International Conference on Acoustics, Fluid Mechanics and Engineering and the 12th Green's Function Seminar (AFME&GFS 2022)

Journal of Physics: Conference Series Volume 2458

Nanjing, China
25-27 November 2022

ISBN: 978-1-7138-7216-0
ISSN: 1742-6588

Printed from e-media with permission by:

Curran Associates, Inc.
57 Morehouse Lane
Red Hook, NY 12571

Some format issues inherent in the e-media version may also appear in this print version.

No changes have been made to the content of these proceedings. There may be changes to pagination and minor adjustments for aesthetics.

Printed with permission by Curran Associates, Inc. (2023)

For permission requests, please contact the Institute of Physics
at the address below.

Institute of Physics
Dirac House, Temple Back
Bristol BS1 6BE UK

Phone: 44 1 17 929 7481
Fax: 44 1 17 920 0979

techtracking@iop.org

Additional copies of this publication are available from:

Curran Associates, Inc.
57 Morehouse Lane
Red Hook, NY 12571 USA
Phone: 845-758-0400
Fax: 845-758-2633
Email: curran@proceedings.com
Web: www.proceedings.com

TABLE OF CONTENTS

ACOUSTIC MATERIAL'S PERFORMANCE RESEARCH AND NOISE CONTROL

FLUID MECHANICS AND AERODYNAMICS APPLICATION

GREEN'S FUNCTION EQUATION CALCULATION AND NUMERICAL SIMULATION

Preface

This issue of the Journal of Physics: Conference Series contains the proceedings of the 2022 International Conference on Acoustics, Fluid Mechanics and Engineering and the 12th Green's Function Seminar (AFME&GFS 2022). The conference took place in Nanjing, China during December 25th-27th, 2022 (hybrid form), with the attendance of about 100 scholars, experts, researchers and scientists in related areas.

The conference aims at creating an active and stimulating forum where current research results and technical advances can be exchanged and the development of acoustics, fluid mechanics, engineering and Green's function can be promoted. For this reason, specialists, researchers, and other related industry personnel in these fields have come from all over the world to present their scientific research findings and innovative approaches to related issues.

To make this conference on schedule, we have a three-part plenary conference including a keynote speech part delivered by five speakers from different countries. Among them, Prof. Chunyan Ji from Jiangsu University of Science and Technology, China performed a keynote speech on Hydrodynamics of Floating Breakwater System. In her presentation, an innovative cylindrical dual pontoon-net FB was designed to enhance the wave attenuation capacity of the FB, and then dual-module FB system was manufactured as the prototype test model. And a fully nonlinear time-domain model based on a coupled iterative solution of water integral equation, floating body and net dynamic equations, was proposed to simulate the interaction of waves and the FB system. We are very optimistic that all participants will acquire new ideas and knowledge from this conference to enhance the capability toward acoustics, fluid mechanics, engineering and Green's function.

We have received tremendous paper submissions from all over China and abroad. Through a rigorous peer-review process, all submissions were performed double blind review to check their quality of content, level of innovation, significance, originality and legibility. Based on the expertise review comments, some excellent papers were accepted and the corresponding authors were invited to submit their papers for final publication. The accepted papers included topics on: Linear Acoustics, Hydrodynamics, Gas Dynamic, Engineering Fluid Mechanics, Green's function in Water Wave Theory, etc.

We would like to thank the Conference Chair and all the Committee members for the procedures that lead to the organization of this event, to our colleagues who contributed to its realization, to the members of the Journal of Physics: Conference Series, for their patience while waiting for the final versions of the papers and to every author who has contributed with an article to these proceedings. We hope you will benefit from all the manuscripts of the proceedings and learn some new and novel ideas.

The Committee of AFME&GFS 2022

Journal of Physics: Conference Series **2458** (2023) 011001 doi:10.1088/1742-6596/2458/1/011001

Committee member

Conference Chair
Prof. Zhi Zong, School of Shipbuilding Engineering, Dalian University of Technology, China

Organizing Committee Chair
Prof. Chunyan Ji, Vice President, Jiangsu University of Science and Technology, China

Academic Committee Chair
Prof. Decheng Wan, Shanghai Jiao Tong University, China

Publication Chair
Prof. Sanjeev Sanghi, Applied Mechanics Department, Indian Institute of Technology, India

Committee Members
Professor. Xiaobo Chen, BV France, France
Professor. Yuzhi Dai, Harbin Institute of Technology, China
Professor. Wenyang Duan, Harbin Engineering University, China
Professor. Nobles Francis, Former employee of Taylor Basin, USA
Professor. Weixi Huang, Tsinghua University, China
Professor. Dongqiang Lu, Shanghai University, China
Professor. Zhiming Yuan, University of Strathclyde, UK
Professor. Papior Nick, Technology University of Denmark, Denmark
Professor. Trebst Simon, University of Cologne, Germany
Professor. Zhen Wang, Beihang University, China
Professor. Chaobang Yao, Huazhong University of Science & Technology, China
Professor. Zhu Renchuan, Shanghai Jiao Tong University, China
Researcher. Zhan Wang, Institute of Mechanics, Chinese Academy of Sciences, China

Peer Review Statement

All papers published in this volume have been reviewed through processes administered by the Editors. Reviews were conducted by expert referees to the professional and scientific standards expected of a proceedings journal published by IOP Publishing.

- **Type of peer review:** Single Anonymous
- **Conference submission management system:** Morressier
- **Number of submissions received:** 115
- **Number of submissions sent for review:** 103
- **Number of submissions accepted:** 52
- **Acceptance Rate (Submissions Accepted / Submissions Received × 100):** 45.2
- **Average number of reviews per paper:** 2
- **Total number of reviewers involved:** 25
- **Contact person for queries:**
  Name: Xuexia Ye
  Email: xx.ye@keoaeic.org
  Affiliation: AEIC Academic Exchange Information Centre

Study of the Hydraulic Characteristics of Slug Flow in the Microfluidic Y-shaped Junctions of Three Different Cross Angles

Yu Han*

R&D Institute of Fluid and Powder Engineering, Dalian University of Technology, Dalian, 116024, China.

e-mail address: 2292135325@qq.com

Abstract: The slug flow in microchannels has been applied for microdroplet/bubble production in recent years. The comparison investigation of hydrodynamic properties of slug flow in three different Y-shaped microchannels was conducted by numerical methods. The numerical simulation results show that the VOF model can describe the velocity and pressure field in the Y-shaped microchannel. It was found that the pressure in the microchannel from upstream to downstream showed an overall decreasing trend, and the liquid phase pressure between the two gas slugs was less than the internal pressure of the adjacent gas slug. The midline pressure of the bubble generation moment is less than the rest of the state. Among the three Y-shaped junctions, the absolute value of the pressure difference at the moment of slug bubble formation is the smallest when the angle of the Y-junction is 90°. Compared to 120° and 90° Y-junctions, the pressure and velocity field of the 60° Y-junction is the most unstable when the slug flow generates.

1. Introduction

With the development of medical treatment, ultrasound diagnosis technology in the field of disease diagnosis is recognized as the safest, fastest, and low-cost detection method [1]. As the most widely used ultrasonic contrast medium, the preparation of microbubbles and related applications has also made great progress. Microfluidic technology is one of the most applicable approaches to preparing microbubbles [2]. For the preparation of microbubbles, the slug flow is preferred appearing in the microchannel because of its controllability to produce monodisperse and uniform microbubbles [3]. Therefore, the formation mechanisms of the slug flow still need deep investigation.

Experiment and numerical simulation are commonly used methods for the study of gas-liquid two-phase flow in microchannels, through which different flow patterns and their change processes with experimental conditions can be observed, but the physical field information such as velocity and pressure of gas-liquid two-phase in microchannels cannot be accurately obtained by experimental methods alone. With the rapid development of computer technology, numerical simulation has become an important means to study the gas-liquid two-phase flow in microchannels. Computational fluid dynamic CFD analyzes physical phenomena such as fluid motion and heat transfer through computer calculations and solutions. At present, the commonly used numerical methods for studying phase interface motion include the fluid volume function method (VOF), the level set method, and the phase field method. These methods provide an effective way to study the two-phase flow in microchannels, in which the VOF method has become the main means to simulate the elastic flow generation process in microchannels [4]. Gupta et al. [5] applied the VOF method to study the liquid film between the slug bubble and the T-shaped microchannel. They demonstrated the important role of wall contact angle in the simulation of the slug bubble generation. Guo and Chen [6] using the VOF method studied the formation process of slug bubbles in a T-junction and found that the viscosity and surface tension influenced the slug bubble interface. Padoin et al. [7] investigated the wettability of the microchannel on the gas-liquid flow patterns. The slug bubble formation process in the microchannel was investigated by Dang et al. [8] using the VOF method. The influence of surface tension, contact angle, and viscosity on the formation of bubbles was studied.

At present, the numerical simulation of the slug flow generation process in the microchannel mainly adopts the VOF method. It can not only obtain basic parameters such as two-phase flow pattern, bullet flow length, and generation period but also deeply analyze the speed field, pressure field, and other information in the flow process to make up for the shortcomings of pure experimental research. In this paper, the speed and pressure field in the flow process of the slug flow in three Y-shaped microchannels with different cross angles were investigated using the VOF method. Comparative investigations were conducted between the three Y-shaped junctions. The mechanism of the slug flow formation in Y-shaped junctions was analyzed and obtained here.

2. Numerical Simulation

To improve computational efficiency, this paper makes reasonable assumptions in the process of establishing a numerical model under the condition of ensuring the accuracy of the results:

(1) The flow in the microchannel is processed according to laminar flow.

It can be seen that the size of the microchannel is at the micron level, and the Reynolds numbers calculated under different experimental conditions in this paper are much less than 2.3×10^3, indicating that the effect of inertial force is much less than the viscous force, so it can be treated according to laminar flow.

(2) The gas-liquid two-phase flow is incompressible.

For liquids, their compressibility can be completely ignored. Because the Mach number of the gas satisfies $Ma=v/c<0.3$ and the variation of gas density is less than 3%, it can be regarded as an incompressible fluid. In this paper, the gas phase flow rate in the microchannel is less than 1 m/s, and the Mach number is much less than 0.3, which is an incompressible fluid.

(3) The physical parameters of the gas-liquid two-phase remain unchanged during the flow process, mainly the density, viscosity, and surface tension of the two phases remain unchanged.

(4) Ignore the influence of gravity.

Hazel et al. [9] showed that the effect of gravity on the system will be significant only if the Bond number is much larger than 1. Under the condition of this simulation parameter, the gravity of the liquid is much less than the surface tension, and the influence of gravity can be ignored during the simulation.

(5) Set the simulation process to normal temperature conditions. There is no heat source inside the research system, no heat exchange with the outside world, and no heat changes are considered.

2.1. Control equations

2.1.1. Continuity equation

From a macroscopic point of view, a fluid can be considered a continuous medium when the flow behavior of individual molecules is not considered. The thermodynamic parameters and mechanical states of the continuum are distributed continuously. The continuity equation is:

$$\frac{\partial\rho}{\partial t}+\nabla(\rho\vec{v})=0 \tag{1}$$

$$\nabla=\vec{i}\frac{\partial}{\partial x}+\vec{j}\frac{\partial}{\partial y}+\vec{k}\frac{\partial}{\partial z} \tag{2}$$

where t is time, ρ is the fluid density, $\vec{v}$ is the velocity vector, and ∇ is the Hamiltonian operator. In Equation (1), from left to right, the unsteady term and the dispersion phase are respectively.

2.1.2. Momentum equation

In addition to the continuity equation, the momentum equation must also be satisfied. The momentum equation is:

$$\frac{\partial\left(\rho\vec{v}\right)}{\partial t}+\nabla(\rho\vec{v}\vec{v})=-\nabla p+\nabla\cdot\left[\mu\left(\nabla\vec{v}+\nabla\vec{v}^{\mathrm{T}}\right)\right]+\rho\vec{g}+\vec{F} \tag{3}$$

where P represents pressure, v represents viscosity, $\vec{g}$ represents gravitational acceleration, and $\vec{F}$ represents external force as surface tension in this simulation.

Surface tension plays a very important role in the formation of slug flow, and the presence of surface tension is also an important reason for maintaining the shape of bubbles, so the influence of surface tension cannot be ignored when discussing the gas-liquid two-phase flow in microchannels. This paper adopts the continuous surface tension model proposed by Brackbill et al. [10] (Continuum Surface Force Model), and expresses the surface tension as a volume force and becomes the source phase in the momentum Equation (3) by the divergence theorem:

$$\vec{F}=\sigma\frac{2\rho K\nabla\alpha_q}{\rho_{\mathrm{L}}+\rho_{\mathrm{G}}} \tag{4}$$

$$K=\nabla\cdot\hat{n} \tag{5}$$

where K is the interface curvature, and $\hat{n}$ in units of positive vectors.

2.1.3. Volume fraction equation

In the VOF method, the tracking phase interface is achieved by solving a continuous equation for the volume fraction of a phase or multiphase fluid, and the volume fraction equation satisfies the following:

$$\frac{\partial\alpha_q}{\partial t}+\vec{v}\cdot\nabla\alpha_q=0 \tag{6}$$

where α_q represents the volume fraction of the fluid. L represents the liquid phase, and G represents the gas phase. In the VOF method, the two-phase fluid is not interpenetrating, and the sum of the phase content in each controlled volume is 1, that is:

$$\alpha_{\mathrm{L}}+\alpha_{\mathrm{G}}=1 \tag{7}$$

The VOF method obtains the position of the phase interface by calculating the volume fraction of each phase in the control volume. The VOF method solves only one set of control equations in the entire solution domain, so the velocity field and pressure field are shared. The physical parameters in the control equation are also determined by the physical parameters of each phase in the control volume. Its mixing density and mixing viscosity are calculated as:

$$\rho=\alpha_{\mathrm{L}}\rho_{\mathrm{L}}+\alpha_{\mathrm{G}}\rho_{\mathrm{G}} \tag{8}$$

$$\mu=\alpha_{\mathrm{L}}\mu_{\mathrm{L}}+\alpha_{\mathrm{G}}\mu_{\mathrm{G}} \tag{9}$$

2.2. Microchannel meshing

The Y-shaped microchannel width is in the micron range, and the physical model is shown in Figure 1. The lowest point on the left side of the graph is taken as the coordinate origin, the horizontal direction is set to the x coordinate, and the vertical direction to the y coordinate.

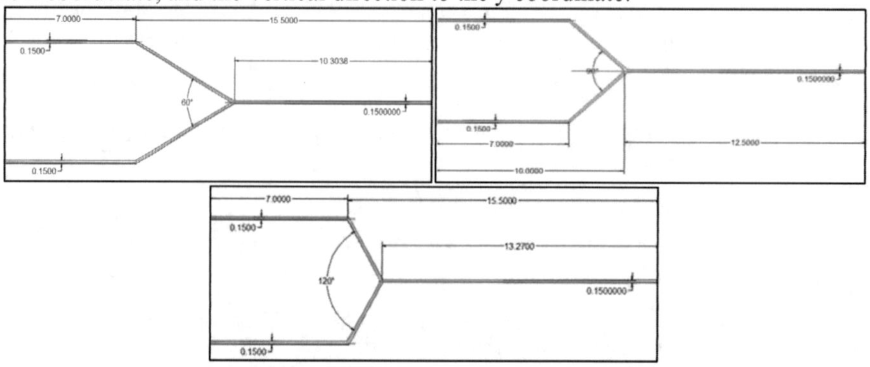

Figure 1. Schematic diagram of microchannels.

After many simulation tests, a two-dimensional model is finally adopted. By using the software Gambit 2.4.6, the physical model of the Y-shaped channels is established and divided into structured quadrilateral meshes with a size of 10 μm, and the resulting mesh is shown in the figure below.

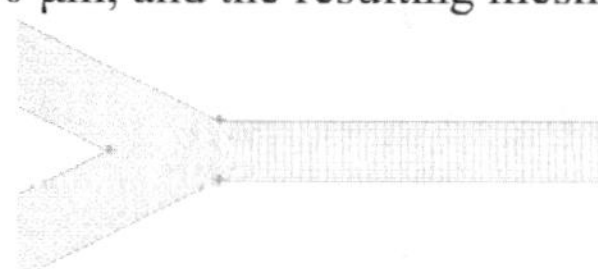

Figure 2. The example of computational mesh domain.

3. Results and discussion

3.1. The formation of slug flow in the Y-shaped microchannel

The change process of the phase interface can be directly observed in the simulation process. A momentary gas phase pinching-off during the flow process is selected as the t=0 moment, and the gas slug generation process is shown in Figure 3. The blue part in the figure represents the air, and the red part represents the aqueous phase. After the slug pinches off, the end of the gas slug continues to grow under the condition of continuous gas feeding, and the liquid phase continuously shears the air thread. When the gas volume accumulates to a certain extent, the gas phase neck near the upper wall of the channel is getting thinner and thinner, and the gas phase cannot support the force of the liquid phase at a certain moment. At this moment, a breakpoint appears on the wall surface of the Y-shaped channel and causes the generation of a new gas slug. The time consumed by the above whole process is the generation cycle of the gas slug.

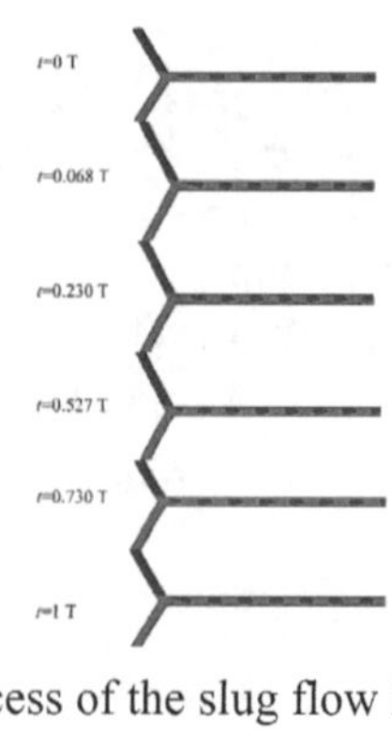

Figure 3. The generation process of the slug flow in the Y-shaped microchannel.

3.2. The pressure field characteristics in the formation process of slug flow

The physics information in the channel changes along with the change of phase interface overtime during the flow process, the most typical of which is the pressure field. This section studies the variation of the pressure field during the flow process. The pressure field in Y-60°, Y-90°, Y-120°, and Y-shaped junctions is shown in Figure 4. According to the legend on the right, it is found that the pressure of the Y-shaped device from upstream to downstream is generally decreasing, and the contour of the gas slug can be observed. It could be seen that the liquid phase pressure between the two gas slugs is less than the internal pressure of the adjacent gas slug.

Figure 4. Pressure program of Q_L=50 μL/min, Q_G=50 μL/min: (a) Y-60°; (b) Y-90°; (c) Y-120°.

The pressure in the horizontal channel midline could represent the pressure variation trend and the data is easy to extract, so the pressure of the midline is obtained for further quantitative analysis. By setting the horizontal channel midline in fluent software and extracting the midline pressure value file, the pressure change △P (the horizontal direction is the x coordinate) in one period is obtained and shown in Figure 5. The comprehensive analysis of the midline pressure and pressure program shows that the part of the solid line with the constant value of △P is the central segment of the gas slug, and the difference between the corresponding coordinates at both ends is the characteristic length of the gas slug. The solid line with the largest absolute slope on both sides of the gas slug is the phase interface. The phase interface represents the transition of pressure from the gas phase to the liquid phase, and a large slope indicates a significant change in pressure. The part where the slope of the interface of the continuous phase is always negative represents the liquid slug, and the corresponding coordinate difference between the two ends is the length of the liquid phase. The variation of the pressure on the midline shows that the "pinching-of pressure" (the pink line in Figure 5) is less than the rest of the state. After the above analysis, the gas phase forms a breaking point near the Y-crossed junction, and the midline pressure value of the "pinching-of moment" has a sudden decline on the right side of the crossed junction, indicating that the occurrence of the pinching-off is accompanied by a significant change in the entire pressure field, and the downstream of the pinching-off location still shows the above-mentioned regularity changes. Compared to the other two angles, the absolute value of pressure difference at the "pinching-of moment" is the smallest in Y-90° junction.

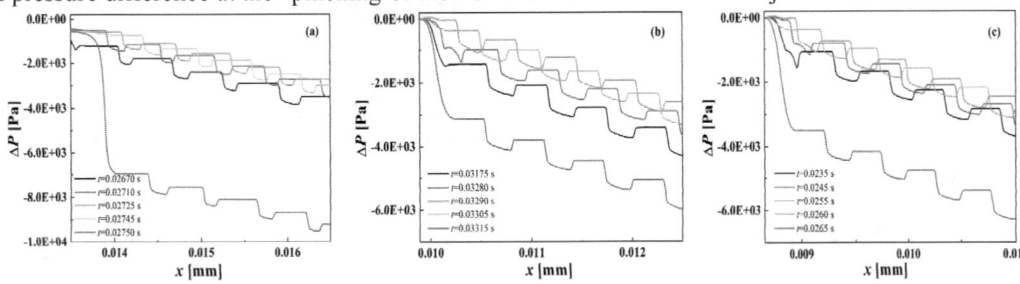

Figure 5. Variation of pressure on the midline over time of Q_L=50 μL/min, Q_G=50 μL/min: (a) Y-60°; (b) Y-90°; (c) Y-120°.

To facilitate the study of the pressure variations at specific locations, monitoring points are set in the model to record the pressure variations during the slug flow in real-time. The monitoring points can record the pressure values of each calculated step of the monitoring position. Comparing the pressure change of the monitoring points in the microchannel with the angle of 60°, 90°, and 120°, it can be found in Figure 6 that the pressure leap of the monitoring point at the "pinching-off moment" is the smallest in Y-90° junction. At the same time, in the Y-shaped microchannel of the above three angles, the pressure variation of the monitoring point within one cycle is: the pressure rises suddenly at the pinching-off time, and then suddenly decreases; the pressure decreases slowly in the bubble growth period. In Y-60° junction, the pressure variation of every cycle is the most unstable.

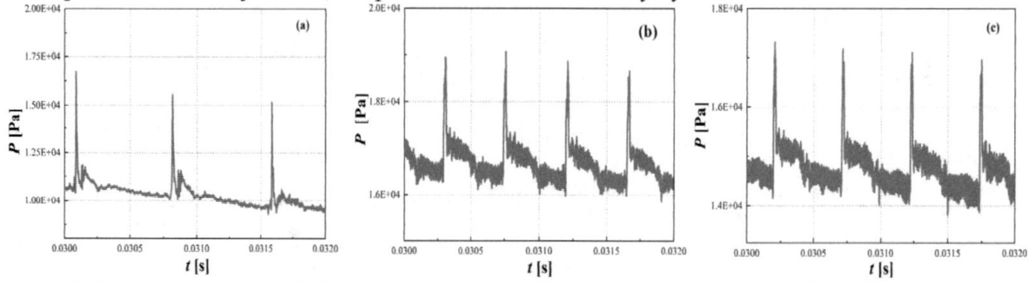

Figure 6. Pressure variation of the monitor points at Q_L=50 μL/min, Q_G=50 μL/min: (a) Y-60°; (b) Y-90°; (c) Y-120°.

3.3. *The velocity field characteristics in the formation process of slug flow*

Figure 7 gives an enlarged view of the velocity vector at the Y-shaped interface during the flow process, which can be seen from the phase diagram at this time as a continuous bubbling stage, and the black arrow in the figure indicates the velocity vector. Under the same liquid phase and gas phase inlet parameters, it can be seen that the smaller the cross angle, the longer the gas slug, and the larger the velocity gradient in the gas slug. During the generation of gas slugs, two vortices are formed between the interior and the wall of the gas slug, and the vortex position changes as the angle increases. Compared to 120° and 90°, the velocity field in Y-60° junction is the most unstable and disorder during the generation of slug flow.

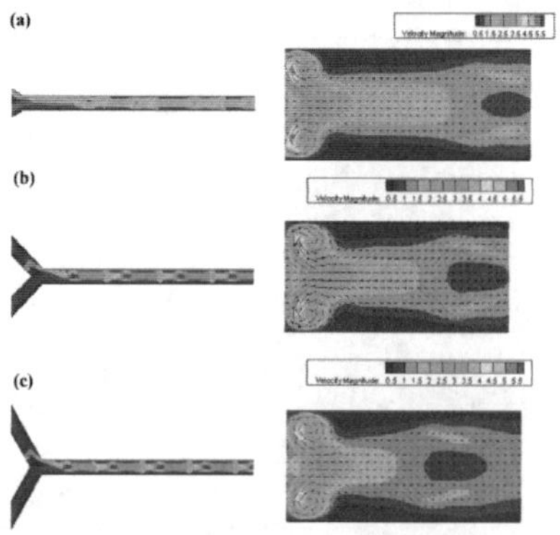

Figure 7. Velocity vector program in the Y-shaped junctions at Q_L=60 μL/min, Q_G=55 μL/min: (a) Y-60°; (b) Y-90°; (c) Y-120°.

4. Conclusion

This paper calculates based on the established numerical model and performs post-processing analysis on the calculation results of Y-shaped microchannels with different angles, and the main conclusions are as follows: Two methods were used to study the variation of pressure field during the flow process in the Y-shaped microchannel. According to the pressure change of the midline and the monitoring point, it can be seen that the pressure near the pinching-off position at the "pinching-off moment" changes abruptly, forming a local low-pressure area, and the pressure value changes exponentially. Among three junction angles, the absolute value of the pressure difference at the moment of bubble generation is the smallest in the Y-90° junction microchannel; The velocity field in the channel changes continuously with time during the flow process, and the pinching-off position and phase interface are the areas with the most obvious speed difference. During the generation of gas slugs, two vortices are formed between the interior and the wall of the gas slug, and the vortex position changes as the angle increases.

Acknowledgments
This work was supported by the Fundamental Research Funds for the central universities [DUT16QY39].

Reference

[1] Zhou W and Zhou L 2022 Ultrasound for the Diagnosis of Biliary Atresia: From Conventional Ultrasound to Artificial Intelligence *Diagnostics* **12** 51.
[2] Li Y K, Wang K, Xu J H and Luo G S 2016 A capillary-assembled micro-device for monodispersed small bubble and droplet generation *Chem. Eng. J.* **293** 182-8.
[3] Fu T and Ma Y 2015 Bubble formation and breakup dynamics in microfluidic devices: A review *Chem. Eng. Sci.* **135** 343-72.
[4] He K, Wang S and Zhang L 2017 Transient split features of slug flow at an impacting micro-T-junction: A numerical study *Int. J. Heat. Mass. Tran.* **112** 318-32.
[5] Gupta R, Fletcher D F and Haynes B S 2009 On the CFD modeling of Taylor flow in microchannels *Chem. Eng. Sci.* **64** 2941-50.
[6] Guo F and Chen B 2009 Numerical Study on Taylor Bubble Formation in a Micro-channel T-Junction Using VOF Method *Microgravity. Sci. Tec.* **21** 51-8.
[7] Padoin N, Souza A Z D, Ropelato K and Soares C 2016 Numerical simulation of isothermal gas−liquid flow patterns in microchannels with varying wettability *Chem. Eng. Res. Des.* **109** 698-706.
[8] Dang M, Yue J and Chen G 2015 Numerical simulation of Taylor bubble formation in a microchannel with a converging shape mixing junction *Chem. Eng. J.* **262** 616-27.
[9] Hazel A L and Heil M 2002 The steady propagation of a semi-infinite bubble into a tube of elliptical or rectangular cross-section *J. Fluid. Mech.* **470** 91-114.
[10] Brackbill J U, Kothe D B and Zemach C 1992 A Continuum Method for Modeling Surface Tension *J. Comput. Phys.* **100**(2) 335-354.

The CFD Investigation of General FPSO Green Water

Min Li, Peilin Dou*, Yue Qi

School of Naval Architecture and Ocean Engineering, Jiangsu University of Science and Technology, Zhenjiang, China.

*Corresponding author's email: yxdpl@just.edu.cn

Abstract: The main research is to investigate the seakeeping effect resulting from general FPSO hull bow flare. CFD-FV (Computational Fluid Dynamics-Finite Volume) method is applied with fluent software to simulate in the 3D time domain. Simulation objects are two different bow flare models exposed to the green water phenomenon, which is under the sea state of the Brazilian sea 20 years of return extreme sea state. Seakeeping analysis includes green water. The results indicated that a 3.5 m vertical bow flare would reduce about 35.09% level of green water on deck. A comprehensive comparison indicated that bow flare improved the global seakeeping performance, which can be used as a reference to the design of general FPSO hull bow flare.

1. Introduction

Green water is one of the seakeeping of ships, which is an important index to evaluate ship performance. ABS, DNV, and other classification societies have given different design specifications for floating structures and their equipment [1-2] to ensure the safety of equipment, and the structural safety of bow floating in the process of waves and prevent equipment corrosion and fatigue damage. Although increasing freeboard height and strengthening bow protection can reduce the degree of wave and prevent structural damage, it will also reduce the ship's stability. Therefore, scholars have made many contributions to FPSO ship design and research methods. GRECO et al. [3] studied the seakeeping of FPSO by changing the turret's position. Based on the seakeeping solver of potential flow theory, the research and analysis method of seakeeping of the FPSO deck is provided. YANG et al. [4] predicted the deck wave risk of single-point berthing ships. They proposed an air gap coupling dynamic analysis method to determine the wave position on the deck and further predict the failure probability of waves on the deck. GATIN et al. [5] used the ocean hydrodynamic data packets in the open-source software foam-extend. They studied the wave loads on the deck by the frequency domain numerical simulation method based on the finite volume method. SILVA et al. [6] carried out the pool model test and numerical simulation of FPSO [7] and studied the relative wave height and impact load of the FPSO deck under the action of a transverse wave and oblique wave. Many scholars have also studied the drainage problems such as falling waves on free water surfaces, outboard waves, and time-domain simulation of outboard waves, providing corresponding solutions and prediction methods to cope with the adverse effects of falling waves on ships [8-9].

In this paper, based on the CFD-FV method, the application software Ansys Fluent is used to simulate the deck wave phenomenon by the 3D time domain method. In the study, the VOF (Volume of Fluid) method of multiphase flow [10] tracks the free surface, mainly used to simulate the green water on the deck of two bow flare models. The focus of the study is the seakeeping performance of the universal FPSO. In the seakeeping analysis, the control variable method is mainly used to exclude the influence of other factors on the ship's seakeeping, and only the influence of the bow flare deformation on the ship's seakeeping is considered. By comparing the changes of seakeeping parameters of two different bow flare models, the variation curves of deck wet surface area and relative wave height under the condition of bow flare are obtained to analyze the ability of bow flare to resist waves and evaluate the seakeeping performance of general FPSO.

2. Bow flare and sea condition setting

2.1 Model parameter setting

The general-purpose FPSO is jointly designed by 708 Institute of China Shipbuilding Group and Waigaoqiao Shipbuilding Co., Ltd., with a total length of 333 m, a typical width of 60 m, a typical depth of 33 m, a displacement of up to 460,000 tons, and a storage capacity of 2.3 million barrels of crude oil. FPSO is in the positive floating state of turret single-point mooring at work. The model establishes the A model without bow flare and the B model with bow flare. The B model flare is stretched from the bow inclined wall, and its vertical distance is 3.5 m. The detailed design parameters are shown in Table 1. The bow flare diagram is shown in Figure 1.

Table 1 General FPSO design parameters

Parameter	Model A (no bow flare)	Model B (bow flare)
Height (*H*/m)	34	37.5
Width (*B*/m)	72.2	75.1
Length (*L*/m)	332.75	334.22
Average draft (*D*/m)	25	25

In the following, models A and B will be cited to replace the no-bow flare model and the bow flare model, respectively. General purpose FPSOs usually operate continuously in the same sea area for up to 20 to 30 years. The probability of extreme sea conditions in Brazil's sea area is high. Therefore, this paper selects the wind wave parameters with a return period of 20 years in Brazil's sea area. The detailed wind wave parameters are shown in Table 2. In this study, Ansys Space claim is used to establish the model, Enclosure is used to establish the external flow field, and Oxyz three-dimensional rectangular coordinate system is established. The origin of the coordinate is located at the intersection point O of the base plane and the stern vertical line, and the intersection line of the base plane and the midline plane is the x-axis, pointing to the bow is positive. The Intersect of the base plane and the middle station is the y-axis. Pointing to the starboard is positive. The intersection of the midline and mid-station planes is the z-axis, which is positive upward.

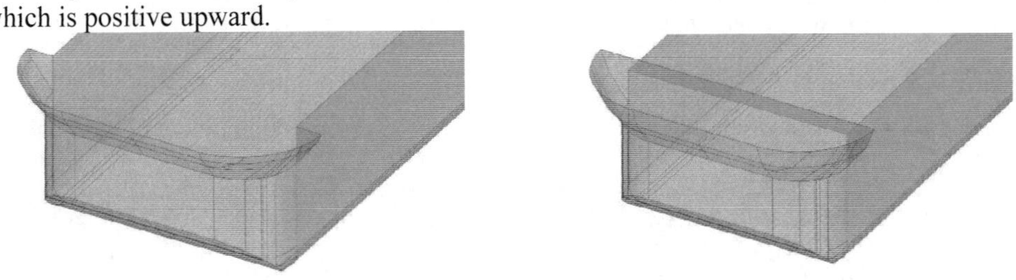

(a) Model A (no bow flare) (b) Model B (bow flare)

Figure1 General FPSO bow flare

2.2 Sea state parameter setting

The wave direction in Table 2 is defined as the angle between the main wave direction and the northerly direction of the Earth, and yr is the number of years required for an extreme wind wave to occur.

Table 2 Wind wave parameters in Brazil's sea area

Wave parameter	Numerical value
Return period/yr	20
Seadirection/m	225
Stormy velocity (m/s)	29.76
Free surface velocity (m/s)	1.57

Significant wave elevation/m	9.74
Maximum wave elevation/m	18.13
The peak period of spectrum/m	14.5
Zero-crossing period/m	10.59

The SW (South-West) wind wave Hs in the Brazilian sea area is the largest, and the wind wave parameters are selected under this extreme sea condition. Because the numerical simulation of wind waves requires a precise and large number of grids to capture accurately, it will cause a great amount of calculation to the numerical solution. And 389.6 m is selected as the design wavelength, and the maximum wave height Hmax, which is most prone to green water, is selected as the design wave height. Considering the continuous effect of wind under extreme sea conditions, the corresponding air flow rate is set, and the design sea conditions are shown in Table 3.

Table 3 Design sea state

Sea state	Design sea conditions
Design wavelength/m	389.6
Design wave height/m	18.13
Average wind speed (m/s)	1.57
Average discharge velocity (m/s)	29.76

3. Seakeeping analysis

3.1 Numerical simulation setting

Previously, the influence of different mesh sizes of the cutting volume on the accuracy and computational efficiency of the free surface has been studied, which reduces the grid correlation of the wave on the FPSO deck. The appropriate grid model has been selected and verified by comparing the pool model test. As the research basis of this numerical simulation, the numerical simulation of this study also applies the Ansys Meshing cutting body grid division method to ensure that the fluid domain is divided into structural grids, increasing the grid quality while controlling the number of grids. The principle of meshing is that the regional grids with small variable gradients are sparsely distributed, and the regional grids with large variable gradients and important regions are densely distributed. The grid model is shown in Figure 2. To ensure the effective control of the number of grids and improve the accuracy of numerical calculation, the grid model parameters are shown in Table 4.

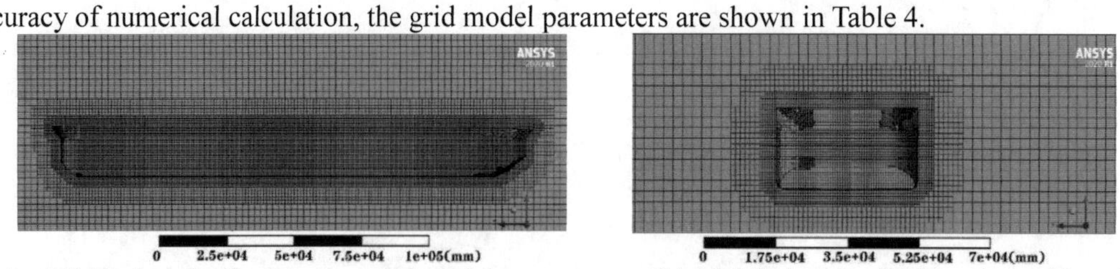

(a) Middle longitudinal section grid model (b) Middle longitudinal section grid model

Figure 2 Grid model of FPSO fluid domain

Table 4 Design sea conditions

Local grid	First floor/1.25 m	Second floor/2.5 m	Outflow field/5 m
Hull surface distance $\pm x$/m	(0, 7.5]	(7.5, 20]	(20, x_{max}]
Hull surface distance $\pm y$/m	(0, 7.5]	(7.5, 22.5]	(22.5, y_{max}]
Hull surface distance $\pm z$/m	(0, 7.5]	(7.5, 20]	(20, z_{max}]

The numerical calculation uses the Fluent internal multiphase flow code Open Channel and Open Wave BC code to set the wave-making, wave-eliminating, and wave boundary. The semi-implicit algorithm of the pressure coupling equations is used to solve the problem, and the boundary conditions of air and seawater are set separately.

3.2 Analysis of green water

Relative Wave Elevation (RWE) refers to the position of the highest free surface of the wave relative to the upper deck of the FPSO. Through RWE, we can know whether the wave on the deck occurs and when it occurs. Figure 3 (a) shows the variation of the RWE maximum of the two models with time. By observing the curves, it can be found that the waves of the two models climbed rapidly at about 18 s, the waves ended at 26.5 s, and then the deck waves occurred. During the 18 s ~ 26.5 s full wave process, the extreme value of the relative wave height in model B is greater than that in model A, which means that the wave will continue to climb with the bow drifting and climb to the extreme value for a longer time.

After analyzing the time of the wave, the severity of the wave can be judged by the wet surface area of the deck, as shown in Figure 3. The wet surface area of the two groups of models increased rapidly at 19 s and 20 s, respectively. During the whole deck surfing process, model A's wet surface is continuously higher than model B's, indicating that the bow flare can effectively reduce the severity of deck surfing. The key parameters describing the wet surface are shown in Table 5. The MAX function in Excel software deals with the maximum wet surface area during the wave process. The maximum wet surface area of model B is 1184.77 m^2, and the maximum wet surface area of model A is 1600.53 m^2. The time corresponding to the maximum wet surface area of the two groups of models is selected, which is 22.76 s and 22.16 s, respectively. The strong nonlinear effect in the wave-floating structure model produces a slight difference between the wave time of the two groups of models. Because the two sets of models control all variables except the bow flare model are the same, the main reason for this nonlinear difference is caused by the spatial term integration in the control equation applied by the CFD-FV method.

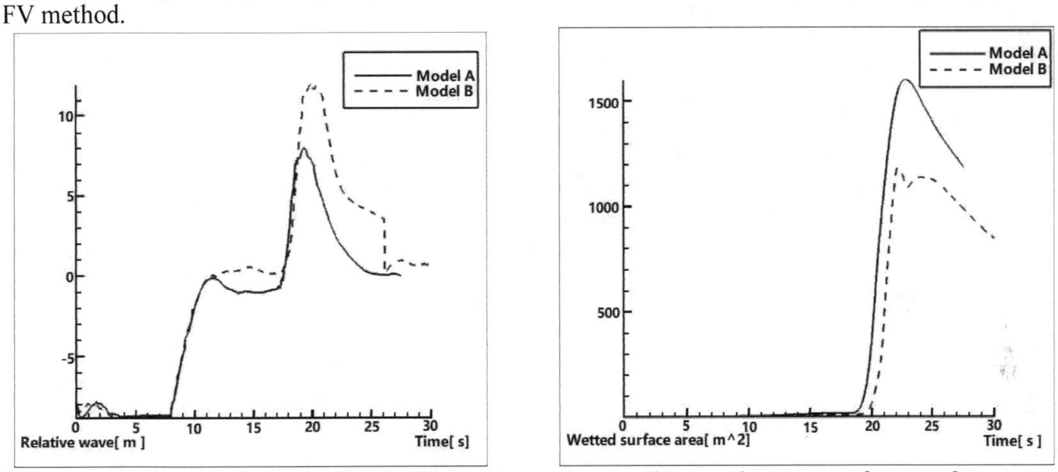

(a) The maximum relative wave height transient curve (b) Transient curve of wet surface area

Figure 3 The transient curve of wave flow on the deck

Table 5 Key parameters of wet surface

Sea state	Model A	Model B
Green water time/s	19.38	20.18
Maximum wet surface time/s	22.76	22.16
Maximum wet surface area/m^2	1600.53	1184.77

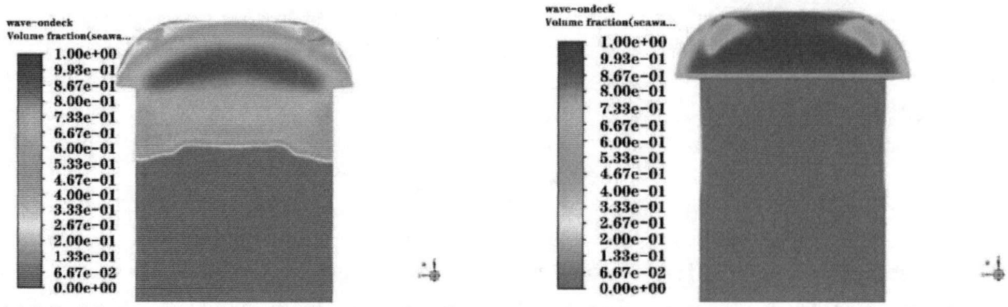

(a) Model A wet surface distribution cloud map (b) Model B wet surface distribution cloud chart

Figure 4 Wet surface distribution cloud

Figure 4 shows the contours of the maximum wetted surface. The maximum wetted surface distribution of FPSO without bow flare can be seen as much larger than that with bow flare. It is proved that the bow flare can resist the wave on the deck, which reduces the maximum wet surface area by 35.09 % and obtains the main areas where the FPSO deck equipment needs to have the waterproof ability.

It can be seen from Figure 4 (b) that the wet surface is mainly distributed on the bow drift plate, which is because the outflow flow of the upper wave water at the drift position is greater than the inflow flow. It can be seen from the streamlined trajectory in Figure 5 (b): 1) the streamlined shape of the wave at the bow of the ship is spiral, and the wave is constantly rising, and the wave area is A1; 2) since the width of the floating plate is greater than the deck width, the upper water layer is continuously lost from both ends, and the loss area is A2 ; 3) Due to the relative level of the outer drifting plate relative to the sea level, the water layer leaving the outer drifting plate performs parabolic motion, and the stagnant water layer is not included in the wet surface area. The parabolic water layer area is A3; 4) the Slamming water layer area of the stagnant water layer after impact deck is A4.

Combined with Figure 3 (b), it can be seen that since 22.16 s is the peak of wet surface area, the total wet surface area corresponding to 22.16 s is less than 22.16 s, indicating that the sum of the area of the lost water layer and the area of the parabolic water layer must be greater than the area of the upper water layer, that is, $A_2+A_3>A_1$. After the parabolic motion of the first water layer hits the deck, the first wave is over because the sum of the slamming water layer area and the rising water layer area must be greater than the loss water layer area and the parabolic water layer area, that is, $A_2+A_3>A_1+A_4$.

From the perspective of wave-up time, the total wave-up time of Model A from climbing to slamming deck is 3.38 s, while that of Model B is 1.98 s. The wave-up time is shortened by 41.42 %, and the time of wave squeezing the bow is shorter, which reduces the fatigue damage of the bow structure and relatively increases the service life. From the wave shape point of view, wave climbing and deformation in the process of A wave is slow and long, and B wave is high and steep.

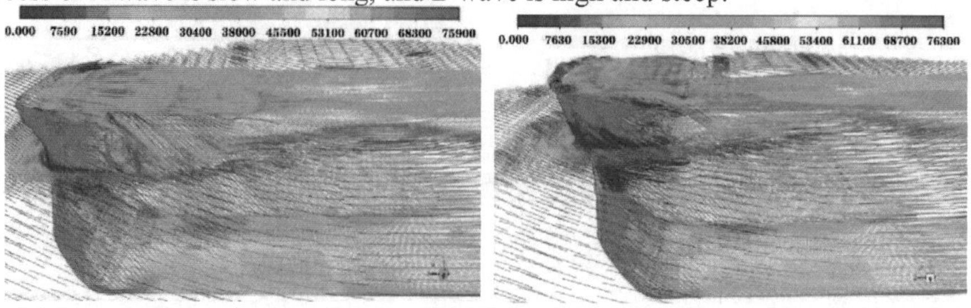

(a) Model A streamline trajectory diagram (22.76 s) (b) Model B streamline trajectory diagram (22.16 s)

Figure 5 Streamline trajectory

4. Conclusions

The seakeeping performance of one of the six major performances of FPSOs must be checked after the first ship is manufactured to review and improve the design of future general-purpose FPSOs. In this paper, the CFD deck wave numerical simulation of two different bow flare models of general FPSO is carried out in the three-dimensional time domain by Fluent software. The extreme wind and wave parameters with a return period of 20 years in the Brazilian sea area are selected for simulation. The control variable method is used to analyze the seakeeping of the bow and deck. In the analysis of the wave on the deck, the two models' relative wave height and wet surface area were compared and analyzed. The relative wave height of model B with the outer wave is increased by 50.88 %, the wet surface area is decreased by 35.09 %, and the wave time is decreased by 41.42 %. Under the same sea conditions, its performance in resisting waves has increased by 35.09 %. Model B with bow flare is more seaworthy under the mooring working condition of FPSO. By analyzing the seakeeping performance of the general-purpose FPSO, it is shown that the bow flare has a significant effect on resisting deck waves, which provides a certain reference value for the subsequent FPSO design. Although the ability to resist rising waves is increased, the ship's stability may be reduced, and the stability impact of bow flare should be further analyzed in future research.

References

[1] Flis L 2016 Application of the rules of the American Bureau of Shipping (ABS) to design mobile offshore structures *Sci. J. Pol. Nav. Acad* **205** 17-28.

[2] Wang S, Wang X and Woo W 2019 Effects of the asymmetric riser and bilge keel arrangements on FPSO green water assessment *Appl. Ocean. Res* **86** 166-76.

[3] Greco M, Lugni C and Faltinsen O 2015 Influence of motion coupling and nonlinear effects on a parametric roll for floating production storage and offloading platform *Philos. Trans. A. Math. Phys. Eng* **373** 1-18.

[4] Yang W, Zhang P, Dong Y J, Yuan Z L and Zhang Z G 2021 Risk analysis of waves on the deck of a single mooring ship *J. Ship. Res* **16** 81-89.

[5] Gatin I, Vladimir N and Malenica S 2019 Green sea loads in irregular waves with finite volume method *Ocean. Eng* **171** 554-64.

[6] Silva D, Coutinho A and Esperanca P 2017 Green water loads on FPSOs exposed to beam and quartering seas, part I: experimental tests *Ocean. Eng* **140** 419-33.

[7] Silva D, Esperanca P and Coutinho A 2017 Green water loads on FPSOs exposed to beam and quartering seas, part II: CFD simulations *Ocean. Eng* **140** 434-52.

[8] Cai S Y, Bao J and Shen S H 2018 Solution of water drainage problem after wave on the deck of surface vessel *Ship. Ocean. Eng* **47** 120-22.

[9] Gong C, Zhu R C, Miao G P and Fan J 2014 Engineering calculation method of high-speed ship deck wave load based on CFD *J. Mech* **18** 524-31.

[10] Hirt C and Nichols B 1981 Volume of fluid (VOF) method for the dynamics of free boundaries *J. Comput. Phys* **39** 201-25.

Study of Near-field Acoustic Pressure in Near-coastal Ice-water Media Based on Angular Spectrum Method

Chen Wang*, Hua Zhang

Zhejiang Ocean University, School of Shipping and Maritime Studies

thankpc@126.com*

Abstract: For seawater detection, acoustic means is currently one of the most promising technical means of development. According to the physical properties of acoustic waves, they can be effectively propagated in seawater and ice, and at the interface of different media, there will be reverse acoustic scattering. Using the theory of acoustic propagation in solids, the propagation velocity and refractive index of transverse and vertical, and taking the longitudinal wave as an example, we use the angular spectrum method to solve the fluctuation equation and obtain the beam angle spectrum. Numerical software simulation calculates the time domain characteristics of acoustic scattering at different incidence angles. The effect of evanescent waves on solid sound transmission is also considered. Considering the effect of apocalyptic waves on thin ice on the ocean surface, it provides a theoretical basis for further research on detecting offshore ice using acoustic methods.

1. Introduction

Sea ice covers about 3% of the seawater area and is a necessary pathway for exchanging materials between seawater and the atmospheric system. The Arctic Ocean has permafrost waters, but even so, it has the Northern Passage and a large amount of fossil energy, which has led scientists from all over the world to devote themselves to it. How to accurately detect the environment under the ice has become a hot topic. Yang and Votaw [1] measured the low-frequency (less than 1 kHz) acoustic reflection coefficients of the sea surface ice north of Ellesmere Island in the Arctic. They found that the reflection coefficients were closely related to the acoustic frequency and grazing angle for grazing angles of 14°-26°. Liu and Li [2] introduced the power spectral density to analyze the coherent reflection coefficient and scattering coefficient numerically and obtained the coherence coefficient for each grazing angle. Several experts have analyzed the sound propagation properties of Arctic ice from experimental and theoretical aspects, respectively.

The northern sea of China also encounters seawater ice in winter, although not as large as the Arctic ice, yet seriously affects ship navigation and underwater detection [3]. For example, in the Bohai Sea region of Dalian, the night air temperature can reach about -20°C during December and January every year, and seawater often freezes rapidly at night. Still, the sea ice roughness is small because of the fast-freezing speed, low salinity at the offshore end, and low thickness of the freezing layer [4]. Bishop G. C. [5] characterized ice crests and flat ice sheets, respectively [5]. Maksimov A.O. [6] studied the characteristics of sound propagation in ice sheets containing air bubbles [6]. So, in the relatively flat case, either experimental measurements or numerical analysis using power spectral density will be more troublesome.

According to previous studies, the frozen areas in offshore waters are mostly fixed ice, compared to flatter ice ridges, which makes theoretical calculations simpler and with less error. In this paper, we focus on the acoustic propagation characteristics of thin ice with low roughness in offshore waters. We treat the offshore thin ice and seawater coupled system as a solid-liquid double-layer structure. We optimize the fluctuation equations in this structure using the theory of elastic dynamics in solids and the theory of acoustic propagation in liquids, and solve them numerically. Further, the finite element method and time domain analysis were used to verify the correctness of the equation derivation and numerical solution. On this basis, the effects of ice thickness, icing time on the porosity of ice, and the choice of empirical equations for the speed of sound in water are investigated. In addition, by decomposing the angular spectrum of the fluctuation equation, the linear superposition solution of the plane and evanescent waves can be obtained to better solve the fluctuation problem in layered media. The results of this paper have a theoretical reference value for the analysis of acoustic propagation signals under near-coastal sea ice and in the ice layer.

2. Water-ice double-media sound propagation characteristics

2.1. Equation of acoustic wave in solid

Compared to the fluid, the situation in the solid is not so plain. Generally, the solid medium can produce both volume and tangential deformation so that the solid medium can propagate not only the longitudinal waves of expansion and compression but also the tangential waves. In an isotropic solid, the direction of vibration of such a shear mass is perpendicular to the direction of wave propagation and is called a transverse wave.

Since two different types of waves are generated in solids, two different reflections and refractions are bound to occur when these waves are incident from one medium to the other. To obtain some basic propagation properties of sound waves in solids. Suppose a plane sound wave comes from a fluid medium ① and is directed to a solid medium ② on an infinitely large flat surface at an angle of incidence θ_i. Since the medium in ① is fluid, the incoming wave must be longitudinal. The medium ② is solid, which can produce transverse waves in addition to longitudinal waves, as shown in Figure 1. There are two types of refracted waves: a longitudinal wave refracted at a θ_{tL} angle, and the other is a transverse wave refracted at a θ_{tT} angle. In the case of two-dimensional xOz plane motion, when the time factor $e^{j\omega t}$ is not considered respectively, the scalar potentials of medium ① and medium ② can be expressed as

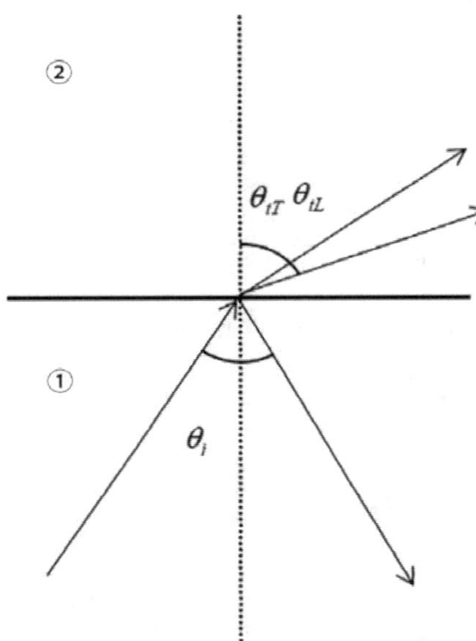

Figure 1 Solid-liquid interface scattering diagram

$$\Phi_1 = \Phi_i e^{-j(k_{1L}\cos\theta_i x + k_{1L}\sin\theta_i z)} + \Phi_r e^{-j(-k_{1L}\cos\theta_r x + k_{1L}\sin\theta_r z)}$$
$$\Phi_2 = \Phi_t e^{-j(k_{2L}\cos\theta_{tL} x + k_{2L}\sin\theta_{tL} z)} \tag{1}$$

For the vector potential in medium ②, since the motion is just in the *xOz* plane, the vector potential only has a component in the *y* direction, so that it can be written as

$$\Psi_2 = \psi_y = \psi_t e^{-j(k_{2T}\cos\theta_{tT} x + k_{2T}\sin\theta_{tT} z)} \tag{2}$$

The boundary condition of normal velocity continuity should be satisfied at the fluid-solid partition interface

$$\left(\frac{\partial\Phi_1}{\partial x}\right)_{x=0} = \left(\frac{\partial\Phi_2}{\partial x} - \frac{\partial\Psi_2}{\partial z}\right)_{x=0} \tag{3}$$

substituting Equation (1) and Equation (2) into Equation (3), we get Equation (4)

$$-\Phi_i k_{1L}\cos\theta_i e^{-jk_{1L}\sin\theta_i z} + \Phi_r k_{1L}\cos\theta_r e^{-jk_{1L}\sin\theta_r z} = -\Phi_t k_{2L}\cos\theta_{tL} e^{-jk_{2L}\sin\theta_{tL} z} - \psi_t k_{2T}\cos\theta_{tT} e^{-jk_{2T}\sin\theta_{tT} z} \tag{4}$$

Equation (4) should hold for all z, so the exponential part should be constant, as

$$k_{1L}\sin\theta_i = k_{1L}\sin\theta_r = k_{2L}\sin\theta_{tL} = k_{2T}\sin\theta_{tT} \tag{5}$$

The law of refraction can be obtained from Equation (5) as

$$\begin{cases} \dfrac{\sin\theta_i}{\sin\theta_{tL}} = \dfrac{k_{2L}}{k_{1L}} = \dfrac{c_{1L}}{c_{2L}} \\ \dfrac{\sin\theta_i}{\sin\theta_{tT}} = \dfrac{k_{2T}}{k_{1L}} = \dfrac{c_{1L}}{c_{2T}} \end{cases} \tag{6}$$

From Equation (6), it can be concluded that the law of refraction between solids and liquids is the same as between fluids [7]. In addition to refracted longitudinal waves, solid media also produce refracted transverse waves. For solids, in general, the longitudinal wave velocity is faster than in fluids, and the longitudinal wave refraction angle in solids is also larger than the angle of incidence. In addition, the propagation speed of longitudinal waves in solids is always faster than that of transverse waves. For the incident from the fluid to the solid, there is always a longitudinal refraction angle bigger than the transverse refraction angle. Hence, the longitudinal wave always reflects the refraction properties better than the transverse wave.

2.2 Solving the fluctuation equation based on the angular spectrum method

There are a variety of solutions to the acoustic wave equation, although most of them separate variables under the simple harmonic solution to the Helmholtz equation of the fluctuation equation, such as the normal wave model, the ray model, and the parabolic equation model [8]. They are both steady-state acoustic fields in the frequency domain, with defects under evanescent waves. The angular spectrum method is based on the analytical solution of the acoustic wave Equation (1), a linear superposition solution of the plane and evanescent waves [9]. In this paper, the longitudinal wave is used as an example, and the fluctuation Equation (7) is shown

$$\frac{\partial^2 p}{\partial x^2} + \frac{\partial^2 p}{\partial y^2} + \frac{\partial^2 p}{\partial z^2} = \frac{\partial^2 p}{c^2\partial t^2} \tag{7}$$

Equation (7) in *p* is the sound pressure of the acoustic wave is a scalar function of space and time. According to the Fourier transform theory, we can obtain Equation (8)

$$p(x,y,z,t) = \frac{1}{2\pi}\int_{-\infty}^{\infty} P(x,y,z,\omega)\exp(-i\omega t)d\omega \tag{8}$$

where *P* is the spectrum in Equation (7), we take its real part P (x, y, z) corresponding to the sound field. It is the steady state sound field with angular frequency ω. P (x, y, z) satisfies the steady state Equation (9)

$$\frac{\partial^2 P}{\partial x^2}+\frac{\partial^2 P}{\partial y^2}+\frac{\partial^2 P}{\partial z^2}+k^2 P=0 \tag{9}$$

In Equation (9), $k=\omega c^{-1}$ is the wave number. c is the speed of sound. Solving the equation by the separation of variables method. Suppose P can be written as the product of three coordinate functions

$P=X(x)Y(y)Z(z)$ substituting into the Equation (9) yields

$$\frac{d^2X(x)}{Xdx^2}+\frac{d^2Y(y)}{Ydy^2}+\frac{d^2Z(z)}{Zdz^2}+k^2=0 \tag{10}$$

The first three terms of Equation (10) are functions of different variables, so they can only be constants, which are denoted as $-k_x^2$, $-k_y^2$ and $-k_z^2$ (They satisfy that $k_x^2+k_y^2=k_z^2$), respectively, to obtain the solution of Equation (7) as $P=P_0\left[i(k_x x+k_y y+k_z z)\right]$ expressed in vector form as

$$P=P_0\exp(i\vec{k}\cdot\vec{r}) \tag{11}$$

In Equation (11), $\vec{r}=(x,y,z)^T, \vec{k}=(k_x,k_y,k_z)^T$. The general sound field can be expressed as a linear superposition of Equation (11) with different ***k***. Formally, it can be expressed as an integral with weighted coefficients in the complex ***k***-space of the form Equation (11). It can be expressed as

$$P_r=\iiint_{\Omega_{\vec{k}}} P'(\vec{k})\exp(i\vec{k}\cdot\vec{r})d\vec{k} \tag{12}$$

where $\Omega_{\vec{k}}$ is the region in ***k***-space. Different issues require different integration regions and weighting factors $P'(\vec{k})$. The weighting factor is the angular spectrum.

The angular spectrum method is mainly used to solve fluctuation problems in layered media. If the sound field is independent of y, the angular spectrum sound field of the 2D acoustic field can be simplified as

$$P(x,y,\omega)=P_1(x,y,\omega)+P_2(x,y,\omega) \tag{13}$$

In Equation (13) P_1 and P_2 are the waves in $\pm k_z$ two directions, respectively,

$$\begin{aligned}P_1(x,z,\omega)&=\int_{-\infty}^{\infty}P_1'(k_x,\omega)\exp[i(k_x x+k_z z)]dk_x\\ P_2(x,z,\omega)&=\int_{-\infty}^{\infty}P_2'(k_x,\omega)\exp[i(k_x x-k_z z)]dk_x\end{aligned} \tag{14}$$

Equation (13) can also be expressed as

$$P(x,z,\omega)=\int_{-\infty}^{\infty}\widetilde{P}(k_x,z,\omega)\exp(ik_x x)dk_x \tag{15}$$

The total sound spectrum in Equation (15) can be expressed as

$$\widetilde{P}(k_x,z,\omega)=\widetilde{P}_1(k_x,z,\omega)+\widetilde{P}_2(k_x,z,\omega) \tag{16}$$

$$\begin{aligned}\widetilde{P}_1&=P_1'(k_x,\omega)\exp(ik_z z)\\ \widetilde{P}_2&=P_2'(k_x,\omega)\exp(-ik_z z)\end{aligned} \tag{17}$$

Equation (17) shows two angular spectra with direction z, respectively. Combining Equation (24) with Equation (25), Equation (23) can be rewritten as

$$P_1(x,z,\omega)=F_1^{-1}[P_1'(k_x,\omega)\exp(ik_z z)]=F_1^{-1}\left[\widetilde{P}_1(k_x,z,\omega)\right] \tag{18}$$

The $F_1^{-1} = \int_{-\infty}^{\infty} dk_x \exp(ik_x x)$ is a one-dimensional Fourier inverse commutator. The results are similar for waves in the *-z* direction, in Equation (17). The transient acoustic field expression is

$$p(x,z,t) = \frac{1}{2\pi}\int_{-\infty}^{\infty} P(x,z,\omega)\exp(-i\omega t)d\omega$$

$$p(x,z,t) = \frac{1}{2\pi}\int_{-\infty}^{\infty}\int_{-\infty}^{\infty}[P_1'(k_x,\omega)\exp(ik_z z) + P_2'(k_x,\omega)\exp(-ik_z z)]\exp(i(k_x x - i\omega t))dk_x d\omega \quad (19)$$

3. Numerical Simulation

According to Equation (15-17), the incident acoustic beam can be represented by the angular spectrum as

$$P_i(x,z,\omega) = \int_{-\infty}^{\infty} P_i'(k_x,\omega)\exp(ik_z z)\exp(ik_x x)dk_x \quad (20)$$

In the case where the sound field is 2D, the angular spectrum is the Fourier transform of the Gaussian envelope modulation function on the $z = 0$ interface for a known sound pressure of x [10], from which it can be deduced that

$$P'(k_x) = P_0 \exp\left[-\left(\frac{k_x - k_{x0}}{k_{xb}}\right)^2 - ik_x x_s\right] \quad (21)$$

In Equation (21), P_0 is a constant and can be directly taken as 1 in the calculation. k_{xb} is proportional to the width of the angular spectrum and $kxb=2x_b^{-1}$. Using numerical simulation software to calculate the sound field, we obtain Figure 2.

The propagation characteristics of transverse and longitudinal waves in solids are similar, and the longitudinal speed of sound is usually greater than the transverse speed of sound. Only the longitudinal waves are taken for the calculation to simplify the calculation. As shown in Figure 3, the one marked with i is the sound beam propagating in the *x-z* direction, and he is incident on the interface called the incident beam, and the expression of the incident beam is the same as Equation (20). Taking the angular spectrum according to Equation (21) will result in a finite wide acoustic beam. However, the analysis reveals that the cross-section of the acoustic beam calculated at oblique incidence is not symmetrical, and the beam's cross-section is not the same for different incidence angles. To solve this problem, the incident acoustic beam is calculated using the rotated coordinate method, i.e., $x_p = x_s + x_n\cos\theta$, $z_p = z_s + x_n\sin\theta$, where (x_s, y_s) is the coordinates of the new coordinate origin in the original coordinate system.

Using the relationship between the old and new coordinate systems, the acoustic field on the x-axis in the new coordinate system can be obtained from the positively incident acoustic beam in the old coordinate system.

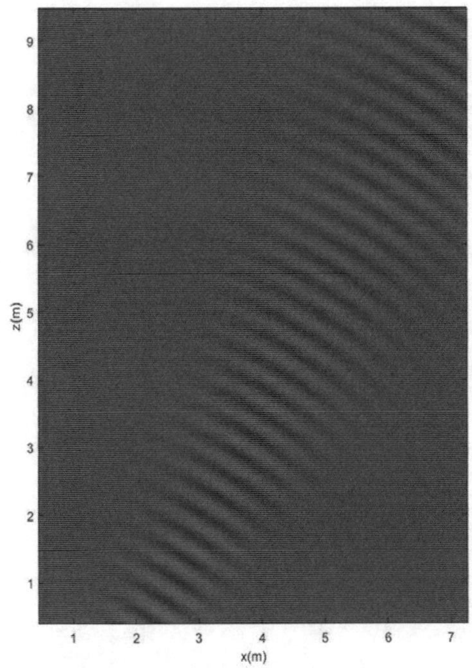

Figure 2 Finite wide beam

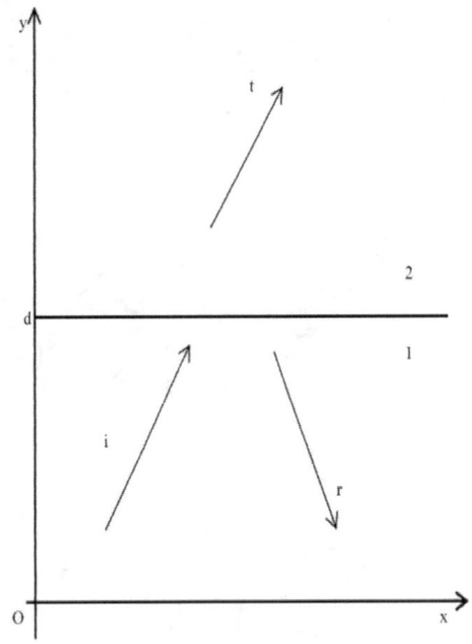

Figure 3 Reflection and Refraction

According to the available historical literature and information data [10-12], we can obtain data on water features in the south of Dalian. The Wilson empirical formula for the most appropriate seawater sound velocity and the empirical formula for sea ice based on porosity can be selected from data such as oven temperature. They can be used to obtain coefficients that are closely related to the acoustic properties of the materials, which are involved in double layers media that are used as input parameters for numerical modeling, as shown in Table 1.

Table 1 Double layer media material parameter setting table

Water density kg/m^3	Speed of sound in water m/s	Ice Density kg/m^3	Speed of sound in ice m/s	Ice Young's modulus MPa	Ice Poisson ratio
1101	1390	915	4175	50	0.31

According to Equation (21) and Equation (6), using it to calculate at different incidence angles, the calculation results are obtained in Figure 4-7. Figures 4-5 show the refraction and reflection of the wave beam at angles greater than the critical angle. Figure 6-7 shows the case where the incident angle is less than the critical angle, and the refracted acoustic wave is evanescent (exponentially decreasing fading wave). A schematic diagram of a two-dimensional model is shown in Figure 8. It has an ice thickness of only 0.5 cm. Figure 9 shows the real-time sound pressure at the detection point for a 500 Hz simple harmonic at a 15-degree oblique incidence angle. Figure 10 shows the evanescent wave's sound pressure at the same parameter.

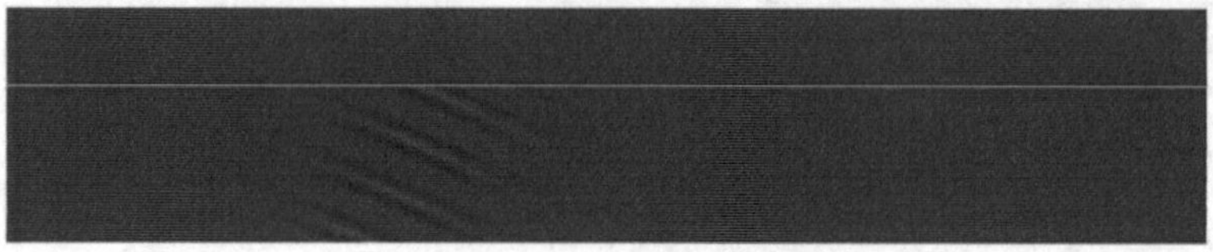

Figure 4 15° incident beam

Figure 5 15° reflection and refraction

Figure 6 45° incident beam

figure 7 45° reflection and refraction

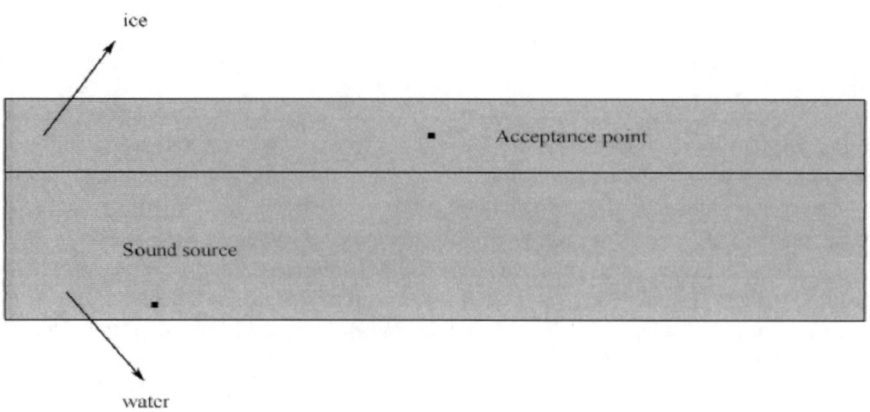

Figure 8 Diagram of the model

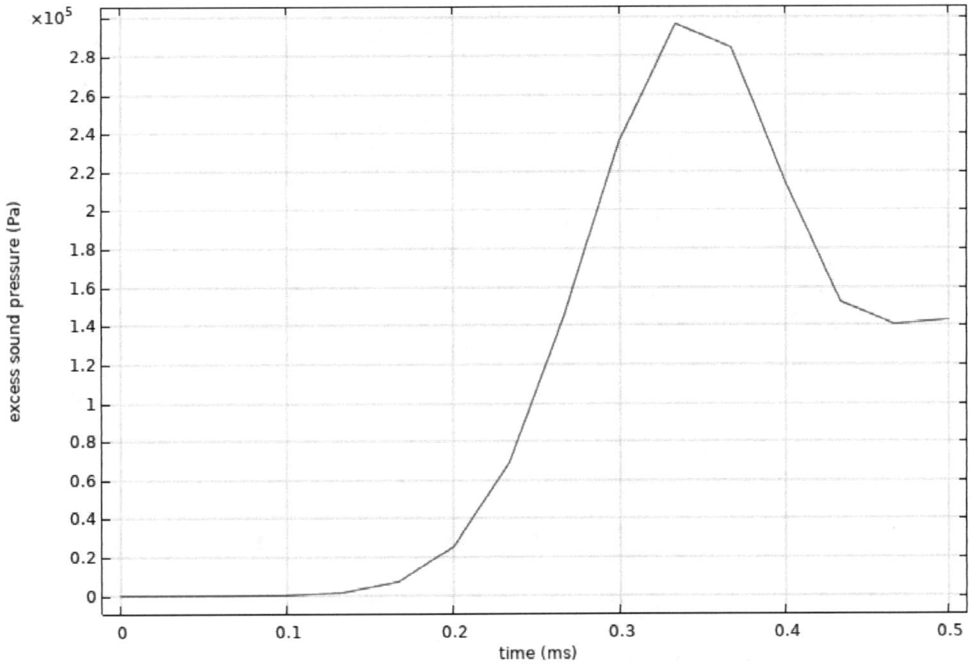

Figure 9 15° Peak sound pressure map of the detection point

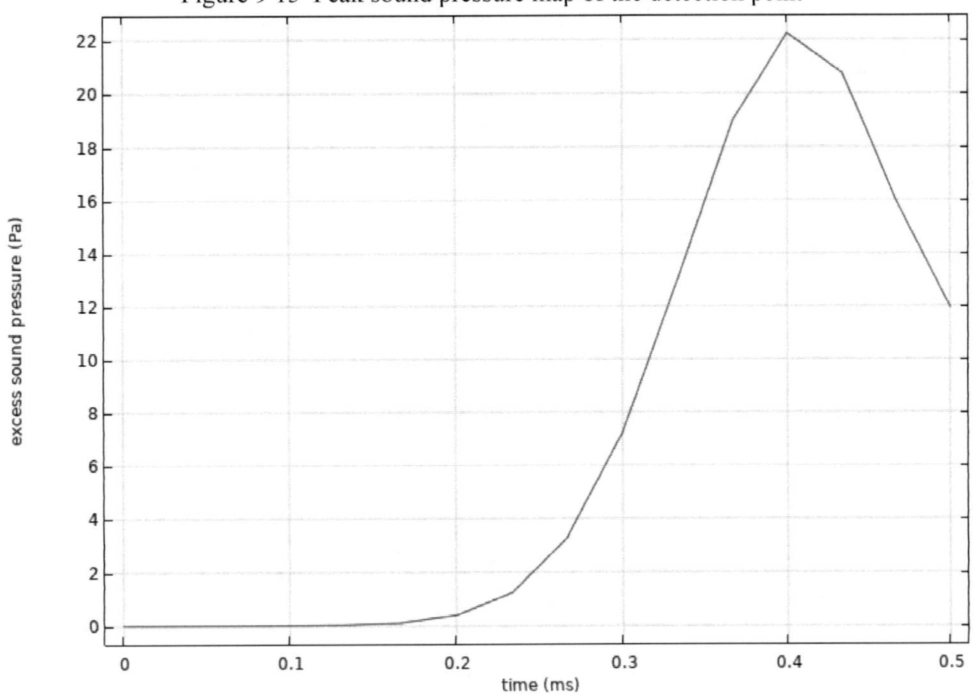

Figure 10 45° Peak sound pressure map of the detection point

4. Conclusion

In this paper, the propagation characteristics of transverse and longitudinal waves are obtained by solid sound transmission theory, and the faster propagating longitudinal waves are used as the object of study. The simulation calculation method proposes the time domain characteristics of the acoustic field of the thin ice layer on the water. The acoustic propagation modes of evanescent and simple harmonic waves are compared so that the refraction theory model can be applied to arbitrary incidence angles. This model allows a more accurate simulation of the acoustic propagation mode under thin offshore ice. Further analysis can be performed to obtain the propagation characteristics within the ice, save costs and make it easier to obtain physical parameters such as ice thickness. The technology provides a theoretical basis for further use of acoustic methods and sea surface ice detection.

Reference

[1] Yang T C, Votaw C W 1981 J. Acoust. Soc. Am. 70 841.
[2] Liu Sh X, Li Zh L 2017 A. Reflecting and Scattering of Acoustic Wave from Sea Ices. Acta Phys. Sin. 66 (23) 234301-234301.
[3] Yu J A, Zhang B, Liu Q Zh, Chen W B, Wang R Sh 1999 S. Oceanologia Et Limnologia Sinica. 30 (5) 552-557.
[4] Li X X, Zhang X, Bao M, Meng J M, Liu G W 2021 Apr. Marine Sciences. 40 (1) 90-101
[5] Bishop G C, Mellberg L E, Albanese C J, et al. 1985 A Simulation Model for High-frequency Under-ice Acoustic Back Scattering[J]. The Journal of the Acoustical Society of America. 77 (S1) S56-S56.
[6] Maksimov A O. 2011 Acoustic Manifestation of Bubbles Frozen in an Ice Sheet[J]. Acoustical Physics 57 (3) 391-400.
[7] Du G H, Zhu Zh M, Gong X F, 2012 Fundamentals of Acoustics. Nanjing University Publisher. Nanjing
[8] Etter P C. 2013 Underwater Acoustic Modeling and Simulation[M]. CRC Press.
[9] Zhang Y Sh, Ma H W, Wang H T, Wang X. 2018 Study on Near Field Sound Pressure of Circular Piston Sound Source Based on Spatial Fourier Transfor. Piezoelectrics and Acoustooptics. 40 (2) 257-261
[10] Lv H B, Sha H F. 2020 Determination of The Calculation Formula of Optimal Submarine Sound Velocity in Haizhou Bay. Journal of Jiangsu University. 29 (1) 34-39.
[11] Feng Ch H, Liu Q, Zhang J. 2017 Porosity Based Ice Sound Velocity Model and Its Empirical Formula. Technical Acoustics. 36 (6) 509-515.
[12] Wu B, Chen Ch A, Lin L. 2014 Analysis of Applicable Scope of Empirical Equation for Sound Velocity. Technical Acoustics. 33 (6) 504-507.

Investigation of Mixing Characteristics of the Transverse Jet in a Supersonic Combustor at Low Mach Inflow Condition

Yi Wang[1], Lin Zhang[1*], Jianhan Liang[1*]

[1] College of Aerospace Science and Engineering, National University of Defense Technology, Changsha, Hunan, 410073, China

*Corresponding author's e-mail: zhlin151@163.com

Abstract: LES is used in the present work to simulate and unfold unsteady flow dynamics and mixing characteristics in a cavity-based supersonic combustor at low Mach numbers. In the combustor, a transverse jet is injected ahead of the cavity. Two inflow Mach numbers (*Ma*=1.2 and *Ma*=2.5) are numerically investigated. Fine instantaneous and average mixing flow structures are well identified and analyzed by LES. The mixing performance is also studied and compared. The simulation results show that penetration is dominated by jet/freestream momentum flux ratio, even though the inflow Mach number varies. It is also demonstrated that for a jet at low Mach inflow, a higher coefficient of pressure is attained. While around the jet location, a much higher peak of the mixing vorticity efficiency is obtained for low Mach inflow, which is of fundamental importance to the near-field mixing.

1. Introduction

Supersonic mixing and combustion are critical problems to develop scramjets for air-breathing hypersonic cruise flight, which have been wide and long-term concerns [1-4]. However, most works are limited to flows at moderate and high flight Mach numbers (e.g., *Ma*>5). The study of flow, mixing, and combustion at low Mach numbers is of great significance for expanding the lower boundary of scramjet and developing the TBCC engine.

The effects of inlet *Ma* in a cavity-based supersonic combustor was numerically investigated by Mengistu et al. [5]. They found that the mixing performance was affected by the inclination angle of reattachment shock. Their simulations were conducted by a RANS solver.

To reveal the unsteady flow dynamics and mixing characteristics at lower Mach numbers, LES is performed to simulate a transverse jet in a cavity-based supersonic combustor at two different inflow Mach numbers. The mixing performance is also studied and compared.

2. Numerical methods and setups

2.1. Governing equations

The governing equation used in LES is given as:

$$
\begin{gathered}
\frac{\partial \bar{\rho}}{\partial t}+\frac{\partial \bar{\rho} \tilde{u}_i}{\partial x_i}=0 \\
\frac{\partial \bar{\rho} \tilde{u}_i}{\partial t}+\frac{\partial}{\partial x_j}\left[\bar{\rho} \tilde{u}_i \tilde{u}_j+\bar{p} \delta_{ij}-\bar{\tau}_{ij}+\tau_{ij}^{sgs}\right]=0 \\
\frac{\partial \bar{\rho} \tilde{e}_t}{\partial t}+\frac{\partial}{\partial x_i}\left[\bar{\rho} \tilde{u}_i \tilde{e}_t+\bar{q}_i+e_i^{sgs}\right]+\bar{\rho} \tilde{S}_{hs}=0 \\
\frac{\partial \bar{\rho} \tilde{Y}_\alpha}{\partial t}+\frac{\partial}{\partial x_i}\left[\bar{\rho} \tilde{Y}_\alpha\left(\tilde{u}_i+\tilde{V}_{i,\alpha}\right)+Y_{i,\alpha}^{sgs}\right]=\bar{\rho} \tilde{S}_\alpha, \alpha=1, \cdots, N_{sp}.
\end{gathered}
\tag{1}
$$

The definition of variables and the closure of sub-grid scale terms can be found in our previous work, and thus not introduced here for clarity.

2.2. Numerical setups in LES solver

To efficiently solve the LES equation, the time discretization is performed by the dual time-step method. The inviscid and viscous fluxes are solved separately for spatial discretization. Specifically, for inviscid flux term, the WENO-CU6 scheme is used.

2.3. Computational details

The simulation domain and basic boundary conditions are illustrated in Figure 1. Considering the low Mach flows, the air enters the combustor at *Ma* = 1.2 and *Ma* = 2.5. The combustor is 255 mm in length, 56 mm in width, and 56 mm in height. A cavity is set 50 mm downstream of the combustor entrance. The cavity length is 90 mm, and the depth is 15 mm, resulting in a length/depth ratio of 6. The rear angle of the cavity is 45°.

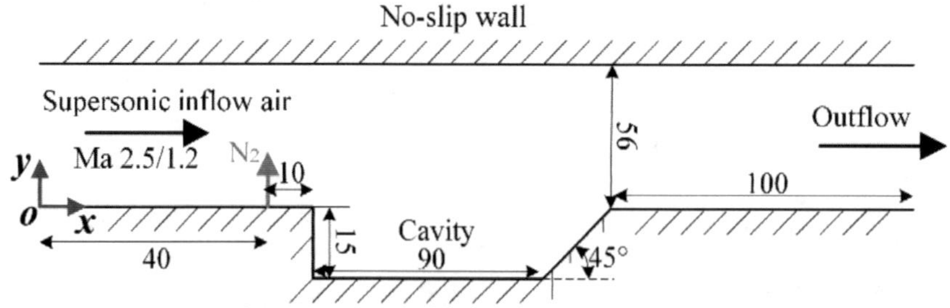

Figure 1. Simulation configuration details.

Table 1 shows the inflow and jet conditions. The mean inflow profile with a thickness of 3.6 mm is imposed onto the combustor entrance. For inflow at both low and high Mach numbers, the stagnation pressure is 100 kPa. The nitrogen is transversely injected at 10 mm in front of the cavity. The orifice diameter of the injector is 4 mm, with a stagnation pressure of 300 kPa. For Mach 1.2 inflow, the Reynolds number is 6.1×10^4. While for Mach 2.5 inflow, this value is 3.8×10^4. The jet/freestream momentum flux ratio $J \equiv \left(\rho V^2\right)_j / \left(\rho V^2\right)_\infty$ is 2.6 and 4.3, respectively.

Table 1. Inflow and jet conditions.

	Low Mach inflow (Air)	High Mach inflow (Air)	Jet (N_2)
Ma	1.2	2.5	1.0
P_t (kPa)	100	100	300
T_t (K)	300	300	300
γ	1.4	1.4	1.4
d (mm)			4
J	2.6	4.3	

δ (mm)	3.6	3.6	
Re	6.1×10^4	3.8×10^4	

Both cases use the same grids. For the main flow, the grid point numbers are 871×257×151. Around the cavity, the grid points are 401×129×151. The grids are densified around the wall and the injector, which is necessary to capture the fine flow structures. For *Ma*=2.5 inflow, the dimensionless distance y^+ is about 1, while this value is 3.4 for *Ma*=1.2 inflow. The total grid numbers are 41.6 million. The grid resolution is presented in Table 2.

Table 2. Grid resolution in computation.

Region	Inflow Mach numbers	N_x	N_y	N_z	Δx^+	Δy^+	Δz^+	Total points (million)
Main flow	*Ma*=1.2	871	257	151	8.5-476	3.4-238	8.5-238	33.8
	Ma=2.5				2.5-140	1-70	2.5-70	
Cavity flow	*Ma*=1.2	401	129	151	8.5-204	3.4-102	8.5-238	7.8
	Ma=2.5				2.5-60	1-30	2.5-70	

3. Results and discussions

3.1. Instantaneous flow features

The dimensionless density gradient magnitude plotted in Figure 2 elaborately depict the fundamental flow features at different Mach inflows. An interesting finding is that the jet penetrated deeper into the supersonic core flow at higher Mach inflow, which can be unraveled by the fact that the jet penetration is mainly affected by jet/freestream momentum flux ratio. Although the Mach number is higher, the inflow momentum flux is smaller, which gives rise to a a deeper jet penetration. The shock waves and vortices structures are also revealed by the density gradient magnitude. The boundary layer, separation regions, separation shock, bow shock, reflection shock, reattachment shock, compression waves, and unsteady large eddy structures induced by velocity shear are clearly shown. It should be noted that as the inflow Mach number is slightly larger than 1 for *Ma*=1.2 inflow, the reattachment shock is merged with the reflection shock.

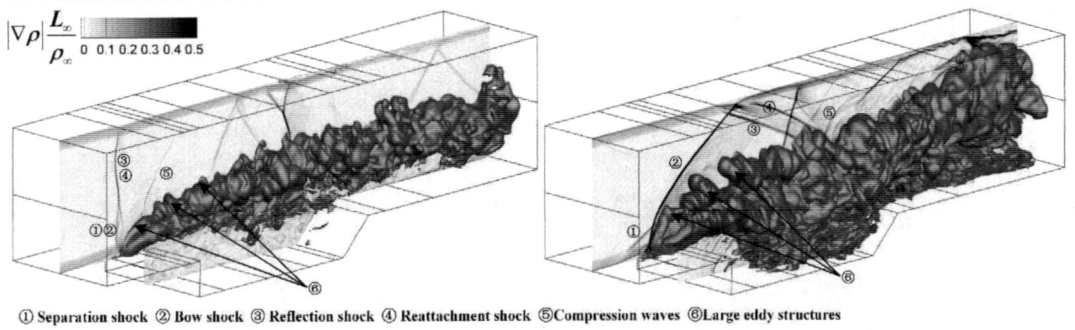

Figure 2. Instantaneous flow structures and iso-surface of N_2 mass fraction (Y_{N2}=0.05) at *Ma*=1.2 (left) and *Ma*=2.5 (right).

3.2. Mixing performance

To quantitatively evaluate the mixing characteristics at low Mach flows, several performance parameters are introduced and numerically compared: the total pressure recovery coefficient σ_{rec}, the coefficient of pressure C_p, and the mixing vorticity efficiency η_w. The definition of these parameters is omitted here for clarity.

The coefficient of pressure indicates the extent of the inflow air entrainment. It is clear from Figure 3 that a lower *Ma* inflow leads to a higher value of the coefficient of pressure, which implies a better jet/freestream mixing. Both around the jet and the ramp, a sudden decrease of the coefficient has been noticed, which is supposed to be caused by the bow shock of jet.

The mixing vorticity efficiency plays a fundamental role in jet/freestream mixing, especially around the near field. Compared with the streamwise circulation, this parameter can effectively exclude the vorticity which is of little help for mixing when the jet is not yet mixed or it is mixed well. The development of the mixing vorticity efficiency is plotted for *Ma*=1.2 and *Ma*=2.5 inflows in Figure 4. It is interesting to find that around the jet location, a higher peak ($\eta_w \approx 2.0$) is attained in the *Ma*=1.2 freestream, whereas a much lower peak ($\eta_w \approx 0.48$) is obtained in the *Ma*=2.5 freestream. Downstream the jet location, mixing vorticity efficiency reduces to a similar level. The peak value implies that near-field mixing is dominated by the streamwise vortices. Meanwhile, another peak of the mixing vorticity efficiency is produced at the cavity aft ramp.

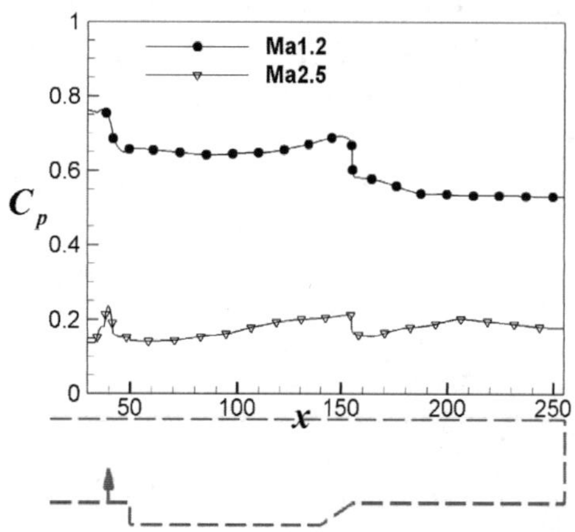

Figure 3. The average coefficient of pressure along the mainstream direction.

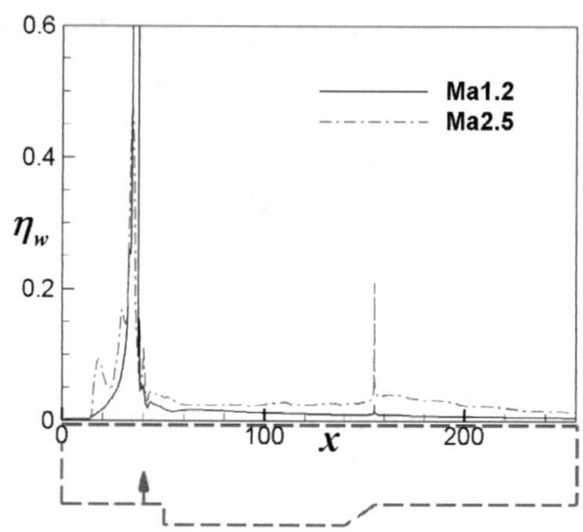

Figure 4. Average mixing vorticity efficiency along the mainstream direction.

The total pressure recovery coefficient progression plotted in Figure 5 displays a somewhat similar manner to the coefficient of pressure. Because the blockage effect of the jet for *Ma*=2.5 is more obvious, it seems that the total pressure loss is larger for higher *Ma* inflow around the cavity. Downstream the cavity, the total pressure is unchanged for *Ma*=1.2, as the flow loss is very small.

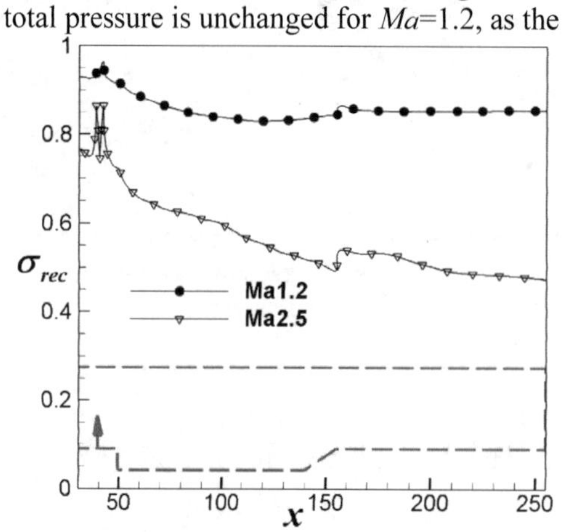

Figure 5. Average total pressure recovery coefficient along the mainstream direction.

4. Conclusions

In this work, LES is employed to unfold the unsteady flow dynamics and mixing characteristics for a transverse jet. Special attention is focused on the mixing performance at lower Mach number flows. Two inflow Mach numbers (Ma=1.2 and Ma=2.5) are numerically investigated. Fine instantaneous and average mixing flow structures are well identified and analyzed by LES. The mixing performance is also studied and compared. Based on the simulation, several important conclusions can be drawn:

(1) The penetration of the jet is dominated by jet/freestream momentum flux ratio, even though the inflow Mach number varies within an appropriate range.

(2) For a jet in a combustor at low Mach inflow, a higher coefficient of pressure is attained. Around the jet, a much higher peak of the mixing vorticity efficiency is obtained in low Mach inflow, which is of fundamental importance to near-field mixing.

Acknowledgments

This work is funded by the National Natural Science Foundation of China (Grant Nos. 12202487 and 11972367) and the Key Laboratory Foundation in China (Grant No. 6142703200310).

References

[1] M.R. Gruber, R.A. Baurle, T. Mathur, and K.Y. Hsu, Fundamental Studies of Cavity-Based Flameholder Concepts for Supersonic Combustors, Journal of Propulsion and Power 17 (1) (2001) 146-153.

[2] Q. Liu, D. Baccarella, and T. Lee, Review of combustion stabilization for hypersonic airbreathing propulsion, Progress in Aerospace Sciences 119 (2020) 100636.

[3] T. Nilsson, S. Zhong, and C. Fureby, LES of H2-air jet combustion in high enthalpy supersonic crossflow, Phys. Fluids 33 (3) (2021) 035133.

[4] W.O. Landsberg, D. Curran, and A. Veeraragavan, Experimental flameholding performance of a scramjet cavity with an inclined front wall, Aerospace Science and Technology 126 (2022) 107622.

[5] Y.G. Mengistu, D.P. Mishra, and V. Hariharan, Numerical characterization of 3D nonreacting supersonic cavity combustor with inlet Mach number variation, Int. J. Hydrogen Energ. 45 (16) (2020) 10130-10144.

Study on the Influence of Low-Frequency Flow Oscillation on the Atomization Characteristics of Gas-Centered Coaxial Centrifugal Nozzle

Jinxin Kang[1], Yiheng Tong[1] ,Chuanjin Jiang[1], Songjiang Feng[1]* , Mingxue Shao[1]

[1] Space Engineering University, Beijing, 101400, China

*Jinxin kang's e-mail: 1104417655 @qq.com
*Yiheng tong's e-mail: 616980682@qq.com
* Chuanjin Jiang's e-mail: 2486829152@qq.com
* Songjiang Feng's e-mail: 2930635318@qq.com*
*Mingxue Shao's e-mail: 2514884580@qq.com

Abstract: The dynamic atomization characteristics of gas center coaxial centrifugal nozzle under the condition of pressure oscillation in the supply system were studied by applying external excitation through the flow pulsator. Obtaining spray image by high-speed photography while using hydro-mechanical pulsator to get generate of supply system. The results show that the spray width changes periodically with low-frequency pressure oscillation. For self-pulsation sprays, it is challenging to maintain a "Christmas tree" shape with low-frequency oscillation. Pressure oscillation can effectively suppress spray self-pulsation. However, the pressure oscillation amplitude may be much larger than that of the self-pulsation itself. Therefore, the premise of applying this method is that the applied external excitation cannot stimulate or amplify the combustion instability. The gas post-pressure drop determines the amplitude and frequency of the self-pulsation, while the excitation frequency will slightly affect the amplitude and frequency of the self-pulsation.

1. Introduction

A closed-loop coupling will be formed, causing severer damage[1]. Bazarov[2, 3] made a pioneering contribution and designed hydro-dynamic pulsators which have been widely used by researchers to carry out experimental research. Fu[4] found that an open-end swirl injector with large geometric parameter A can suppress flow oscillation. Chung[1] reported that the amplitude of the liquid film thickness at the injector orifices decreases with the increase in the diameter, length, and tangential entry number of the swirl chamber. Furthermore, for open-ended swirl injector, the tangential entry number was found to have a greater impact on the dynamic characteristics. Xue[5] pointed out that when the oscillation frequency is high, the thicker liquid film is less sensitive to short disturbance waves. When the mass flow of the supply system oscillates, the shape of the gas core will also change accordingly. Khil[6] conducted that the air core of the chain shape is caused by the integration of the Klystron effect and swirl in the swirl chamber. Cheng[7] investigated that air core oscillates propagates in a long wave manner in the injector. In addition, the wavelength of the oscillation is about 11 times larger than the injector length.

Under some operating conditions, the spray and pressure will shock periodicly. The important physical phenomenon Self-pulsation can be observe. So many of studies have been explored the reasons behind the phenomenon in the recent period. These studies mainly aimed to find out the generation mechanism of self-pulsation and its influence on atomization characteristics. Bazarov[8] Through compareing under the same conditions ,found the self-pulsation spray has a better atomization effect. Thus, it can make spray quality bater and improve propellant mixing uniformity .So reduce the probability of combusion instability. Zhou[9, 10] pointed out that screaming has an intensifying effect on atomization. Kang[11] found that the gas core size significantly affects the shape of the self-pulsation spray. When the gas core size is large, a "Christmas tree" shape spray is formed. And when the size of the gas core is small, the spray is in a "string" shape. Moreover, self-pulsation causes an increase in the spray angle. Im[12] reported that self-pulsation may weaken the atomization performance of the spray and amplify combustion instability.

In summary, many studies have been devoted to exploring the periodic flow pulsation and the injector self-pulsation. Most of these studies focused on single injectors and LCSC injectors. However, few studies many studies have been devoted to exploring the coupling of self-pulsation and periodic flow oscillation of the GCSC injector. In this study, through analyzing the experimental data under high-speed photography the characteristics of injection pressure, the a GCSC injector oscillation frequency and amplitude on the spray pattern is explored. In addition, the relationship between self-pulsation and periodic flow oscillation is discussed.

2. Experimental methods

2.1 Experimental facilities

A schematic of the GCSC injector is shown in Figure 1. The coaxial swirl injector is composed of an inner and an outer injector. four tangential entries, a swirl chamber and orifices compose the outer injector. The liquid (water) inflow the swirl chamber. On the effect of centrifugal force the, The liquid forms a rotating liquid film. The gas is injected through the inner nozzle. The the injector geometrical parameters are writed in Table 1.

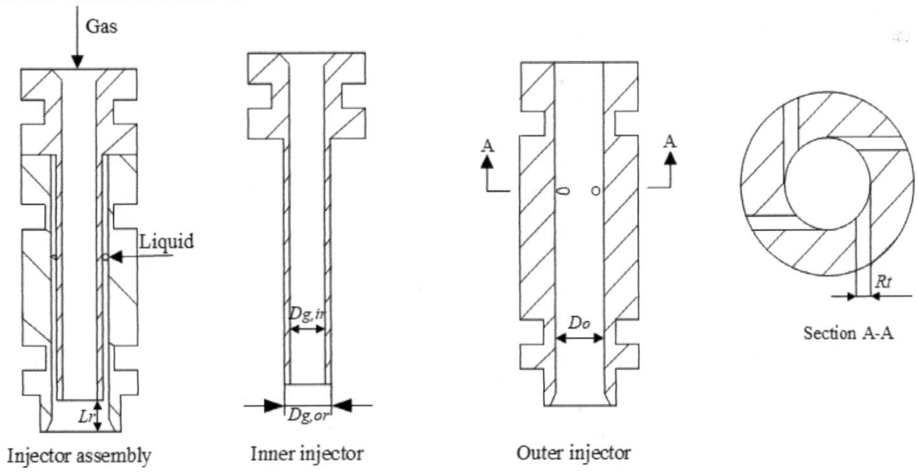

Figure 1 Injector geometry

Table 1 The injector Dimensional parameters

Parameters	Values (mm)
Do	5.5
Dg	3.5
Hgap	0.5
Dt	0.6
Lr	0, 2, 7

As shown in Figure 2, the experimental apparatus used in this study is comprised of a hydro-mechanical pulsator, a propellant feed system, an image acquisition system, a spray collector, a measurement and control system and a GCSC injector. The hydro-mechanical pulsator is designed based on CHENG's study[7], and its principle is shown in Figure 3.

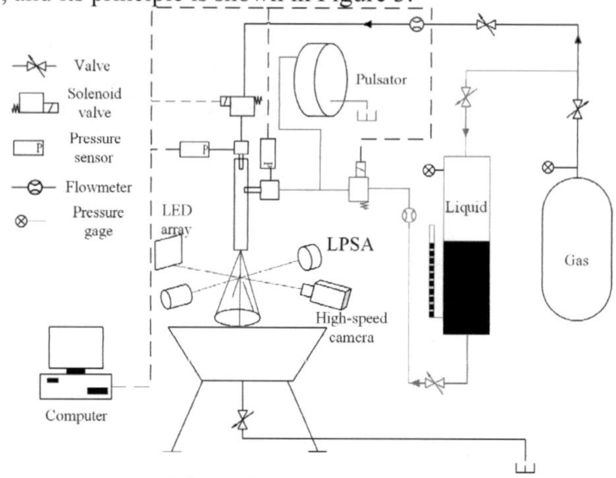

Figure 2 system

Figure 2 shows the experimental system, which is consisted of a hydro-mechanical pulsator, a propellant feed system, an image acquisition system, a spray collector, a measurement and control system and a GCSC injector

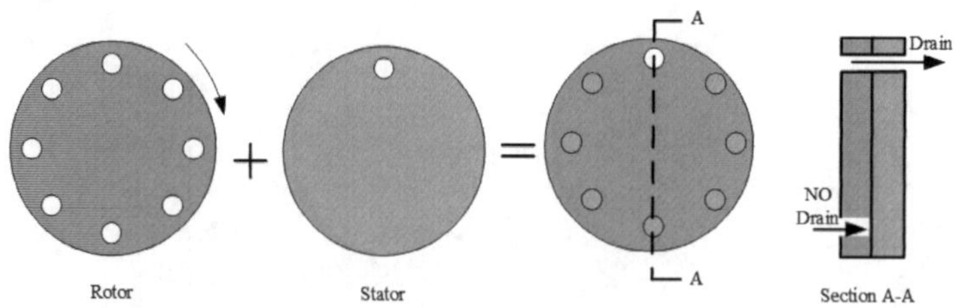

Figure 3 Principle of the hydro-mechanical pulsator

2.2 Experimental system

Pressure sensors (the 4730 diffused silicon pressure sensors) were used to measure the pressure in the liquid and gas manifold. The accuracy of the pressure sensors is 0.25% FS. A high-speed camera was used to get spray images. To capture clear spray images, a backlighting imaging technology was adopted for instantaneous spray images. Meanwhile LED light (Scienploer Light Tech, HLS-30, Maximum power: 250 W) was employed to illuminate the spray. The resolution of the camera was 800 × 600 pixels with a frame rate of 31784, and the exposure time was set to 2 μs.

2.3 Experimental conditions

Table 2 shows the experimental conditions of the GCSC injector used in this study. The environment is at the atmosphere pressure.The water is filtered by filter to avoid impurities blocking the inlet dried air supplied through a propellant feed system.The The hydro-mechanical pulsator is set with three excitation frequencies of 1 Hz, 10 Hz, and 15 Hz. And the manifold pressure of the liquid and its frequency spectrum are shown in Figure 4. It shows that when the propellant feed system oscillates, the pressure in the GCSC injector manifold oscillates periodically. Moreover, the excitation frequency of the pulsator is consistent with the frequency of the pressure oscillation in the manifold.

Table 2 Operating conditions in experiments

Test NO	ΔPl /Mpa	ΔPg /Mpa	Oscillation frequency/Hz
1	0.27	0-0.5	0
2	0.27	0-0.5	1
3	0.27	0-0.5	10
4	0.27	0-0.5	15

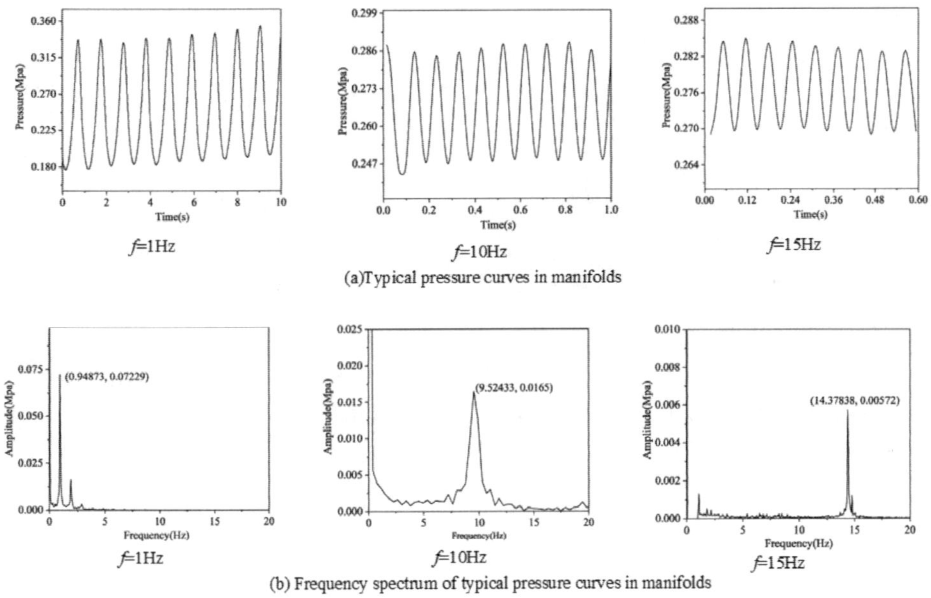

Figure 4 Manifold pressure of the liquid and its frequency spectrum

2.4 Data processing

2.4.1 Measurement of self-pulsation frequency

In this study, the oscillation spray frequency is obtained by processing the transient spray images according to a time series. The procedures are as follows. First, the background interference is eliminated by calculating the gray gradient difference between two adjacent raw images. Then the image is binarized without the background information. Third, the sum of gray values along a detecting line are extracted. Finally, the frequency and amplitude of the spray oscillation can be obtained by summing up the gray values through FFT (fast Fourier transform). Figure 5 (a) shows the frequency and amplitude of the sum of the gray values of pulsating spray extracted by the FFT method when the excitation frequency is 10 Hz. We can see from Figure 5(b) that its dominant frequency is consistent with the excitation frequency. This clearly shows that the frequency and amplitude of the oscillation spray can be extracted in this way. In this paper, FFT is performed on the spray gray value information at different positions from the nozzle. The results are given in Figures 5(c) and 5(d).

At the position of 270 pixels, 310 pixels, 350 pixels, and 390 pixels downstream of the nozzle, the frequency measurement results are all 9.5352 Hz. The amplitude measurement results are 183.84 pixels,

186.40 pixels, 197.39 pixels, and 192.77 pixels. And the relative error is 3.8%. Therefore, the measurement results will not vary accordingly to the location.

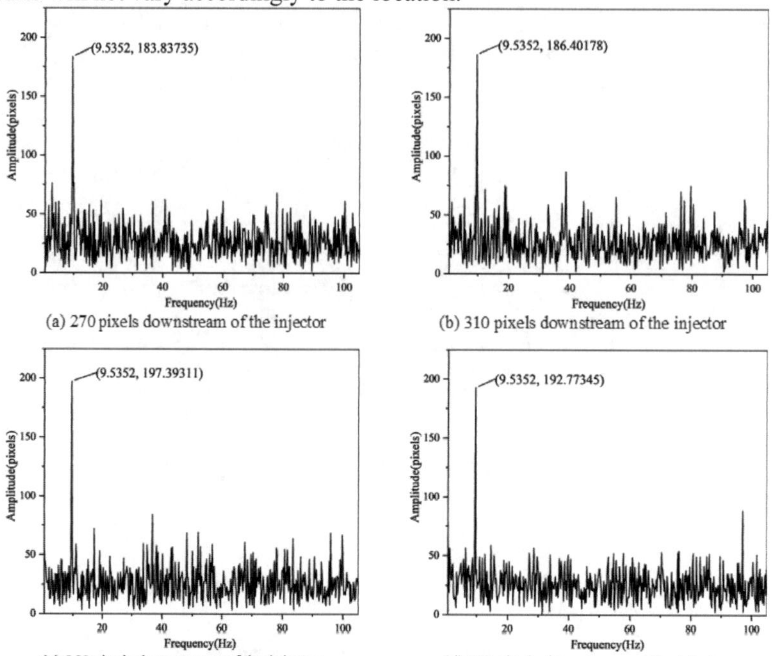

(a) 270 pixels downstream of the injector (b) 310 pixels downstream of the injector

(c) 350 pixels downstream of the injector (d) 390 pixels downstream of the injector

Figure 5 The sum of spray gray value and frequency during pressure oscillation

2.5 Proper Orthogonal Decomposition

Proper Orthogonal Decomposition (POD) is the earliest proposed method for modal decomposition. This method decomposes the flow field into several spatial orthogonal modes. Then the various modes are sorted according to the magnitude of energy to select the primary flow mode. Orthogonal modes reflect the spatial distribution characteristics of oscillation energy. Since the first orthogonal mode is the mode with the largest energy in the process of liquid film breaking. In this study, it is the only mode selected for the analysis of the spatial distribution of oscillating spray energy.

3. Results and Discussion

3.1 Spray Pattern

The self-pulsation spray pattern without pressure oscillation can be found in Figure 6. It can be clearly seen from the figure that when the self-pulsation is generated by the injector, the spray oscillates periodically, showing a "Christmas tree" shape as a whole. The distance between the "branches" of the "Christmas tree" shape self-pulsation spray is more uniform. It can be seen from the whole image that the self-pulsation spray forms an obvious "Christmas tree" shape. This shows that the self-pulsation spray has stable periodicity.

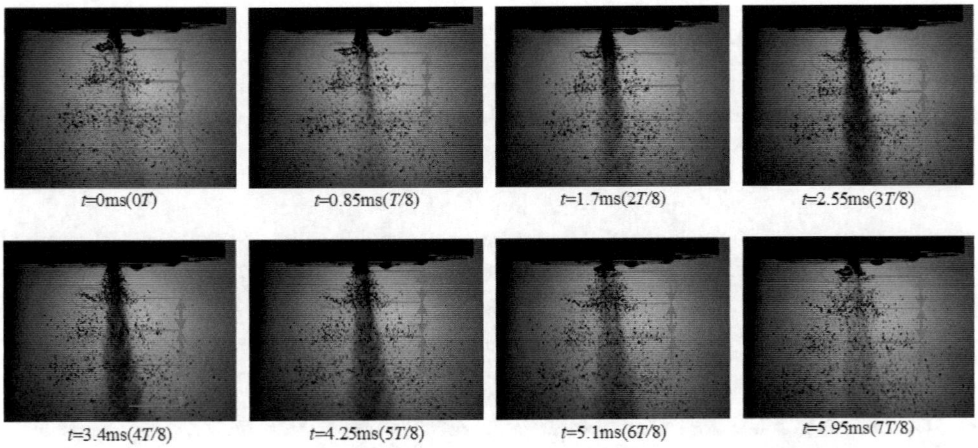

Figure 6 Variation of self-pulsation spray pattern with time, f=0 Hz, ΔPg=0.237 Mpa

Figure 7 shows the variation of spray pattern at different time points when the excitation frequency of the hydro-mechanical pulsator is 10 Hz. The spray shows a "bell" shape, and the cone-shaped liquid film is not fully opened. The reason is that the liquid post-pressure drop is slightly smaller, and the surface tension of the liquid film is greater than the inertial force of the liquid film. Therefore, the liquid film gradually converges downstream, forming a "bell" shape. It can be seen from the figure that the width of the spray changes periodically. In this study, no Klystron effect was found. Xue[5] points out that the Klystron phenomenon of the injector takes place only when the excitation frequency of the mass flow increases. Before this, the spray only showed periodic changes in the spray angle. Therefore, the excitation frequency in this study is low (≤15 Hz), and the Klystron phenomenon will not be generated. The spray oscillates with the spray width and the spray angle varying periodically. However, when ml is low, the liquid film of the spray is comparatively long due to the weakness of the centrifugal forces. The location of the first breakup of the film is about 50 mm downstream of the injector/ dump plane. Also, because of the surface tension, the spray shrinks before the film breaks up into droplets. For those reasons, the spray shows a "bell" shape. The cone-shaped liquid film is not fully opened, and hence the change in the spray cone angle is not very obvious. Nevertheless, the oscillation of the spray is shown by the spray width in this paper since the variation of the spray width is more obvious than that of the spray angle.

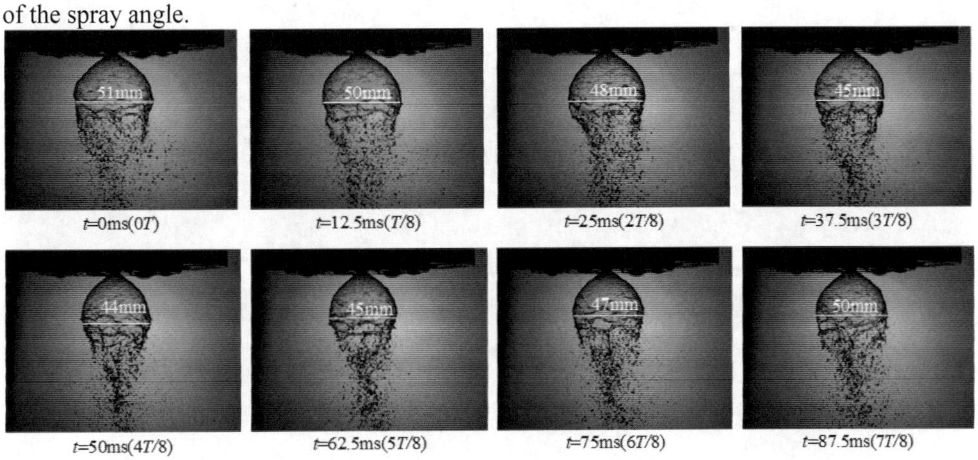

Figure 7 Variation of spray pattern with time under pressure oscillation

Journal of Physics: Conference Series **2458** (2023) 012005 doi:10.1088/1742-6596/2458/1/012005

Figure 8 shows that the variation of the self-pulsation spray pattern changes with time when the supply system pressure oscillates. This is significantly different from the spray pattern when the supply system is stable (Figure 6). The pressure oscillation of the supply system has a significant influence on the self-pulsation spray pattern. In Figure 8, only the spray images of T/8 and 6T/8 are similar to the spray images of self-pulsation when the pressure of the supply system is stable. The spray pattern at other times is significantly different from the self-pulsation spray pattern when the pressure of the supply system is stable. At this time, the oscillation of the spray can only be observed near the injector orifice, while downstream far away from the injector orifice is difficult to capture the oscillation of the spray. The distribution of self-pulsation spray is more uniform than that when the pressure is stable.

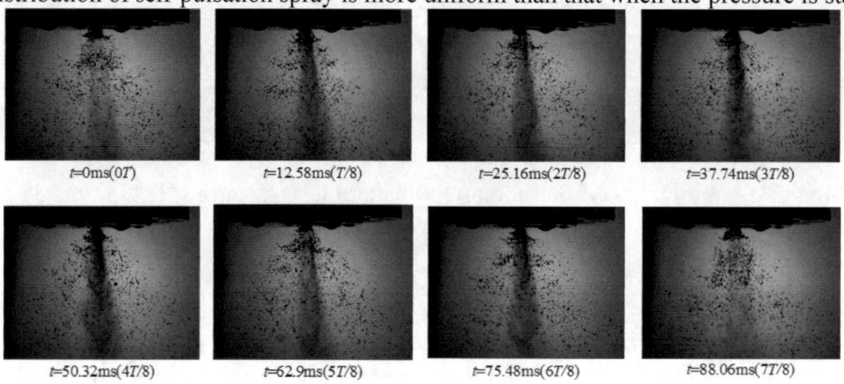

Figure 8 Self-pulsation spray pattern in one cycle with pressure pulsation

3.2 Response Characteristics

The method introduced in Section 2.4 is adopted to extract the frequency of the spray oscillation. Figure 9 shows the spray oscillation frequency when the gas post-pressure drop is 0.237 Mpa and the hydro-mechanical pulsator excitation frequency is 0 Hz, 1 Hz, 10 Hz, and 15 Hz. Figure 11 indicates that when the supply system is stable, the frequency of the spray self-pulsation is 139.94 Hz. However, when the supply system produces pressure oscillations, the spray produces two oscillation frequencies. One is low-frequency oscillation, and the other is high-frequency oscillation. The frequency of the low-frequency oscillation is the frequency of the supply system oscillation, and the frequency of the high-frequency oscillation is the frequency of the spray self-pulsation. The comparison of Figure (a), Figure (b), Figure (c), and Figure (d) shows that the widening of the frequency spectrum of the self-pulsation frequency is caused by the supply system pressure oscillation . When supply system pressure oscillates, the self-pulsation is suppressed. However, the amplitude of oscillation generated in some operating conditions by supply system pressure is much higher than that of self-pulsation, which may cause combustion instability.

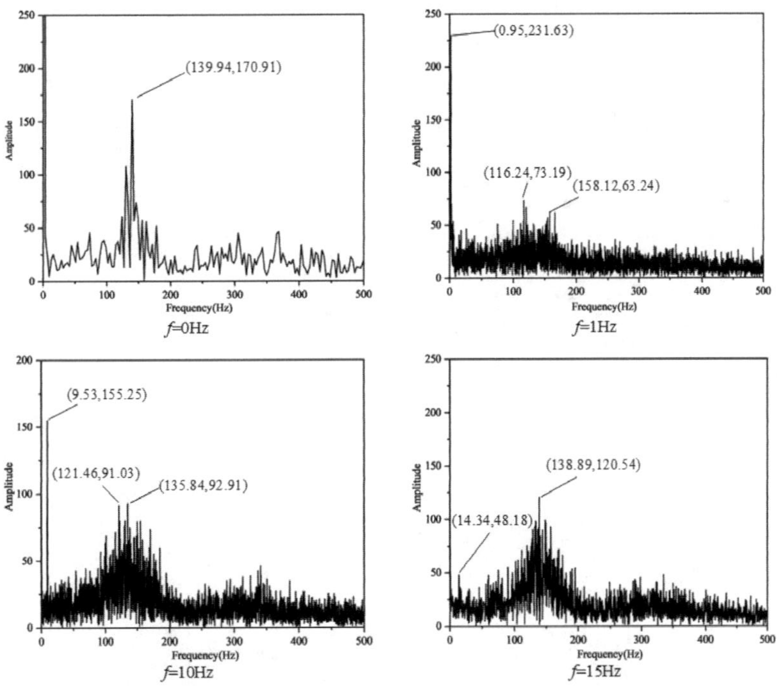

Figure 9 Amplitude frequency diagram of the self-pulsation spray

Figure 10 reveals that as the excitation frequency increases, the spray pressure oscillation amplitude decreases. However, the spray self-pulsation amplitude will decrease significantly at first and then increase slowly with the excitation frequency increase. The intensity of self-pulsation is the greatest with stable pressure from the supply system. When excitation is applied to the supply system, self-pulsation is suppressed. However, as the excitation frequency increases, the degree of suppression of self-pulsation decreases.

Figure 11(a) shows the influence of gas post-pressure drop on the frequency of self-pulsation. The black line represents the self-pulsation frequency with the stable pressure from supply system. The red line represents self-pulsation frequency with pressure oscillation from supply system. It can be seen that the frequency of the self-pulsation with the stable pressure from the supply system is consistent with that with pressure oscillation. Therefore, the gas post-pressure drop determines the self-pulsation frequency. The pressure oscillation of supply system has relatively little influence on the self-pulsation frequency. From Figure 11(b), we can see that when the supply system pressure oscillates, the amplitude of the excitation frequency can be divided into three sections: the initial rising section, the decreasing section, and the rising section as the post-pressure drop increases. The reason for the sharp increase in the amplitude in the initial rising section is that the spray is in a wave-assisted breakup. The liquid film has a longer breakup length at this period of time. Increasing the gas post-pressure drop will make the liquid film unstable. This will increase the liquid film fluctuation and amplify the liquid film oscillation of itself. The decreasing section is driven by the significant reduction of the liquid film breakup length. The liquid film breakup mode is changed from wave-assisted breakup to perforated breakup, segmented breakup, and self-pulsation. Moreover, when the spray is in self-pulsation, the amplitude of the excitation frequency is the smallest. The gas pressure drop continues to increase, and the amplitude of the excitation frequency rises again. The spray breakup mode changes to the air-blasted breakup. Xu[13] pointed out that the spray breakup mode is an air-blasted breakup, and the spray angle decreases when the gas-liquid ratio increases. Therefore, when the supply system oscillates, the gas-liquid ratio will also oscillate. When spray is an air-blasted breakup, the spray angle is more sensitive to the gas-liquid ratio. Moreover, when the supply system pressure oscillates, the amplitude of self-pulsation is reduced compared to that with a stable supply system.

In summary, applying external excitation to the supply system can indeed suppress the self-pulsation of the spray. So it seems that a weaker low-frequency external excitation can be applied to suppress combustion instability. The premise of the application of this method is that the applied external excitation cannot cause or amplify combustion instability.

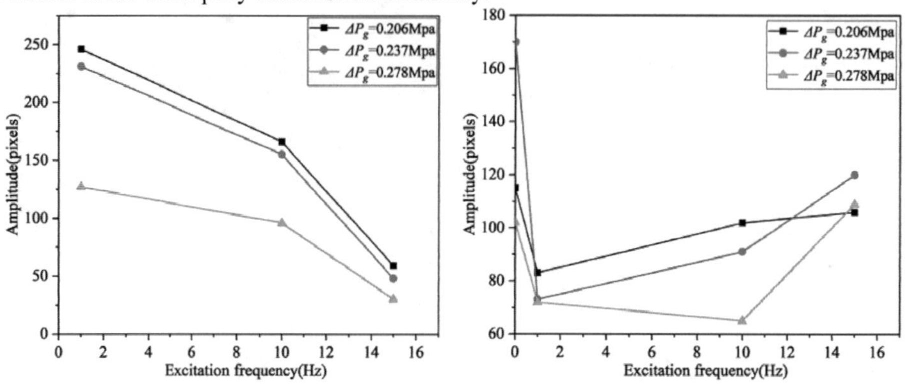

(a) Amplitude of excitation (b) Amplitude of self-pulsation with pressure oscillation

Figure 10 Influence of excitation frequency on amplitude

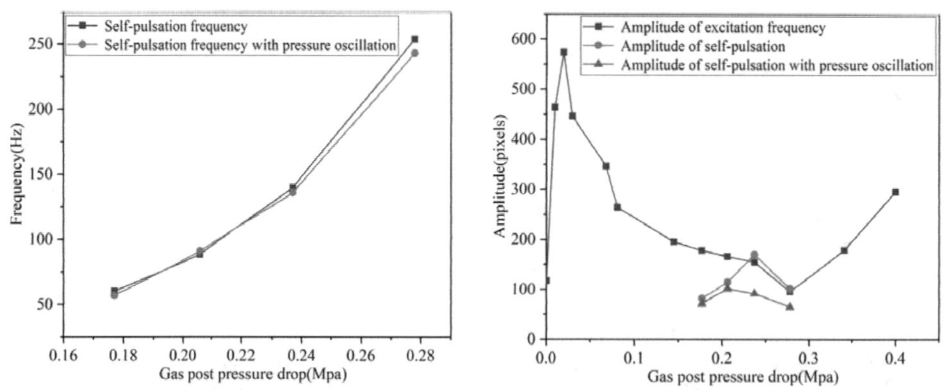

(a)The effect of gas post pressure drop on response frequency (b)The effect of gas post pressure drop on response amplitude

Figure 11 The effect of gas post-pressure drop on response frequency

4. Conclusion

The characteristics of gas-centered swirl coaxial injectors with low-frequency flow pulsation were investigated experimentally. The effects of gas post-pressure drop and excitation frequency on the spray pattern and spray response characteristics are analyzed. The following conclusions can be drawn from this study:

1. The structure of the self-pulsation spray shows an apparent “Christmas tree” shape. The distance between the branches of the “Christmas tree” is roughly the same. But it can be destroyed by pressure oscillation from the supply system. The oscillation of the spray can only be acquired near the injector orifice. And it is challenging to keep the oscillation of the spray downstream stable at other points far away from the injector orifice.

2. The spray responds well to excitation, and the supply system pressure oscillation can suppress the self-pulsation. Therefore, it is possible to apply external excitation to suppress self-pulsation without causing or amplifying combustion instability.

References

[1] CHUNG Y, KIM H, JEONG S, et al. Dynamic Characteristics of Open-Type Swirl Injector with Varying Geometry [J]. Journal of Propulsion and Power, 2015, 32(3): 583-91.
[2] V.G B. Hydro-Mechanical Pulsator: USSR. 1967-.
[3] BAZAROV V, LEE E, LINEBERRY D, et al. Pulsator Designs for Liquid Rocket Injector Research; proceedings of the Aiaa/ASME/SAE/ASEE Joint Propulsion Conference & Exhibit, F, 2013 [C].
[4] FU Q, YANG L. Theoretical Investigation on the Dynamics of a Gas-Liquid Coaxial Swirl Injector [J]. Journal of Propulsion and Power, 2011, 27(1): 144-50.
[5] SHUAIJIE X, HONGJUN L, LIU H, et al. Experimental on dynamic characteristics of an open-end swirl injector with thick liquid film %J Acta Aeronautica et Astronautica Sinica [J]. 2018, 39(12): 122534.
[6] KHIL T, KIM S, KIM H, et al. Spray Characteristics of a Single Simplex Injector with Low Hydrodynamic Disturbance generated by Pressure Fluctuation in Feed Line [M]. ICLASS06. Kyoto, Japan. 2006.
[7] CHENG P, LI Q, KANG Z, et al. Response of inner flow and spray characteristics of a pressure swirl injector to pressure oscillation in supply system [J]. Acta Astronautica, 2018, 154(82-91).
[8] BAZAROV V. Self-pulsations in coaxial injectors with central swirl liquid stage; proceedings of the 31st Joint Propulsion Conference and Exhibit, F, 1995 [C].
[9] JIN Z, XIAOPING H, YUHUI H, et al. A Study of the Mixing Characteristics of Gas-Liquid Coaxial Swirling Injector of Liquid Rocket Engine [J]. JOURNAL OF NATIONAL UNIVERSITY OF DEFENSE TECHNOLOGY, 1997, 19(4): 9-13.
[10] JIN Z, XIAOPING H, YUHUI H, et al. AN EXPERIMENTAL STUDY ON ACOUSTIC CHARACTERISTICS OF GAS-LIQUID COAXIAL INJECTOR OF LIQUID ROCKET ENGINE [J]. JOURNAL OF PROPULSION TECHNOLOGY, 1996, 17(4): 37-41.
[11] ZHONGTAO K. The Unsteady Atomization Mechanism and Combustion Characteristics of Gas-Liquid Swirl Coaxial Injector [D]. National University of Defense Technology, 2016.
[12] IM J H, KIM D, HAN P, et al. SELF-PULSATION CHARACTERISTICS OF A GAS-LIQUID SWIRL COAXIAL INJECTOR [J]. Atomization and Sprays, 2009, 19(1): 57-74.
[13] SHUN X. Research on the dynamic spray characteristics of the gas-centered swirl coaxial injector [D]. National University of Defense Technology, 2016.

Study on vibro-acoustic performance of composite laminates under transient impact excitation

Hongfu Wang[*a], Hanhui Guo[a], Yu Wang[a], Yuan Du[b], Wangqiu Cai[a]

[a] Marine Design and Research Institute of China, Shanghai, China, 200011

[b] College of Shipbuilding Engineering, Harbin Engineering University, Harbin, China, 150001

[*] Corresponding author: hongfuw@foxmail.com

Abstract. With the widespread use of composite structures in ship superstructures, their vibro-acoustic response under transient impact loads is of increasing concern. This paper studies the vibro-acoustic behavior of a carbon fibre reinforced composite laminate under transient impact excitation based on the time-domain hybrid FEM/BEM method. Firstly, the numerical method is verified through the model test of a homogeneous plate. Subsequently, the vibration and sound radiation characteristics of composite laminates are studied. It was found that the impact noise energy mainly concentrated in the low-frequency range of 10~200Hz, and the peak noise spectrum was 9dB lower than that of the steel plate. The variation of lay-up orientations causes differences in the structural properties of the laminates, resulting in differences in their vibro-acoustic radiation performance. In this study, the laminate with the $[0°/90°]_{5s}$ orthogonal lay-up scheme has better acoustic radiation performance under the impact load.

1. Introduction

With a range of advantages unmatched by traditional materials, composite materials are widely used in ship superstructures as an effective means of achieving lightweight design, while their excellent noise reduction properties also effectively enhance the stealth characteristics of the ship and the comfort of the personnel[1]. Jia et al.[2] and Gao et al.[3] investigated the vibro-acoustic performance of composite superstructures of underwater vehicles based on the finite element method. Vergassola et al.[4,5] studied the acoustic characteristics of large yacht superstructures through numerical and experimental methods. As steady-state noise is effectively suppressed, the impact of noise induced by transient loads on the comfort of superstructures comes to the fore. When the ship's sailing state changes, transient loads on the propeller and rudder are transmitted through the hull structure to the superstructure, triggering transient noise, or the impact of combat ship weapon firing loads and airwave pressure on superstructures[6], etc. Pang et al.[7] numerically investigated the impact response of an offshore rocket launch system. Abdelkhalek et al.[8] studied the acoustic response of a 35,000 DWT bulk carrier under propeller excitation through the time domain boundary element method.

As a fundamental component of the superstructure, the vibro-acoustic performance of composite laminated structures is of paramount importance for the comfort of ships. This study researches the vibro-acoustic behavior of a carbon fibre reinforced composite laminate under the transient impact excitation based on the time-domain hybrid FEM/BEM method. Firstly, the numerical method is verified through the model test of a homogeneous steel plate. Subsequently, the vibration and sound radiation characteristics of the composite laminate are studied, which aims to provide support for the design of vibration and noise reduction for superstructures.

2. Numerical methods

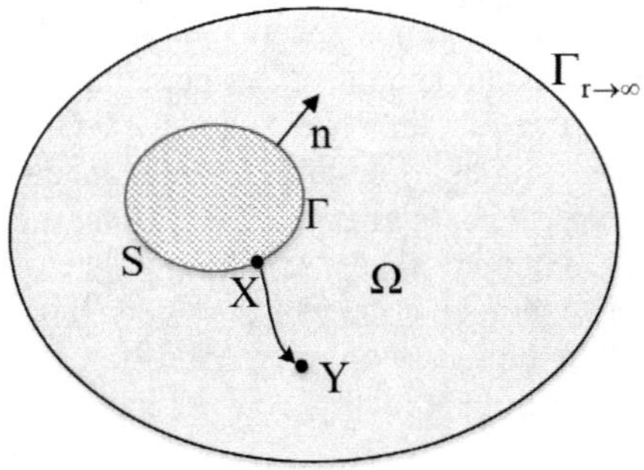

Figure 1. Solution domain of acoustic propagation

In this study, the boundary element method based on the acoustic wave equation is used to solve the transient sound propagation of composite laminates. The acoustic wave equation is

$$\nabla^2 p(Y,t) - \frac{1}{c^2}\ddot{p}(Y,t) = 0 \tag{1}$$

Where $\ddot{p}(Y,t)$ is the second derivative of sound pressure p with respect to t, c is the speed of sound, and ∇^2 is the Laplace operator. The structural vibration leads to the propagation of sound wave in acoustic medium, so the structural acceleration is regarded as the boundary condition, expressed as

$$q = -\rho \ddot{u}_n \tag{2}$$

Where q is the sound flux, ρ is the density, and $\ddot{u}_n$ is the normal acceleration of the structural surface. The acoustic boundary integral equation is:

$$C(Y)p(Y,t) + \int_\Gamma \int_0^t q^*(Y,t;X,\tau)p(X,\tau)d\tau d\Gamma = \iint_\Gamma \int_0^t p^*(Y,t;X,\tau)q(X,\tau)d\tau d\Gamma \tag{3}$$

Where $C(Y)$ is the boundary shape parameter, $p_0^*(Y,t;X,\tau)$ is the basic solution of the sound pressure of the field point Y, as shown in Figure 1. Then, the sound pressure at each field point in the domain Ω can be calculated by the sound pressure and sound flux on the boundary Γ:

$$C_p^n + \sum_{m=1}^{\bar{n}}\sum_{\alpha=1}^{n_\alpha}[\alpha(m)]H^{(n-m+1)}p^{(\alpha_g)} = \sum_{m=1}^{\bar{n}}\sum_{\alpha=1}^{n_\alpha}[\alpha(m)]G^{(n-m+1)}q^{(\alpha_g)} \tag{4}$$

Where p^n is the sound pressure vector corresponding to time step n, $p^{(\alpha_g)}$ and $q^{(\alpha_g)}$ are the overall vectors formed by the sound pressure and sound flux of all nodes, respectively. Then,

$$\hat{p}^n = -\sum_{m=n_i}^{\bar{n}}\sum_{\alpha=1}^{n_\alpha}[\alpha(m)]H^{(n-m+1)}p^{(\alpha_g)} + \sum_{m=1}^{\bar{n}}\sum_{\alpha=n_i}^{n_\alpha}[\alpha(m)]_G^{(n-m+1)}q^{(\alpha_g)} \tag{5}$$

Where $n_i = \max(1, \bar{n} - n_{\max} + 1)$ and $n_{\max} = r_{\max} / c\Delta t + 2$.

In this study, the structural transient response of the laminate under impact loading was obtained firstly based on the Abaqus/Explicit solver, and then the results of the structural response were extracted as the boundary conditions for the boundary element analysis. Eventually, the transient radiation noise prediction of the laminate was carried out using the Virtual.Lab Acoustics solver.

3. Validation of transient noise prediction methods

In order to verify the hybrid FEM/BEM method adopted in this study, the transient acoustic vibration response of a homogeneous thin plate is calculated in this section and compared with the results of Pang et al.'s experiment[9]. As shown in Figure 2 and Figure 3, the calculational parameters, boundary conditions and impact loads used in the calculations were kept consistent with those in the test.

The time domain vibration and noise response of the plate under impact excitation is calculated numerically and the vibration acceleration and sound pressure spectra were obtained by Fast Fourier Transform (FFT). As shown in Figure 4 and Figure 5, the reference values of vibration acceleration level and sound pressure level were 1×10^{-6}m/s^2 and 2×10^{-5}Pa respectively. In the range of 10Hz to 4kHz, the results calculated by the numerical method are consistent with the general trend of test results. The total horizontal error of vibration acceleration at the location of the vibration measurement point does not exceed 2.2dB, and the radiated noise at the location of the noise measurement point is less than 3dB, which meets the engineering accuracy requirements.

Figure 2. Transient radiated noise test

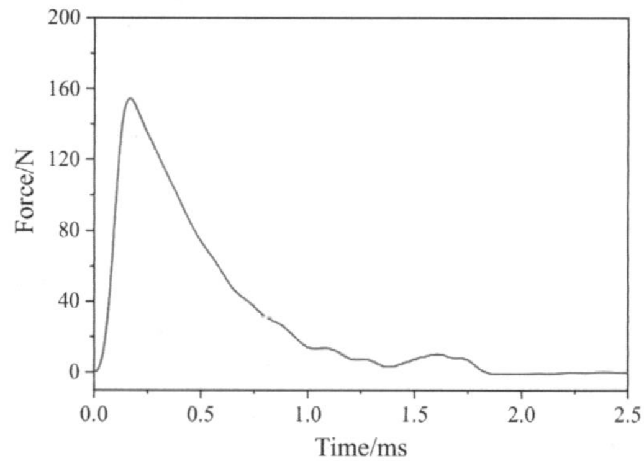

Figure 3. Time history curve of impact load

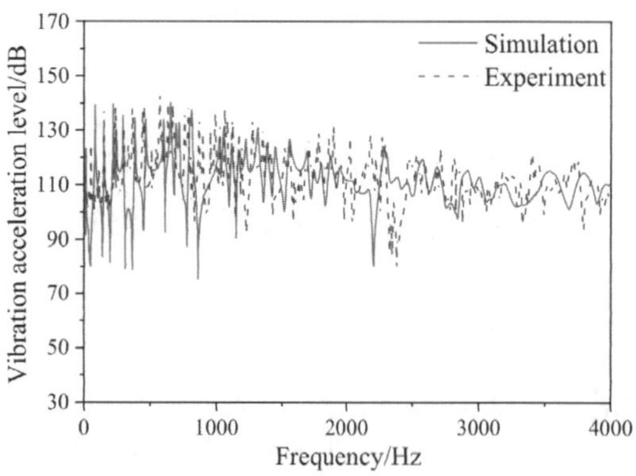

Figure 4. Comparison of vibration acceleration level at measurement point c

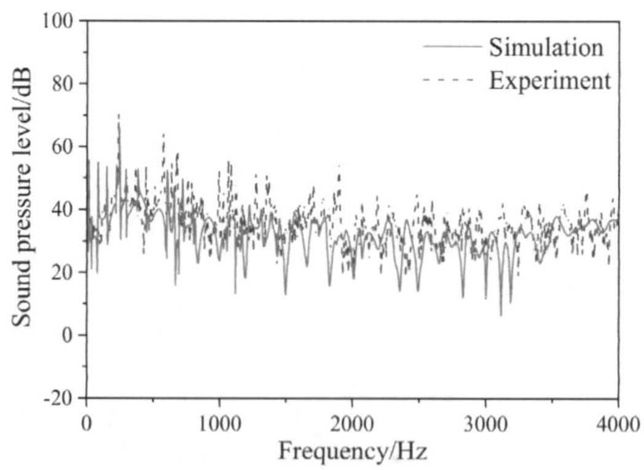

Figure 5. Comparison of sound pressure level at 2# microphone

4. Computational model and Numerical setups

4.1 Numerical model and mesh

The T800/3200 carbon fibre reinforced composite laminate with dimensions of 1000mm×1000mm ×6mm, divided into 20 layers, each with the same thickness and material parameters, was used for the study, and the material parameters of the model are shown in
Table 1.

Table 1. Fibre layer mechanical performance parameters

ρ / kg·m^{-3}	E_{11} /GPa	E_{22} /Gpa	E_{33} /Gpa	v_{12}	v_{23}	v_{13}	G_{12} /Gpa	G_{13} /Gpa	G_{23} /Gpa
2700	160	8.97	8.97	0.016	0.24	0.016	6.21	6.21	3.45

The structural finite element model of the composite laminate was established in ABAQUS, and the conventional shell was used to simulate the laminate with S4R cells[10]. The boundary conditions were set as the four-sided simple support. Considering the calculation accuracy and efficiency, the mesh size was first analyzed for convergence, and the final finite element mesh size was chosen as 25mm×25mm, as shown in Figure 6. The Rayleigh damping coefficient of the laminate is given as $\alpha = 6$, $\beta = 1.5\times10^{-6}$. Furthermore, the acoustic boundary element mesh was created in LMS Virtual lab, with dimensions satisfying $\Delta x \le c/6f_{\max}$, where c is the speed of sound in air, $f_{\max}$ is the maximum analysis frequency. Eventually, the acoustic grid is presented in Figure 7 with the size of 60mm×60mm.

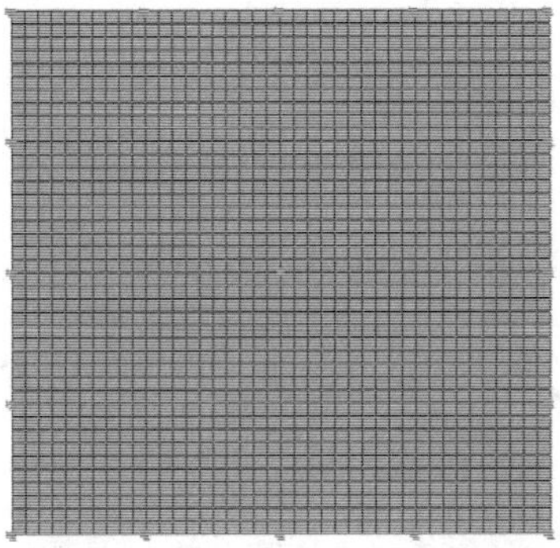

Figure 6. Finite element model of the laminate

Figure 7. Acoustic boundary element model

4.2 Description of the pulse load

Ship superstructures are subjected to extremely complex forms of transient loads, which are generally simplified by equating them to symmetrical impulse loads. In this paper, a symmetrical triangular waveform is used as the transient load (Figure 8), with T =2ms and $P_{\max}$ =100N. The normal excitation force is applied to the centre of the laminate at the coordinates of (0.5,0.5,0)m, as shown in Figure 9.

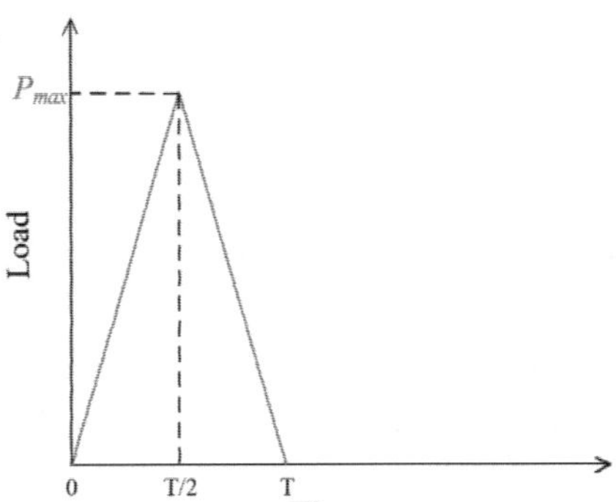

Figure 8. Triangular pulse load

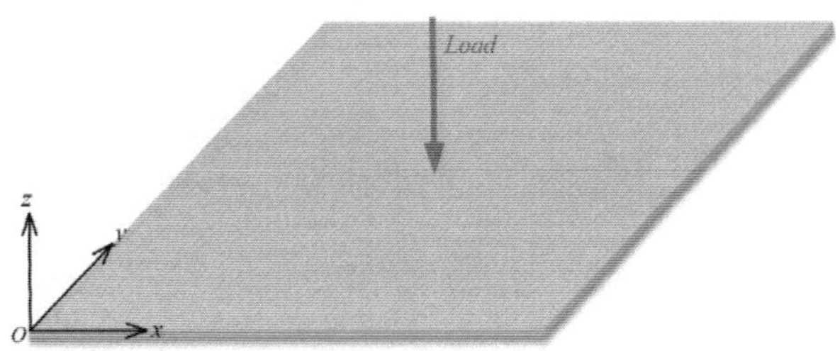

Figure 9. Schematic diagram of the excitation position

5. Results and discussion

5.1 Influences of composite materials on transient vibration noise

In order to study the influence of composite laminates in ship superstructure on vibration and noise control, this section first compares the vibro-acoustic performance of a homogeneous steel plate and a composite laminate, the lay-up orientation of which is [0°/45°/90°/-45°]$_5$. Modal analysis is an important method to study structural vibration behavior[11]. And the Lanczos algorithm is used for the eigenvalue solution. The natural frequencies of the preceding six modes of laminates and steel plates are shown in

Table 2. It can be found that each modal order of the laminate is slightly smaller than those of the steel plate, which indicates that the overall stiffness of the laminates with the lay-up orientations of [0°/45°/90°/-45°]$_5$ is slightly weaker than that of the steel plate.

Table 2. Natural frequencies comparison results

Modal order	Laminated plate	Steel plate
1	26.78	29.48
2	66.61	73.84
3	106.79	118.05
4	133.49	148.22
5	173.08	192.25
6	227.61	253.26

As given in Figure 10, the vibration responses of plates made of different materials are compared at the measurement point of (0.5,0.3,0)m. With the loading of the impact excitation, the vibration acceleration reaches the peak value within a short period time. At the initial moment, the acceleration amplitude of the laminates is larger than that of the steel plate, which is due to its low rigidity and its inferior deformation resistance. However, the vibration's attenuation rate of the laminate is obviously higher than that of steel plates, which is caused by the high damping of composite laminates. The maximum peak of the vibration acceleration spectrum of the steel plate is 5.1dB higher than that of the laminate, which indicates that the composite material can effectively restrain the vibration response under transient impact loading. It can be seen from Figure 11 that the transient noise induced by the impact excitation has an obvious fluctuation process. The calculation results of the sound pressure at 1.5m above the plate were extracted for comparison. As shown in Figure 12, the variation tendency of the radiated noise is similar to the vibration response results. The maximum value of noise spectrum of steel plate is 9dB higher than that of laminates. This indicates that the composite material substantially reduces the radiated noise of the flat structure under transient impact loading.

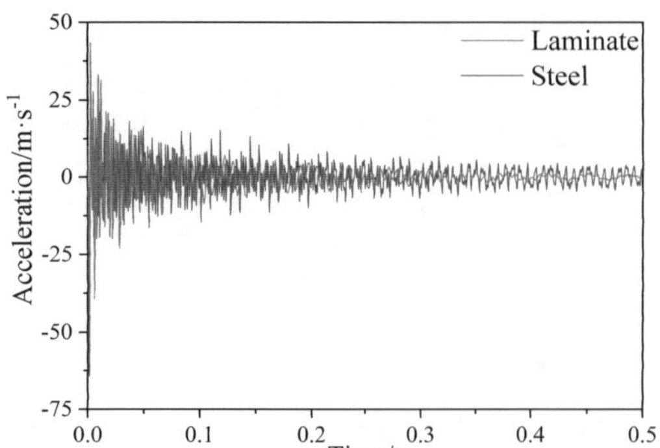

(a) Time domain response

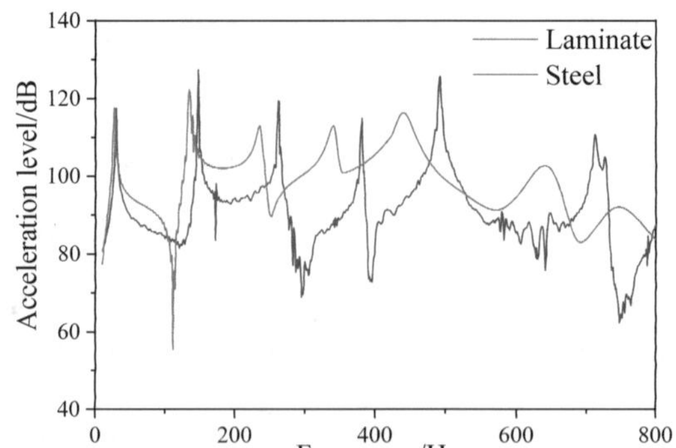

(b) Frequency domain response

Figure 10. Comparison of vibration acceleration response at the measurement point (0.5,0.3,0)m

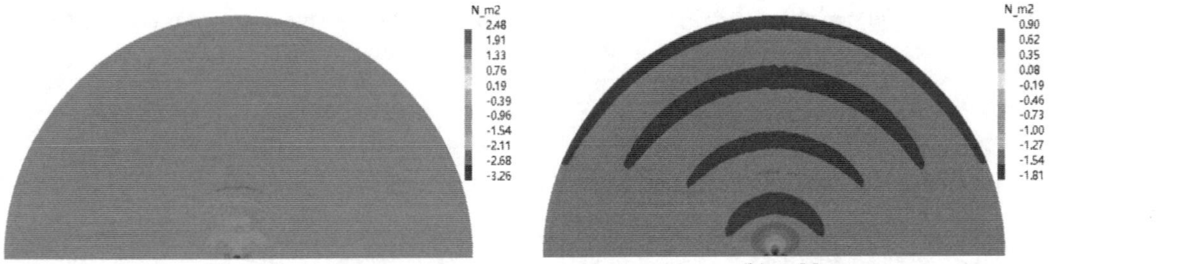

(a) t=10ms (b) t=20ms

Figure 11. Transient radiation noise propagation nephogram

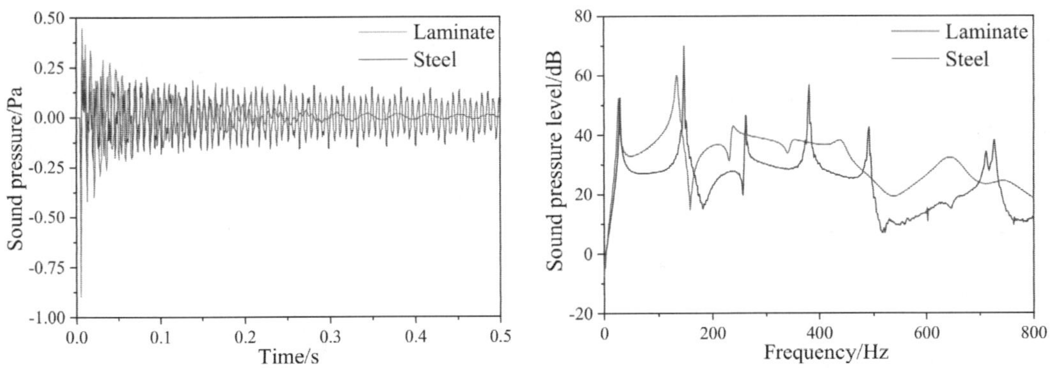

(a) Time domain response (b) Frequency domain response

Figure 12. Comparison of radiation noise at 1.5m above the plate

5.2 Influence of ply angle on impact noise

As shown in Figure 13, in order to investigate the effect of fibre lay-up direction on the vibration and acoustic characteristics of carbon fibre composite panels under impact loading, three typical lay-up sequences with lay-up angles of $[0°/90°]_{5s}$, $[0°/45°]_{5s}$ and $[0°/45°/90°/-45°]_5$ were selected.

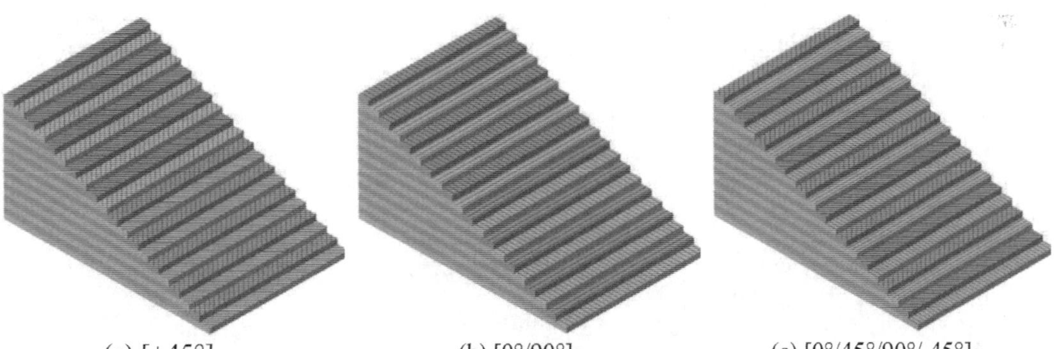

(a) $[\pm45°]_{5s}$ (b) $[0°/90°]_{5s}$ (c) $[0°/45°/90°/-45°]_5$

Figure 13. Diagram of laying angles

The acoustic responses of the laminates with different lay-up orientations under the impact load are shown in Figure 14. The impact noise energy under triangular pulse wave loading is mainly concentrated in the low frequency region in the range of 10~200Hz. After 200 Hz, the noise energy decays rapidly under the effect of damping. The amplitude of the source level of $[0°/45°/90°/-45°]_5$ is slightly larger than that of $[0°/45°]_{5s}$, and $[0°/90°]_{5s}$ is the smallest. As shown in Figure 14(b), the sound source level spectrum curve is converted to A-weighted band sound pressure level. It can be found that the source level trends of laminates with different ply angles are close to each other.

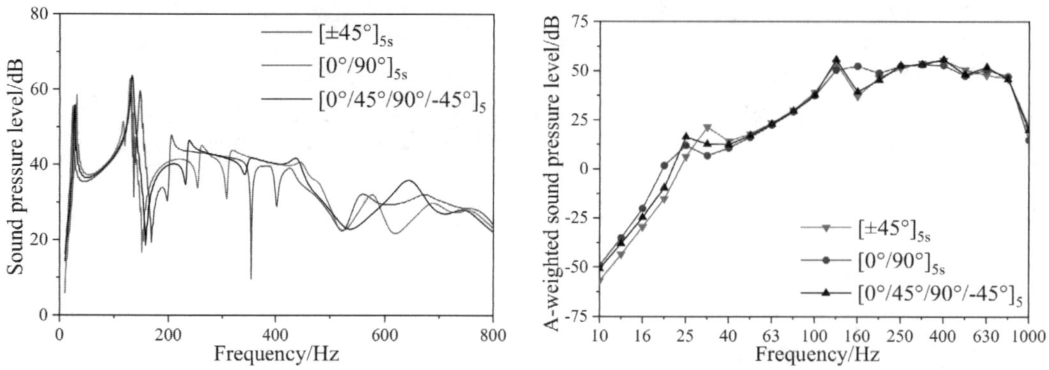

(a) Spectrum of sound source level (b) A-weighted sound pressure level in band

Figure 14. Comparison of sound source levels of different ply angles

Table 3. Results of the overall sound pressure level in band

Frequency band/Hz	$[\pm45°]_{5s}$	$[0°/90°]_{5s}$	$[0°/45°/90°/-45°]_5$
10~100	63.1	60.8	62.1
100~200	70.3	69.6	71.5
200~300	61.6	63.7	62.7
300~400	61.6	60.0	60.9
400~500	58.6	55.2	58.5
500~600	48.2	49.2	46.1
600~700	46.6	50.7	53.2
700~800	47.1	48.4	46.6

The total band levels of the radiated noise of laminates with different ply angles are shown in Table 3. The influence of ply angles on the radiated noise characteristics of the laminate under impact loading varies in different frequency band ranges. In the range of 10~800Hz, the total level of $[0°/45°/90°/-45°]_5$ is the largest, $[\pm45°]_{5s}$ is the second largest and $[0°/90°]_{5s}$ is the smallest. This indicates that the laminate with the lay-up orientations of $[0°/90°]_{5s}$ has better acoustic performance.

6. Conclusions

In view of the insufficient research on the vibro-acoustic response law of composite superstructures under transient impact loads, this study explores the vibro-acoustic characteristics of a carbon fibre reinforced composite laminate under transient impact excitation based on the time-domain hybrid FEM/BEM method. It was found that the acoustic performance of the carbon fibre reinforced composite laminate is significantly better than that of the steel plate under triangular wave loading, with the peak vibration spectrum 5.1dB lower than that of the steel plate, and the peak noise spectrum 9dB lower than that of the steel plate. The decay rate of vibration acceleration and noise amplitude of laminated plate is significantly faster than that of steel plate, which is caused by the high damping of composite material. The impact noise energy is mainly concentrated in the low-frequency region in the range of 10~200Hz. Different lay-up directions lead to differences in the structural properties of the laminates, resulting in differences in their vibro-acoustic radiation performance. Among the three lay-up schemes, the laminate with $[0°/90°]_{5s}$ orthogonal lay-up scheme has better acoustic radiation performance under impact loading. This research can provide support for the acoustic improvement of ship superstructures and the reduction of vibration and noise under transient loads in cabins. In subsequent studies, we will calculate more complex structures and impact loads, and then analyse the impact noise characteristics of composite structures in detail.

References

[1] Jiang Q, Nan L, Wenqiang S. (2015). The application and development of composites for foreign naval warships' super-structure. *Ship Science and Technology*;**37**(1):233-7.

[2] Di J, Xueren W, Xuhong M. (2018). Simulation study on underwater vibration noise of composite material structures. *Transducer and Microsystem Technologies*;**37**(04):80-3.

[3] Cong G, Xichun H, Guobing H, Huiling D. 2019. Study on Vibration and Acoustic Radiation Characteristics of Composite Material Superstructures of Underwater Vehicles.**In:** *IOP Conference Series: Materials Science and Engineering*: IOP Publisher. pp. 012-48.

[4] Vergassola G, Pais T, Boote D. (2019). Low-Frequency analysis of super yacht free vibrations. *Ocean Engineering*;**176**:199-210.

[5] Tatiana P, Dario B, Gianmarco V. (2018). Vibration analysis for the comfort assessment of a superyacht under hydrodynamic loads due to mechanical propulsion. *Ocean Engineering*;**155**:310-23.

[6] Xiaowen L, ZhaoYi Z, Yan L, Zhe H, Ping L. (2018). Concept Design and Mechanical Behavior Research of Composite Superstructure for Ships. *Ship Engineering*;**40**(05):88-93.

[7] Fuzhen P, Yuxuan Q, Haichao L, Yao T, Qingtao T. (2021). Study on impact resistance of composite rocket launcher. *Reviews on Advanced Materials Science*;**60**(1):615-30.

[8] Abdelkhalek H, Duanfeng H, Liangtian G. 2015. Computational fluid dynamics and Numerical acoustic response for ship accommodation areas due to propeller excitation, towards a human factors recommendations.**In:** *Applied Mechanics and Materials*: Trans Tech Publ. pp. 406-11.

[9] Fuzhen P, Hongfu W, Xuhong M, Haichao L, Yuan D. 2021. Study on underwater acoustic radiation of double stiffened cylindrical shells under transient impact loading.**In:** *The 18th Ship Underwater Noise Symposium*; Kunming. pp. 378-88.

[10] Singh S, Köpke UG, Howard CQ, Petersen D. (2014). Analyses of contact forces and vibration response for a defective rolling element bearing using an explicit dynamics finite element model. *Journal of Sound and Vibration*;**333**(21):5356-77.

[11] Chen Z, Wang A, Qin B, Wang Q, Zhong R. (2020). Investigation on free vibration and transient response of functionally graded graphene platelets reinforced cylindrical shell resting on elastic foundation. *The European Physical Journal Plus*;**135**(7):582.

Sound Insulation Performance of the Stiffened Laminated Plate with Different Reinforcement Forms

Y L Bu, L Chen*, and H F Wang

Marine Design and Research Institute of China, Shanghai, China

*Corresponding author: chldysh@163.com

Abstract: A semi-analytical methodology on the sound transmission of stiffened laminated plate under different reinforcement forms is developed to explore its sound insulation performance under the action of a plane sound wave load, in accordance with the classical laminated plate theory and in consideration of stiffener flexion and torsional motions. The formula of acoustic transmission loss is obtained by utilizing spatial harmonic expansion and the virtual work principle. Subsequently, the predicted values of proposed methodology are validated by the existing models. Eventually, some characteristic parameters are considered to investigate their effects on sound insulation behavior.

1. Introduction

Composite laminated plates with significant advantages of sound insulation, lightweight, and high strength are widely used in the aerospace, shipbuilding, military industry, construction, and other engineering fields [1, 2]. Due to the development of composite technology and the increasing acoustic requirements of people's surroundings, the design and optimization of low-noise composite structures with excellent sound insulation performance have become a research hotspot in recent years.

There are many pieces of literature about the problems [2-6,11,14] of sound insulation of periodically stiffened plates. The existing models can be broadly grouped into two categories: the equivalent simplified model [2, 3] and the plate-beam combined model [4]. In Lee and Kim's study [2,14] on the behavior that sound propagation in thin stiffened plates, the two-dimensional panel of periodical stiffeners was modeled as a one-dimensional periodically supported beam by simplifying stiffeners into translational motional and rotational springs and lumped masses. For the purpose of resolving the vibroacoustic performance of a stiffened platter, Maxit [4] approximated stiffeners as the Euler-Bernoulli beam. Through the theory of classical laminated composite plate and Mace's stiffened uniform isotropic plate, Yin and Gu [5] investigated the sound radiation of the composite laminate reinforced by double periodic stiffeners. Nevertheless, the sound insulation of laminated plates with different periodical stiffeners is rarely concerned. Those theoretical models accurately describing the stiffener's moveable influence and reflecting the stiffened laminated plate's sound propagation mechanism are not yet well developed.

In this paper, with full consideration of the impacts of the stiffeners' bending and torsional motion on laminated plates, a semi-analytical methodology on the stiffened laminated plate's sound propagation with different reinforcement forms is developed. Firstly, the formula of acoustic propagation loss is secured by having access to the nonplanar harmonic expansion and virtual work principle. Subsequently, by comparing subsistent models[12,14], the methodological availability is verified. Ultimately, employing numerical calculations studies various parametric impacts on acoustic insulation performance.

2. Descriptions of the semi-analytical methodology

A system of right-handed Cartesian coordinates is defined as exhibited in Figure 1. Plane $z=0$ where the laminated plate is located separates the spatial sound field into the incident ($z<0$) and transmission($z>0$) field. The b_x and b_y are the stiffeners' widths in x- and y-direction respectively, and the d_x and d_y are the stiffeners' heights in x- and y-direction respectively. There exists a harmonic plane sound wave P_i with pitch angle φ and azimuth θ on the stiffened laminated plate's surface by means of obliquely incident, and this then causes the bending wave to propagate along the plate. The bending wave is transmitted to the space $z>0$, causing the transmitted sound pressure P_t. Simultaneously, the plate reflects the incident wave into space $z<0$, which induces the reflected sound wave P_r.

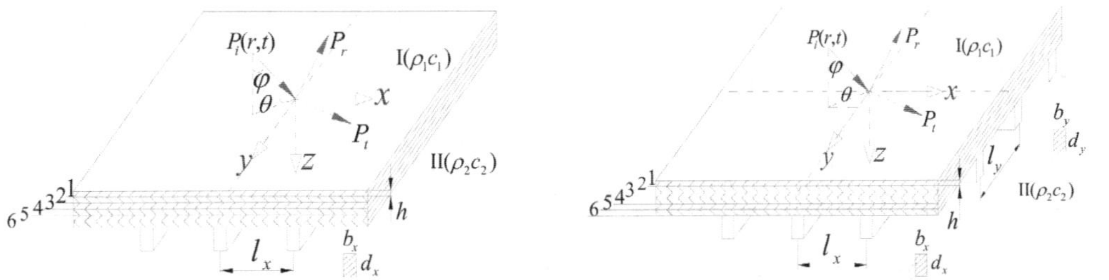

(a) A single periodically stiffened laminated plate (b) An orthogonally stiffened laminated plate

Figure 1. Sound transmission process through stiffened laminated plates

2.1. Establishment of vibroacoustic coupling equation

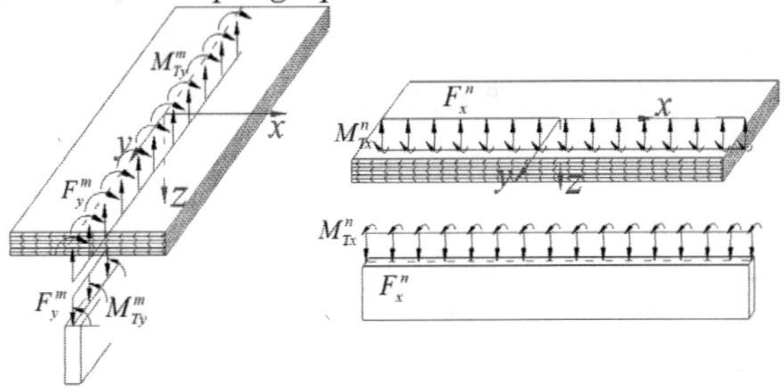

Figure 2. Force diagram of the basic elements of a stiffened laminated plate

As shown in Figure 2, the forces along the y-axis acting on the laminated plate are F_y^m and M_{Ty}^m . (F_x^n, M_{Tx}^n) replaces (F_y^m, M_{Ty}^m) in the x-direction. According to the classical laminated plate theory, According to the classical laminated plate theory, the orthogonal stiffened laminated plate's dynamic governing equation can be expressed as follows [7]:

$$D_*[W(x,y,t)] = -f_x - m_x - f_y - m_y + P_1(x,y,0) - P_2(x,y,0) \quad (1)$$

where D_* is the general linear differential operator form, W represents the laminated plate's lateral displacement, P_1 and P_2 mean the aggregate sound pressure in the incident and transmission field $f_x = \sum_{n=-\infty}^{+\infty} F_x^n \delta(y - nl_y)$, $f_y = \sum_{m=-\infty}^{+\infty} F_y^m \delta(x - ml_x)$, $m_x = \sum_{n=-\infty}^{+\infty} M_{Tx}^n \delta'(y - nl_y)$, $m_y = \sum_{m=-\infty}^{+\infty} M_{Ty}^m \delta'(x - ml_x)$, and δ is the Dirac delta function.

The flexural displacement of a metallic plate can be described by a degree of freedom:

$$D_* = D(\frac{\partial^2}{\partial x^2} + \frac{\partial^2}{\partial y^2})^2 - \omega^2 \rho h \tag{2}$$

where D, ρ and h represent the flexural rigidity, the density, and the thickness of the plate.

However, with regard to composite laminated plates with thin plies, its bending motion can be described by the degrees of freedom raising to three. Utilizing theory of the classical laminated plate here, the expression D_* is a 3-order square matrix, which is proposed by Yin et al. [5].

The governing equations of stiffeners' bending and torsional motions in the x- and y- direction are:

$$E_s I^n \frac{\partial^4 W}{\partial x^4} + \rho_s A^n \frac{\partial^2 W}{\partial t^2} = F_x^n(x,t), \quad -G_s J^n \frac{\partial^3 W}{\partial y \partial x^2} - \omega^2 \rho_s I_0^n \frac{\partial W}{\partial y} = M_{Tx}^n(x,t) \tag{3}$$

$$E_s I^m \frac{\partial^4 W}{\partial y^4} + \rho_s A^m \frac{\partial^2 W}{\partial t^2} = F_y^m(y,t), \quad -G_s J^m \frac{\partial^3 W}{\partial x \partial y^2} - \omega^2 \rho_s I_0^m \frac{\partial W}{\partial x} = M_{Ty}^m(y,t) \tag{4}$$

where $(E_s I^n, E_s I^m)$, $(\rho_s A^n, \rho_s A^m)$, $(G_s J^n, G_s J^m)$, and $(\rho_s I_0^n, \rho_s I_0^m)$ are the bending stiffness, the line mass, the torsional stiffness, and the moments of inertia of stiffeners per unit length, respectively. With regard to an infinite orthogonally stiffened laminated plate motivated via a harmonic plane acoustic wave, the response can be expressed by spatial harmonic expansion [10,14], as

$$W(x,y,t) = \sum_{m=-\infty}^{+\infty} \sum_{n=-\infty}^{+\infty} A_{mn} e^{-j\left[(k_x + 2m\pi/l_x)x + (k_y + 2n\pi/l_y)y - \omega t\right]} \tag{5}$$

where A_{mn} is the vibration amplitude and ω is the incident sound wave's circular frequency.

2.2. *Wave equation and boundary condition*

Sound pressure P_1 and P_2 can be expressed as [9]:

$$P_1(x,y,z,t) = Ie^{-j(k_x x + k_y y + k_z z - \omega t)} + \sum_{m=-\infty}^{+\infty} \sum_{n=-\infty}^{+\infty} B_{mn} e^{-j\left[(k_x + 2m\pi/l_x)x + (k_y + 2n\pi/l_y)y - k_{z,mn} z - \omega t\right]} \tag{6}$$

$$P_2(x,y,z,t) = \sum_{m=-\infty}^{+\infty} \sum_{n=-\infty}^{+\infty} C_{mn} e^{-j\left[(k_x + 2m\pi/l_x)x + (k_y + 2n\pi/l_y)y + k_{z,mn} z - \omega t\right]} \tag{7}$$

where I is amplitude of P_i, B_{mn}, and C_{mn} are respectively the (m,n)th spatial harmonic amplitude of the positive wave and negative wave amid the transmitted field. What is more, (k_x, k_y, k_z) represent are the wavenumber constituents of x-, y-, and z-directions, and $k_{z,mn}$ represents the (m,n)th Spatial harmonic wavenumber in the z-direction. Both P_1 and P_2 satisfy the wave equation:

$$\nabla^2 P_i - \frac{1}{c_i^2} \frac{\partial^2 P_i}{\partial t^2} = 0 \qquad (i = 1,2) \tag{8}$$

The coupling of the fluid and laminated plate gratifies the equation of momentum in the z-axis is:

$$\left.\frac{\partial P_i}{\partial z}\right|_{z=0} = \rho_0 \omega^2 W \qquad (i = 1,2) \tag{9}$$

Equations (5), (6), and (7) are substituted into Equation (9), and the relations among B_{mn}, C_{mn}, and A_{mn} can be obtained:

$$B_{mn} = \begin{cases} 1 + \omega^2 \rho_0 A_{00} / jk_z & n = 0 \\ \omega^2 \rho_0 A_{mn} / jk_{z,mn} & n \neq 0 \end{cases}, \quad C_{mn} = -\frac{\omega^2 \rho_0 A_{mn}}{jk_{z,mn}} \tag{10}$$

2.3. Solution of the equations

By employing the virtual work's principle [13], the following expression can be obtained:

$$\{D_* - \frac{\rho_0\omega^2}{jk_{z,kl}}\}l_x l_y A_{kl} + \sum_{n=-\infty}^{+\infty}\{(\rho_s A^n\omega^2 - E_s I^n\alpha_k^{\ 4}) + \beta_l\beta_n(-G_s J^n\alpha_k^{\ 2} + \rho_s I_0^n\omega^2)\}\times l_x A_{kn}$$
$$+\sum_{m=-\infty}^{+\infty}\{(\rho_s A^m\omega^2 - EI^m\beta_l^{\ 4}) + \alpha_k\alpha_m(-G_s J^m\beta_l^2 + \rho_s I_0^m\omega^2)\}l_y A_{ml} = \begin{cases}2Il_x l_y & k=0 \text{ and } l=0\\ 0 & k\neq 0 \text{ or } l\neq 0\end{cases} \tag{11}$$

where $\alpha_k = k_x + 2k\pi/l_x$, $\alpha_m = k_x + 2m\pi/l_x$, $\beta_l = k_y + 2l\pi/l_y$, $\beta_n = k_y + 2n\pi/l_y$.

Once the unknown A_{mn} is obtained, the bare laminated plate and (P_1, P_2) in the fluid domain adjacent to the laminated plate are obtained as well, then enabling the acoustic transmission solution to stiffened laminated plate of the fluid-loaded orthogonally. With the same method, the acoustic transmission analysis of the stiffened laminated plate with single period can also be obtained.

3. Determination of sound transmission loss (TL)

A function that relates the pitch angle φ and azimuth θ can be defined as the acoustic transmission coefficient $t(\varphi,\theta)$. It is equivalent to an intensity ratio of the transmitted and the incident wave [14]:

$$t(\varphi,\theta) = \sum_{m=-\infty}^{\infty}\sum_{n=-\infty}^{\infty}|C_{mn}|^2 Re(k_{z,mn})\Big/(|I|^2 k_z) \tag{12}$$

The sound wave transmission coefficient $t(\varphi,\theta)$ can be obtained by substituting the acquired coefficient C_{mn} into Equation (12). Ultimately, the transmission loss (*TL*) is described in dB scale:

$$TL = -10\log_{10}\left(t(\varphi,\theta)\right) \tag{13}$$

4. Results and discussion

4.1. Convergence check and validation

For the purpose of ensuring the convergence of results, Sufficient terms must be adopted. Previous studies [14] showed that the required quantity of series terms depends on the maximum analysis frequency, which is set as 10 kHz in this study. In the convergence analysis, the incident sound wave has a 45° degree pitch angle and a 0° degree azimuth. The material of stiffeners is aluminum, and the used parameters are as follows: l_x=l_y=0.3 m, b_x=b_y=0.3 m, d_x=d_y=0.6 m, ρ_s =2700 kg/m^3, Young's modulus E_s =71 GPa, Poisson's ratio μ_s=0.33, and damping dissipation coefficient η=0.02. The sound velocity and density of air as a fluid medium are respectively 342 m/s and 1.21 kg/m^3. The laminated plate consists of six layers, each with a thickness of 0.001 m. The ply angles are ($45^\circ/30^\circ/15^\circ/15^\circ/30^\circ/45^\circ$). See Table 1 for further details.

Table 1. A composite laminated plate's material parameters

Layer	Material	E_1 (GPA)	E_2 (GPA)	G_{12} (GPA)	v_{12}	ρ (kg/m^3)
1、6	Case 1	138	8.96	7.1	0.3	1600
2、5	Case 2	207	20.7	6.9	0.3	2000
3、4	Case 3	131	10.3	6.9	0.22	1500

Figure 3 indicates that TL_1 of the single periodically stiffened laminated plate converges when $n_1 \geq 23$, and TL_2 of the orthogonally stiffened laminated plate converges when $n_2 \geq 30$.

Next, the validation model's parameters refer to the published literature [12,14]: the stiffness of the spring $K_t = 3.6\times10^9$ N/m, the stiffness of the torsion spring $K_r = 60$ Nm/rad, and stiffeners with width *b*=0.3 m, and height *d*=0.6 m. As shown in figure 4, the predicted results obtained by the plate-beam combined model [13] are in good agreement with those obtained by the presented model with orthogonal stiffeners. However, obvious differences are observed when $f < f_2$ by comparing the results of the simplified model [14] with that of the presented model with parallel stiffeners, which is because the simplified model fails to treat stiffeners as an integration of translational and torsional springs. In sum, the proposed method is reasonable and reliable.

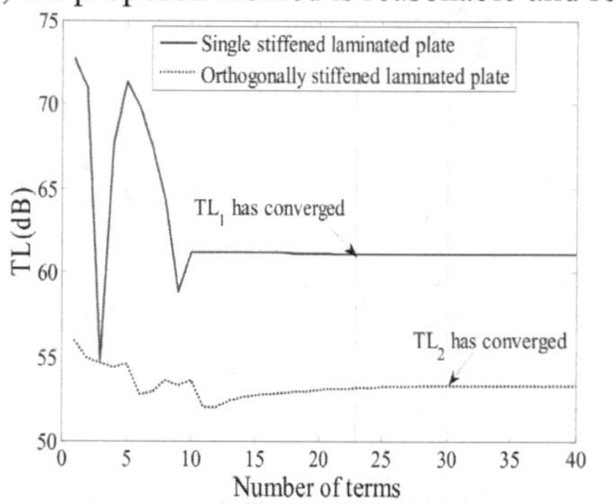

Figure 3. Convergence check of TL under an incident sound wave at 10 kHz

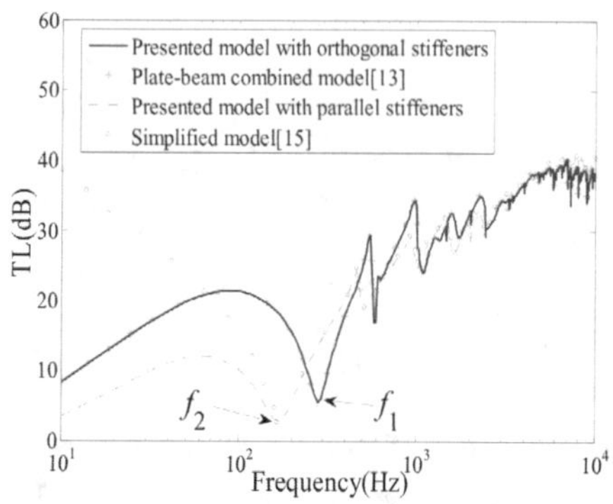

Figure 4. Comparison of transmission loss of an infinite single stiffened plate

4.2. *Effects of angle-ply schemes*

Subsequently, the TL curves of the orthogonally stiffened laminated plate with an antisymmetric angle-ply (45°/30°/15°/-15°/-30°/-45°) and a symmetrical angle-ply (45°/30°/15°/15°/30°/45°) are shown in figure 5 for evaluating the influences of different angle-ply schemes on the sound transmission. Angle-ply schemes have different effects on the orthogonally stiffened laminated plate's acoustic transmission. The peaks and troughs of TL are shifted to high frequencies as the angle-ply scheme from antisymmetrical to symmetrical, implying that the symmetrical angle-ply scheme leads to increasing the stiffness in the case of unchanging mass of the laminated plate.

4.3. *Effects of stiffening forms*

Three different stiffening forms, including the laminated plates without stiffeners, with single parallel stiffeners, and with orthogonally stiffeners, are compared in figure 6. It can be found that the overall trend of TL curves for the three kinds of laminated plates are like each other. The existence of parallel and orthogonal stiffeners complicates the mode shapes of the laminated plate, so their TL curves oscillate around the TL curve of bare laminated plates. Moreover, the orthogonal stiffened laminated plates have a higher amplitude of oscillation than parallel stiffened plates. But three TL curves will tend to uniformity when the exciting frequency exceeds a certain value, and thus stiffened laminated plates can be regarded as bare laminated plates.

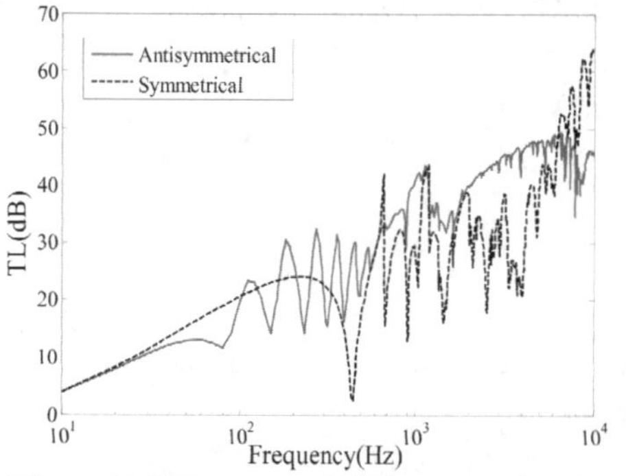

Figure 5. Influences of different angle-ply schemes on sound insulation

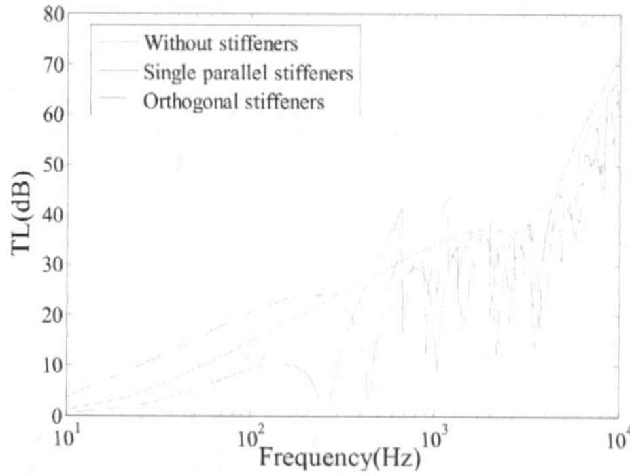

Figure 6. Influences of different stiffening forms on sound insulation

4.4. Effects of damping loss factor

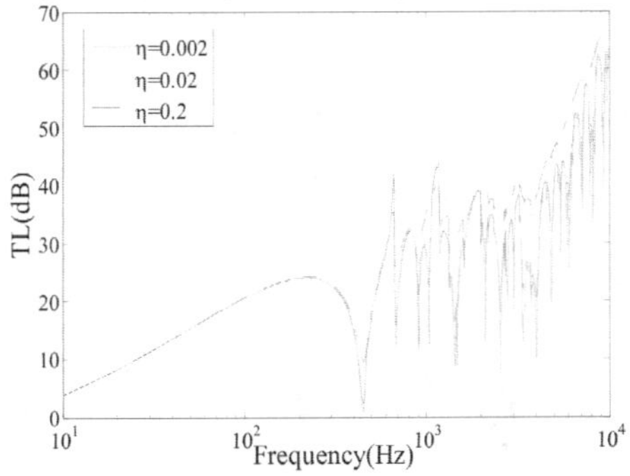

Figure 7. Impacts of damping loss factor on sound insulation performance

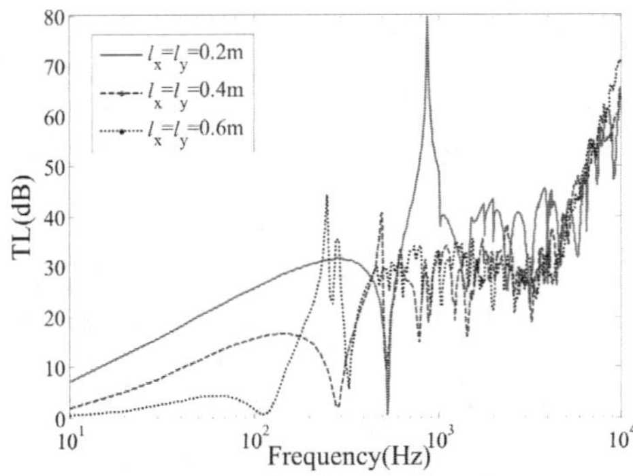

Figure 8. Impacts of stiffeners spacing on sound insulation performance

In this section, the damping dissipation coefficients are taken as 0.02, 0.1, and 0.2, respectively. As shown in Figure 7, the increase of damping loss factor will reduce both the peaks and troughs of the TL curve at the resonance frequencies, but has have little influence at the frequency without peaks and troughs. This indicates that growing material damping dissipation coefficients can enhanced the acoustic insulation performance at resonance frequency.

4.5. Effects of stiffener spacings

To study the impacts of stiffener spacings on sound insulation performance, the spacings are taken as 0.2m, 0.4m, and 0.6m, respectively. The comparison results in Figure 8 show that, as the spacing increases, the TL curvilinear peaks and troughs will shift towards the lower frequency, but the trends of curves are like each other. Overall, increasing the number of stiffeners can improve the sound insulation behavior of the laminated plate.

5. Conclusion

With full consideration of the impacts of the stiffeners' bending and torsional motion on laminated plates, a semi-analytical methodology on the stiffened laminated plate's the acoustic transmission under different reinforcement forms is developed to explore its sound insulation performance under the action of a plane sound wave load.

Subsequently, Comparison of predicted values with the results obtained by the plate-beam combined model and the simplified model can bear out the efficacy and practicability of proposed methodology in this research.

The research shows that the peaks and troughs of the TL curves shift towards the higher frequency as the angle-ply scheme from antisymmetric to symmetrical, implying that the symmetrical angle-ply scheme will increase the stiffness of the laminated plate. Besides, the overall trend of TL curves for the three kinds of laminated plates with different forms are like each other. When the excited frequency exceeds a certain value, the three TL curves will tend to uniformity. The increment of damping will decrease both the peaks and troughs of the TL curve at the resonance frequencies while having little effect at the frequency where peaks and troughs do not occur. With the increase of the spacing of stiffener, The peaks and troughs in the TL curves shift towards the lower frequency. In the low-frequency region, with the augmented spacing of stiffener, the sound insulation performance decreases significantly, while the sound insulation performance is improved in the whole frequency range.

References

[1] Shen G L, Hu G K 2006 Mechanics of composite materials. Tsinghua University Press, Beijing.
[2] Lee J H, Kim J 2002 Sound transmission through periodically stiffened cylindrical shells J. Journal of Sound and Vibration, 251(3): 431-456.

[3] Mead D J 1970 Free wave propagation in periodically supported infinite beams J. Journal of Sound and Vibration, 11 (2): 181-197.
[4] Maxit L 2009 Wavenumber space and physical space responses of a periodically ribbed plate to a point drive: A discrete approach J. Applied Acoustics, 70(4): 563-578.
[5] Yin X W and Gu X J 2007 Acoustic radiation from a laminated composite plate reinforced by doubly periodically parallel stiffeners J. Journal of Sound and Vibration, 306: 877-889.
[6] Legault J and Atalla N 2010 Sound transmission through a double panel structure periodically coupled with vibration insulators J. Journal of Sound and Vibration, 329: 3082-3100.
[7] Berhelot J M 1999 Maté riaux composites-Comportement mécanique at analyse des structures. Editions TEC & DOC, Paris.
[8] Reddy J N 1997 Mechanics of laminated composite plates: theory and analysis. CRC Press, Boca Raton.
[9] Mace B R 1981 Sound radiation from fluid loaded orthogonally stiffened plates J. Journal of Sound and Vibration, 79 (4): 439-452.
[10] Xin F X and Lu T J 2010 Analytical modeling of fluid loaded orthogonally rib-stiffened sandwich structures: Sound transmission J. Journal of the Mechanics and Physics of Solids, 58:1374-1396.
[11] Xin F X and Lu T J 2010 Analytical modeling of sound transmission across finite aeroelastic panels in converted fluids J. Journal of the Acoustical Society of America, 128 (3): 1097-1107.
[12] Jin Y Q, Pang F Z and Yao X L 2012 Research on the shell-beam theoretical model of sound transmission through the orthogonally stiffened plate J. Acoustics Journal, 37 (6): 611-620.
[13] Mead D J and Pujara K K 1971 Space-harmonic analysis of periodically supported beams: response to convected random loading J. Journal of Sound and Vibration, 14 (4): 525-541.
[14] Lee J H and Kim J 2002 Analysis of sound transmission through periodically stiffened panels by space harmonic expansion method J. Journal of Sound and Vibration, 251 (2): 349-366.

Research on Acoustic Reflection Characteristics of Typical Concave Body based on COMSOL

Shisong Zhang*, Xiaofang Sun, Zhongshan Yao

Department of Military Oceanography and mapping, Dalian Naval academy

*394403007@qq.com

Abstract: In the design of an underwater acoustic imaging system, to improve the signal-to-noise ratio (Snr) of the signal receiver and achieve good acoustic imaging results, it is necessary to discuss the acoustic focusing problem of the concave body. Based on COMSOL finite element analysis, this paper makes a comparative analysis of concave structures in several typical acoustic reflection modules, takes plane waves as the background sound field, analyses the acoustic reflection performance of typical concave structures at different wavelengths and angles of incidence, and studies the acoustic convergence effect under different concave structures, providing reference and help for further optimizing the design of acoustic imaging systems.

1. Introduction

As for the acoustic reflection of concave bodies, it was primarily concentrated in the acoustic structure design of halls and other buildings [1]. However, with the continuous development of underwater imaging technology in recent years, to improve the signal-to-noise ratio of the receiver and improve the imaging quality, the problem of acoustic energy convergence at the signal receiver has gradually attracted people's attention [2]. The reflective interface of a concave body can be a sphere or a sphere-like surface, and the sphere-like surface includes paraboloids, ellipsoids, etc. The relevant expressions are shown in Table 1 [3]:

Table 1. Spherical expression

Interface shape	Equation	Curvature
Spherical interface	$(x - x_0)^2 + y^2 = R^2$	0
Oblate elliptic interface	$\frac{(x - x_0)^2}{A^2} + \frac{y^2}{B^2} = 1(A > B)$	$\frac{\sqrt{A^2 - B^2}}{A}$
Oblate elliptical interface	$\frac{(x - x_0)^2}{A^2} + \frac{y^2}{B^2} = 1(A < B)$	$\frac{\sqrt{B^2 - A^2}}{A}$
Parabolic interface	$y^2 = 2p(x - x_0)$	1

In this paper, we will choose typical spherical, ellipsoidal, and parabolic surfaces as research objects to analyze their acoustic reflection characteristics.

2. Research content

2.1 Reflection law of concave mirror

The sound reflection of a sphere is studied by using its semi-circular section, the incident sound wave is assumed to be a plane sound wave, and its incident direction is perpendicular to the diameter of the semicircle.

2.1.1 Primary acoustic ray reflection focusing

It is defined that the sound ray incident on the sphere and leaving the hemisphere space after only one reflection is called a primary sound ray. As shown in Figure 1, according to the reflection law, among the sound rays' incident on the arc, the primary sound ray with the largest distance from the main axis OZ should be the sound ray that reaches the end point of the semicircle diameter after the arc is reflected once. In Figure 1, we can see that the incident sound ray is CB. Then the incident sound ray of the primary sound ray is concentrated in the circle with OC as the radius. According to the geometric relationship, the distance between the sound ray CB and the main shaft can be obtained, so the sectional area of the hemisphere where the primary sound ray is located is$OC = R/2$

$$S_{11} = \pi R^2/4。 \quad (1)$$

Then the spherical crown obtained by rotating the arc BZ around the main shaft OZ is the reflecting surface of the primary sound ray. Its area is

$$S_{21} = 2\pi R(R - CB) = (2 - \sqrt{3})\pi R^2 \quad (2)$$

The hemispherical space with OC as the radius contains all the primary sound rays. It can be seen in Figure.1 that a conical focus area is formed in front of the sphere after the primary sound ray is reflected.

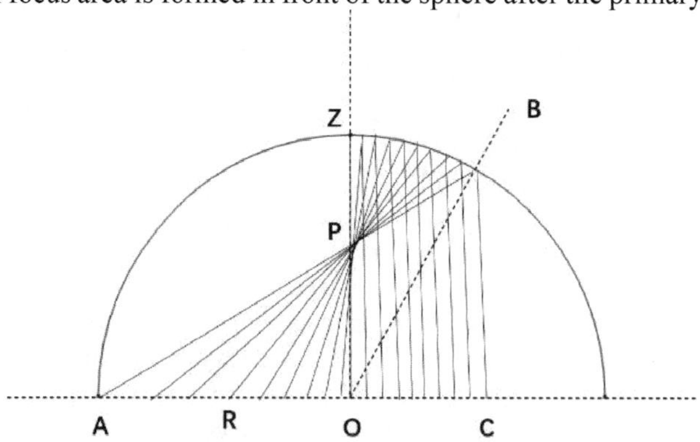

Figure.1 Focus area of primary sound ray

2.1.2 Secondary acoustic ray reflection focusing

The definition of the sound ray incident on the sphere and leaving the hemispherical space after secondary reflection is called the secondary sound ray. As shown in Figure 2, according to the reflection law, among the sound rays' incident on the arc, the second sound ray farthest from the main axis should be the sound ray that reaches the end of the diameter after the second reflection of the arc. In Figure 2, we can see that the incident sound ray is FE. Then the incident sound line of the secondary sound line is concentrated in the hollow hemisphere formed by the rotation of the CF segment around the OZ axis in the figure. According to the geometric relationship, the distance between FE and the main shaft can be obtained as follows:

$$OF = Rsin(3\pi/10) \quad (3)$$

Then the area of the hollow hemisphere section where the secondary sound line incident sound line is located is:

$$S_{12} = \pi(r_2^2 - r_1^2) = \pi R^2(sin^2\left(\frac{3\pi}{10}\right) - \frac{1}{4}) \quad (4)$$

The spherical defect obtained by rotating the circular arc EB around the main axis OZ is the reflecting surface of the secondary sound ray. Its area is:

$$S_{22} = 2\pi R(CB - FE) = (\sqrt{3} - 2\cos(3\pi/10))\pi R^2 \quad (5)$$

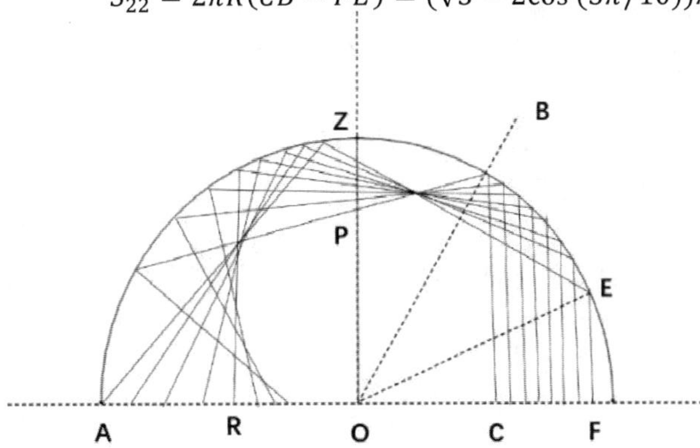

Figure. 2 Focus area of secondary sound ray

Combined with the previous analysis, it is not difficult to see from Figure 2 that the "acoustic focusing" is formed at the place where the secondary sound line is closer to the sphere, and there are two focusing areas. But obviously, the focusing ability is not as obvious as that of the primary sound ray.

In the same way, we can get the reflection focusing on the third and fourth sound rays. Under the assumption that the sound energy is uniform, the sum of the energy contained in the primary and secondary sound rays accounts for 65.5% of the total energy, while the total area of the first reflecting surface corresponding to them only accounts for 41.2% of the hemispherical surface. The energy contained in more than four times sound rays accounts for 34.5% of the total energy, and the area of the first reflecting surface corresponding to them accounts for 55.5% of the total area. Their energy distribution is mainly reflected in the creeping along the spherical direction, which does not affect the acoustic focusing in the central area. Therefore, the study of the acoustic focusing phenomenon of the hemisphere is mainly to analyze the reflection characteristics of the primary and secondary acoustic lines [4].

From the above analysis, it can be found that the sound lines of plane wave incident on the sphere do not intersect at a point, and they form multiple focus areas in a certain space area in front of the sphere, which will also change with the change of incidence angle.

For ellipses and parabolas, they have a clear focus, which can be determined by the concave mirror reflection law. In the case of an ellipse, the sum of the distances between two focal points and any point on the ellipse is a constant. The sound ray radiated from one focus, regardless of its direction, will focus on the other focus after reflection. In the case of a paraboloid, the plane acoustic wave will focus on the focal point of the paraboloid after being reflected by the paraboloid [5].

Next, we will use COMSOL to conduct finite element analysis for spherical structures with different geometric shapes to explore their corresponding acoustic reflection characteristics.

2.2 Finite element simulation analysis

COMSOL Multiphysics is a large-scale advanced numerical simulation software. It is widely used in scientific research and engineering calculation in various fields and is called "the most professional multi-physical field full coupling analysis software" by scientists in the world today. By simulating various physical processes in the scientific and engineering fields, COMSOL Multiphysics has achieved highly accurate numerical simulation with efficient computing performance and outstanding multi-field bidirectional direct coupling analysis capability. Applications include acoustics, biological sciences, chemical reactions, electromagnetics, fluid dynamics, and optics. This paper will use COMSOL to simulate and analyze several typical concave bodies discussed earlier. First, a two-dimensional spherical model in COMSOL is built, where the inner radius is 1m, the outer radius is 1.2m, the PML thickness is 0.2m, and the internal filling area is seawater medium. The pressure acoustics in COMSOL, the frequency domain module, and the plane wave with the initial background pressure of 1Pa are selected. Assume that the equation is as follows [6]:

$$\nabla \cdot \left(-\frac{1}{\rho_c}(\nabla p_t - q_d)\right) - \frac{k_{eq}^2 p_t}{\rho_c} = Q_m \tag{6}$$

Wherein, $p_t = p + p_b$ $k_{eq}^2 = \left(\frac{\omega}{c_c}\right)^2 - k_z^2; c_c = c\rho_c = \rho$

To reduce the calculation error, the maximum cell of the grid is set to 1/5 of the shortest wavelength. The maximum frequency calculated in this verification is 10kHz, and the maximum size of the finite element grid is 0.03m. The number of elements of PML is 5, the growth formula is an isochromatic sequence, and the boundary is selected as the axial direction outward along the center of the circle.

2.2.1 Acoustic scattering characteristics of the rigid hemisphere

As shown in Figure 3, a rigid hemisphere with a radius of 0.15 m and a layer thickness of 0.01 m is constructed by COMSOL and placed in an infinite uniform and stable liquid seawater.

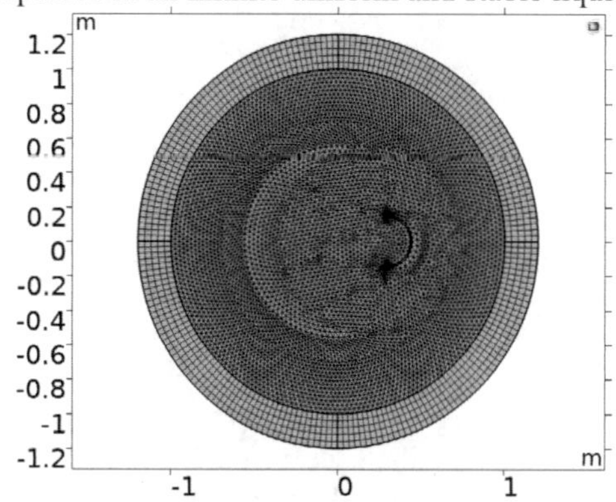

Figure. 3 Rigid Hemisphere Model Grid in Infinite Loud Field

The plane wave with the initial value of 1Pa and pressure amplitude of 10Pa is selected as the incident background sound field. The incident directions are 0°, 30°, and 60° respectively. The frequency range of the incident sound wave is 5kHz-10kHz.

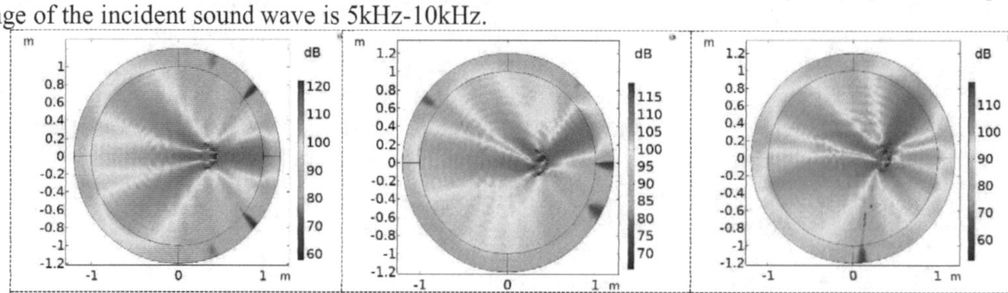

(a) 0° scattering pressure field corresponding to plane wave in the direction	**(b)** 30° scattering pressure field corresponding to plane wave in the direction	**(c)** 60° Scattering pressure field corresponding to plane wave in the direction

Figure. 4 Distribution of sound scattering pressure of rigid hemisphere

It can be found from the analysis in Figure 4 that the acoustic focus area of a rigid hemisphere changes with the change of the incident angle, and there are multiple positions in the focus area. The following is to further analyze the scattering pressure field distribution characteristics of the focus area, establish a two-dimensional section line along the incident direction of the plane wave, and study the pressure acoustic field distribution along the section line.

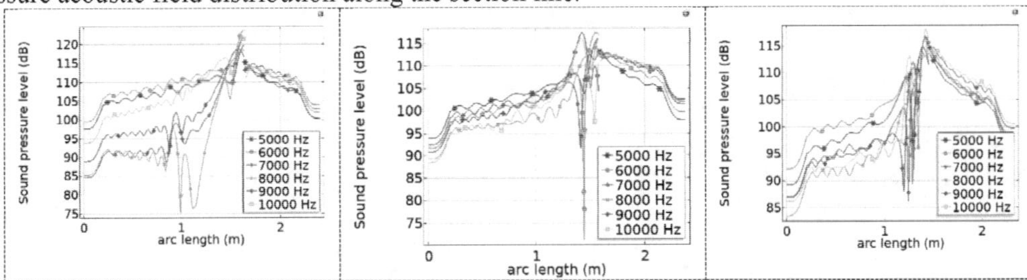

(a) 0° directional two-dimensional intercept scattering pressure field	(b) 30° directional two-dimensional intercept scattering pressure field	(c) 60° directional two-dimensional intercept scattering pressure field

Figure. 5 Scattering Pressure Field of Rigid Hemisphere in two-dimensional Section Direction

According to the analysis results in Figure 5, it is not difficult to find that for the hemispherical structure, the convergence area of the acoustic wave is uncertain in different incident directions, and multiple areas will appear. At the same time, for the acoustic lines with different frequencies, the scattering pressure field along the two-dimensional intercept direction has also changed greatly.

2.2.2 Acoustic scattering characteristics of the rigid paraboloid

Similarly, as shown in Figure 6, the COMSOL structure is used to construct a rigid paraboloid with a layer thickness of 0.005m and place it in an infinite uniform and stable liquid seawater.$x^2 = -2y(y \in (-0.3m, 0.3m))$

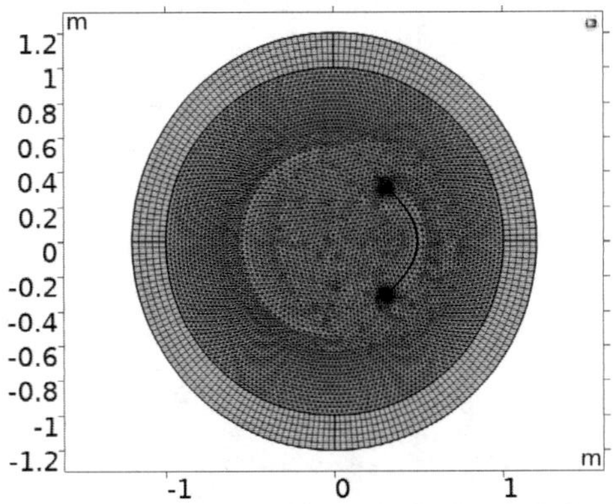

Figure. 6 Mesh of rigid paraboloid model in the infinite loud field

The plane wave with the initial value of 1Pa and pressure amplitude of 10Pa is selected as the incident background sound field. The incident directions are0^o, 30° and60^o respectively. The frequency range of the incident sound wave is 5kHz-10kHz.

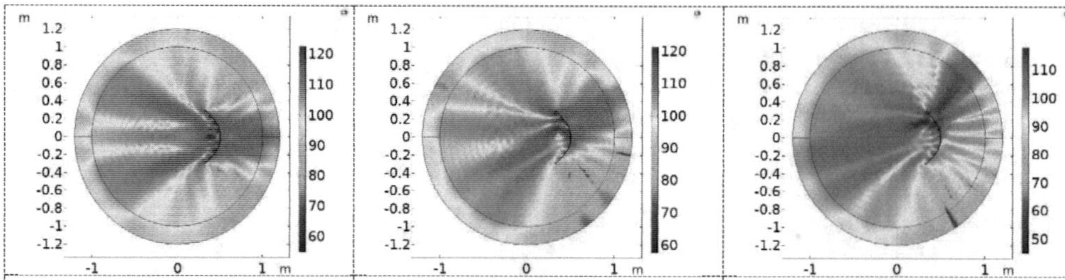

(a) 0° scattering pressure field corresponding to plane wave in the direction	(b) 30° scattering pressure field corresponding to plane wave in the direction	(c) 60° Scattering pressure field corresponding to plane wave in the direction

Figure. 7 Distribution of sound scattering pressure of Rigid paraboloid

It can be found from the analysis in Figure.7 that the acoustic focusing area of the rigid paraboloid changes with the change of the incident angle, but there is only one main focusing area along the incident direction of the plane wave. The following is a further analysis of the scattering pressure field distribution characteristics of the focusing area. A two-dimensional section line is established along the incident direction of the plane wave to study the pressure acoustic field distribution along the section line.

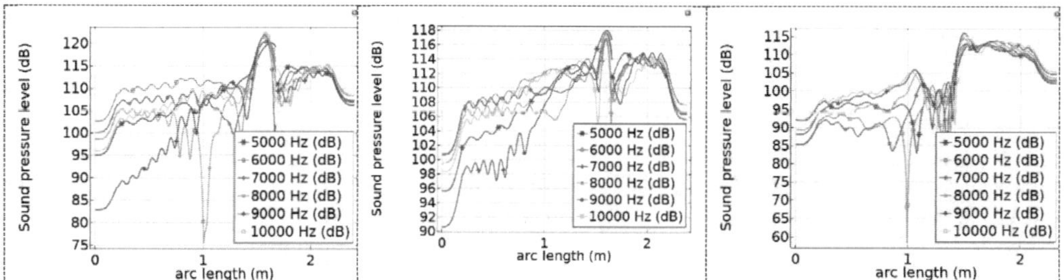

(a) 0° directional two-dimensional intercept scattering pressure field	(b) 30° directional two-dimensional intercept scattering pressure field	(c) 60° directional two-dimensional intercept scattering pressure field

Figure. 8 Scattering pressure field of rigid paraboloid in a two-dimensional cross-section direction

Through the simulation analysis shown in Figure 8, it is found that for paraboloidal structures, the convergence area of sound waves will change with the incident angle in different incident directions, but the focus area is relatively concentrated. At the same time, for sound rays of different frequencies, the change law of the scattering pressure field along the two-dimensional intercept direction is the same, and the location where the maximum value of the pressure field appears is the same, so the convergence effect is better.

2.2.3 Acoustic scattering characteristics of rigid semi ellipsoid

As shown in Figure 9, COMSOL is used to construct a rigid semi-ellipsoid with half-axis A of 0.1m, half-axis B of 0.15m, and layer thickness of 0.01m, and placed in an infinite uniform and stable liquid seawater.

Journal of Physics: Conference Series **2458** (2023) 012008 doi:10.1088/1742-6596/2458/1/012008

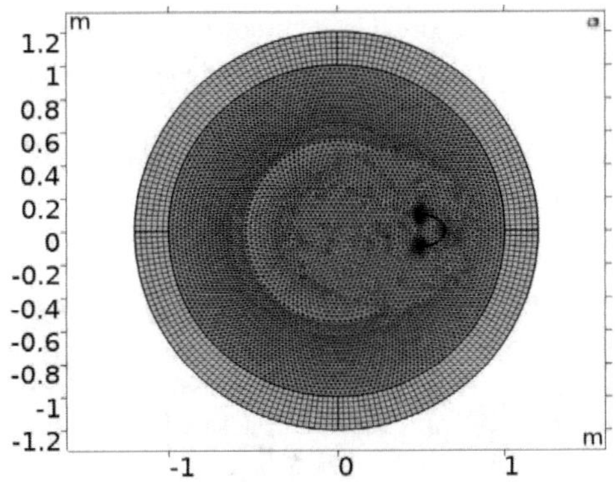

Figure. 9 Mesh of rigid semi-ellipsoid model in the infinite sound field

The plane wave with the initial value of 1Pa and pressure amplitude of 10Pa is also selected as the incident background sound field. The incident directions are 0°, 30° and 60° respectively. The frequency range of the incident sound wave is 5kHz-10kHz.

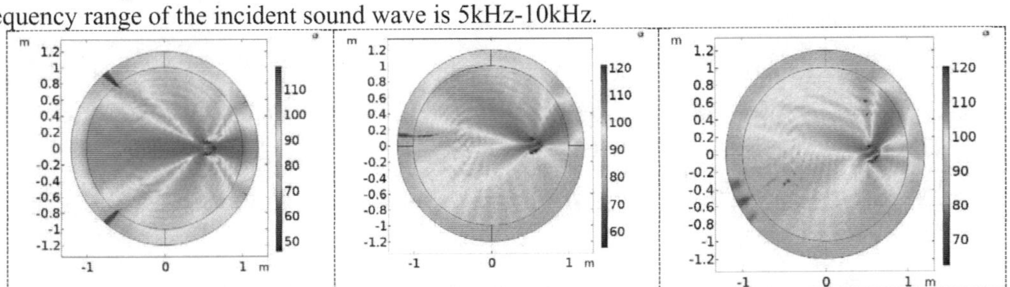

(a) 0° scattering pressure field corresponding to plane wave in the direction

(b) 30° scattering pressure field corresponding to plane wave in the direction

(c) 60° Scattering pressure field corresponding to plane wave in the direction

Figure. 10 Sound scattering pressure distribution of rigid semi ellipsoid

It can be found from the analysis in Figure 10 that the acoustic focusing area of a rigid semi-ellipsoid changes with the change of the incident angle, and the focusing area has several positions as well as that of a hemispherical body. To further analyze the distribution characteristics of the scattering pressure field in the focusing area, a two-dimensional section line is established along the incident direction of the plane wave, and the pressure acoustic field distribution along the section line is studied.

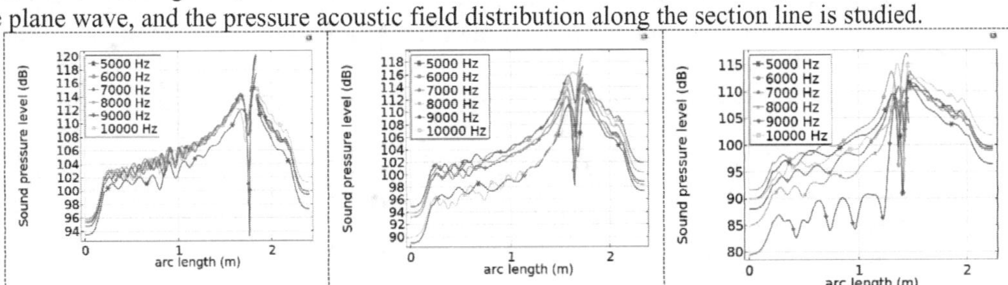

(a) 0° directional two-dimensional intercept scattering pressure field

(b) 30° directional two-dimensional intercept scattering pressure field

(c) 60° directional two-dimensional intercept scattering pressure field

Figure. 11 Scattering pressure field from a rigid semi-ellipsoid in the direction of a two-dimensional cross-section

Through the simulation results in Figure 11, it is not difficult to find that, for the semi-ellipsoidal structure, the convergence area of the acoustic wave is uncertain in different incident directions, and there will be multiple areas. However, for the acoustic lines with different frequencies, the variation rules of the scattering pressure field along the two-dimensional intercept direction are the same, but there is no strong sound convergence area.

3. Conclusion

Based on the COMSOL finite element analysis, this paper analyzes the acoustic focusing characteristics of different typical concave structures. Through comparison, it is found that the paraboloid structure has better convergence ability for acoustic waves in all directions, and is least affected by frequency changes. It is the most suitable reflector structure for underwater acoustic imaging systems. However, the research process of this paper also has the following shortcomings:

First, when using COMSOL to simulate the marine environment, the influence of the inhomogeneity of ocean elements such as the mobility of seawater, temperature, and salinity density on the acoustic convergence effect is not considered.

Secondly, during the simulation, the rigid structure is assumed for each concave structure, without considering the acoustic reflection and absorption loss, and the coupling effect of the actual concave structure in the acoustic field.

Because of this, the research content should be refined in the future, the impact of marine environmental elements on acoustic focusing should be studied in detail one by one, and the impact of each element should be analyzed qualitatively and quantitatively; In addition, targeted selection of concave structure materials and coupling effect can be considered, to provide a reference for actual application selection. At the same time, it can be considered to add a concave sound convergence test in an anechoic tank for comparative verification.

References

[1] Kumar N, Mitra U, Narayanan S S. Robust object classification in underwater side scan sonar images by using a reliability-aware fusion of shadow features[J]. IEEE Journal of Oceanic Engineering,2015,40(3):592-606.

[2] LeHardy P K, Moore C. Deep Ocean search for Malaysia airlines flight 370[C]. Proceedings of the 2014 Oceans-St John's,2014:1-4.

[3] Wu Ju. Research on key technologies of acoustic lens imaging [D]. Harbin: Harbin Engineering University, June 40, 2013.

[4] Jin Difeng. Research on spherical structure acoustic reflection [D]. Chongqing: Chongqing University, 2003.5:6, 16-17.

[5] Luo Songqin. Research on Acoustic Design of Spherical Interior Space [D]. Changsha, Hunan: Hunan University, 2006.4:9.

[6] Zhang Ranyue, Zhang Minghui. Vibration and Sound Foundation [M]. Harbin: Harbin Engineering University Press, 2016.12:211-213.

Vibro-acoustics time domain FEM-BEM analysis of a Titian Stradivari violin: the role of the bridge

Evaggelos Kaselouris[1,2], Vasilis Dimitriou[1,2,*]

[1]Hellenic Mediterranean University, Department of Music Technology & Acoustics, Rethymnon, GR-74100, Greece

[2]Hellenic Mediterranean University Research Centre, Institute of Plasma Physics & Lasers, Rethymnon, GR-74100, Greece

Email: dimvasi@hmu.gr

Abstract. The acoustic behaviour of the violin is defined during manufacturing since the vibrational acoustic properties of the instrument are related to the top plate's vibrational modes in a direct way. An important component of the violin assembly is the bridge. The bridge transfers the excitation forces from the strings to the soundboard via the bridge feet and acts as a sound filter. In this study the Titian Stradivari violin CAD geometry of the soundboard, including the f-holes and bass bar, and the bridge is modelled, and time domain coupled finite element method-boundary element method, vibro-acoustic, simulations are performed to analyse the interference of the instrument's sound with its structural components. A comparison of the vibroacoustic behaviour for the soundboard with and without the bridge enlightens the crucial role of the bridge. This initial computational approach will be extended, with similar studies on the role of every structural component of the violin individually, aiming to decode the full violin soundbox vibroacoustic behaviour and provide key insights for violinmakers on how structural changes affect the quality of the emitted sound.

1. Introduction

The violin is a well-known stringed musical instrument and has been studied experimentally via modal analysis [1] and holographic [2] methods, as well as via numerical modelling and simulations, especially via the finite element method [3-6]. The vibro-acoustic properties of the violin and the construction of the instrument are related to the top plate's vibrational modes in a direct way [7]. Nevertheless, the violin's vibro-acoustic behaviour is still not fully understood since it is significantly affected by various parameters such as the geometrical dimensions, the thickness distribution, the existence of f-holes and the existence of a bass bar. One of the most important components of the violin is the bridge since it transfers the excitation forces from the strings to the soundboard via the bridge feet. The bridge acts as a sound filter, and the material properties and geometric configuration of the bridge may have a significant influence on the vibroacoustic behaviour of the instrument [8].

Various studies exist in the literature that have numerically studied the vibrational characteristics of the violin's soundboard, providing crucial information to violinmakers on the factors that affect the top plate frequencies and modal shapes before assembling the whole instrument. In [9] the tuning of free violin plates was investigated via frequency domain FEM simulations. The assembly of several components of a violin body were numerically investigated via the FEM method in [7]. Recently, it was shown [10] that the frequency modes of a violin soundboard may be predicted by a neural network based on their geometric parameters, while other studies [11] investigated the possibility to substitute the wooden material of the soundboards by using composite materials. The influence of the bridge and the area between the f-holes on the acoustic behaviour of the violin was investigated in [12], while the bridge hill was simulated by a supported plate with the f-holes and bridge in [13]. Moreover, the numerical studies of the violin interaction with the surrounding environmental air are limited in the literature due to large computational needs. In [14] a vibroacoustic FE model was developed to study the behaviour of the Titian Stradivari violin. The automatically matched layer (AML) property was adopted to handle the interior/exterior acoustic problem and a very good agreement, between simulation and experimental results, on the predicted structural modal frequencies was found.

In this study the 1715 Titian Stradivari violin CAD geometry of the soundboard, including the f-holes and bass bar, and the bridge developed in [4] is modelled and simulated via time domain coupled finite element method-boundary element method vibro-acoustic simulations that determine the relationship between the instrument's sound and its structure. This violin acknowledged to Antonio Stradivari is selected since it still considered the finest in the produced sound quality. The structure is subjected to time-dependent loading condition that considers inertia and damping effects. A comparison of the vibroacoustic behaviour for the soundboard with and without the bridge is performed. This initial computational approach will be further extended to every structural component of the violin individually, to decode the full violin soundbox vibroacoustic behaviour and its relation to the final sound emission of the instrument and to provide these crucial manufacturing details to the violinmakers.

2. Mathematical description

FEM is a powerful method that numerically solves differential equations which concern engineering and Multiphysics problems [15]. It has the capability to predict the response of music instruments to different loading conditions and vibrations. Variations in the environmental conditions (such as relative humidity, temperature) may also be considered in vibro-acoustic simulations [15]. In this study the vibro-acoustic response of the violin top plate with and without the bridge during is investigated via FEM-BEM time domain transient numerical simulations. To perform the time domain FEM-BEM analysis LS-DYNA software [16] is used, by coupling the FEM dynamic solver with a BEM acoustic solver. BEM models the environment surrounding the vibrated structure and calculates the pressure of the emitted sound. For the time domain transient FEM-BEM analysis, initially the time domain FEM analysis is carried out, and the dynamic structure's response is transformed to the frequency domain via the use of the Fast Fourier Transform. Then, the computed boundary velocities are coupled to the following BEM acoustic computations [16].

The fundamental equation of motion for the transient mechanical analysis is:

$$[M]\left\{\frac{\partial^2 U}{\partial t^2}\right\} + [C]\{\dot{U}\} + [K]\{U\} = \{F\} \tag{1}$$

where *[M]* is the matrix of mass, *{U}* is the vector of displacement, *[C]* is the matrix of damping, *[K]* is the matrix of stiffness, and *{F}* is the load vector. The damping is defined by a stiffness matrix multiplier, and it holds $[C]=\beta[K]$, where β is the Rayleigh damping constant. For the BEM frequency domain acoustic analysis, the propagation of acoustic waves in an ideal fluid is governed via the Helmholtz equation, neglecting any volume source:

$$\nabla^2 p + k^2 p = 0 \tag{2}$$

where p is the acoustic pressure and k is the wavenumber. Equation 2 is transformed into an integral equation by using Green's theorem and the pressure at any fluid medium point is the integral provided by the expression:

$$P_Q(\omega) = \int_S (G\frac{\partial p(\omega)}{\partial n} - p(\omega)\frac{\partial G}{\partial n})dS \tag{3}$$

where $P_Q(\omega)$ is the sound pressure at field point Q and $p(\omega)$ is the surface pressure, S is the structure surface, n is the normal vector on surface S and G is the Green's function, which is expressed as:

$$G = \frac{e^{-ikr}}{4\pi r} \tag{4}$$

where r is the distance between the surface point of integration and Q. It also applies that:

$$\frac{\partial p(\omega)}{\partial n} = -i\rho\omega v_n \tag{5}$$

where v_n is the normal velocity and ρ is the acoustic fluid density.

3. Modeling and simulation

The CAD geometry of a Titian Stradivari soundboard and bridge is adopted from the work in [4]. The model of the structure includes the bass bar and the f-holes and is shown in figure 1. A total number of approximately 20000 elements is considered to model the violin parts, as shown in figure 1, while a mesh independent study was performed to conclude to this number. Tetrahedral shell elements are used to create the mesh of the FEM model. Furthermore, under free conditions the vibrational variations of the plate's response are computed, except for the bridge's top, where out-of-plane displacements at four points were not permitted resembling the limitations that the strings provide. The structure was modelled using orthotropic wooden material properties that can be found in [4]. The Rayleigh damping constant has a value of 2×10^{-6} [4]. The transient mechanical analysis was carried out to determine the structure's response when it is subjected to a time-dependent loading that considers damping and inertia effects. A delta time dependent force function was applied at the bridge's bass corner like the work in [4]. In the case where the bridge was not modelled the force was applied on the soundboard in the location where the bridge foot touches the soundboard.

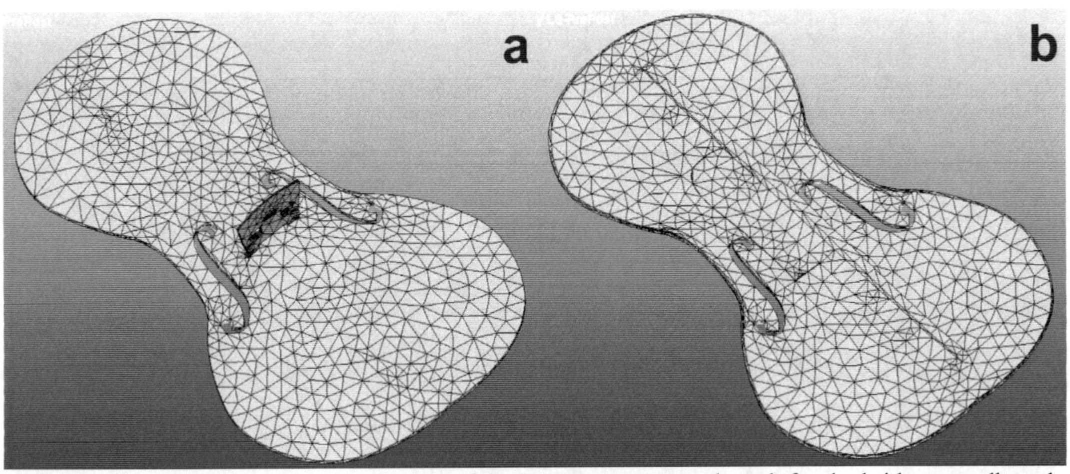

Figure 1. a) Front view and b) Back view of the CAD geometry and mesh for the bridge-soundboard system

Regarding the BEM acoustic numerical simulations, a frequency range between 0 to 1000 Hz is considered. A resolution of 1 Hz is considered. The modelled acoustic medium is air, having at room temperature a speed of sound of 340 m/s and a density of 1.21 kg/m^3, while a pressure of 20 μPa is also considered as reference. 2000 BEM elements model the spherical geometry of the domain of the air, which contains the structure. An acoustic node with no mass is set 0.5 m above and 0.5 m in front of the structure's centre and its location corresponds to an assumed microphone position in the acoustic field, where the sound pressure is measured. The total runtime of the vibroacoustic simulation is 1 s. Like in our previous works [11,16] the High-Performance Computer (HPC) ARIS for parallel processing is used to run our simulations.

4. Results and Discussion

Figure 2 shows time domain FEM-BEM results of the normalized sound pressure for a) the violin soundboard and b) the violin soundboard and bridge system. Comparing the two cases it is evident that the inclusion of the bridge is responsible for filtering frequencies especially in the frequency range of 150 Hz to 250 Hz and between 0 Hz to 50 Hz. For the simulation with no modeled bridge the maximum frequencies at 182 Hz and at 389 Hz agree with the modal frequencies found in [4] and correspond to the X-mode and ring mode respectively. The second mode (X mode) and the fifth mode (ring mode) are considered the most important by violinmakers, to determine the acoustical properties of the whole assembled violin [11]. For the simulation with modeled bridge the maximum computed frequencies at 278 Hz and at 480 Hz correspond to the Helmholtz f-hole resonance [4] and to the B_1^- signature mode. More detailed analysis for the well-known signature modes of the violin soundbox will take place in future studies. The full violin soundbox will be simulated, while experimental measurements via the electronic speckle pattern interferometry will validate the vibroacoustic simulations of the structure.

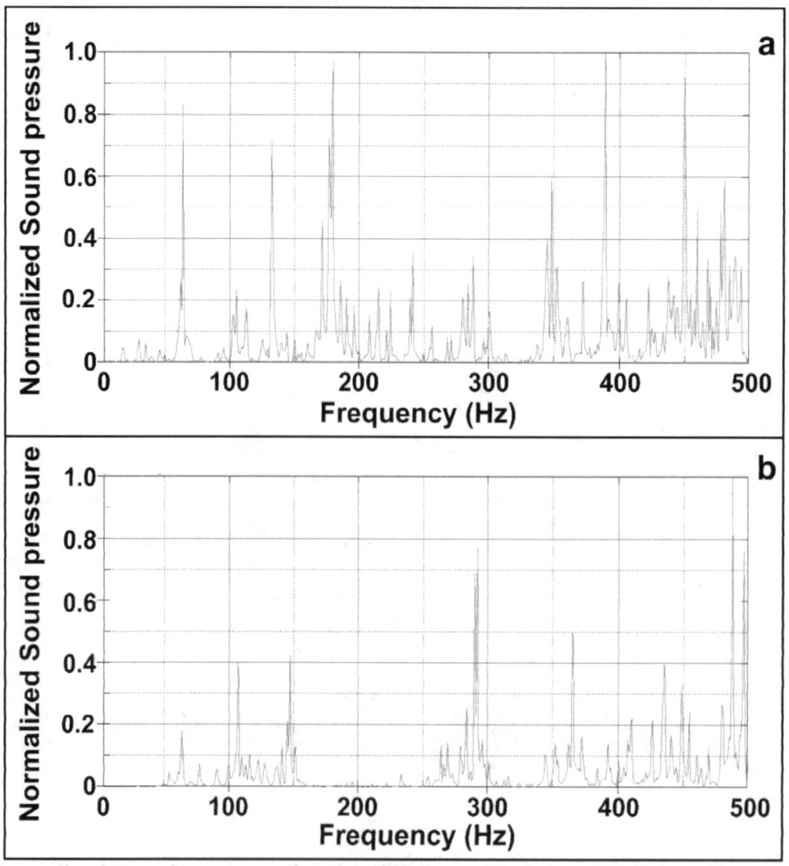

Figure 2. Normalized Sound pressure of a) the violin soundboard and b) the violin soundboard and bridge system

Figure 3 demonstrates time domain FEM-BEM results of the pressure spectrogram for a) the violin soundboard and b) the violin soundboard and bridge system. The harmonics developed on the air can be depicted. Likewise, it is evident that the existence of the bridge filters specific frequency ranges as previously mentioned.

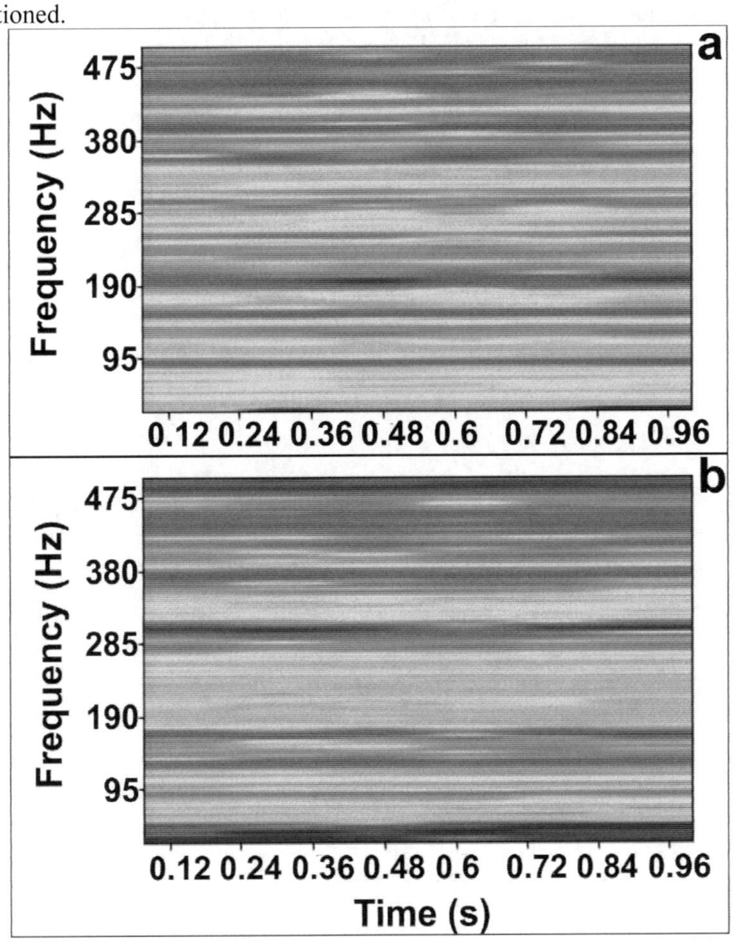

Figure 3. Sound pressure spectrograms for the violin soundboard a) without the bridge and b) with the bridge

5. Conclusions

FEM-BEM time domain simulations of a musical violin soundboard, with and without the bridge are presented and analysed. The model can provide key insights on the vibroacoustic behaviour of the instrument. The existence of the bridge filters a specific frequency range, and the simulation results agree well with literature data of the vibrating structure. In future works more simulations will focus on the mobility of the bridge, while the full violin soundbox will be simulated, and experimental measurements via the electronic speckle pattern interferometry will validate the vibroacoustic simulations of the instrument. This first vibroacoustic computational investigation on the role of the bridge, will be further extended to every structural component of the violin, to release crucial manufacturing details related to the sound emission of the instrument to the violinmakers.

Acknowledgments

This work was supported by computational time granted by the Greek Research & Technology Network (GRNET) in the National HPC facility ARIS-under project ID pr011027-LaMPIOS.

References

[1] Bissinger G 2008 *J Acoust. Soc. Am.* **124** 1764
[2] Bakarezos E, Orphanos Y, Kaselouris E et al 2019 Laser-Based Interferometric Techniques for the Study of Musical Instruments *Computational Phonogram Archiving. Current Research in Systematic Musicology* ed R Bader (Cham: Springer) pp 251-268
[3] Gough C 2018 *J. Acoust. Soc. Am.* **143** 1
[4] Torres J A, Soto C A, Torres-Torres D 2020 *J. Acoust. Soc. Am.* **148** 1496
[5] Efstathopoulos N et al 2010 *Injury* **41** 259
[6] Kaselouris E et al 2018 *Plasma Phys. Control. Fusion* **60** 014031
[7] Gough C 2015 *J. Acoust. Soc. Am.* **137** 139
[8] Woodhouse J 2005 *Acta Acust. United Acust.* **91** 155
[9] Bretos J, Santamaría C, Moral J A 1999 *J. Acoust. Soc. Am.* **105** 1942
[10] Gonzalez S, Salvi D, Baeza D et al 2021 *Sci. Rep.* **11** 9455
[11] Kaselouris E, Orphanos Y, Bakarezos M et al 2021 *Proceedings of the INTER-NOISE 2021—2021 International Congress and Exposition of Noise Control Engineering (Washington)* vol 263 (Reston: Institute of Noise Control Engineering) pp 3369-3377
[12] Gough C 2018, *J. Acoust. Soc. Am.* **143** 1
[13] Jansson E V, Kabała A 2018 *Vib. Phys. Syst.* **29** 2018027
[14] Pyrkosz M A, Van Karsen C 2014 Coupled vibro-acoustic model of the Titian stradivari violin *Topics in Modal Analysis I* vol 7 (Berlin/Heidelberg: Springer) pp 317–332
[15] Kaselouris E, Bakarezos M, Tatarakis M, Papadogiannis N A, Dimitriou V 2022 *Acoustics* **4** 183
[16] Kaselouris E, Alexandraki C, Bakarezos M et al 2022 *Int. J. Circuits Syst. Signal Process.* **16** 948

Method and Experiment Study of the Array Shape Calibration for Underwater Vertical Line Arrays

Xuegang Li[1,*], Shuyou Wu[1], Shengli SU[1], Lan Ding[1] and Guoqing Jiang[2]

[1] Wuhan Second Ship Design and Research Institute, Wuhan 430205, China

[2] School of Marine Science and Technology, Northwestern Polytechnical University, Xi'an, 710072, China

[*] Email: xuegang608@126.com

Abstract: Vertical line arrays have been widely used in underwater target location and noise measurement, and the attitude and shape of the vertical array have a significant influence on detection accuracy. The paper presents an array shape calibration method for underwater vertical acoustic array-based measurement systems. Firstly, a standard acoustic source was used to transmit underwater acoustic signals at a certain distance from the acoustic array, and the target location based on the received data of the acoustic array was estimated concerning distance and depth to obtain the two-dimensional ambiguity function of the depth and distance of the array. Then, the ambiguity function was used to calculate the center depth and the inclination angle of the vertical acoustic array and to further obtain the shape and location information of the acoustic array. Finally, a lake experiment study was carried out. The results showed that the calibrated array shape was in good agreement with the actual location, and the sound source location estimated by using the calibrated array shape was consistent with the actual sound source location, thereby effectively reducing the impact on the measurement accuracy due to the inclination of the measurement system under the influence of water current.

1. Introduction

The radiated noise of an underwater target is one of the important indexes of its stealthy and is used to evaluate the danger of causing the enemy's sonar detection and underwater weapon attack by the underwater target. Accurate measurement and evaluation of the radiated noise levels of underwater targets is an important basis for testing the realization of the technical indexes and design finalization of a product. Array signal processing technology has a wide range of applications in sonar, radar, and other fields [1][2][3]. With the development of vibration and noise reduction technology, the radiated noise of underwater targets is gradually reduced [4][5]. Under the same background noise condition, the traditional single-hydrophone measurement method cannot meet the requirement of the signal-to-noise ratio (SNR) in the testing of silent underwater targets. Compared with the single-hydrophone measurement method, the array measurement method can improve the SNR, reduce measurement fluctuation, and improve the accuracy of the measurement results [6][7][8].

The underwater vertical array will become inclined under the influence of water current [9], which will affect the detection accuracy and the measurement results of the measurement system. To reduce the measurement error and realize the accurate measurement of the radiated noise of silent underwater targets, it is necessary to perform array shape calibration for the vertical arrays in the underwater attitude. With underwater target detection, radiated noise measurement, and evaluation as the application background, the paper proposes an array shape calibration method for underwater acoustic array-based measurement systems. Using array signal processing, the array shape calibration of vertical arrays underwater current disturbance has been realized, laying a good foundation for improving the accuracy of the underwater target detection and the underwater target radiated noise measurement.

2. Method for the Vertical Array Shape Calibration

Vertical acoustic arrays are usually arranged underwater using survey ship suspension and underwater anchoring. In an ideal state where there are no water currents and wind waves, a vertical array is in a vertical state by gravity, as shown by the dotted line in Figure 1. In an actual ocean environment, the underwater attitude of the vertical array will become inclined under the influence of water currents and wind waves. For a vertical array in a suspension arrangement, the inclination may be reduced by hanging a counterweight at the bottom of the array. Because of the influence of the counterweight, the inclination of the lower part of the vertical acoustic array is usually smaller than that of the upper cable, so the vertical array shape can be simplified to the inclination model as shown by the solid line in Figure 1.

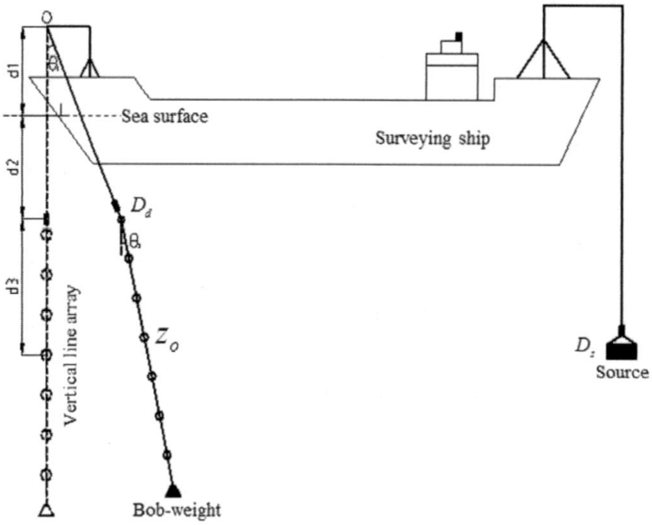

Figure 1. Schematic Diagram of the Shape of an Underwater Acoustic Array

Assuming that the height of the suspension point O of the acoustic array in the vertical state from the water surface is d_1, the distance between the top of the acoustic measurement section of the vertical array and the water surface is d_2, the half-array length of the vertical array is d_3, and the top-end depth of the acoustic segment under the condition that the acoustic array is inclined is D_d, then, the inclination angle in the vertical direction of the upper cable of the vertical array is:

$$\theta_1 = \cos^{-1}\left(\frac{d_1 + D_d}{d_1 + d_2}\right) \tag{1}$$

The inclination angle in the vertical direction of the acoustic segment of the vertical array is:

$$\theta_2 = \cos^{-1}\left(\frac{z_o - D_d}{d_3}\right) \tag{2}$$

where z_o is the center depth of the acoustic segment of the vertical array.

Then, the target location is estimated concerning the distance and depth dimensions. For an acoustic array with M elements, the data of the time domain of the array element domain can be expressed as:

$$x(t)=[x_1(t), x_2(t),\cdots x_n(t),\cdots x_M(t)] \tag{3}$$

where $x_n(t)$ is the time domain data of the n-th sampling point, and M is the number of the elements of the acoustic array.

Fourier transform on Formula (3) is performed to obtain the frequency domain data:

$$X(f)=\int_{-\infty}^{+\infty} x(t)e^{-j2\pi ft}dt \tag{4}$$

The following formula is calculated to obtain the covariance matrix of the measured data of the acoustic array:

$$\hat{R}(f)=X^{\mathrm{H}}(f)\cdot X(f) \tag{5}$$

where the superscript H stands for conjugate transposition.

The target localization function with the horizontal distance r and depth z between the acoustic array and the sound source can be expressed as:

$$\Phi(r,z)=\boldsymbol{w}(r,z)^{\mathrm{H}}\hat{R}(f)\boldsymbol{w}(r,z) \tag{6}$$

where $\boldsymbol{w}(r,z)$ represents the manifold vector of the near-field array, the superscript H represents conjugate transposition, and the central array element of the acoustic array is taken as the reference array element, which is expressed as:

$$\boldsymbol{w}_i(r,z)=\frac{d_{M/2}(r,z)}{d_i(r,z)}e^{j2\pi f(d_i(r,z)-d_{M/2}(r,z))/c}, i=1,\ldots,M \tag{7}$$

where c is the sound speed in the water, f is frequency, and $d_i(r,z)$ is the distance between the location point (r,z) and the i-th array element, which is expressed as:

$$d_i(r,z)=\sqrt{r^2+(z-z_i)^2} \tag{8}$$

where z_i represents the depth of the i-th array element.

When Formula (3) is used to obtain the two-dimensional ambiguity function of the array depth and distance, the depth z_o corresponding to the maximum value point of the target function is the center depth of the vertical array, and the location of each array element can be found by further combining the inclination angles θ_1 and θ_2 of the acoustic array. Taking the hanging point 0 of the vertical array as the original point, the location of post-calibration array elements can be found by calculation as follows:

$$P_i(x,z)=\begin{cases}(d_1+d_2)\cdot\sin\theta_1+(i-1)\cdot h\cdot\sin\theta_2 \\ (d_1+d_2)\cdot\cos\theta_1+(i-1)\cdot h\cdot\cos\theta_2\end{cases}, i=1,\ldots,M \tag{9}$$

where h is the spacing between the elements of the acoustic array in the vertical state.

Journal of Physics: Conference Series **2458** (2023) 012010 doi:10.1088/1742-6596/2458/1/012010

3. Experiment Study in Lake

A lake experiment study was carried out in October 2020. First, the length parameters of the acoustic array in the vertical state were obtained by measuring with a meter ruler, which was 5.0 m for d_1, 9.6 m for d_2, 15.36 m for d_3, and 5.0 m for h; the top-end depth D_d of the acoustic array under the condition of vertical array inclination was recorded as 8.34 m by a depth finder; and the depth D_s of the standard sound source was recorded as 24.65 m by a depth finder, which was then substituted into Formula (1) to calculate the inclination angle θ_1 of the upper cable of the vertical array under this working condition, which was found to be 24°.

The time domain data $x(t)$ of the acoustic array with M elements acquired by using the standard sound source to transmit linear frequency modulation signals are shown in Figure 2, and the time-frequency diagram of the received signals obtained by time-frequency analysis is shown in Figure 3.

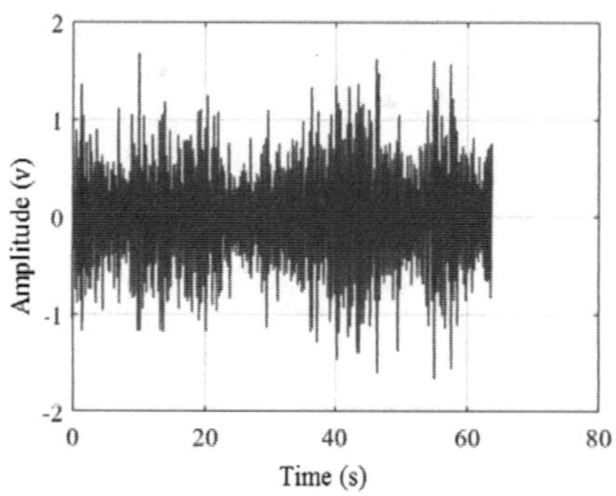

Figure 2. Time-Domain Waveform of Received Signals.

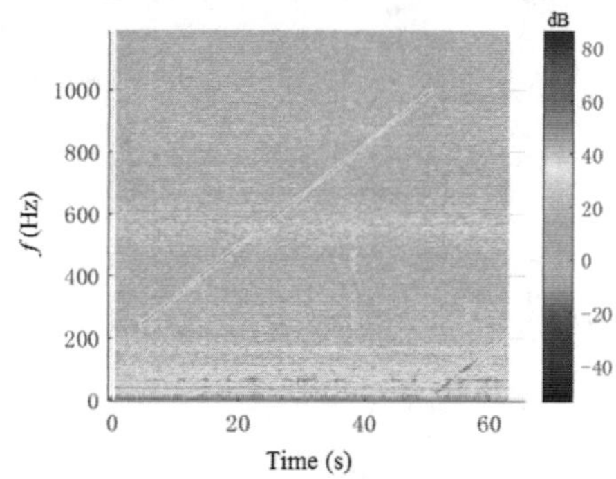

Figure 3. Time-Frequency Diagram of Received Signals

Formula (4) was used to perform Fourier transform on the time domain data $x(t)$ of the vertical array with M elements to obtain the frequency domain data $X(f)$, and Formula (5) was used to find the covariance matrix $\hat{R}(f)$ of the frequency domain data. Formula (7) and Formula (8) were used to calculate the array manifold vector $\mathbf{w}(r,z)$ of the vertical array, and the result was substituted into Formula (6) to calculate the target location function $\Phi(r,z)$. The two-dimensional pseudo-color picture of the target location function $\Phi(r,z)$ concerning the distance r and depth z is shown in Figure 4. The depth corresponding to the maximum value point (23.69, 39.6) is the center depth of the array, that is, the center depth of the vertical array z_o is 23.69 m.

Substituting the estimated center depth z_o of the acoustic array into Formula (2) for calculation, we got an inclination angle of the nested array as 2° and substituting it into Formula (9) for calculation, and the actual array element location was obtained. Using the calibrated vertical array shape to estimate the location of the standard sound source, a two-dimensional diagram of distance and depth as shown in Figure 5 was obtained. The estimated sound source location and the actual sound source location (39.6, 24.56) are consistent, indicating that the calibrated array shape is in good agreement with the actual location.

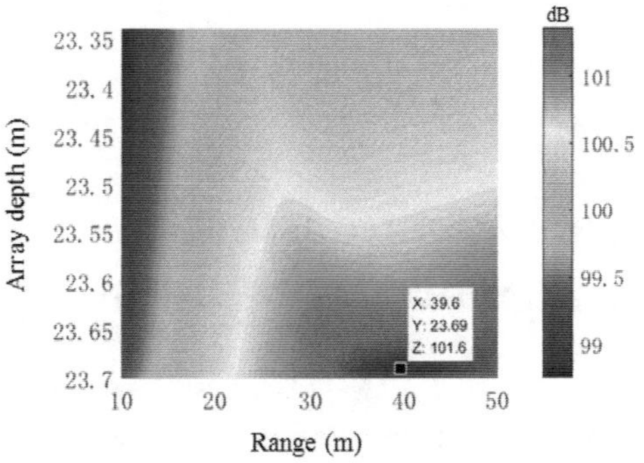

Figure 4. Estimation Results of the Center Depth and Distance of the Acoustic Array

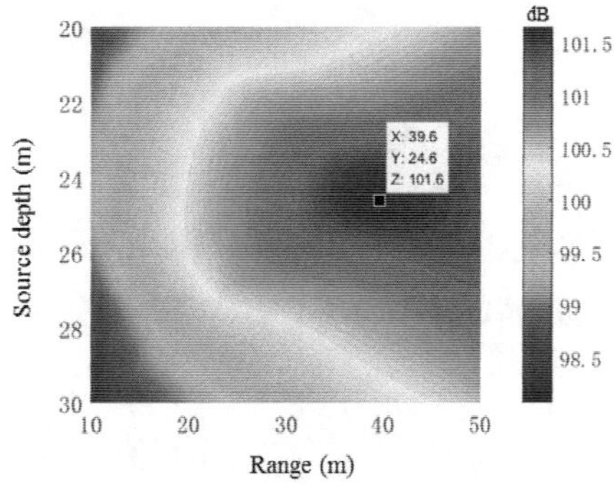

Figure 5. Inversion Results of the Sound Source Depth and Distance

4. Conclusions

Given the problem that the detection accuracy of an underwater vertical array-based measurement system is affected by the inclination under the influence of water current, this paper presents an array shape calibration method for underwater vertical acoustic array-based measurement systems. By using a standard sound source to transmit underwater acoustic signals at a certain distance from the acoustic array, and in combination with the information of the top-end depth of the acoustic array recorded by a depth finder, the target location based on the data received by the acoustic array was estimated in the distance and depth dimensions, and the two-dimensional ambiguity function of the array depth and distance was obtained. The depth corresponding to the maximum value point of the ambiguity function was the center depth of the acoustic array, and the shape and location information of the acoustic array were found by further calculation. The results of a lake experiment showed that the calibrated array shape was in good agreement with the actual position, and the sound source location estimated by using the calibrated array shape was consistent with the actual sound source location, thereby effectively reducing the impact on the measurement accuracy due to the inclination of the measurement system under the influence of water current. In the next step, in combination with a lake sound field measurement test, by using the array shape calibration method proposed in this paper, an underwater target radiated noise measurement test based on a vertical array will be carried out, and the uncertainty of the measurement results will be analyzed and evaluated.

Acknowledgments

This work was supported by the National Natural Science Foundation of China (Grant No. 51409199).

References

[1] Yan S F, Ma Y L, and Hou C H 2007 *J. Acoustic. Soc. Am.*, **122** 2680
[2] Yan S F and Ma X C 2008 *Acta Acustica*, **33** 316
[3] Gan T and Wang Y M 2012 *Acta Acustica*, **37** 18
[4] Su Q, Wang G B and Zhu P F 2014 *Ship Science and Technology*, **36** 1
[5] He L 2006 *Ship Science and Technology*, **28** 9
[6] Jiang G Q, Sun C and Liu X H 2020 *Journal of Applied Acoustics*, **39** 81
[7] Wu G Q, Wang M G, Chen S H and Ma L 2007 *Acta Acustica*, **32** 398
[8] Yan S F 2005 *Acta Acustica*, **30** 309-316
[9] Du J L, Ma Z C, An T S and Li A 2015 *Journal of Test and Measurement Technology*, **29** 309

Design and Realization of Ultra-wideband Constant Beamwidth Beamforming in a Waveguide Environment

Xuegang Li[1, *], Jie Zhu[1], Guoqing Jiang[2], Runan Hua[1] and Wuping Yao[1]

[1] Wuhan Second Ship Design and Research Institute, Wuhan 430205, China

[2] School of Marine Science and Technology, Northwestern Polytechnical University, Xi'an, 710072, China

[*] xuegang608@126.com

Abstract: Array signal processing technology has been widely used in signal detection and underwater target radiation noise measurement. A design and realization method of an ultra-wideband and high-robustness constant beamwidth beamformer based on a vertical nested array is proposed. Using the nested array multiplexing, the ultra-wideband constant beamwidth beamforming was realized by using fewer array elements; the main lobe of each sub-band beam in the designed band was made to approach the desired constraint value based on the convex optimization method; the beamforming robustness and array gain were improved through the constraint of the weight vector two-norm. In the paper, the influence on beamforming due to inconsistent array element amplitude-phases in a measured waveguide environment is analyzed. The computer simulation was used to realize the measurement of constant beamwidth beamforming within a 100-octave band range in the measured waveguide environment and verify the effectiveness of the proposed method.

1. Introduction

Array signal processing technology has been widely used in sonar and radar fields [1][2][3]. With the gradual reduction of the radiation noise level of underwater targets, improving the output signal-to-noise ratio by array signal processing has become a vital problem [4][5]. Beamforming is a very important means of array signal processing. In the beamforming, the spatially distributed sensor array data are combined to get a scalar beam output, and then the signal in a specified direction is selectively enhanced while the signals in other directions are inhibited by designing an appropriate weight value, thereby improving the output signal-to-noise ratio, inhibiting the interference from the water surface and underwater, and reducing the measurement fluctuation [6][7]. The performances of a beamformer can be examined in terms of correlative indicators such as array gain, robustness, sidelobe level, main lobe width, and response. Beam optimization is designed to find an optimal trade-off among these conflicting performances, thereby obtaining a satisfactory beamformer with the best overall performance.

Wideband constant beamwidth beamforming methods may be classified into frequency domain beamforming ones and time domain beamforming ones. In the frequency domain approach, the time domain data is converted to that of the frequency domain by Fourier transform, and then beamformed in each sub-band to keep the beam patterns of the sub-bands consistent, thereby realizing a constant beamwidth. In the time domain processing approach, the inter-array time delay is compensated by setting a time delay filter to each array element, and finally, the signals of various array elements are summed up and output to achieve the wideband constant beamwidth processing. The frequency domain processing is simple and feasible, so the paper uses the frequency domain beamforming method to achieve a constant beam width.

The paper proposed an ultra-wideband space array information processing method based on a vertical nested array under an application background of underwater target radiation noise measurement. The method improved beamforming robustness and array gain through the constraint of the weight vector two-norm. The paper also analyzed the influence on beamforming results due to inconsistent array element amplitude-phases in the measured waveguide environment to lay a foundation for the effective measurement of underwater target radiation sound field.

2. High-robustness constant beamwidth beamforming method

The radiation noise of underwater vehicles is a wideband signal with frequency components in a range from tens of hertz to tens of kilohertz. Generally, the overall wideband signal is measured by using several subarrays of a linear array to measure different frequency band signals. In this paper, an ultra-wideband space array information processing method based on a vertical nested array was used to lay a foundation for measuring the wideband radiation sound field of an underwater target.

According to the constant beamwidth beamforming in the assumed plane wave, a similar design method was adopted for the wideband constant beamwidth beamforming in the waveguide environment. The weight vectors at different sub-band frequencies were solved by a convex optimization algorithm [8]. To improve the beam robustness, the weight vectors were constrained by two-norm to achieve constant beamwidth beamforming, which is expressed as follows:

$$\begin{aligned} &\min_{\boldsymbol{w}} \quad \left|B_d\left(D_q\right)-\boldsymbol{w}^H(f)\boldsymbol{H}_f\left(D_q\right)\right|, \quad D_q \in Z_{ML} \\ &\text{subject to} \quad \left|\boldsymbol{w}^H(f)\boldsymbol{H}_f\left(D_s\right)\right| \le \varepsilon, \quad D_s \in Z_{SL} \\ &\qquad\qquad \left\|\boldsymbol{w}(f)\right\|_2 \le \xi \end{aligned} \tag{1}$$

where $B_d(D_q)$ represents the main lobe of the desired beam pattern; Z_{ML} represents the depth range corresponding to the main lobe, and Z_{SL} represents the depth range corresponding to the sidelobe. The weight vectors were supplemented with a two-norm constraint condition to ensure the beamformer robustness.

The array gain is expressed as

$$AG = \frac{AG_s}{AG_n} = \frac{S_A/S_E}{N_A/N_E} = \frac{S_A/N_A}{S_E/N_E} = \frac{SNR_A}{SNR_E} \tag{2}$$

where AG_s is the signal gain representing the ratio of the array-weighted output signal power S_A to the signal power S_E received by an individual array element; AG_n is noise gain representing the ratio of the array-weighted output noise power N_A to the noise power N_E received by an individual array element; SNR_A represents the signal-to-noise ratio at the array output end; and SNR_E represents the signal-to-noise ratio received by an individual array element.

The noise gain may be expressed as

$$AG_n = \boldsymbol{w}^H \boldsymbol{Q} \boldsymbol{w} \tag{3}$$

where $\boldsymbol{w}$ represents the weight vector of the array; and $\boldsymbol{Q}$ represents the normalized noise cross-spectral density matrix, of which the (m, n)-th element in an isotropic noise field is expressed as follows:

$$\boldsymbol{Q}_{mn} = \frac{\sin\left[2\pi d(m-n)/\lambda\right]}{2\pi d(m-n)/\lambda} \tag{4}$$

where d represents array element spacing and λ represents the wavelength. Then, the noise gain formula is expanded to:

$$\begin{aligned} AG_n &= \sum_{m=1}^{M}\sum_{n=1}^{M}\left(w_m^* w_n \frac{\sin\left[2\pi d(m-n)/\lambda\right]}{2\pi d(m-n)/\lambda}\right) \\ &= \sum_{i=1}^{M}\left|w_i\right|^2 + 2\sum_{m=1}^{M-1}\sum_{n=m+1}^{M}\left(w_m^* w_n \frac{\sin\left[2\pi d(m-n)/\lambda\right]}{2\pi d(m-n)/\lambda}\right) \end{aligned} \tag{5}$$

For wideband signals, when analyzing the sub-band frequency f_0, the output signal of the i-th hydrophone is expressed as:

$$X_i(r,f_0) = H_i(r,f_0)S(f_0), \quad i = 1,2,\mathrm{K},M \tag{6}$$

The array weighted output signal power is:

$$\begin{aligned} \left|Y(f_0)\right|_A^2 &= \left|\boldsymbol{w}^H \boldsymbol{X}(r,f_0)\right|^2 = \left|\boldsymbol{w}^H \boldsymbol{H}(r,f_0)S(f_0)\right|^2 \\ &= \left|S(f_0)\right|^2 \left|\boldsymbol{w}^H \boldsymbol{H}(r,f_0)\right|^2 \end{aligned} \tag{7}$$

Let the average power of the output signals of the hydrophones be the output power of the reference hydrophone:

$$\begin{aligned} \left|Y(f_0)\right|_E^2 &= \frac{1}{M}\sum_{i=1}^{M}\left|X_i(r,f_0)\right|^2 = \frac{1}{M}\sum_{i=1}^{M}\left|H_i(r,f_0)S(f_0)\right|^2 \\ &= \frac{1}{M}\left|S(f_0)\right|^2 \sum_{i=1}^{M}\left|H_i(r,f_0)\right|^2 \end{aligned} \tag{8}$$

Therefore, the signal gain is:

$$AG_s = \frac{\left|S(f_0)\right|^2 \left|\boldsymbol{w}^H \boldsymbol{H}(r,f_0)\right|^2}{\frac{1}{M}\left|S(f_0)\right|^2 \sum_{i=1}^{M}\left|H_i(r,f_0)\right|^2} = \frac{M\left|\boldsymbol{w}^H \boldsymbol{H}(r,f_0)\right|^2}{\sum_{i=1}^{M}\left|H_i(r,f_0)\right|^2} \tag{9}$$

Assuming that the parameters of the sound field are known, the channel transfer function $\hat{H}_i(r,f_0)$ can be calculated by the Scooter model based on the wavenumber integration technique or the Kraken C model based on the normal mode. Substitute it into Formula (1) to obtain a signal gain as follows:

$$\dot{A}G_s = \frac{M\left|\boldsymbol{w}^H \hat{\boldsymbol{H}}(r,f_0)\right|^2}{\sum_{i=1}^{M}\left|\hat{H}_i(r,f_0)\right|^2} \tag{10}$$

The array gain is expressed as

$$AG = \frac{\dot{A}G_s}{AG_n} = \frac{\dfrac{M\left|\boldsymbol{w}^H \hat{\boldsymbol{H}}(r,f_0)\right|^2}{\sum_{i=1}^{M}\left|\hat{H}_i(r,f_0)\right|^2}}{\sum_{i=1}^{M}\left|w_i\right|^2 + 2\sum_{m=1}^{M-1}\sum_{n=m+1}^{M}\left(w_m^* w_n \dfrac{\sin\left[2\pi d(m-n)/\lambda\right]}{2\pi d(m-n)/\lambda}\right)} \tag{11}$$

3. Design of the vertical nested array

The vertical array is proposed to reduce the influence of interference noise and multipath interference, which is essential to the radiation noise measurement of silent targets. The beam direction of the vertical array can be accurately controlled vertically, so that the reflected waves and interference noise sources from the water surface and lakebed can be suppressed by adjusting the beam direction, thereby obtaining a high signal-to-noise ratio. In the design of a hydrophone array, factors such as target radiation noise source level, background noise level, measuring distance, and signal-to-noise ratio (generally not lower than 6 dB) should be considered in addition to ease of implementation and cost control of the project. The vertical linear array measurement system works by using vertical directivity to improve the signal-to-noise ratio.

The length is inversely proportional to the operating frequency for linear arrays under the same beamwidth. According to the spectrum characteristics of underwater target radiation noise, the nested multiplexing method was used to design the wideband array with consideration of project implementation ability, in which the linear array could realize wideband measurement in the 100Hz-10kHz frequency band range, and was divided into seven subarrays with the operating frequency of each subarray at one octave for beamforming. The seven subarrays covered the 100Hz-10kHz operating frequency band range, so that the signal-to-noise ratio of measurement in the frequency band range improved by about 7 dB, thereby improving the measurement signal-to-noise ratio of the system. The seven sub-bands covered the ranges of 100 Hz-200 Hz, 200 Hz-400 Hz, 400 Hz-800 Hz, 800 Hz-1600 Hz, 1.6 kHz-3.2 kHz, 3.2 kHz-6.4 kHz, and 6.4 kHz-12.8 kHz, respectively. To reduce the number of array elements, array elements with the same size and performance were used in the seven subarrays to ensure the phase and frequency consistency among the elements, thereby facilitating the formation of a nested array by the common parts of the array elements. The array elements were distributed in a way where there were small spacings between the middle elements (corresponding to the bands with higher frequencies) and gradually increasing spacings between the elements at both ends (corresponding to the bands with gradually decreasing frequencies), and there was a total of 31 elements in the seven subarrays. The cable was 80 m in total length (30.72 m of acoustics section and 50 m of extension cable), and covered seven subarrays, with the 1st subarray being a 5-element array and the other subarrays being 9-element arrays). The elements in the seven subarrays were spaced apart from each other by 60 mm, 120 mm, 240 mm, 480 mm, 960 mm, 1920 mm, and 3840 mm, respectively. As shown in Figure 1, the schematic diagram of the vertical linear array (semi-array) illustrates that the array is symmetrical concerning Element 0 and has the minimum element spacing d=60mm.

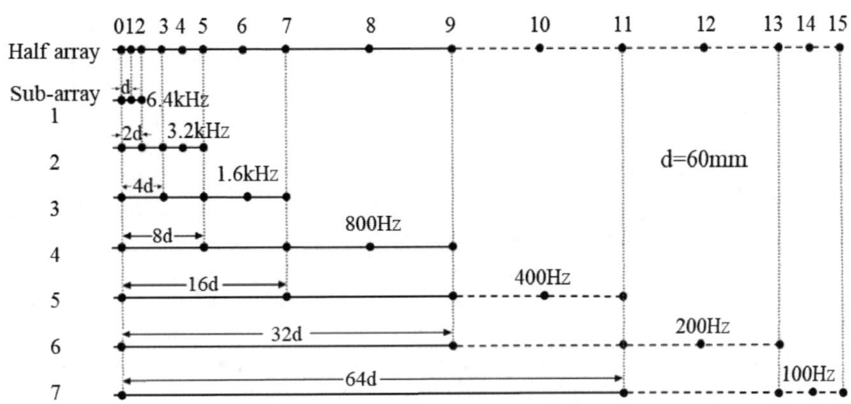

Figure 1. The diagram of a vertical nesting array

4. Analysis of the beamforming results in a waveguide environment

Sound propagation is more complex in a waveguide environment, so it is necessary to verify the effect of applying the near-field constant beamwidth weight vector calculated under the assumed plane wave to the waveguide environment. A freshwater lake was used as the test area. The sound velocity profile calculated after using a temperature-depth measuring instrument to acquire the temperature and depth data is shown in Figure 2. The water depth at the measured point was about 190 m, and there was an approximate iso-sound-velocity layer on the lake surface. The iso-sound-velocity layer was 20 m in depth, in which the sound velocity decreased rapidly with a temperature drop; increased slowly with an increase of depth when the temperature was constant or kept unchanged when the depth was greater than 116 m. The sampling measurement on the depth distribution of the test lake area by a depth measuring instrument showed uniform distribution of the longitudinal depth and an average water depth of about 190 m. In addition, sampling analysis on the lakebed sediment demonstrated the sediment was silt sediment, for which the density and the sound velocity were taken as 1.42 g/cm^3 and 1520 m/s, respectively. For the basement below the silt sediment, the density and sound velocity were taken as 2.0 g/cm^3 and 1800 m/s, respectively.

The real vertical nested array is shown in Figure 3. Since high error precision of the array elements was required for the mid-high frequency bands, a fixed bracket was specially designed and manufactured for the hydrophone elements at the small spacing between the middle elements in the nested array to effectively reduce the position error of the array elements.

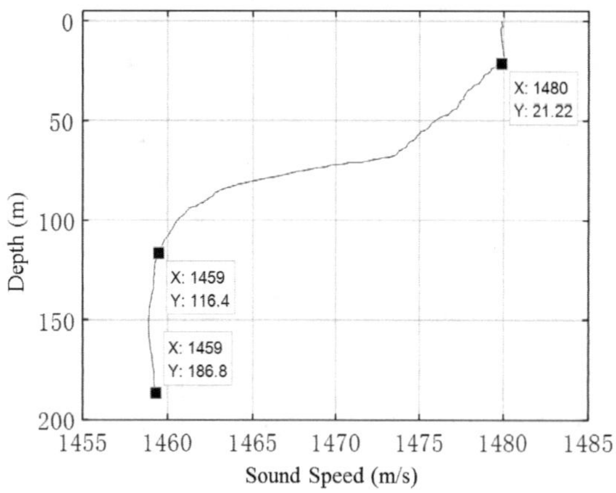

Figure 2. The sound speed profile in the water.

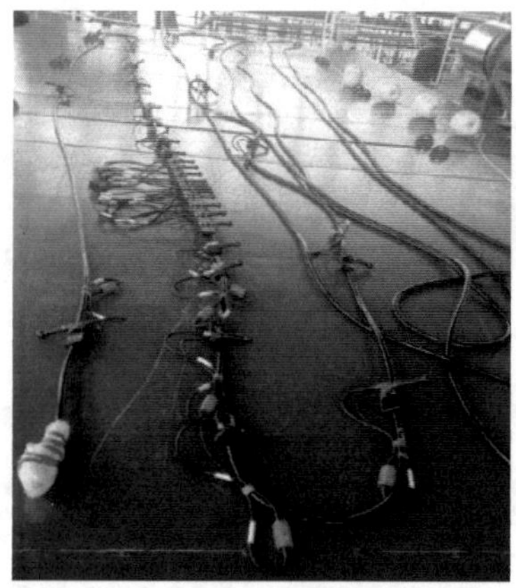

Figure 3. The vertical nesting array.

For the nested array used in the experiment, when the near-field constant beamwidth beamforming was weighted, the beam focusing distance was adjusted according to the actual working conditions, and a weighting coefficient of beamforming was solved by convex optimization, so that the beam patterns of different frequencies were identical, thereby achieving a constant beamwidth. The convex optimization was conducted by allowing the beam patterns in the main lobe region to approach the desired beam pattern and simultaneously constraining the beam patterns in the sidelobe region to be smaller than or equal to a fixed value, thereby meeting the requirement of iso-sidelobe. In a test with the condition including a sound source depth of about 24 m, a center depth of about 25m for the nested array, and a distance of about 33m between the sound source and the nested array, the desired beam was shown in Figure 4, and the calculated constant beamwidth beam pattern and the array gain were shown in Figure 5 and Figure 6, respectively. It was shown that the constant beamwidth weight vector was still valid in the measured waveguide environment, and the array gain was almost close to 7 dB at all frequencies.

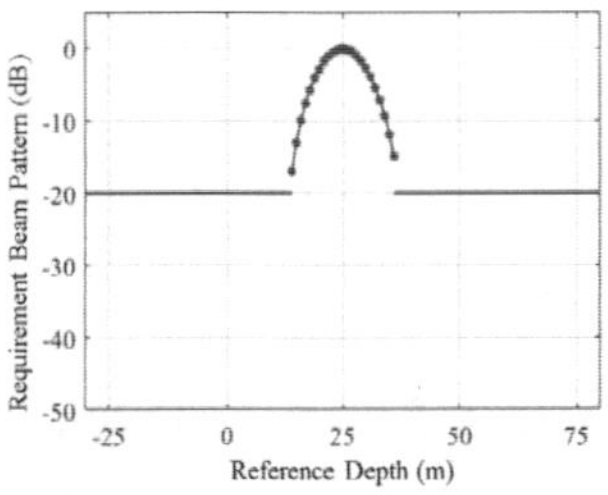

Figure 4. The desired beam patterns.

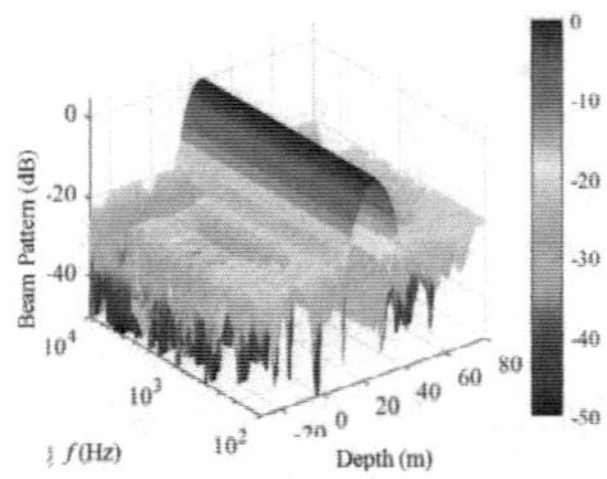

Figure 5. The constant width beam pattern

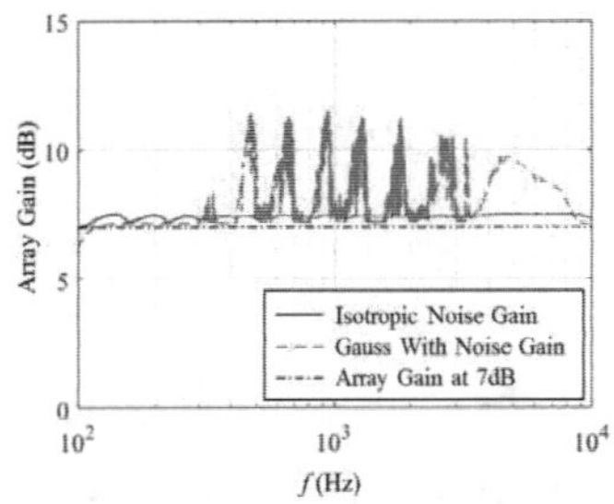

Figure 6. The array gain

5. Conclusions

This paper summarizes the design and realization of an ultra-wideband and high-robustness constant beamwidth beamformer based on a vertical nested array. Using nested array multiplexing, constant beamwidth beamforming at a 100-octave band (100 Hz-10 kHz) was realized by using fewer array elements. Based on the convex optimization method, the main lobe of each sub-band beam in the designed band was made to approach the desired constraint value, and then the beamforming robustness and array gain were improved through the constraint of the weight vector two-norm. In the paper, the influence on beamforming due to inconsistent array element amplitude-phases in the measured waveguide environment is analyzed. The proposed method was verified for correctness by computer simulation combined with the measured waveguide environmental parameters, which lays a foundation for the measurement of underwater target radiation sound field. The experiment designed to confirm the array manifold of the vertical nested array will be the future work.

References

[1] Yan S F, Ma Y L, and Hou C H 2007 *J. Acoust. Soc. Am.* **122** 2680
[2] Yan S F and Ma X C 2008 *Acta. Acustica.* **33** 316
[3] Gan T and Wang Y M 2012 *Acta. Acustica.* **37** 18
[4] Su Q, Wang G B and Zhu P F 2014 *Ship Science and Technology* **36** 1
[5] He L 2006 *Ship Science and Technology* **28** 9
[6] Jiang G Q, Sun C and Liu X H 2020 *Journal of Applied Acoustics* **39** 81
[7] Wu G Q, Wang M G, Chen S H and Ma L 2007 *Acta Acustica.* **32** 398
[8] Yan S F 2005 *Acta. Acustica.* **30** 309

Numerical Simulation Analysis of Flow Noise Characteristics of Underwater Vehicle

Kaijing Leng[1]*, Leilei Deng[1], Haozheng Zhang[1], Shijia Hao[1]

[1]Kunming Shipborne Equipment Research and Test Center, Kunming, Yunnan, China

*Corresponding author e-mail: 1123890565@qq.com

Abstract: Based on the standard large eddy simulation and k-epsilon model, the steady flow field and navigation flow noise of the underwater vehicle shell are simulated. In this paper, the underwater vehicle model is established by referring to classic SUBOFF model. The standard k-epsilon method is used to study the pressure distribution characteristic, velocity distribution characteristic, and velocity vector distribution characteristic on and around the shell of the underwater vehicle at different speeds. The results indicate that the pressure increases with the increase of the speed and in the bow of the shell is the largest. A large eddy simulation was used to analyze the noise characteristics of sailing flow at different speeds. The results show that the sound pressure level of navigation flow noise is higher in the low-frequency band, and will increase with the increase in speed.

1. Introduction

Flow noise is an important part of the self-noise of underwater vehicles, and one of the main factors which influence the detection range of sonar. Compared with the noises caused by mechanical vibration and propeller rotation, flow noise has the characteristics of low frequency and slow attenuation. Therefore, it is of great significance to go deeply into the flow noise characteristics of underwater vehicles to guide the placement of sonar arrays to reduce the impact of flow noise on detection.

In recent years, researchers have carried out lots of research on numerical simulation and experimental data analysis of the characteristics of convective noise. In the numerical study of flow noise, large eddy simulation (LES) is the most extensively used means[1,2]. The LES method can predict the unsteady flow. For large-scale separation areas, LES simulation is more accurate than RANS. Researchers have conducted numerous studies both on the cavity flow noise characteristics and the optimization of the underwater vehicle model reducing the navigation flow noise [3-4]. It is necessary to analyze the stream noise of the entire underwater vehicle to find the placement of the sonar array that is less affected by the flow noise. The first step in the analysis process is to use standard k-epsilon method to simulate the steady flow field around the simplified underwater vehicle model. Then, the flow characteristic noise of underwater vehicle is studied by using LES and Ffowcs Williams-Hawkins (FW-H) equations. Through these numerical simulation experiments, the distribution characteristic of flow noise during the navigation of the underwater vehicle are analyzed. The numerical results provide a reference for sonar array placement.

2. Mathematical model

2.1. Underwater vehicle model

In this paper, a simplified model of an underwater vehicle is established based on the classic SUB OFF model [5]. The simplified model of an underwater vehicle mainly includes three parts. The first part is the axisymmetric bare vehicle shell. The second part is four tail fins. The third part is the annular rib inside the shell. The shell of the simplified underwater vehicle model is an axisymmetric revolving body structure, with a total length of 2.422m and a maximum diameter of 0.324m. In addition, there are 5 ring ribs inside the shell. Figure 1 shows the underwater vehicle model which is simplified.

Figure 1. The simplified model of an underwater vehicle

2.2. Standard k-epsilon model

The steady flow field around the underwater vehicle model is simulated by the standard k-epsilon (k-ε) model. In this model, we need to solve dissipation rate ε equation and the turbulent kinetic energy k equation. The dissipation equation ε equation is an empirical equation obtained through physical reasoning and mathematical simulation. The turbulent kinetic energy k equation is an accurate equation obtained through accurate derivation.

The standard k-ε model is applicable to the case where the fluid is completely turbulent. Under the condition of complete turbulence, dissipation rate ε equations and the turbulent kinetic energy k equation are expressed as Equation (1) and Equation (2).

$$\frac{\partial}{\partial t}(\rho\varepsilon)+\frac{\partial}{\partial xi}(\rho\varepsilon u_i)=\frac{\partial}{\partial x_j}\left[\left(\mu+\frac{\mu_t}{\sigma_\varepsilon}\right)\frac{\partial\varepsilon}{\partial x_j}\right]+C_{1\varepsilon}\frac{\varepsilon}{k}(G_k+C_{2\varepsilon}G_b)-C_{3\varepsilon}\rho\frac{\varepsilon^2}{k}+S_\varepsilon \tag{1}$$

$$\frac{\partial}{\partial t}(\rho k)+\frac{\partial}{\partial t}(\rho k u_i)=\frac{\partial}{\partial x_j}\left[\left(\mu+\frac{\mu_t}{\sigma_k}\right)\frac{\partial k}{\partial x_j}\right]+G_k+G_b-\rho\varepsilon-Y_M+S_k \tag{2}$$

where, G_k expresses turbulence kinetic energy which is caused throungh laminar velocity gradient. Y_M is the wave caused by the diffusion of the transition in compressible turbulence. G_b expresses the turbulent kinetic energy which is caused througn buoyancy. C is a constant. S_k/S_ε are user-defined data. σ_ε and σ_k are turbulent Prandtl numbers of the ε equation and k equation. The calculation equation of the turbulent viscosity coefficient μ_t in the above equation which is expressed as:

$$\mu_t=\rho C_\mu\frac{k^2}{\varepsilon} \tag{3}$$

2.3. Large eddy simulation method

The first step of LES is to establish the filter function. The filter function divides the flow variable into two parts which are large-scale variables and small-scale variables. Then the unsteady Navier Stokes (N-S) equation is used for filtering to obtain the control equation of LES.

The filtering function and continuous function of LES are Equation (4) and Equation (5). The N-S equation is shown in Equation (6).

$$G(x,x)=\begin{cases}\frac{1}{V},x\in V\\ 0,x\notin V\end{cases} \tag{4}$$

$$\frac{\partial\rho}{\partial t}+\frac{\partial}{\partial xi}\left(\rho\bar{u}_i\right)=0 \tag{5}$$

$$\frac{\partial u_i}{\partial t}+\frac{\partial u_i u_j}{\partial x_j}=v\frac{\partial^2 u_i}{\partial x_j^2}-\frac{1}{\rho}\frac{\partial p}{\partial x_i} \tag{6}$$

where V is the control volume, p expresses pressure, $v=\mu/p$ expresses the viscosity coefficient of fluid motion, ρ expresses the undisturbed fluid density.

Therefore, τ_{ij} is defined as the stress tensor of the sub grid, the filtered N-S equation can be expressed as:

$$\frac{\partial}{\partial t}\left(\rho\bar{u}_i\right)+\frac{\partial}{\partial x_j}\left(\rho\overline{u_i u_j}\right)=\frac{\partial}{\partial x_j}\left(\mu-\frac{\sigma_{ij}}{\partial x_j}\right)-\frac{\partial\bar{p}}{\partial x_i}-\frac{\partial\tau_{ij}}{\partial x_j} \tag{7}$$

2.4. *Ffowcs Williams and Hawkings equations*

Ffowcs Williams-Hawkings (FW-H) equations can be used to calculate sound generation of moving solid walls. The FW-H [6] equations can be expressed as:

$$\frac{1}{c_0^2}\frac{\partial^2 p'}{\partial t^2}-\nabla^2 p'=\frac{\partial^2}{\partial x_i\partial x_j}T_{ij}H(f)-\frac{\partial}{\partial x_i}\left[p_{ij}n_j\delta(f)\right]+\frac{\partial}{\partial t}\left[\rho_0 u_n\delta(f)\right] \tag{8}$$

where $H(f)$ and $\delta(f)$ are the Hayside and Dirac function function respectively. $p'=p-p_0$, p' expresses far-field sound pressure. v_n represents the corresponding velocity component. f is a wall function. u_n expresses the velocity component which come from the x direction of the vertical wall. .

The sound source of Equation (5) consists of three parts. The first part of the sound source is caused by velocity fluctuations (four dipoles) in the flow field. The second part of the sound source is caused by the interaction between fluid and solid boundary (dipole source). The third part of the sound source is caused by the interaction between the object's movement and the flow field (monopole source). Because the FW-H equation contains different types of typical noise sources, it can solve most problems encountered in stream noise.

3. Simulation experiment

3.1. *Flow field characteristics*

We use standard k-ε model to simulate steady flow field of the underwater vehicle model through Fluent software. Figures 2 and 3 show the static pressure of the flow field on the surface of the underwater vehicle at 10m/s and 20m/s respectively.

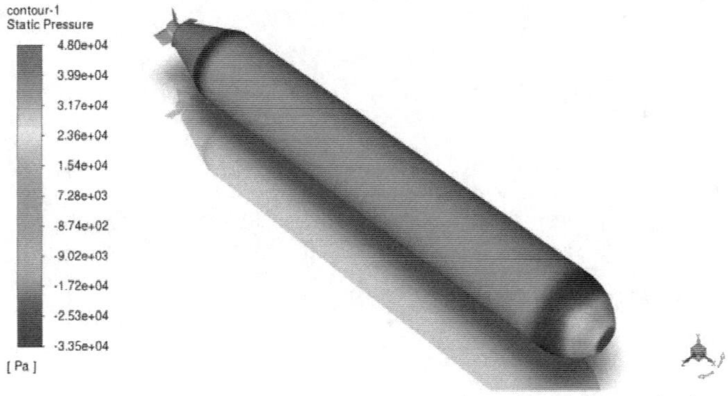

Figure 2. Cloud map of surface static pressure with an incoming velocity of 10m/s

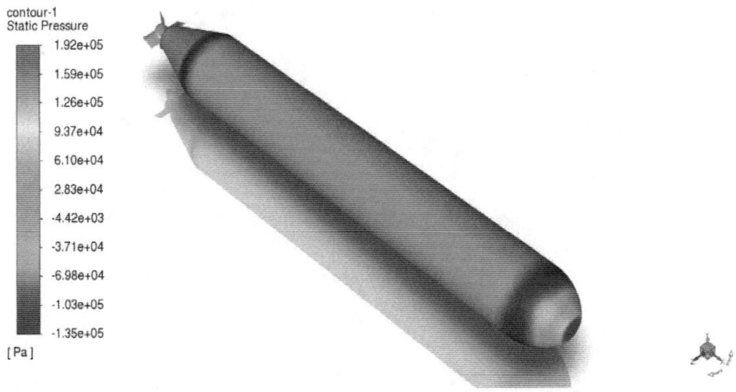

Figure 3. Cloud map of surface static pressure with an incoming velocity of 20m/s

Two conclusions can be drawn by analyzing the results. First, the static pressure to which the model has been subjected increases with the incoming flow velocity increases. Second, the maximum static pressure appears at the bow of the model. The simulation results indicate that the distribution characteristics of the surface pressure of the underwater vehicle are the same under different inflow velocities.

We choose the condition of 10m/s inflow velocity to further analyze the flow field details of underwater vehicle. Figure 4 expresses the pressure distribution cloud atlas on cross-section of flow field under this condition. Figure 5 and Figure 6 respectively show the velocity distribution cloud diagram and velocity vector diagram of flow field cross-section under these conditions.

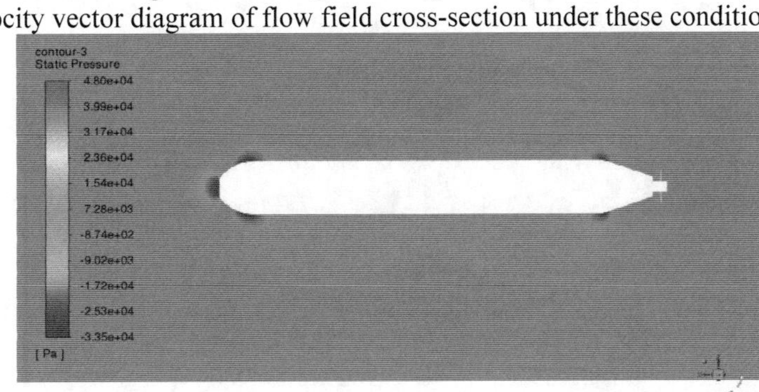

Figure 4. Cloud atlas of pressure distribution in intersecting surface of the flow field

Figure 5. Cloud atlas of velocity distribution in intersecting surface of the flow field

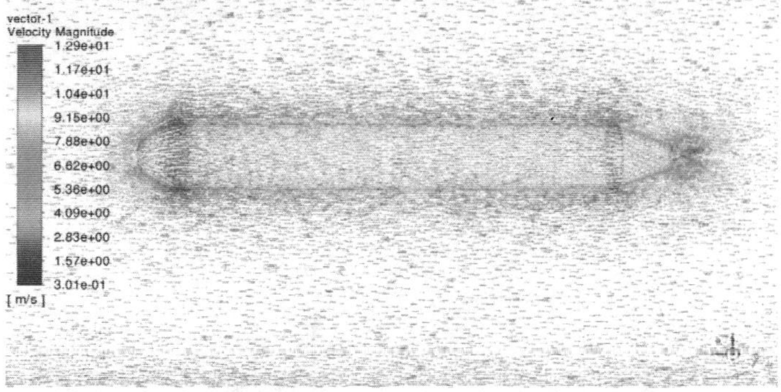

Figure 6. Flow field cross-section velocity vector diagram

The results in Figures 4 and 5 show that the flow field around the underwater vehicle has a pressure variation pattern opposite to the velocity variation pattern. This conforms to the law reflected in the Bernoulli equation. Figure 6 shows that the flow field at aft of an underwater vehicle will generate eddy currents.

3.2. Flow noise characteristics

The flow noise characteristics of this model are analyzed by using the FW-H equation with Fluent software. First, we use LES to calculate the flow field. Then, we install a virtual hydrophone (30cm below the middle of the model) in the flow field. Next, we calculate sound pressure frequency curves of characteristic points under 10m/s and 20m/s respectively. Figure 7 shows the calculation results.

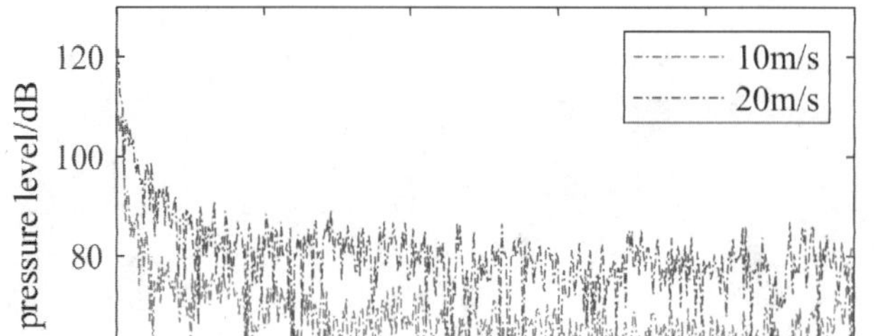

Figure 7. Sound pressure frequency response curve at characteristic points at speeds of 10m/s and 20m/s

Figure 7 shows that there is noise in the whole frequency band, and the noise sound pressure level within 1000Hz is higher. In addition, the amplitude of the sound pressure frequency response curve increases with the increase in sailing speed. The changing trend of the sound pressure frequency curve under different speeds is consistent.

4. Conclusion

In this paper, we carried out simulation experiments based on Fluent software to study the flow field on the surface of the underwater vehicle model and distribution characteristics of flow noise around this model. Through the simulation experiment, we found that the pressure brought by the fluid was mainly concentrated in the bow section of the model. The pressure increases as the speed increases. In addition, the flow noise caused by the model navigation is distributed throughout the analysis band and is higher in the lower frequency bands. The flow noise increases with increasing velocity.

Reference

[1] Benson R and Persson T 2004 Large eddy simulation of the viscous flow around submarine hulls (Canada: 25th Symposium on Naval Hydrodynamics).

[2] Fu H P and Miao G P 2009 Application of LES method to flow noise calculation. Journal of Shanghai Jiaotong University, China, 43.

[3] Zhang B, Shang D J, Liu J and Song Hao 2016 Numerical simulation of underwater flow noise from the open cavity under different attack angles *2016 OES China Ocean Acoustics (COA).*

[4] Li Y L, Ye J X, Wu J Y, Zhang Tao, Zhang Z H and Liu Y W 2021 The hydrodynamic noise suppression of a scaled submarine model by trailing-edge serrations *2021 OES China Ocean Acoustics (COA).*

[5] Sarakinos S S, Lygidakis G N and Nikolos I K 2015 Assessment of the academic CFD code galatea-I with the DARPA SUB OFF test case *ASME 2015 International Mechanical Engineering Congress and Exposition.*

[6] Ffowcs Williams J E and Hawkings D L. 1969 Sound generation by turbulence and surface in arbitrary motion. Phil Trans Roy Soc 264: pp 321-342.

A Low-Frequency Unidirectional Underwater Acoustic Transducer

Qiumu Wang[1], Yu Lan[2,3,4]

[1]State Key Laboratory of Ocean Engineering, School of Naval Architecture, Ocean & Civil Engineering, Shanghai Jiao Tong University, Shanghai 200240, China

[2]Acoustic Science and Technology Laboratory, Harbin Engineering University, Harbin 150001, China

[3]Key Laboratory of Marine Information Acquisition and Security (Harbin Engineering University), Ministry of Industry and Information Technology; Harbin 150001, China

[4]College of Underwater Acoustic Engineering, Harbin Engineering University, Harbin 150001, China

Electronic mail: lavendermu@sjtu.edu.cn

Abstract: A unidirectional underwater acoustic transducer has the advantage of narrower beam width. In the low-frequency field, a single transducer is usually omnidirectional because its dimension is much smaller than the wavelength of the sound waves. Here, a low-frequency unidirectional underwater acoustic transducer, the Single-Sided Radiation Flextensional Transducer (SSR FT), which is driven by a stack with the longitudinal vibration, was designed by the finite-element method. The vibration mode of the transducer was first studied and analyzed. Then, an underwater harmonic response analysis of the transducer was conducted. We found that the second resonance, corresponding to the fourth vibration mode, could be unidirectional based on a specific S_{br}, which is the ratio of r_{b} (the radius of the tail mass) to l_{maj} (the inner length of the semi-major axis of the radiation surface). Finally, the effect of S_{br} on the vibration velocity and directivity of the SSR FT was investigated. We established that the vibration velocity amplitude and phase of the shell were different at the same vibration mode but at different S_{br}, which resulted in different directivity.

1. Introduction

Unidirectional underwater acoustic transducers possess the typical advantages of narrower beam width and a higher gain. The use of a reflector is one of the approaches to achieving unidirectionality.[1-2] The other common method is to use the multimode vibration excitation of the transducer[3-5]. Unidirectional flextensional transducers were previously designed to synthesize several modes based on the Class IV flextensional transducer [3-4]. Moreover, the Class V and Class VII flextensional transducers were also driven by the stack with the flexure mode to synthesize the vibration mode[5-6]. A model projector was also designed by synthesis of modes[7-10].

In this paper, a low-frequency unidirectional transducer was designed based on the special geometric structure and a single vibration mode of the shell of the Single-Sided Radiation (SSR) Flextensional Transducer (FT)[11]. The first-order vibration mode, which is omnidirectional, was studied and analyzed before. [11]Here, the research focus of our study was placed on the effect of a new structural parameter, which is the key for the unidirectionality of the transducer in the higher-order resonance (mode).

2. The finite element analysis of the transducer

As can be seen in Figure 1, a geometric parameter s_{br} (the subscript "b" represents "bottom mass," and the subscript "r" represents "radiation surface"), which is the ratio of r_b to l_{maj}, was introduced for the analysis of the second resonance of the SSR FT. Here, r_b is the inner radius of the tail mass (bottom mass), and l_{maj} is the inner length of the semi-major axis of the radiation surface.

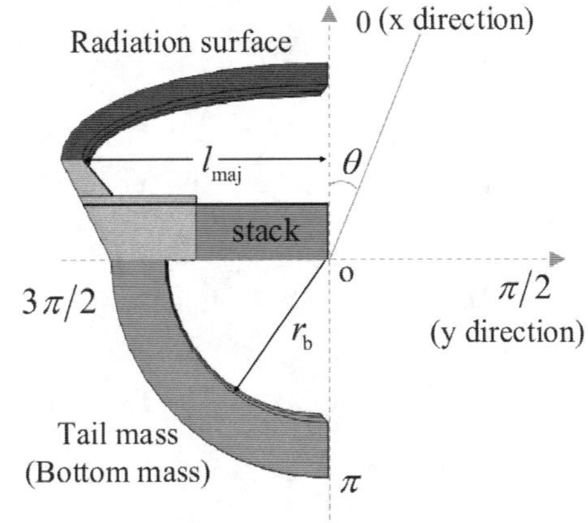

Figure 1 The model of the SSR FT[11]

2.1 Vibration mode analysis

This study's first and fourth modes correspond to the first two vibration modes of the shell, as illustrated in Figure 2(a) and (d). A stack can drive them with longitudinal vibration. The second mode, depicted in Figure 2(b), requires a driven stack with a flexure vibration. Additionally, the third mode, illustrated in Figure 2(c), is the vibration mode referring to the direction of the height, which requires a more complicated driven stack.

(a) (b) (c) (d)
Figure 2 The four vibration modes

2.2 Underwater harmonic response analysis

A finite element model of the SSR FT was designed to perform underwater harmonic analysis, which is presented in Figure 3(a). The model possessed 10,944 elements and 3649 nodes. The thickness of the Perfect Matched Layer (PML) was 0.07 m, and the radius of the water domain was 0.28 m.

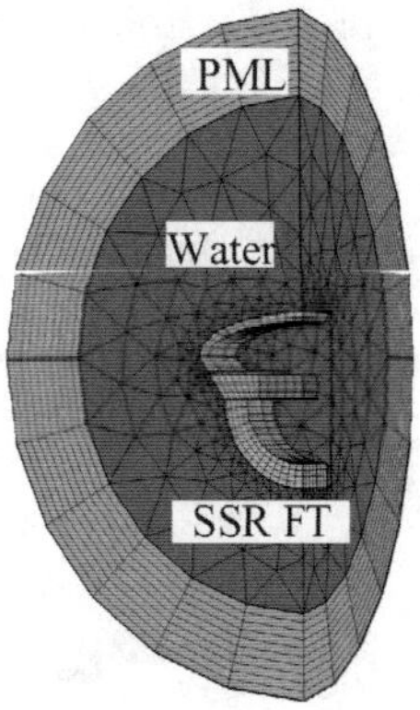

(a)

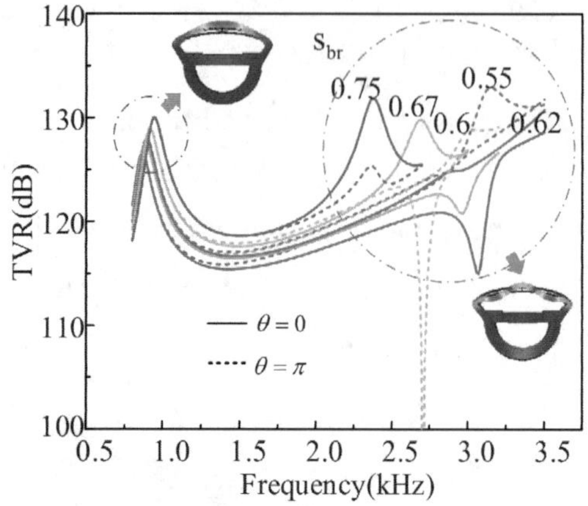

(b)

Figure 3 Underwater harmonic response analysis of SSR FT

Figure 3(b) displays the Transmitting Voltage Response (TVR) variation of the SSR FT, where two resonances are generally observed in the frequency range of 0.5–3.5 kHz. The first resonance, which corresponded to the first mode, was observed within the frequency range 0.8–1 kHz, corresponding to the s_{br} range 0.55–0.75. As can be observed in Figure 4, the directional beam pattern of the first resonance was omnidirectional. The results of the study of the first resonance are not presented here.

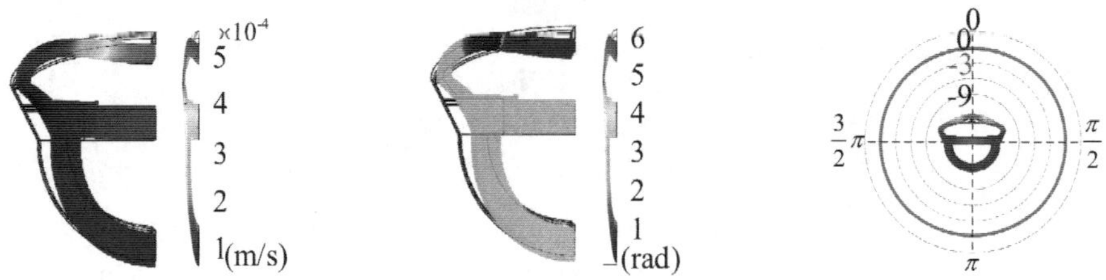

(a) (b) (c)

Figure 4 Vibration and radiation performance correspond to the first resonance

As visible in Figure 3(b), the second resonance, which corresponded to the fourth vibration mode, was observed within the frequency range of 2.0–3.5 kHz. The dotted lines represent the TVR variations at the direction of $\theta=\pi$, while the solid lines represent the TVR variations at the direction of $\theta=0$. Hence, the main lobe occurred at the direction of $\theta=\pi$ in case of $0.55 < s_{br} < 0.62$, whereas it occurred at the direction of $\theta=0$ when $0.62 < s_{br} < 0.75$ for the second resonance. The second resonant peak approximately vanishes under the case of $s_{br} = 0.62$. As illustrated in Figures 5(a)–(c), although the transducers operated in the same radiation mode, the amplitude and phase differed. Thus, the second resonance of the SSR FTs with different s_{br} radiated different beam patterns.

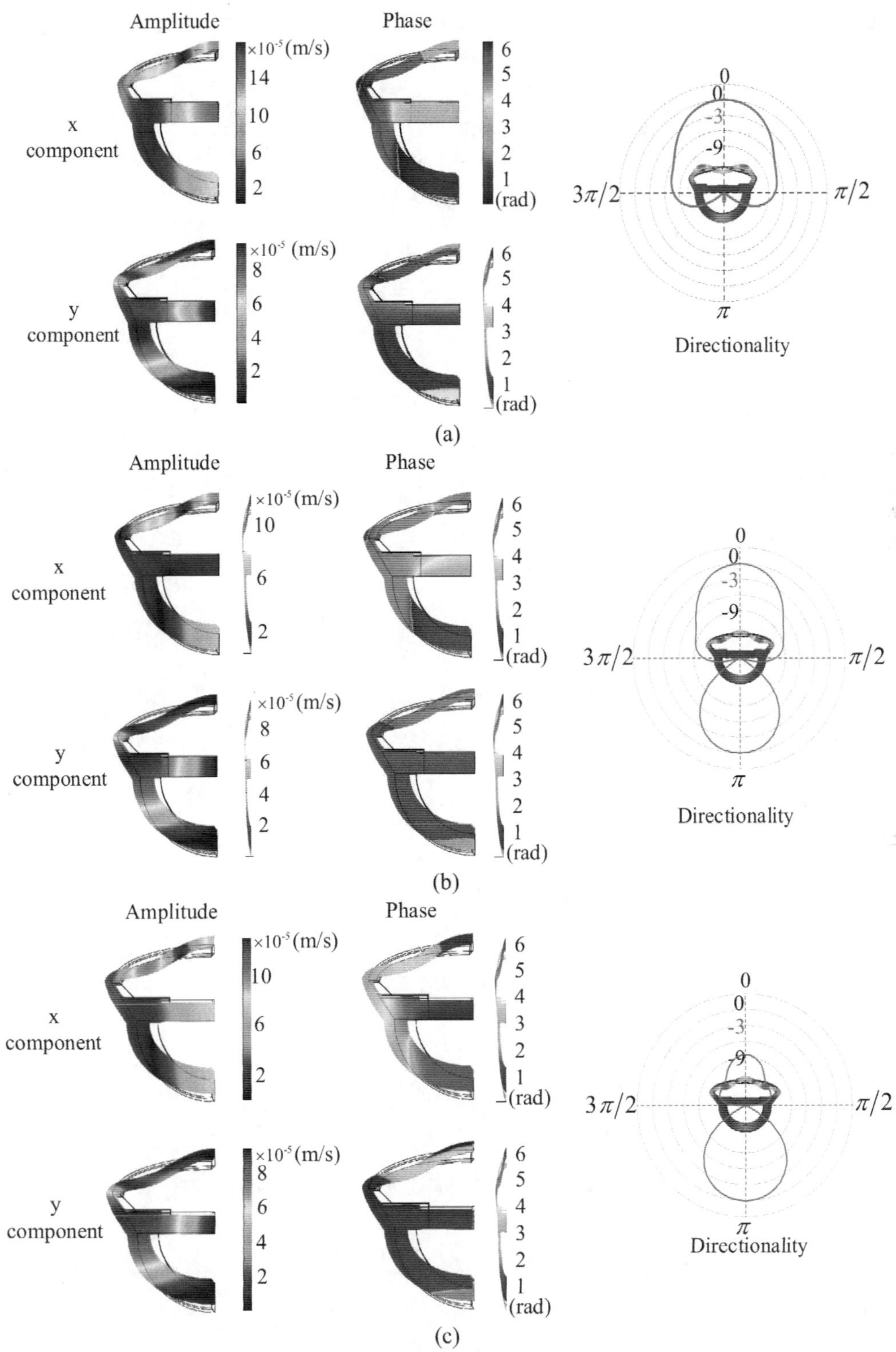

Figure 5 Vibration and radiation performance correspond to the second resonance

As visible in Figure 6, the SSR FT was approximately equivalent to the model composed of point sources. The amplitude and phase of point sources corresponding to different s_{br} are presented in Table 1. As depicted by Eq. 1–Eq. 7, the beam pattern of the transducer was approximately achieved. The beam patterns of the equivalence model were approximate with the beam patterns of SSR FT when they were the same s_{br} (Figure 7). The analysis results of the approximate modal confirmed that the beam pattern of the transducer was related not only to the vibration mode but also to the amplitude and the phase of the vibration velocity of its shell.

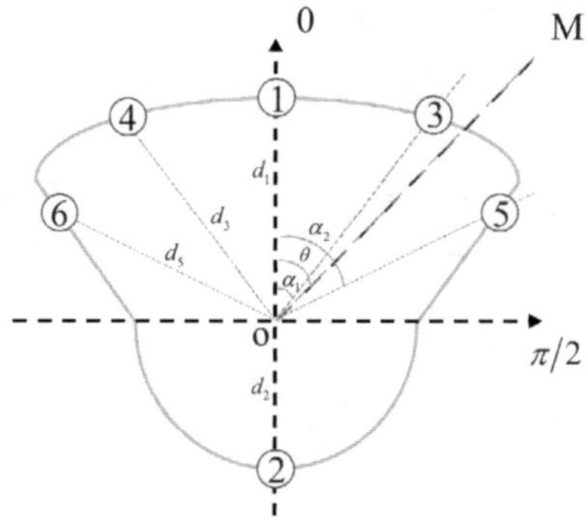

Figure 6 Schematic diagram of an approximate model of the SSR FT (at the second resonance)

Table 1 The amplitude and phase of point sources correspond to different s_{br}

No.	$s_{br}=0.66$		$s_{br}=0.62$		$s_{br}=0.6$	
	a_n	φ_n (rad)	a_n	φ_n (rad)	a_n	φ_n (rad)
1	13.55	2.45	9.64	1.77	8.77	0.95
2	10.71	3.83	6.80	4.49	7.18	5.02
3(4)	11.87	5.65	8.51	4.80	10.21	3.88
5(6)	9.45	1.12	9.92	1.52	9.22	1.37

$$p_1(\theta)=p_0\left(a_1\exp\left(-i\varphi_1-ikd_1\cos(\theta)\right)\right) \tag{1}$$

$$p_2(\theta)=p_0\left(a_2\exp\left(-i\varphi_2+ikd_2\cos(\theta)\right)\right) \tag{2}$$

$$p_3(\theta)=p_0\left(a_3\exp\left(-i\varphi_3-ikd_3\cos(\theta-\alpha_1)\right)\right) \tag{3}$$

$$p_4(\theta)=p_0\left(a_3\exp\left(-i\varphi_3+ikd_3\cos(\theta+\alpha_1)\right)\right) \tag{4}$$

$$p_5(\theta)=p_0\left(a_5\exp\left(-i\varphi_5-ikd_5\cos(\theta-\alpha_2)\right)\right) \tag{5}$$

$$p_6(\theta)=p_0\left(a_5\exp\left(-i\varphi_5+ikd_5\cos(\theta+\alpha_2)\right)\right) \tag{6}$$

$$p_M(\theta)=\sum_{n=1}^{6}p_n \tag{7}$$

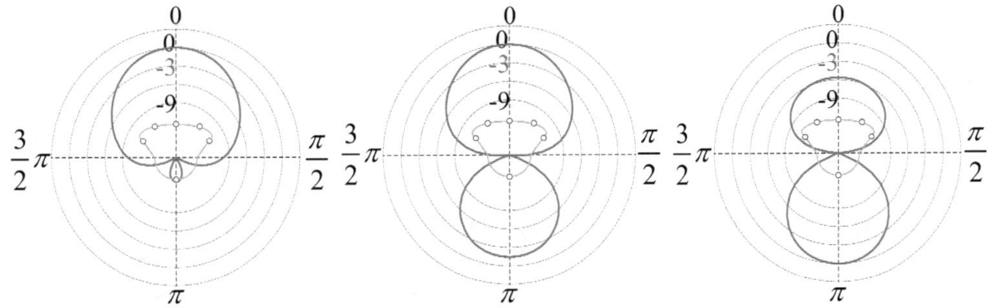

(a) (b) (c)

Figure 7 The beam pattern of the approximate equivalent model (at the second resonance)

3. Conclusion

The SSR FT with the fourth vibration mode can be unidirectional at $0.55 < s_{\mathrm{br}} < 0.62$ and $0.62 < s_{\mathrm{br}} < 0.75$. In the case of $s_{\mathrm{br}} = 0.62$, the beam pattern represents the “8”. This result indicates that although the transducer operates with the same vibration mode, the beam pattern is related to the structure of the transducer shell.

Acknowledgments

Financial support for this work was provided by the National Nature Science Foundation of China (Grant No. 10804023). COMSOL Multiphysics provided the finite element method for this work.

References

[1] K. P. B. Moosad, G. Chandrashekar, M.J. Joseph and R. John. (2011) “Class IV flextensional transducer with a reflector,” Appl. Acoust. 72:127–131.

[2] D. A. Brown, B. Aronov. (2004) “Baffled ring directional transducers and arrays,” US 6,768,702 B2

[3] S. C. Butler, J. L. Butler and A. L. Butler. (1997) “A low-frequency directional flextensional transducer and line array,” J. Acoust. Soc. Am. 102 (1): 308-314.

[4] S. C. Butler, A. L. Butler and J. L. Butler. (1992) “Directional flextensional transducer,” J. Acoust. Soc. Am. 92 (5): 2977–2979.

[5] S. C. Butler. (2010) “A directional dogbone flextensional transducer,” Proceedings of Meetings on Acoustics, Vol. 11: 1-23.

[6] J. Zhang, R. E. Newnham. (2003) “Class V flextensional transducer with directional beam patterns,” US 6,614,143 B2

[7] J. L. Butler, A. L. Butler, and S. C. Butler. (2011) “The modal projector,” J. Acoust. Soc. Am. 129 (4): 1881–1889.

[8] J. L. Butler, A. L. Butler, and J. A. Rice. (2004) “A tri-modal directional transducer,” J. Acoust. Soc. Am. 115 (2): 658-665.

[9] J. L. Butler, C. H. Sherman, *Transducers and arrays for underwater sound* (ASA Press, New York, 2006), p. 217–218

[10] A. L. Butler, J. L. Butler. (2002) “A synthesized tri-modal directional transducer,” J. Acoust. Soc. Am. 111: 2444.

[11] Q. Wang, J. Fan, Y. Lan. (2022) “Transducer design for low-frequency circular close-packed array and its mutual radiation analysis,” J. Acoust. Soc. Am. 151: 2223-2237

Experimental Study on the Airborne Noise of Rain

Zekun Li and Hengqi Liu*

Shenzhen College of International Education

* henry.liu@scie.com.cn

Abstract: Rainfall noise is investigated experimentally in this paper. The modern high-speed camera and microphone are employed to study both drop dynamics and sound signals. Sound signals produced by single-drop impacts on deep pools, thin liquid films, steel plates, and sandy surfaces were analyzed through a combination of videos and audio. Classical Fourier transforms are performed for converting the original signal to a power spectrum. Results show that in the deep-water cases, the noise is proved to be the entrapment of a tiny breathing bubble that sheds from the dimple. The dominant frequencies of droplets impacting water film, rigid walls, and sandy surfaces are much compared with deep pool cases.

1. Introduction

Raining is one of the most common natural phenomena that can be observed in daily life, and it always produces distinctive sounds. The formation of rainfall noise has been studied for a long time. People have demonstrated that rainfall noise is highly relevant to drop size and impact velocity [1] since the droplet will create underwater bubbles in certain conditions [2] and these oscillation bubbles are the primary source of rainfall noise [4]. This conclusion is widely supported by evidence from numerical simulation [6], hydrophone measurement [4], and analytical models [2].

Although extensive studies have been performed to record the sound of a droplet impacting water, the audio signals and images are not strictly synchronized. Besides, the sounds produced by drops' impact on the rigid or sandy surface are rarely mentioned in previous literature. In this article, acoustical data would be collected through professional microphones and hydrophones with a high sample rate and the droplet impact process would be recorded through a high-speed camera. The influence of the substrate on the sound formation and its acoustic features will be carefully investigated.

2. Experimental set-up

In this experiment, a water droplet falls into a water tank (15cm*15cm*15cm) from a nozzle with an inner diameter of 0.5 mm. The release height H, measured from the nozzle to the water surface, was varied in the range H = 12~20 cm, giving impact velocities in the range of U = 1.5~2.0 m/s. The impact of a droplet on water or other substrate is captured by using a high-speed camera with a resolution of 640 × 640 pixels at a framerate of 1500 per second. The camera synchronizes with a Zoom H1N microphone, with a 96 kHz sample rate. Compulsorily, a LED light source is required to provide enough lighting so the image could be limpidly displayed. A photo of our experimental arrangements is shown in Figure 1.

Journal of Physics: Conference Series **2458** (2023) 012014 doi:10.1088/1742-6596/2458/1/012014

3. Results and Discussion

3.1 Sound induced by droplet impact on deep water pool

Figure 2 presents a series of instantaneous images which correspond to the time stamp. All images are synchronized with the microphone to ensure homogenous information was given at every specific frame. Firstly, the waterdrop collides with the water surface thus creating an air cavity. A thin layer of air does not have enough time to escape the range of the falling drop. This could create a high-speed oscillation but its' frequency might be inaudible [1].

Due to the surface tension, the water body is pulled upwards whilst a dipole is created through the superposition of capillary waves from the initial unstable structure presented in Figure 2 (d). The dipole contracts at halfway, and a bubble will be formed as the dimple finally encloses, as shown in Figure 2 (h). Since the crater is open to the air, we can safely consider that the inner crater pressure is equal to the atmospheric pressure [3, 5, 7].

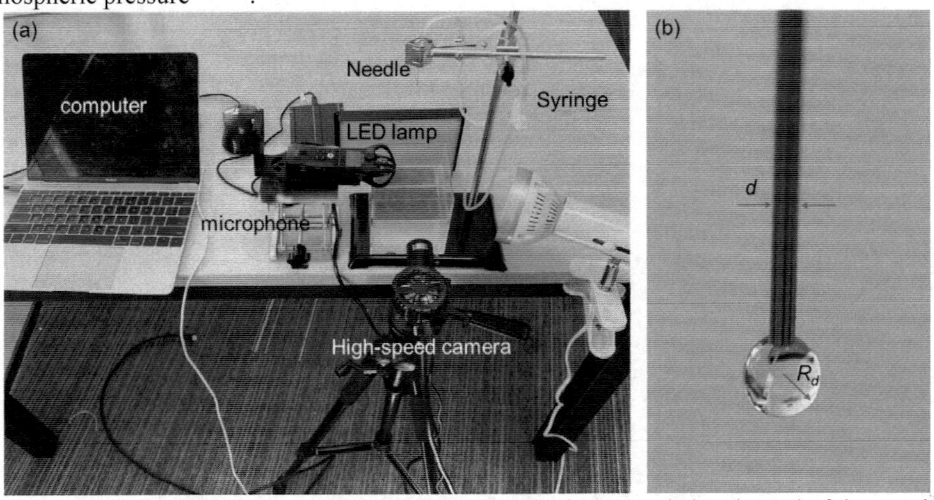

Figure. 1 (a) Photo of the experimental set-up. (b) A droplet is suspended at the end of the nozzle with a diameter of Rd=1.5 mm.

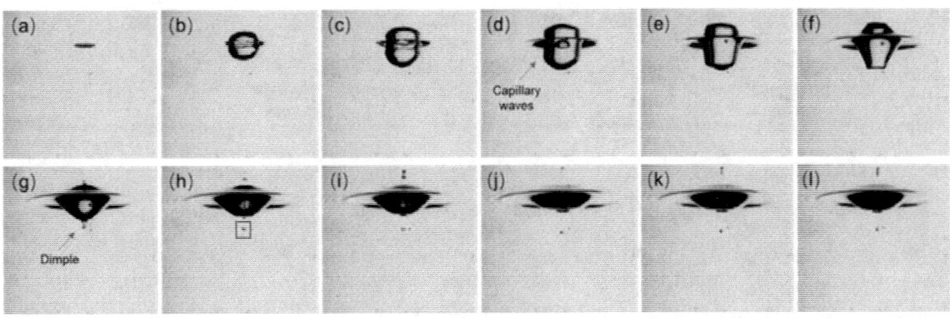

Figure 2. Photos of water drop impact into a deep pool. The time interval between frames is 3μs in (a)~(f) and 0.66 μs in (g)~(i). The red rectangular in (g) indicates the appearance of the underwater bubble.

Figure 3 shows the airborne audio signals recorded by the microphone when a 1.5 mm diameter droplet impacts onto the water at 2m/s. The blue lines indicate the locations of video frames shown in Figure 2. It's not difficult to find two wave packets in Figure 3. The first one locates between (a) and (b) is related to the initial droplet impact. Then, the audio signal reaches its maximum value at Frame (g), where a bubble is separating from the dimple at the bottom of the crater. Thus, it demonstrates that bubble oscillation is the main source of rainfall noise.

3.2 Sound induced by droplet impact on various surfaces

As shown in Figure 4 (a), the crown splash and the crater are observed when dropping impact onto the surface of the liquid layer. The presence of a rigid boundary indeed deters the crater from expanding. Thus, air entrapment cannot be observed in shallow water cases. For the liquid layer, unlike the pool, the emitted sound has a lower characteristic frequency since the existence of a rigid wall constrains the bubble's formation, leading to a featured frequency of 2 kHz, as shown in Figure 5.

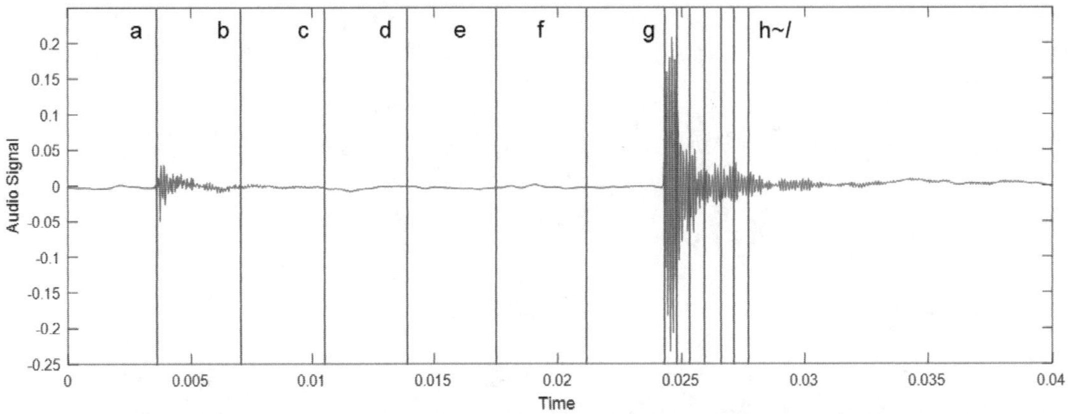

Figure 3. Airborne sound excited by the water entry of a 1.5 mm diameter spherical droplet (red line). Each event takes place at Frames (a)~(l) in Figure 2 (blue line). The acoustic signal was obtained from the microphone with a sample rate of 96000Hz. The x-axis indicates time (s) while the y-axis represents the relative magnitude of droplet sound.

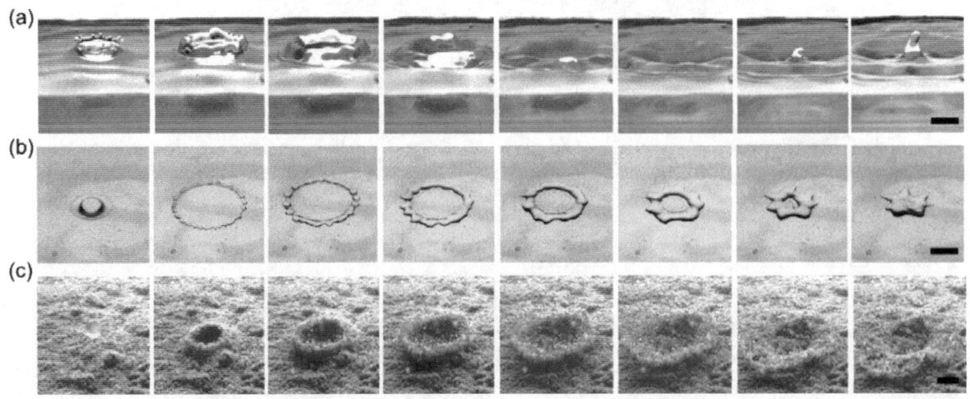

Figure 4. Images capture the impact of a water drop on water film (a), dry smooth steel plate (b), and sandy surface (c). The time intervals between frames are 0.005s, 0.0033s, and 0.005s in (a), (b) and (c). The black scale bar indicates 1mm.

The impacts of a water drop onto the dry steel surface, cause a broadening of the drops into a thin wavy film with rims around, as shown in Figure 4 (b). The lamella with a limited radius suffers a recession due to the effect of surface tension and the surface energy of the steel plate. It's worth mentioning that the bounce of the impacted droplet is not found in this experiment, since the substrate is hydrophilic. The impact of a single droplet impact on to solid and smooth plate leads to the formation of sound with a relatively lower frequency, compared with those cases of water impacts (Figure 6). Previous studies have demonstrated that pressure waves will be excited after the droplet contact with the solid ground. Thus, the droplet begins to oscillate due to the propagation of pressure waves inside the droplet which presumably causes the low-frequency sound heard here for the rigid steel surface drop impact case. Besides, the total impacting period is much longer than that of the impacts of droplets onto a liquid layer. Drops falling on fixed surfaces produce sounds with fixed resonant frequencies. The resulting harmonic impact sounds can be modeled by an impact atom [8] with amplitude a and damping rate α.

$$x(t)=\sum_{m=1}^{M} a_m \sin\left(2\pi f_m t\right) e^{-\alpha t} \tag{1}$$

where FM represents the modal frequencies of impact bodies.

For the sandy surface, the tiny sands eject outward and a crater is formed after droplet impact. Compared with rigid surfaces, the flow pattern drastically shifts toward the chaotic collapse of the drop with a splashing of sand. This phenomenon leads to a relatively wide, but intense, acoustic response with pockets of energy centered around 0.5 and 0.8 kHz. In the case of the sandy surface, the radiated noise is clustered within a millisecond of drop impact because of the instant fluid dynamics of drop breakup and sand ejection. s. Such sounds are efficiently reproduced by a noisy impact atom, which is a sum of 8 contiguous subbands of noise $S(t)$.

$$x_1(t)=\sum_{i=1}^{8} a_i s\left(t\right) e^{-\alpha t} \tag{2}$$

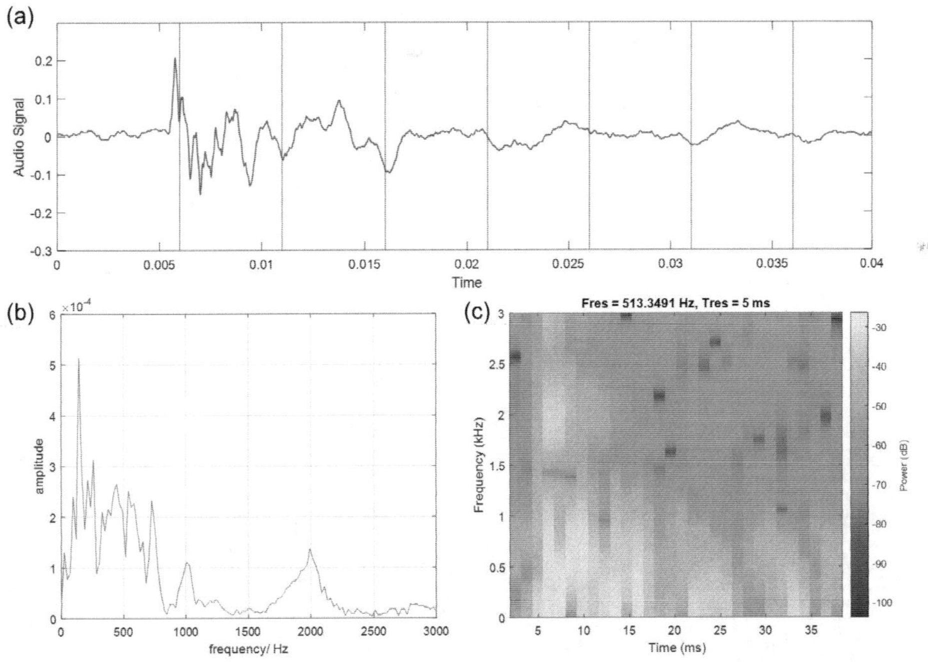

Figure 5. Audio signals (a), frequency spectra (b), and power spectrum (c) for droplet impact on the water layer. The red lines in (a) match the images in Figure 5 (a).

4. Conclusions

The sound generated by the impact of liquid water drops on four surfaces including a deep pool, water layer, rigid plate, and sand has been investigated. We both use high-speed cameras to capture fluid motions and handy microphones to record the impacting acoustic signals.

The audio recordings reveal that time-dependent sounds are related to events, like initial contact and air dimple collapse observed on synchronized high-speed images. This investigation demonstrates that the oscillation of an entrapped air bubble is the key driver for the characteristic airborne 'plink' sound. The damping rate and frequency of the impacting sound can be explained by an analytical model that simplified the breathing bubble to bob on the spring. Besides, the sound signals for droplets' impact on a thin liquid layer are obtained. The sound energy concenters around 2 kHz, much lower than the deep-water cases because the presence of a rigid wall will prevent the air cavity from expanding and deterring the formation of sound bubbles. For the smooth and sandy surfaces, the featured frequency ranges from 50 to 800 Hz.

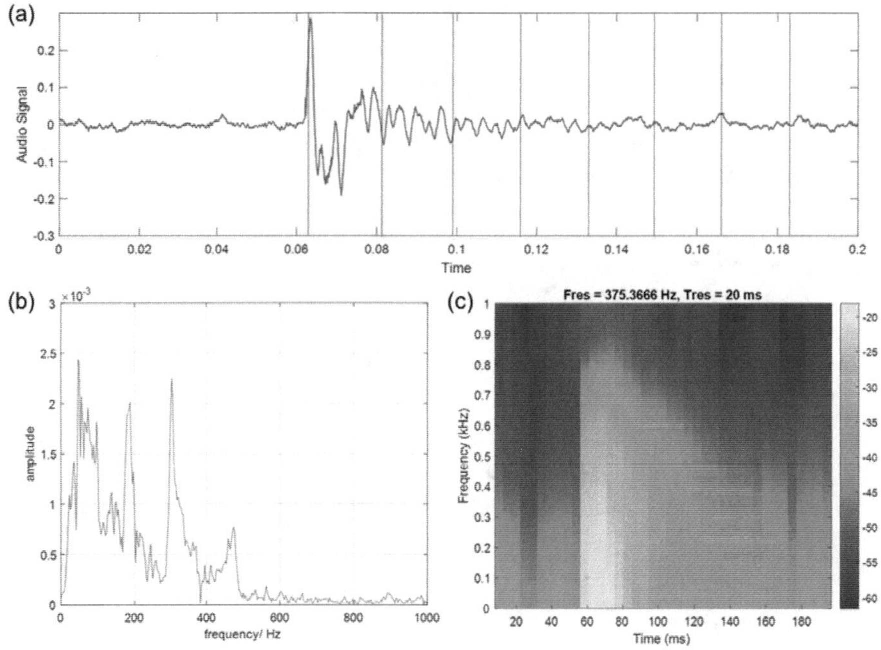

Figure 6. Audio signals (a), frequency spectra (b), and power spectrum (c) for droplet impact on the rigid metal ground. The red lines in (a) match the images in Figure 6 (b).

Reference

[1] Pumphrey, H.C. and Elmore, P.A., 1990. The entrainment of bubbles by drops impacts. *Journal of Fluid Mechanics*, *220*, pp.539-567.

[2] Prosperetti, A. and Oguz, H.N., 1993. The impact of drops on liquid surfaces and the underwater noise of rain. *Annual Review of Fluid Mechanics*, *25*(1), pp.577-602.

[3] Wang, Y., Ye, B., Wang, Z., Huang, J., Wang, Y. and Huang, C., 2020. Vertical water entry of projectiles with surface seal. *Ocean Engineering*, 216, p.107606.

[4] Pumphrey, H.C., Crum, L.A., and BjoRno, L., 1989. The underwater sound is produced by individual drop impacts and rainfall. *The Journal of the Acoustical Society of America*, *85*(4), pp.1518-1526.

[5] Wang, Y., Wang, Z., Du, Y., Wang, J., Wang, Y. and Huang, C., 2022. The mechanism of a surface-seal splash during water entry. *Physics of Fluids*, *34*(4), p.042110.

[6] Oğuz, H.N. and Prosperetti, A., 1991. Numerical calculation of the underwater noise of rain. *Journal of fluid mechanics*, 228, pp.417-442.

[7] Wang, Y., Wang, Z., Du, Y., Wang, J., Wang, Y. and Huang, C., 2022. On the airflow in a cavity during water entry. *International Journal of Multiphase Flow*, *151*, p.104073.

[8] Verron, C. and Drettakis, G., 2012, October. Procedural audio modeling for particle-based environmental effects. In *Audio Engineering Society Convention 133*. Audio Engineering Society.

Research on Robust CS-DOA Estimation Method Based on Signal Sparsity Reconstruction

Peilin Su [1*], Di Huang [1], Jieqi Zhu[1]

Hangzhou Applied Acoustics Research Institute

798970918@qq.com

Abstract. The application of compressed sensing theory in DOA estimation of underwater signals has always been difficult. Aiming at the slow operation speed of the current L1-SVD algorithm and the poor robustness under various conditions such as multi-array elements, this paper proposes an improved CS-DOA estimation algorithm combined with the greedy algorithm. The algorithm sparsely reconstructs the received signal itself, which effectively improves the strict sparsity of the processed signal. The feasibility of the new algorithm is proved through theoretical analysis and simulation experiments. Compared with the L1-SVD algorithm, the measurement performance under different conditions is significantly improved. It has development prospects in academic research and engineering applications.

1. Introduction

Since the theory of compressed sensing was proposed by D. Donoho et al.[1], it has been applied in the fields of medical imaging, radar detection, and wireless communication[2][3][4][5][6]. Up to now, the research of this theory has mainly focused on three aspects: sparse transformation, observation sampling, and reconstruction algorithm. The reconstruction algorithm can be extended into a convex optimization algorithm, greedy algorithm, statistical optimization algorithm, and combination algorithm. With the development of the theory and the improvement of the algorithm, the idea of compressive sensing has also begun to be used in the incoming wave signal estimation algorithm[7]. D. Malioutov and M. Cetin[8] proposed a sparse representation of the signal under the multi-measurement vector observation model. On this basis, various CS-DOA estimation algorithms were developed. Compared with the traditional beamforming algorithm, the CS-DOA algorithm generally has an extremely high resolution but lacks sufficient robustness and operation speed. One of the typical representatives is the L1-SVD algorithm.

One of the difficulties of underwater array signal processing lies in the variable conditions brought by the complex environment, which makes the received signal to be processed differ greatly from the ideal signal in the simulation experiment. The array element fails to transmit signals, and the signal-to-noise ratio between different array elements is inconsistent[9]. This makes the previous CS-DOA estimation methods unable to achieve the performance of high-resolution direction finding.

In this paper, a new CS-DOA estimation algorithm is proposed based on the L1-SVD algorithm. The improved greedy algorithm is used to reconstruct the received signal, which significantly improves the robustness and operation speed of the algorithm. The theoretical analysis and simulation experiments verify the performance of the algorithm. Feasibility, high-resolution direction finding effect can be achieved under various conditions such as the increase of array elements, the decrease of the signal-to-noise ratio of the source, the increase of the difference of the signal-to-noise ratio of different array elements, and the increase of the number of faulty array elements. On the premise of the advantages of the traditional CS-DOA estimation method, the shortcomings have been fully improved, which has the significance of continuous research.

2. Convex optimization class CS-DOA estimation method L1-SVD

CS-DOA estimates the multi-measurement vector model as follows.

$$\mathbf{X} = \mathbf{A}(\bar{\boldsymbol{\theta}})\bar{\mathbf{S}} + \mathbf{E} \tag{1}$$

Where, $\mathbf{X}$ is the received signal data matrix, $\mathbf{E}$ is the noise data matrix captured, $\mathbf{A}(\bar{\boldsymbol{\theta}})$ is the array manifold matrix corresponding to the grid point vector $\bar{\boldsymbol{\theta}}$, and the corresponding sparse coefficient matrix is $\bar{\mathbf{S}}$. The CS-DOA estimation problem under multi-shot conditions can be transformed into a sparse reconstruction problem. The position of the non-zero elements in the sparsity signal is confirmed, and the corresponding angle information is the direction of the incoming wave.

The convex optimization algorithm adopts the idea of approximately transforming the l 0 norm optimization problem into the l 1 norm optimization problem. This idea requires the sensing matrix to comply with the RIP criteria. l The 1-norm optimization problem can be solved by a convex optimization class algorithm. The unconstrained form was rewritten.

$$\bar{\mathbf{S}} = \arg\min \left\| \mathbf{X} - \mathbf{A}(\bar{\boldsymbol{\theta}})\bar{\mathbf{S}} \right\|_f^2 + \lambda \left\| \bar{\mathbf{s}}^{(l_2)} \right\|_1 \tag{2}$$

Among them, λ is the regularization parameter. $\bar{\mathbf{s}}^{(l_2)}$ is a sparse coefficient vector of length N. Since the l 2 norm itself does not have sparseness, the calculation of the l $\bar{\mathbf{S}}$ 2 norm of each row of the matrix will not bring sparseness to the snapshot domain. $\bar{\mathbf{s}}^{(l_2)}$ the sparse structure is guaranteed to $\bar{\mathbf{S}}$ be consistent with each column in the sparse coefficient matrix. The method of solving the above problem can make full use of the joint sparsity between different snapshots and improve the resolution of the algorithm. But the inversion step increases the amount of computation.

In order to reduce the complexity of the algorithm, the L 1 -SVD algorithm uses singular value decomposition to achieve dimensionality reduction.

$$\mathbf{X} = \mathbf{ULV}^H \tag{3}$$

Where, $\mathbf{U}$ and $\mathbf{V}$ represent the signal subspace and the noise subspace respectively. And $\mathbf{L}$ diagonal matrices are formed by arranging the singular values of the matrix from large to small $\mathbf{X}$. Then the signal subspace and denoise the noise subspace are kept.

$$\mathbf{X}_{SV} = \mathbf{ULD}_K = \mathbf{XVD}_K \tag{4}$$

Among them, $\mathbf{D_K} = [\mathbf{I_K}\ \mathbf{0}]$, $\mathbf{I_K}$ is the identity matrix. $\mathbf{0}$ is the zero matrix. K is the sparsity. The received signal matrix is replaced with the newly received signal matrix $\mathbf{X}$. The $\mathbf{X}_{SV}$ dimensionality is $M \times N_p$, which is reduced from to $M \times K$. The same can be obtained through the following equation.

$$\bar{\mathbf{S}}_{SV} = \bar{\mathbf{S}}\mathbf{VD}_K \tag{5}$$

$$\mathbf{E}_{SV} = \mathbf{EVD}_K \tag{6}$$

The equation is converted.

$$\mathbf{X}_{SV} = \mathbf{A}(\bar{\boldsymbol{\theta}})\bar{\mathbf{S}}_{SV} + \mathbf{E}_{SV} \in \mathbb{R}^{M \times K} \tag{7}$$

Then the CS-DOA estimation optimization model is transformed into the following equation.

$$\bar{\mathbf{S}}_{SV} = \arg\min \left\| \mathbf{X}_{SV} - \mathbf{A}(\bar{\boldsymbol{\theta}})\overline{\mathbf{S}_{SV}} \right\|_f^2 + \lambda \left\| \bar{\mathbf{s}}_{SV}^{(l_2)} \right\|_1 \tag{8}$$

Equation (8) can be regarded as a second-order cone programming problem that can be solved by the CVX toolkit.

$$\begin{aligned} &\min p + \lambda q \;\; s.t. \;\; \left\| \mathbf{z}_1^H, \ldots, \mathbf{z}_K^H \right\|_2^2 \le p. \; and \;\; \mathbf{1}^H \mathbf{r} \le q \\ &where \sqrt{\sum_{k=1}^{K} \left(s_i^{SV}(k) \right)^2} \le r_i, \;\; i = 1, \ldots, Ns \\ &and \;\; \mathbf{z}_k = \mathbf{x}^{SV}(k) - \mathbf{A}\mathbf{s}^{SV}(k), \;\; k = 1, \ldots, K \end{aligned} \tag{9}$$

The spatial spectrum obtained by the algorithm has an extremely high resolution, but the robustness is not good. Some studies[10] pointed out that when it is applied to an underwater array receiving signal processing, the simulation experiment with a large number of array elements and a low signal-to-noise ratio of the source signal is conducted. There will always be a large number of "false peaks" and a loss of the direction-finding effect. How to improve the robustness of such high-resolution algorithms under complex conditions is an important topic.

3. Signal sparsity reconstruction algorithm ReGOMP

In Figure 1

, $\bar{\mathbf{S}}$ is a $N \times N_p$ dimensional sparse coefficient matrix. N is the number of grid points and the number N_p of snapshots. The sparsity of a matrix and the sparsity of a vector are calculated slightly differently. $\bar{\mathbf{S}}$ $\{\bar{\mathbf{s}}_1, \bar{\mathbf{s}}_2, \ldots, \bar{\mathbf{s}}_N\}$ corresponds to a different grid point. Under ideal conditions, the default signal wave will only enter the direction corresponding to a very small number of grid points. So the matrix $\bar{\mathbf{S}}$ has row sparseness and is sparse in the longitudinal spatial angle domain. However, it does not have sparsity in the horizontal snapshot domain. Since the source direction information carried by the signal data under different snapshots is substantially the same, the positions and numbers of non-zero elements in the sparse coefficient vectors under different snapshots are also the same.

In the actual scene of receiving signals from the array, the existence of noise interference will inevitably cause some grid points to receive signals other than the source signal. Although the value is not large, it will obviously destroy the sparsity in the spatial angle domain. When the signal itself does not conform to strict sparsity, the effect of the CS-DOA estimation algorithm will inevitably be affected.

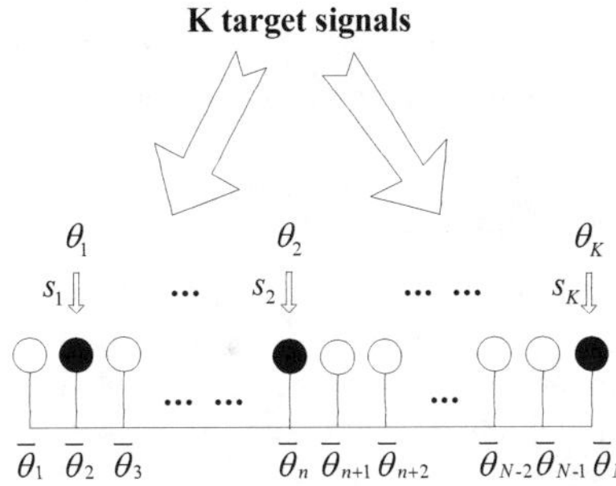

Figure 1 Grid discretization in one-dimensional space (black dots represent grid points that receive signals, and white dots represent grid points that do not receive signals)

To obtain the sparse signal, the method of changing the sparse transformation basis is usually adopted. However, the spatial signal already has approximate sparsity in the one-dimensional angle domain, and changing the sparse transform basis has little effect on further improving the strict sparsity of the signal. It is advisable to use the compressed sensing algorithm to reconstruct the obtained sparse vector to obtain a more sparse array received signal, thereby improving the effectiveness of the conventional CS-DOA estimation method.

There is a class of compressed sensing algorithms called greedy algorithms. The basic idea of the greedy algorithm is to use the iterative method to directly solve the l0 norm optimization problem. In each iteration, a local optimal solution is selected, and the signal is continuously approximated according to the local optimal solution until the optimal approximation is achieved. Although the reconstruction accuracy is sacrificed, the calculation speed is improved, which just meets the need of improving the strict sparsity of the received signal.

The classic OMP algorithm[11] is taken as an example. The received signal matrix is inputted to the sense matrix and sparsity. The signal sparse coefficient matrix is outputted, and the signal matrix is reconstructed. The improved algorithms based on OMP include CoSaMP[12], ROMP[13], gOMP[14], and StOMP[15]. Since it is necessary to use the characteristics of the greedy algorithm to perform the secondary reconstruction of the received signal, the performance of the reconstruction algorithm has higher requirements. This paper mainly combines the algorithm characteristics of ROMP and gOMP to construct the ReGOMP algorithm to achieve a better signal reconstruction effect than OMP. The algorithm steps are as follows.

- The sparsity estimation $K_0 = 1$ is initialized. The residuals are $\mathbf{r}_0 = \mathbf{y}$, index set of column serial numbers. $\mathbf{\Lambda}_0 = \varnothing$ is set as the number of columns selected $\mathbf{A}_0 = \varnothing$ from the sensing matrix according to the index $\mathbf{A}(\bar{\boldsymbol{\theta}})$. The iteration is $i = 1$.
- The first iterative process is started. First, from $\mathbf{A}(\bar{\boldsymbol{\theta}})$ the overcomplete basis of the sensing matrix, the $2K_0$ column with the largest correlation with the current residual (that is, the received signal vector of the original array) is found to calculate $\mathbf{u}=abs\left[\mathbf{A}^T\mathbf{r}_{i-1}\right]$. $\langle \mathbf{r}_{i-1}, \boldsymbol{\alpha}_j \rangle, 1 \le j \le N$ is calculated, and $\mathbf{u}$ the $2K_0$ largest value among them is selected. The column numbers j corresponding to these values in $\mathbf{A}(\bar{\boldsymbol{\theta}})$ is replaced by a set $\mathbf{J}$.
- In this column vector, another selection is made according to the regularization criteria. $2K_0$ is used to find $|\mathbf{u}(i)| \le 2|\mathbf{u}(j)|,\ i,j \in J_0$, a subset of $\mathbf{J}_0$ the set $\mathbf{J}$ that satisfies this requirement. The one with the greatest energy $\mathbf{J}_0$ is chosen. The standard of regularization here refers to selecting a group with the largest energy among the subsets, in which the absolute value of the inner product of each column vector and the residual is less than twice the minimum value.
- It can be seen that $\mathbf{\Lambda}_i = \mathbf{\Lambda}_{i-1} \cup \{\mathbf{J}_0\}$, and $\mathbf{A}_i = \mathbf{A}_{i-1} \cup \boldsymbol{\alpha}_j, j \in \mathbf{J}_0$. This saves the selected column vector and its corresponding column number.
- As $\mathbf{x} = \mathbf{A}_i\mathbf{s}_i$ and $\hat{\mathbf{s}}_i = \arg\min_{\mathbf{s}_i} \left\|\mathbf{x} - \mathbf{A}_i\bar{\mathbf{s}}_i\right\| = \left(\mathbf{A}_i^T\mathbf{A}_i\right)^{-1}\mathbf{A}_i^T\mathbf{x}$, the least squares solution is obtained. The item with the largest absolute value is selected from it $\hat{\mathbf{s}}_i$ K_0, and the K_0

corresponding $\hat{\mathbf{s}}_{iK_0}$ item $\mathbf{\Lambda}_i$ is recorded. And the corresponding $\mathbf{\Lambda}_{iK_0}$ set $\mathbf{\Lambda}_i = \mathbf{\Lambda}_{iK_0}$ of column numbers is recorded. $\mathbf{\Lambda}_{iK_0}$ is the residuals updated as $\mathbf{r}_i = \mathbf{x} - \mathbf{A}_i\hat{\mathbf{s}}_i = \mathbf{x} - \mathbf{A}_i\left(\mathbf{A}_i^T\mathbf{A}_i\right)^{-1}\mathbf{A}_i^T\mathbf{x}$.

- Then $i = i+1$. The previous steps are repeated until $i > K_0$, $\|\mathbf{\Lambda}_i\|_0 \geq 2K_0$, or $\mathbf{r}_i = 0$.

4. Robust CS-DOA estimation method L1-ReGOMP

Through the theoretical analysis in the previous two sections, this paper proposes a new robust CS-DOA estimation method L1-ReGOMP, which completes the sparse reconstruction of the received signal through the ReGOMP algorithm. Then this paper combines the L1-SVD algorithm to achieve high-resolution spatial spectrum estimation. The algorithm flow is shown in the figure below.

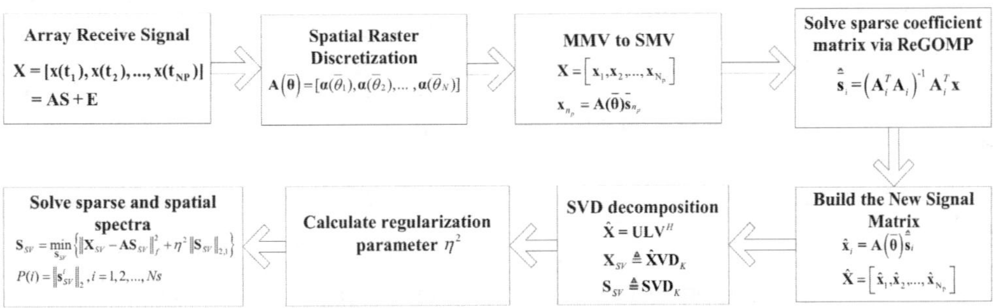

Figure 2 Flow chart of L1-ReGOMP algorithm

Next, further analysis is carried out through two simulation experiments of compressed sensing signal reconstruction and CS-DOA estimation.

4.1 Compressed sensing signal reconstruction

The key step of the L1-ReGOMP algorithm is to reconstruct the received signal in advance by using the compressed sensing theory. The choice of reconstruction algorithm has a great influence on the reconstruction effect of the signal. A linear single-frequency signal is taken as an example, and the signal frequency is set to 1 kHz. The sampling rate is set to 3 kHz, and the number of snapshots is 1000. The noise is Gaussian white noise. The signal and noise are irrelevant, and the signal-to-noise ratio is set to 0 dB. Figure 3 shows the noise-free Linear single frequency signal, analog received signal after noise is added.

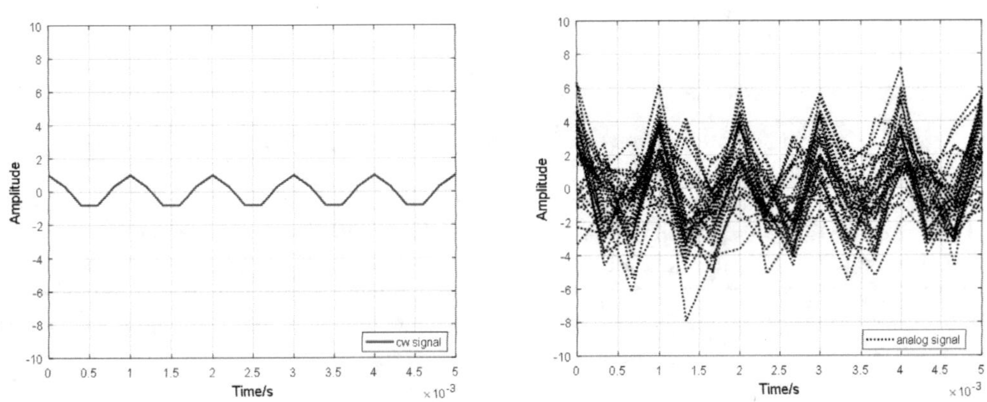

Figure 3 (a) Linear single frequency signal waveform (b) Analog receiving signal waveform

Then, the OMP algorithm and the ReGOMP algorithm are respectively used to reconstruct the analog received signal after the noise is added. And the waveform of the reconstructed signal is shown in Figure 4. It can be seen that the reconstructed signal is smoother than the analog received signal, and the reconstruction effect of the ReGOMP algorithm is closer to the linear single-frequency signal itself than the OMP algorithm.

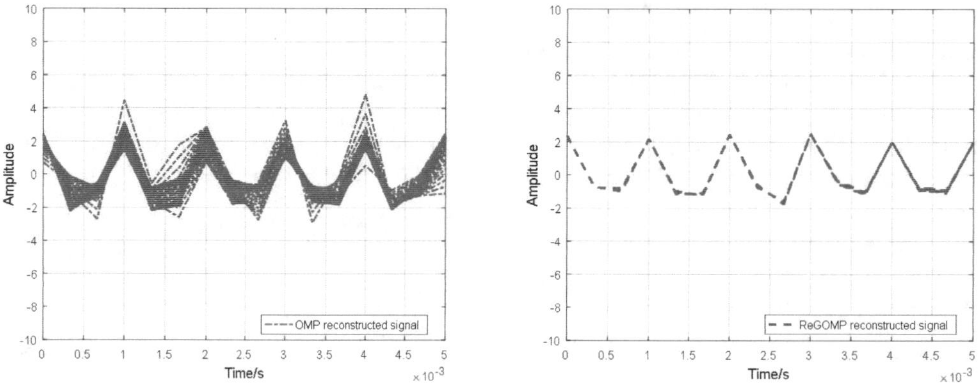

Figure 4 (a) OMP reconstructed signal waveform (b) ReGOMP reconstructed signal waveform

4.2 CS-DOA estimation

Next, the spatial spectrum estimation effect of the L1-ReGOMP algorithm is verified. The basic array is a uniform linear array, and the number of array elements is 36. The array element spacing is a half wavelength. The scanning angle is -60° to 60°, the scanning interval is 0.2°, and the source direction is 0°. Other settings are consistent with the noise-added linear single-frequency signal in Section 3.1. The obtained spatial spectrum is shown in Figure 5.

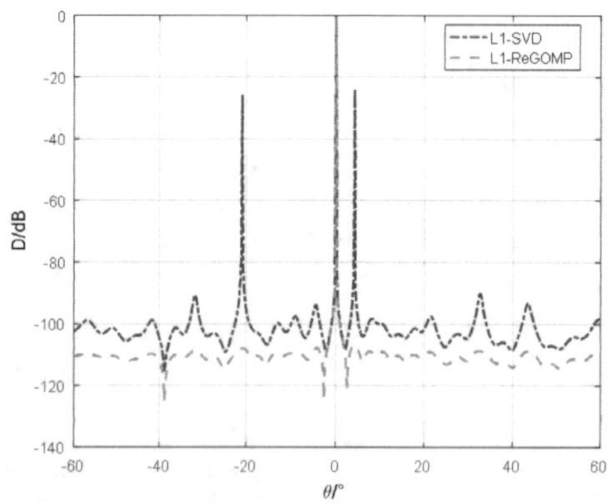

Figure 5 Comparison of direction-finding effects between L1-SVD and L1-ReGOMP

It can be seen that both algorithms accurately determine the source position as 0°. Compared with L1-SVD, L1-ReGOMP has two significant improvements. 1. The directivity is stronger, the maximum sidelobe height is lower, and there is no occurrence that can affect the signal and the "pseudo-peak" of the source direction judgment. 2. The direction-finding speed has been improved, and the algorithm running time has been effectively shortened. It shows that with the increase of the sparsity of the processed signal, the effect of the algorithm is also effectively improved.

5. Conclusion

In the array signal processing in underwater acoustic engineering, research on high-resolution DOA estimation methods has not been used. Among them, research based on compressive sensing theory is one of the important applications of the frontier. The classic L1-SVD algorithm can achieve high-resolution DOA estimation with a small number of array elements, but not so good with a high number of array elements. In this paper, through theoretical analysis and computer simulation experiments, the ReGOMP algorithm for signal sparsity reconstruction is proposed and applied to the improved CS-DOA estimation method L1-ReGOMP. The feasibility of this idea is estimated by CS-DOA, which effectively enhances the robustness of the algorithm and retains the high-resolution direction-finding effect. New signal sparsity reconstruction algorithms can be further explored, and the process of the CS-DOA estimation algorithm can be optimized to further improve the operation speed and robustness, and have broad development in future academic research and engineering applications.

References

[1] Donoho. D. L. Compressed Sensing[J]. IEEE Trans on Information Theory, 2006, 54(4): 1289-1306
[2] Varshney K R, Çetin M, Fisher J W, et al. Sparse representation in structured dictionaries with application to synthetic aperture radar[J]. IEEE Transactions on Signal Processing, 2008, 56(8): 3548-3561.
[3] Bajwa W U, Haupt J D, Sayeed A M, et al. Joint source-channel communication for distributed estimation in sensor networks[J]. IEEE Transactions on Information Theory, 2007, 53(10): 3629-3653.
[4] Elad M. Optimized projections for compressed sensing[J]. IEEE Transactions on Signal Processing, 2007, 55(12): 5695-5702.
[5] Trzasko J, Manduca A. Highly undersampled magnetic resonance image reconstruction via homotopic ℓ0-minimization[J]. IEEE Transactions on Medical Imaging, 2009, 28(1): 06-121.
[6] Baboulaz L, Dragotti P L. Exact feature extraction using the finite rate of innovation principles with an application to image super-resolution[J]. IEEE Transactions on Image Processing, 2009, 18(2): 281-298
[7] Fuchs J J. On the application of the global matched filter to DOA estimation with uniform circular arrays[J]. IEEE Transactions on Signal Processing, 2001, 49(4): 702-709.
[8] Malioutov D, Çetin M, Willsky A S. A sparse signal reconstruction perspective for source localization with sensor arrays[J]. IEEE Transactions on Signal Processing, 2005, 53(8): 3010- 3022.
[9] Takaaki Musha, Tatsuo Kikuchi. Adaptive signal processing for reducing nearby generated flow noise[J]. Applied Acoustics,2005, (66): 1206-1217.
[10] Jihao Yin, Tianqi Chen, Direction-of-Arrival Estimation Using a sparse Representation of Array Covariance Vectors[J]. IEEE Transactions on Signal Processing, 2011, 59(9): 4489-4493.
[11] Tropp J A, Gilbert A C. Signal recovery from random measurements via orthogonal matching pursuits [J]. IEEE Transactions on Information Theory, 2007, 53(12): 4655-4666.
[12] NEEDELL D, TROPP J A. CoSaMP: Iterative signal recovery from incomplete and inaccurate samples [J]. Applied and Computational Harmonic Analysis, 2009, 26(3): 301-321.
[13] NEEDELL D, VERSHYNIN R. Uniform uncertainty principle and signal recovery via regularized orthogonal matching pursuit [J]. Foundations of computational mathematics, 2009, 9(3): 317-334.
[14] Jia Wang, Seokbeop Kwon, Byonghyo Shim. Generalized orthogonal matching pursuit, IEEE Transactions on Signal Processing, vol.60, no.12, pp.6202-6216, Dec.2012
[15] Donoho DL, Tsaig Y, Drori I, et al. Sparse solution of underdetermined systems of linear equations by stagewise orthogonal matching pursuit [J]. Information Theory, IEEE Transactions on, 2012, 58(2): 1094-1121.

An Analytical Method of Beamwidth under Different Angles of Planar Array

Qiang Kong, Jieqi Zhu, Peilin Su*

Hangzhou Applied Acoustics Research Institute

798970918@qq.com

Abstract. It is widely recognized that the calculation of the beam width of the large-aperture planar array through beam pattern simulation is slow and inefficient, and obtaining the ability to distinguish the target from different angles within the observation range quickly is difficult. This paper uses methods such as the array product theorem, discrete array, continuous array beam width equivalence, and coordinate transformation to acquire a more accurate analytical solution of the beam width of the planar array. The analytical value of the beam width of the planar array at different angles and the number of array elements were compared with the simulated value. The results show that with the usage of the new method, the average deviation of the beam width analytical values in different main maxima directions can reach less than 1% when the number of array elements is less than 10 elements. Compared with the simulation method, the planar array beam width can be obtained quickly and accurately by the new method.

1. Introduction

The planar array is a commonly used array method for basic arrays. Because of its simple layout method, uniform array element spacing, good symmetry, two-dimensional directivity, and easy engineering implementation, it is often used in sonar, radar, and other array designs[1]. The plane array is used as the basic array arrangement.

The beam width of the array is an important parameter in the design of the array, which is defined as the two sides of the array whose angular deflection response value is lower than the maximum response value on the main axis by a specified value at a given frequency and a given plane including the main axis. The angle between the two directions is also measured[2]. The planar array has two-dimensional directivity, and its basic array beam pattern is three-dimensional, so the horizontal beam width and vertical beam width are often described separately. The beam width reflects the ability of the acoustic array to distinguish space targets. If this specified value is taken as 0.5 (linear scale) or -3 dB (logarithmic scale) in the directivity map, the beam width is called half-power beam width $\Delta\theta_{3dB}$. And the beam width mentioned in this article refers to the half-power beam width.

It is an important indicator to measure the ability of the acoustic array to resolve multiple targets in different azimuths. The beam width value of the plane array is related to the azimuth and elevation angles of the main maximum value of the beam pattern, which can be calculated by the beam pattern simulation. However, the plane array generally has a large aperture, a large number of array elements and beam patterns are two-dimensional, resulting in a huge budget for simulation calculation[3]. A fast and convenient analytical calculation formula is urgently needed to calculate the beam width of the planar array at different angles.

In this paper, according to the product theorem of the directivity function of the acoustic array[4], the directivity function of the planar array is decomposed by coordinate transformation, and the calculation of the beam width of the planar array is converted into the calculation of the linear array to finally obtain the horizontal and vertical beamwidth analytical formulas in different pitch angles and azimuth angles of the planar array.

2. Formula Derivation

The plane array is located in the XOY plane, and the array lengths in the X-axis direction and the Y-axis direction are L, H, d_1, and d_2 respectively. The array element spacing m and n in the X-axis and Y-axis directions are presented in Figure 1.

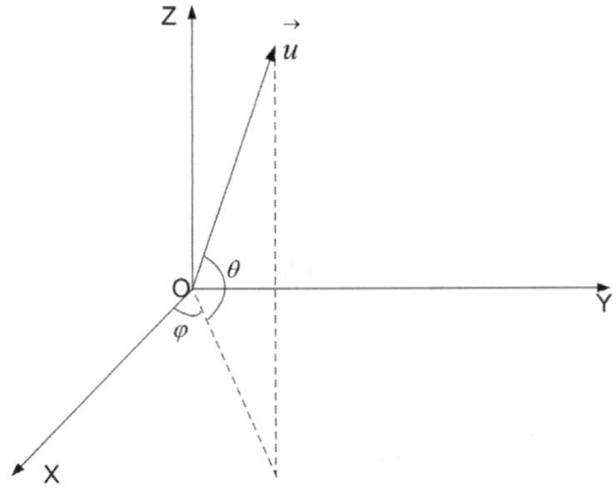

Figure 1 The coordinate system for directivity calculation

According to the array directivity function product theorem[5], the directivity function $D(\theta,\varphi)$ is the product of the linear array directivity functions. When the main maximum direction is located in the positive and lateral directions of the plane array, the directivity function is as follows.

$$D(\theta,\varphi)=\frac{\sin(md_1\pi/\lambda\cos\theta\cos\varphi)}{m\sin(d_1\pi/\lambda\cos\theta\cos\varphi)}\cdot\frac{\sin(nd_2\pi/\lambda\cos\theta\sin\varphi)}{n\sin(d_2\pi/\lambda\cos\theta\sin\varphi)}$$

At this time, since the horizontal and vertical main maximum directions are both located at 0°, the directivity function can be simplified as follows in the calculation of the vertical beam width $\varphi=0^\circ$.

$$D(\theta,\varphi)=\frac{\sin(md_1\pi/\lambda\cos\theta)}{m\sin(d_1\pi/\lambda\cos\theta)}=0.707$$

At this time, the vertical beam width is equal to the linear beam width. When the array element spacing of the linear array is less than half a wavelength and the number of array elements is greater than 10, the beam width is basically the same as that of a continuous linear array. Therefore, the vertical beam width can be calculated by solving the following formula.

$$D(\theta,\varphi)\approx\frac{\sin(L\pi/\lambda\cos\theta)}{L\pi/\lambda\cos\theta}=0.707$$

The above formula is a sinc function, and the horizontal beam width can be obtained by looking up the table.

$$\begin{aligned}\theta_{3dB}&=\arccos(-0.443\lambda/L)-\arccos(0.443\lambda/L)\\&=2\arcsin(0.443\lambda/L)\end{aligned}$$

However, in the calculation of the horizontal beam width, $\theta=0^\circ$, the following formula needs to be obtained to get the horizontal beam width.

$$D(\theta,\varphi)=\frac{\sin(md_1\pi/\lambda\cos\varphi)}{m\sin(d_1\pi/\lambda\cos\varphi)}\cdot\frac{\sin(nd_2\pi/\lambda\sin\varphi)}{n\sin(d_2\pi/\lambda\sin\varphi)}=0.707$$

However, the above formula cannot obtain an effective analytical solution. In addition, Equation (4) is only applicable when the main maximum is located in the positive and horizontal directions of the plane array. If the azimuth of the main maximum is not in the positive and horizontal directions, the main maximum pitch angle is θ_0 azimuth. The angle is φ_0. At this time, the directivity function of the plane array is as follows.

$$D(\theta,\varphi)=\frac{\sin\left[md_1\pi/\lambda(\cos\theta\cos\varphi-\cos\theta_0\cos\varphi_0)\right]}{m\sin\left[d_1\pi/\lambda(\cos\theta\cos\varphi-\cos\theta_0\cos\varphi_0)\right]}\cdot\frac{\sin\left[md_2\pi/\lambda(\cos\theta\sin\varphi-\cos\theta_0\sin\varphi_0)\right]}{n\sin\left[md_2\pi/\lambda(\cos\theta\sin\varphi-\cos\theta_0\sin\varphi_0)\right]}$$

$\varphi=\varphi_0$ and $\theta=\theta_0$, the analytical solution of the horizontal and vertical beam segments, cannot be obtained. When $\theta=\theta_0$, the horizontal beam width is calculated, and the directivity function simplifies as follows.

$$D(\theta,\varphi)=\frac{\sin\left[md_1\pi/\lambda(\cos\theta_0\cos\varphi-\cos\theta_0\cos\varphi_0)\right]}{m\sin\left[d_1\pi/\lambda(\cos\theta_0\cos\varphi-\cos\theta_0\cos\varphi_0)\right]}\cdot\frac{\sin\left[md_2\pi/\lambda(\cos\theta_0\sin\varphi-\cos\theta_0\sin\varphi_0)\right]}{n\sin\left[md_2\pi/\lambda(\cos\theta_0\sin\varphi-\cos\theta_0\sin\varphi_0)\right]}$$

Obviously, Equation (7) cannot form φ an analytical solution to the beam width. In order to obtain the beam width of the plane array when the main maximum is located in the non-orthogonal direction, the coordinates of the plane array are transformed. The plane array is located in the YOZ plane, and the array lengths in the Y-axis and Z-axis directions are L, H, d_1, and d_2 respectively. As for the distance between the array elements m in the Y-axis direction and the Z-axis direction, n is the number of array elements in the Y-axis direction and the Z-axis direction. Y-axis is the horizontal direction and the Z-axis is the vertical direction. Since the aperture of the planar array and the spacing of the array elements have not changed, its beam width should remain unchanged. At this time, the directivity function of the plane array is $D(\theta,\varphi)$ and it is transformed as follows.

$$D(\theta,\varphi)=\frac{\sin\left[md_1\pi/\lambda(\cos\theta\sin\varphi-\cos\theta_0\sin\varphi_0)\right]}{m\sin\left[d_1\pi/\lambda(\cos\theta\sin\varphi-\cos\theta_0\sin\varphi_0)\right]}\cdot\frac{\sin\left[md_2\pi/\lambda(\sin\theta-\sin\theta_0)\right]}{n\sin\left[md_2\pi/\lambda(\sin\theta-\sin\theta_0)\right]}$$

As $\theta=\theta_0$, the horizontal beam width is calculated, and the directivity function is simplified as follows.

$$D(\theta,\varphi)=\frac{\sin\left[md_1\pi/\lambda\cos\theta_0(\sin\varphi-\sin\varphi_0)\right]}{m\sin\left[d_1\pi/\lambda\cos\theta_0(\sin\varphi-\sin\varphi_0)\right]}=0.707$$

Similarly, when the number of array elements is greater than 10, the beam width is equivalent to that of a continuous linear array. So the horizontal beam width is calculated by the following formula:

$$D(\theta,\varphi)\approx\frac{\sin\left(L\pi/\lambda\cos\theta_0(\sin\varphi-\sin\varphi_0)\right)}{L\pi/\lambda\cos\theta_0(\sin\varphi-\sin\varphi_0)}=0.707$$

The horizontal beam width can be obtained by looking up the table:

$$\varphi_{3dB}=\arcsin\left[\sin\varphi_0+0.443\lambda/(L\cos\theta_0)\right]-\arcsin\left[\sin\varphi_0-0.443\lambda/(L\cos\theta_0)\right]$$

In the calculation of the vertical beam width through the coordinate transformation, the plane array is also located in the YOZ plane. But the array length in the Y-axis direction is H, and the array length in the Z-axis direction is L. Namely, the sides of the plane array in the Y-axis and Z-axis are interchanged. Then the vertical beam width of the array can be obtained by Equations (8) to (10).

$$\theta_{3dB}=\arcsin\left[\sin\theta_0+0.443\lambda/(H\cos\varphi_0)\right]-\arcsin\left[\sin\theta_0-0.443\lambda/(H\cos\varphi_0)\right]$$

3. Simulation Analysis

3.1. Accuracy Analysis

In order to verify the accuracy of the horizontal and vertical beamwidth analytical formulas, the simulation method is used to verify the accuracy.

The length and width of the plane array are 8 m and 4 m respectively, and the spacing of the array elements is 0.1 m. The maximum operating frequency is 7.5 kHz. The main maximum pitch angle is 30°, and the azimuth angle is 60°. The two-dimensional beam pattern of the array at 6 kHz is shown in Figure 2. The beam widths in the horizontal and vertical directions are 3.66° and 7.36° respectively. The horizontal and vertical beam widths are calculated by Equations (11) and (12) to be 3.67° and 7.34°, respectively. It can be seen that the two equations are very accurate for the calculation of the beam width of the planar array, and the difference from the simulation value is less than 0.3 %.

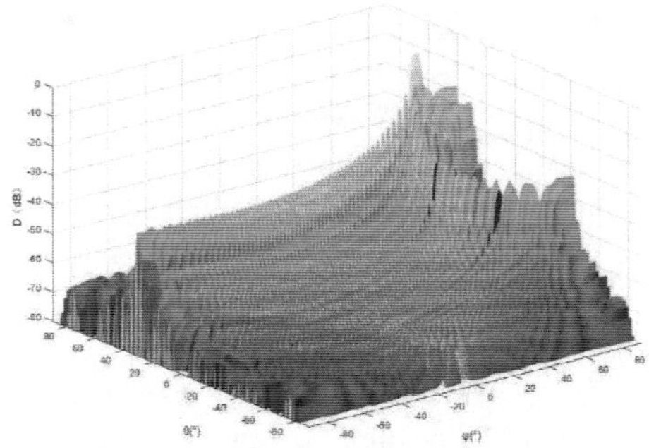

Figure 2 Beam pattern of the planar array

The size of the plane array and the spacing of the array elements remain unchanged. The frequency is also 6 kHz. The main maximum pitch angle is fixed at 30°, and the horizontal and vertical beam widths under different main maximum azimuth angles are compared. The maximum difference between the analytical value of the horizontal beam width and the vertical beam width and the simulated value is 0.02°. The maximum difference between the analytical value of the horizontal beam width and the simulated value is 1.3%, the average difference is 0.7%, and the analytical value of the vertical beam width is the largest. The difference was 0.5%, and the mean difference was 0.3%. The comparison in Figure 3 shows that the analytical value is calculated very accurately, and the difference between the analytical value and the simulated value is smaller when the beam width is wider.

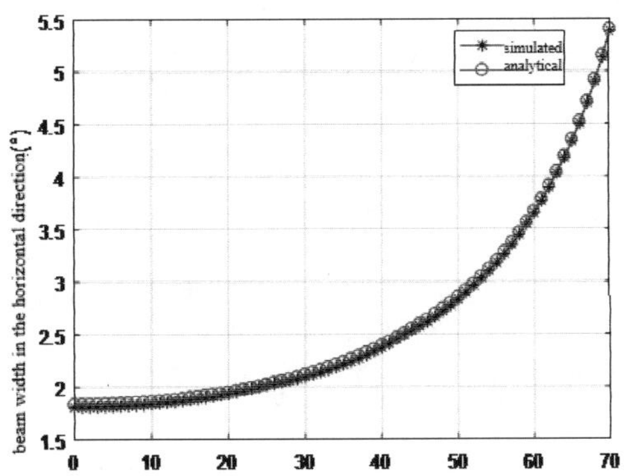

(a) Comparison of analytical and simulated values of beam width in the horizontal direction

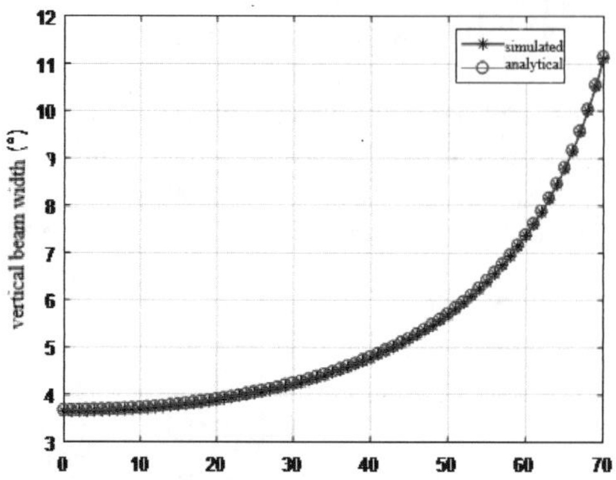

(b) Comparison between the analytical value of the vertical beam width and the simulated value
Figure 3 Comparison between analytical and simulated values of planar array beam width

3.2. Analysis of the influence of the number of array elements

The premise of Equations (11) and (12) for the horizontal and vertical beamwidth analytical formulas of planar arrays is that the beam widths of discrete planar arrays and continuous planar arrays are equivalent. The analytical value of the planar array beam width is compared with the simulated value under different numbers of array elements.

Similarly, the length and width of the plane array are 8 m and 4 m respectively. In order to ensure that the distance between the array elements is less than half a wavelength, the signal frequency is set to 750 Hz. At the same time, to effectively calculate the beam width on the beam pattern in both the horizontal and vertical directions, the main maximum azimuth angle is set to 40°, and the pitch angle is 30°. When the array element spacing is 0.1 m, 0.2 m, 0.4 m, 0.5 m, 0.8 m, and 1 m, the number of array elements in the length and width directions also changes. At this time, the beam widths in the horizontal and vertical dimensions are different for the number of array elements. The comparison between the analytical value and the simulated value is shown in Figure 4.

Figure 4 shows that when the array element spacing is large and the number of array elements is small. The error generated by the analytical value of the beam width gradually. At this time, the horizontal beam width error is 0.1°, and the difference between the analytical value and the simulation is 0.5%. When the vertical beam width error reaches 1.3°, the difference between the analytical value and the simulation is 3.1%. When the array element spacing is 1 m, the accuracy of the analytical value of the horizontal beam width is better. But the error of the analytical value of the vertical beam width is too large. The number of array elements in the width direction is only 4, which is caused by a certain deviation between the discrete directivity function of the matrix array and the directivity function of the continuous array in Equation (10). The result echoes the general number of array elements in the length and width directions mentioned above. When it is more than 10, the calculation of the beam width analytical value is more accurate.

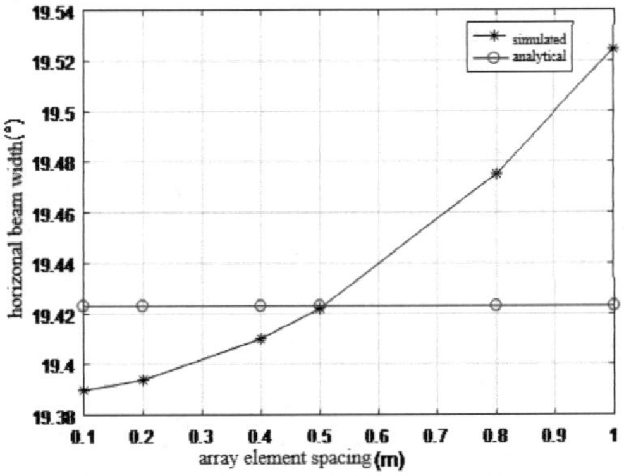

(a) Comparison of analytical and simulated values of beam width in the horizontal direction under different array element spacing (number of array elements)

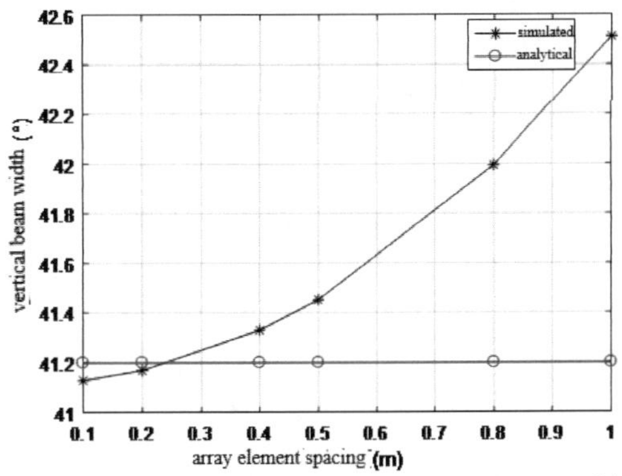

(b) Comparison of analytical and simulated values of vertical beam width under different array element spacing (number of array elements)

Figure 4 Comparison of analytical and simulated values of planar array beam width with different array element spacing (number of array elements)

4. Conclusion

The planar array is the most commonly used array arrangement, and the beam width is an important parameter in the design of the array. Due to the generally large aperture of the planar array, the calculation of the beam width by 2D beamform simulation requires a large amount of computation, slow speed, and low efficiency, and cannot quickly distinguish targets from different angles within the observation range. Aiming at the above problems, this paper deduces the analytical solution of the horizontal beam width when the main maximum of the two-dimensional beam pattern is located in the positive and horizontal directions according to the array directivity function product theorem. It also combines with the situation that the beam widths of the discrete planar array and the continuous planar array are basically equal. . Then, in the case that the main maximum is not in the positive and horizontal directions, the horizontal and vertical beamwidth analytical solutions are obtained respectively through coordinate transformation. Through simulation analysis, it is found that the average deviation between the analytical value of the planar array beam width and the simulation calculation value can be controlled within 1% when the number of array elements in the length or width direction is greater than 10 elements. When the number of array elements in the length or width direction is less than 10 elements, the analysis deviation between the value and the simulation value gradually increases. And the maximum deviation reaches 3.1% when the number of array elements in the width direction is 4 elements. Through analysis, the analytical calculation method of the planar array beam width proposed in this paper can be applied to the calculation of the planar array beam width with the number of long and wide array elements greater than 10 elements, and the beam width value can be obtained quickly and accurately.

Reference

[1] Bruno C, Taufik A, Closed-Form Directivity Expression for Arbitrary Volumetric Antenna Arrays, IEEE Transactions on Antennas and Propagation, 2018, 66 (12): 7443-7448.

[2] Richard PH, Underwater Acoustics Analysis, Design and Performance of Sonar[M]. John Wiley & Sons, 2010: 49-54

[3] William SB, Underwater Acoustic System Analysis[M]. Second edition, Peninsula Publishing, 2002: 303-338

[4] AD Waite, Sonar for Practicing Engineers[M]. Third edition, Beijing: Publishing House of Electronics Industry, 2004: 30-34.

[5] Li Qihu, Introduction to Sonar Signal Processing[M]. Second edition, Beijing: Ocean Press, 2000: 208-210.

Finite Element Analysis of Acoustic Performance of Water-filled Helmholtz Resonator with the Effect of Elastic Cavity Walls

Shengli Su*, Shuyou Wu

Wuhan Second Ship Design and Research Institute, Wuhan 430205, China

*Email: sushenli@163.com

Abstract. Water-filled Helmholtz resonator is a primary means to reduce the hydrodynamic noise of pump and valve in naval vessels. The predicted acoustic performance of water-filled Helmholtz resonators with the effect of elasticity of cavity walls using finite element method is compared with theoretical and experimental results in references, which demonstrated the applicability and accuracy of the finite element numerical method. And then, the effect of thickness and materials of cavity walls on resonance frequency and transmission loss of resonators is discussed, including the comparison of rigid and elastic cavity walls predictions simultaneously. The results of the discussion in this paper show that the acoustic performance of water-filled Helmholtz resonators is influenced by the elasticity of cavity walls to a great extent, which must be taken into consideration in the process of design.

1. Introduction

The Helmholtz resonator has been widely used in the noise control engineering of pipelines of air, water, hydraulic pressure and other media because of its simple structure, strong adaptability, high reliability and ideal noise reduction effect. The Helmholtz resonator consists of a relatively large cavity and a relatively small connecting tube in series. According to the lumped parameter theory, this is equivalent to a simple "mass-spring" vibration system, in which the medium in the connecting pipe provides acoustic quality, and the medium in the cavity is equivalent to an acoustic spring. At the point of resonance frequency of this vibration system, the resonator will produce strong acoustic attenuation.

As is well known, when the medium in the pipeline is air, the elasticity of the cavity wall structure of the Helmholtz resonator is far less than the compressibility of air, and in this case, the effect of the cavity wall elasticity on the acoustic performance of the resonator can be ignored. Because of this, the research of the Helmholtz resonators with respect to the noise control of air medium-based pipelines has been focused on the calculation method of resonant frequency, the optimization and improvement of resonator structure and so on [1-4]. Few people have paid attention to the effect of cavity wall elasticity.

However, when the medium in the pipeline is heavy fluid such as water, because the compressibility of water is extremely poor, it can be compared to the cavity wall elasticity of the Helmholtz resonator. In this case, the hypothesis of rigid walls will be no longer established, and the effect of the elasticity of the cavity wall on the acoustic performance of the resonator must be considered. Internationally, A.N. Norris and G. Wickham, on the basis of the thin shell theory, conducted the study on the effect of thin-wall elasticity of a spherical Helmholtz resonator with a circular hole on its acoustic performance [5]. In China, on the basis of the equivalent lumped parameter theory, Zhou Chengguang et al. studied the effect of the elastic walls of water-filled Helmholtz resonators on their acoustic characteristics [6] and carried out the relevant experimental verification. Wang Zefeng et al. also studied the effect of resonant cavity wall elasticity by using the lumped parameter theory [7, 8]. The difference is that they also considered the effect of the wall radiation impedance of the resonant cavity in the study. Yuan Jianping et al. used the acoustic finite element method to numerically analyze and test the acoustic characteristics of Helmholtz resonators [9, 10]. The effect of the wall elasticity of the resonant cavity on the resonance frequency was not mentioned in the study, which was one of the reasons why the theoretically predicted value of the resonance frequency was inconsistent with the test result. Up to now, the author has not seen the related literature about using the acoustic finite element method to analyze the effect of resonant cavity wall elasticity of Helmholtz resonators on its acoustic performance.

The main purpose of this paper is to study, on the basis of the acoustic finite element method, by using the Virtual. Lab Acoustics numerical software, the effect of the cavity wall elasticity of the Helmholtz resonator on its acoustic performance and compare the results with the experimental results reported in reference [6], so as to verify the correctness of the method adopted herein. On this basis, the effect of the parameters such as thickness and materials of cavity walls on the resonance frequency and transmission loss of the resonator is discussed, and the hypothesis of rigid cavity walls and the test results of elastic cavity walls are compared and analyzed.

2. Principle of calculation

We used the coupled acoustic finite element module of Virtual Lab Acoustics is to calculate the response characteristics of the internal acoustic field of Helmholtz resonator with respect to the cavity wall elasticity, and to further find its transmission loss and resonance frequency.

According to the "Advanced Application Examples of Acoustic Simulation Computation by Virtual Lab Acoustics"[11], when the effect of structural vibration is not considered, the system equations of the acoustic finite element method for acoustic field solutions are as follows:

$$(K_a + j\omega C_a - \omega^2 M_a)\cdot\{p_i\} = \{F_{ai}\} \tag{1}$$

where K_a, C_a and M_a are the stiffness matrix, damping matrix and mass matrix of the acoustic system, respectively, $\{p_i\}$ is sound pressure vector, and $\{F_{ai}\}$ is acoustic excitation vector.

When the density of fluid medium is relatively large, or the size of structure is very large and the structure is relatively soft, it is necessary to consider the coupling relationship between the fluid and the structure, that is, the coupled acoustic finite element method, and its system equations are as follows:

$$\left(\begin{bmatrix} K_s & K_c \\ 0 & K_a \end{bmatrix} + j\omega\begin{bmatrix} C_s & 0 \\ 0 & C_a \end{bmatrix} - \omega^2\begin{bmatrix} M_s & 0 \\ M_c & M_a \end{bmatrix}\right)\cdot\begin{Bmatrix} u_i \\ p_i \end{Bmatrix} = \begin{Bmatrix} F_{si} \\ F_{ai} \end{Bmatrix} \tag{2}$$

where, K_s, C_s and M_s are the stiffness matrix, damping matrix and mass matrix of the structural system, respectively, K_c is the coupling stiffness matrix, M_c is the coupling mass matrix, $\{u_i\}$ is the particle vibration velocity vector, and $\{F_{si}\}$ is the structural excitation vector.

Based on formula (2), by means of the coupled acoustic finite element module of Virtual Lab Acoustics, the internal acoustic field response of Helmholtz resonator with respect to the effect of the elasticity of cavity wall structure is obtained, and then the transmission loss TL is found by using the following formula:

$$TL = 20lg\left|\frac{p_1 + \rho c_0 v_1}{2p_2}\right| \tag{3}$$

where ρ is the density of water medium, c_0 is the speed of sound in water, p_1 and v_1 are the sound pressure and the particle vibration velocity at the inlet of the main pipeline where an Helmholtz resonator is located, and p_2 is the sound pressure at the exit of the main pipeline.

3. Numerical Calculation and Analysis

The Helmholtz resonator in reference [6] was used as the calculation model in this paper, as shown in Fig. 1, where the length of the connecting pipe is l = 87 mm, the diameter of the connecting pipe is d = 34 mm, the resonant cavity is cylindrical, the length is L = 200 mm, the diameter is D = 160 mm, and the diameter of the main pipe is D_P = 80 mm. The resonant cavity wall was composed of circular plate, cylindrical shell and annular plate, all of which were 4 mm thick. The resonant cavity, connecting pipe and main pipe were all made of steel and welded to form a complete resonator with Young's modulus E=2. 16×10^{11}Pa, density ρ=7800 kg/m^3, and Poisson's ratio μ = 0.3. In the calculation process, the density of water is 1000kg/m^3 and the sound speed of water is 1480m/s.

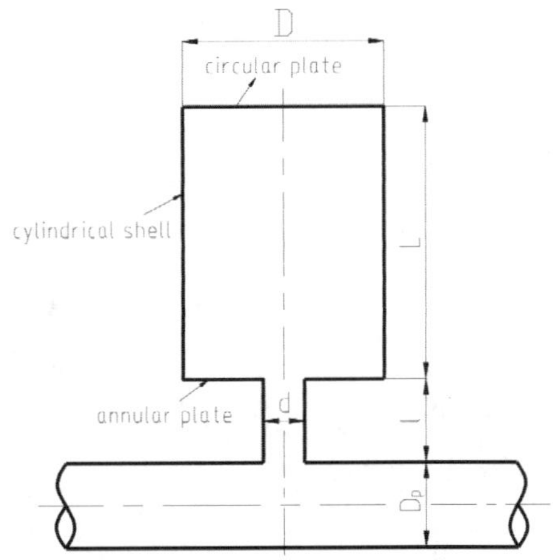

Fig.1 Structure of Water-filled Helmholtz Resonator

Fig.2 Finite Element Model of Water-filled Helmholtz Resonator

According to reference [6], the circular plate, cylindrical shell and annular plate in the resonator shown in Fig. 1 were welded together. The boundary conditions among the above three structures should be between the fixed-end and the simply- support boundary conditions. The results obtained under the fixed-end boundary condition were the upper theoretical limit and the results obtained under the simply-support condition were the lower theoretical limit. On this basis, the finite element software was used to calculate the vibration modes of the above three structures under the fixed-end and simply-supported connection conditions, and then they were introduced into the Virtual Lab Acoustics, and the coupled acoustic finite element calculation model as shown in Figure 2 was established. Then, the transmission loss of Helmholtz resonator was calculated under the fixed-end and simply-supported boundary conditions, with respect to the effect of resonant cavity wall elasticity, as shown in Fig. 3. The result is basically consistent with the theoretical calculation result disclosed in reference [6] as shown in Fig. 4, which proves the accuracy of the calculation method adopted herein. The difference between them may be caused by the following reasons: 1) The lumped parameter method was adopted in reference [6] while the three-dimensional finite element method is adopted in this paper; 2) The viscosity of water medium is not considered in the calculation process of this paper.

It can be known from Fig. 3 and Fig. 4 that the test results are between the upper and lower theoretical limits. As the circular plate, cylindrical shell and annular plate of the resonant cavity are welded together, it is difficult to accurately obtain their boundary conditions, but they must be between the fixed-end and simply-supported conditions. In addition, the resonance frequency of the resonator under the hypothesis of rigidity condition is 330 Hz, which is much higher than the calculated results and experimental results under the fixed-end and simply-supported boundary conditions, which shows that the elasticity of the resonant cavity wall has a great effect on the acoustic characteristics of the Helmholtz resonator.

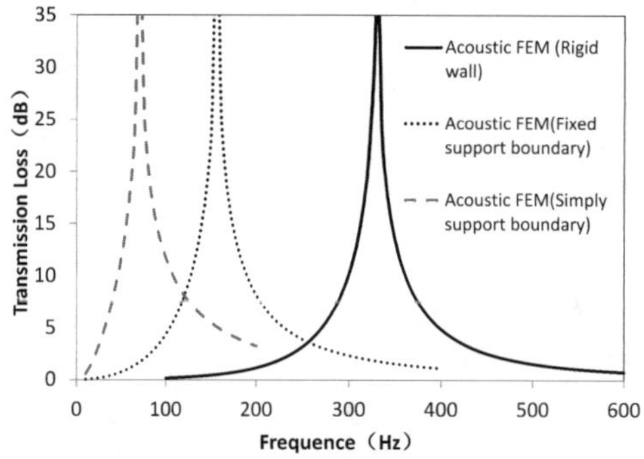

Fig.3 Transmission Loss of water-filled Helmholtz Resonator by Finite Element Method

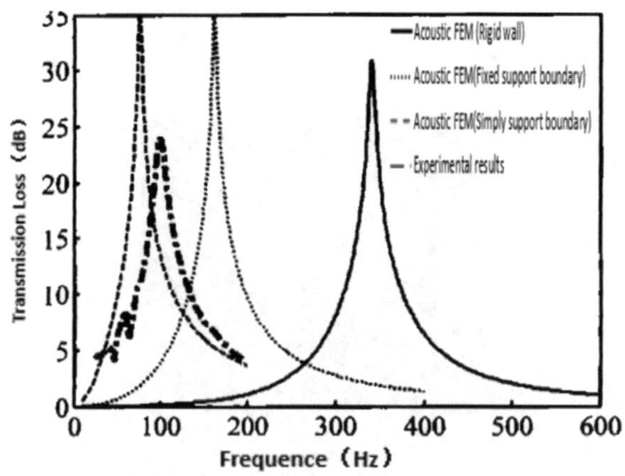

Fig.4 Transmission Loss of Water-filled Helmholtz Resonator in reference [6] （theoretical and experimental results）

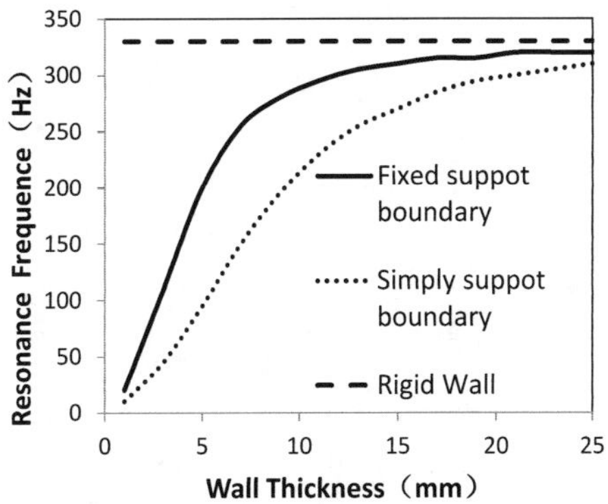

Fig.5 Effect of Thickness of Cavity on Resonance Frequency of Helmholtz Resonator

Fig. 5 shows the effect law of the thickness change of the resonant cavity walls including the circular plate, cylindrical shell and annular plate on the resonance frequency of the water-filled Helmholtz resonator when the resonant cavity wall is made of steel and the boundary conditions are fixed-end and simply-supported conditions. The wall thickness varies from 1 mm to 25 mm. It can be seen from the figure that the resonance frequency of the resonator does not change with the change of wall thickness under the hypothesis of rigidity; and when the elasticity of the resonant cavity wall is considered, the resonance frequency will increase with the increase of wall thickness under either the fixed-end boundary condition or the simply-supported boundary condition. When the thickness is small, the resonance frequency increases rapidly with the increase of thickness. When the thickness increases to greater than 15 mm, the resonance frequency increases slowly with the thickness, and when the thickness increases to greater than 20 mm, the resonance frequency hardly changes. In addition, we can also find that the resonance frequency under the fixed-end boundary condition is always higher than that under the simply-supported boundary condition, which is consistent with the foregoing statement herein, that is, the fixed-end boundary condition is the theoretical upper limit of resonance frequency, while the simply-supported boundary condition is the theoretical lower limit, and the actual resonance frequency should fall between those of the fixed-end boundary condition and the simply-supported boundary condition.

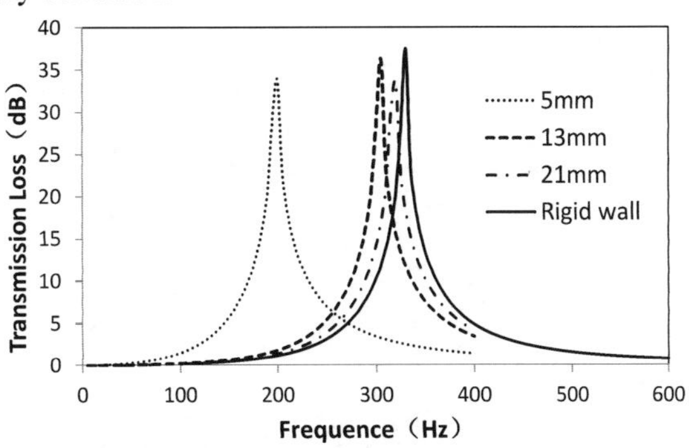

Fig.6 Effect of Thickness of Fixed Cavity on Transmission Loss of Helmholtz Resonator

Fig. 6 shows the effect law of the thickness change of the resonant cavity walls on the transmission loss of the water-filled Helmholtz resonator. It can be seen from the figure that with the increase of thickness, the resonance muffling frequency also increases rapidly. When the thickness is 21 mm, the resonance frequency of the resonator is little different from that under the hypothesis of rigidity. We can also find that when the thickness changes, the muffling bandwidth in the transmission loss curve does not change obviously.

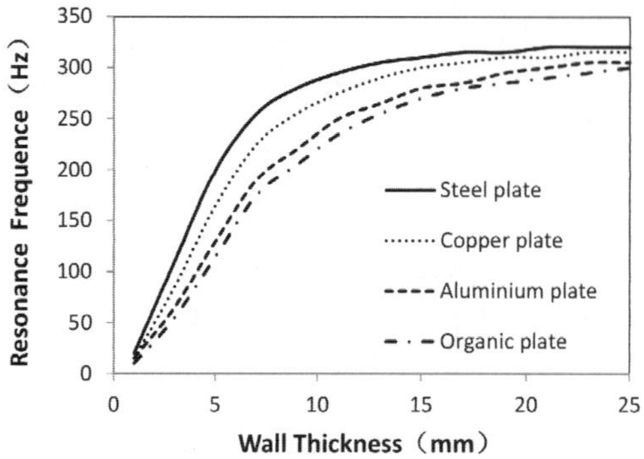

Fig.7 Effect of Thickness of Fixed Cavity of Different Materials on Transmission Loss of Helmholtz Resonator

Table 1 Mechanical Constants of Different Materials

Materials	Density (kg/m^3)	Young's modulus (10^{10}N/m^2)	Poisson's ratio
Steel	7800	21.6	0.3
Copper	8900	12.3	0.35
Aluminum	2700	6.85	0.34
Plexiglass	1200	5.35	0.35

Fig. 7 shows the effect law of the thickness change of the resonant cavity walls made of different materials on the transmission loss of the water-filled Helmholtz resonator. Table 1 gives the mechanical constants of steel, copper, aluminum and plexiglass. It can be seen from the figure that the wall thickness effects of various cavity wall materials on the resonance frequency are roughly the same; however, when the thicknesses are the same while the wall materials are different, the resonance frequencies of the resonator are also different, and the materials arranged in a frequency-based order from large to small are steel, copper, aluminum and plexiglass. This is consistent with the order of Young's modulus of these materials.

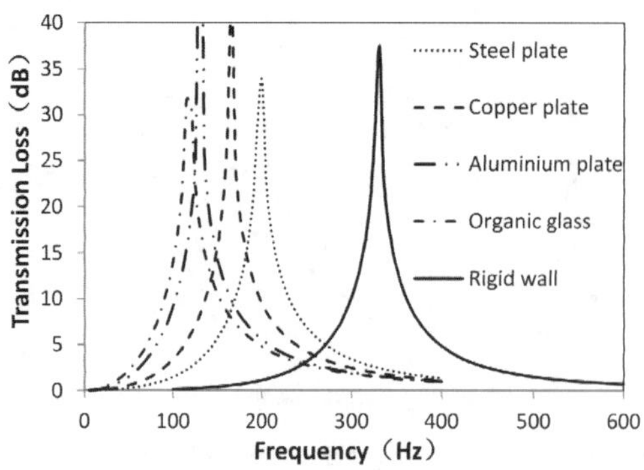

Fig.8 Effect of Fixed Cavity of Different Materials（5mm） on Transmission Loss of Helmholtz Resonator

Fig. 8 shows the effect of different materials on the transmission loss of the resonator when the boundary condition is the fixed-end condition and the wall thickness of the resonant cavity is 5 mm. It can be seen from the figure that when the thicknesses are the same and the materials used on the wall are different, the resonance frequencies of the resonator are also different, which is consistent with the foregoing statement. Moreover, the muffling bandwidth in the transmission loss curve does not change obviously.

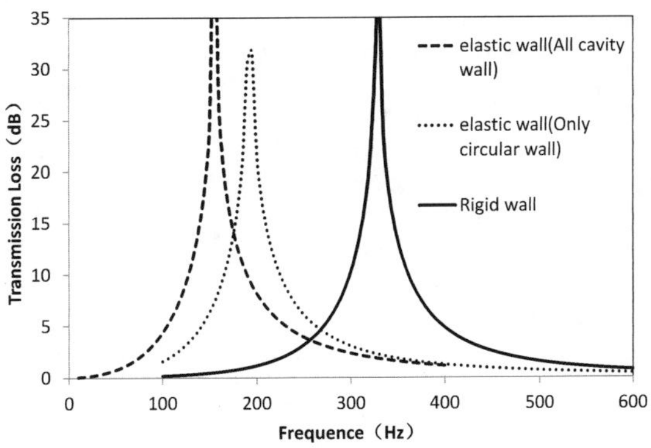

Fig.9 Effect of Different Cavity Hypothesis on Transmission Loss of Helmholtz Resonator

In the above calculation process, the effect of the elasticity of all walls of the resonant cavity, including the circular plate, cylindrical shell and annular plate, on the acoustic characteristics of Helmholtz resonator was considered. In order to investigate the effect of different wall assumptions, the effect of different wall assumptions on the transmission loss of the resonator were calculated when the wall thickness was 4 mm, as shown in Fig. 9. It can be seen from the figure that the resonance frequency of the resonator is obviously lower than that of the hypothesis of rigidity when only the elasticity of the circular plate is considered, and when the elasticity of all the walls of the resonant cavity is considered, the resonance frequency of the resonator becomes even lower with the further increase of the elasticity of the system. This shows that it is best to consider the elasticity of all the walls of the resonant cavity in the design and analysis of the water-filled Helmholtz resonator.

4. Conclusions

In this study, the effect of cavity wall elasticity on the acoustic characteristics of the water-filled cylindrical Helmholtz resonator is analyzed by the acoustic finite element method. The results show that when the internal working medium is water and other heavy fluids, the elasticity of the resonant cavity walls will make the resonance frequency of the resonator shift to the low frequency direction obviously. The smaller the thickness is, the greater the wall elasticity and the lower the resonance frequency will be. When different materials are used, the resonance frequency will be different. The method used herein and the related conclusions can provide reference for the design of water-filled cylindrical Helmholtz resonators, and also provide a feasible means for the acoustic performance analysis of other resonators with more complex structures and shapes. In the future, acoustic finite element method can be used to design more complicated Helmholtz resonator to reduce the hydrodynamic noise caused by pump and valve in piping system.

References

[1] Zhang Z L, Wang S R, Ruan D F and Y C 2004 *Noise and Vibration Control*, **24** 38.
[2] Fan Q W, Shen Y G, Zhang W, Shi L and Chen G S 2007 *Audio Engineering*, **31** 116.
[3] Ji Z L 2005 *Journal of Sound and Vibration*, **283** 1180.
[4] A. Selamet, and Ji Z L 2000 *Journal of the Acoustical Society of America*,**107** 2360.
[5] A.N.Norri, and G.Wickham 2009 *Journal of the Acoustical Society of America*, **93** 617.
[6] Zhou C G, Liu B L, Li X D, and Tian J 2007 *Acta Acustica*, **32** 426.
[7] Wang Z F, Hu Y M, Meng Z, and Ni M 2008 *Acta Acustica*, **57** 7022.
[8] Wang Z F, Hu Y M, Xiong S D, Luo H, Meng Z, and Ni M 2009 *Acta Physic Sinica*, **58** 2507.

[9] Yuan J P, Jin R, and Chen H L 2012 *Journal of Drainage and Irrigation Machinery Engineering*, **30** 141.
[10] Yuan J P, Zhang R C, and Jin R 2013 *Journal of Drainage and Irrigation Machinery Engineering*, **31** 1.
[11] Li Z G, and Zhan F L 2010 A*dvanced Application Examples of Acoustic Simulation Computation by Virtual Lab Acoustics* (Beijing:National Defense Industry Press)

Study on vibro-acoustic characteristics of an adaptive piezoelectric locally resonant structure

Anmin Huang[1,3], Minqing Wang[1,3], Daxiong Liao[2*] and Zhiwei Guo[1,3]

[1] School of Marine Science and Technology, Northwestern Polytechnical University, 710072 Xi'an P. R. China

[2] Facility Design and Instrumentation Institute, China Aerodynamics Research and Development Center, Mianyang 621000, China

[3] Ningbo Institute of Northwestern Polytechnical University, 315103 Ningbo P. R. China

liaodaxiong@cardc.cn

Abstract. Adaptively reducing low-frequency broadband vibration and noise is a challenge in the engineering and needs to be resolved. A piezoelectric locally resonant structure with periodic piezoelectric units and shunt damping circuits of synthetic impedance was studied to provide an option to deal with the problem. The z-scores method was applied to identify the peak frequencies used in the setting of control parameters. The band-gap performance and vibro-acoustic response were analyzed based on the plane wave expansion method and finite element method, respectively. The results show that the considered locally resonant piezoelectric structure with the z-scores peak-detection method can adaptively suppress the vibration and acoustic radiation of the plate, providing an effective method to reduce vibration and noise in time-varying excitation situations.

1. Introduction

The development of piezoelectric materials contributes to the development of piezoelectric locally resonant technology which can suppress low-frequency vibration and noise[1,2]. Using piezoelectricity, mechanical energy can be transformed to electrical energy which can be dissipated or stored by a shunt damping circuit connected to the piezo-element. On this basis, Fleming[3] presented a synthetic impedance circuit, which employed three operational amplifiers and a digital signal processor (DSP) to represent an impedance of arbitrary complexity to be attached to the piezo-element. The DSP circuit of synthetic impedance was proved to be better at low frequency in vibration and noise control than the traditional shunt damping circuit. In recent years, piezoelectric shunt damping circuits have been combined with the locally resonant technique[4], which can broaden the controlled frequency band. In 2017, Christopher[5] studied the piezoelectric locally resonant beam with synthetic impedance shunt circuits and analyzed the electro-elastic bandgap formation of the structure. In recent years, many simulational and experimental researches on piezoelectric locally resonant structures combined with synthetic impedance circuits were carried out[6,7]. In this work, the plane wave expansion method[8] was used to analyze the band structure of the piezoelectric locally resonant plate with synthetic impedance shunt circuits. Vibration response and acoustic radiation were also calculated with the finite element method to evaluate the effect of the designed structure on vibro-acoustic suppression.

The peak frequencies of the vibration are the reference for setting control parameters when the piezoelectric shunt circuit is used for vibration and noise control. Sometimes, the peak frequencies may be changed due to the time-varying external forces and boundary conditions. To deal with these situations, a new adaptive control method that can be programmed in a DSP circuit of synthetic impedance was presented. It can recognize peak frequencies of structural vibration. On this basis, by adjusting the parameters of the control system, the vibration and noise can be effectively controlled.

2. Band structure of an adaptive piezoelectric locally resonant plate

In this work, an adaptive piezoelectric locally resonant plate with two-dimension periodicity was considered, as shown in Figure 1. The piezo-elements were attached to a thin plate. The DSP circuit of synthetic impedance was attached to each patch.

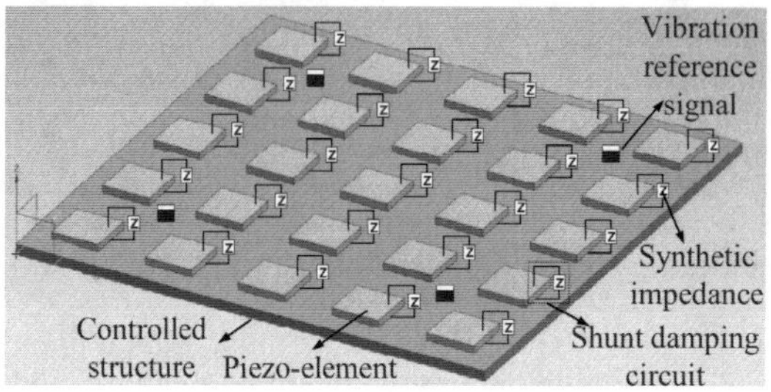

Figure 1. Piezoelectric locally resonant plate

The piezoelectric shunt circuit can be equivalent to a 1-DOF dynamic vibration absorber. Taking the mass of piezo-elements into consideration, the governing equations of the unit can be written as

$$\begin{cases} D\nabla^4 w(x,y) - \omega^2 \rho h w(x,y) \\ = \sum_{s=-\infty}^{\infty} \sum_{t=-\infty}^{\infty} \left[f_{st}\delta(x-x_s)\delta(y-y_t) + mw''(x,y)\delta(x-x_s)\delta(y-y_t) \right] \\ f_{st} = k_{me} V_{\mathrm{R}st} \\ L\ddot{q}_{\mathrm{R}st} + R\dot{q}_{\mathrm{R}st} + \dfrac{1}{C_p}\left[q_{\mathrm{R}st} - k_{me} w(x,y) \right] = 0 \\ L\ddot{q}_{\mathrm{R}st} + R\dot{q}_{\mathrm{R}st} = V_{\mathrm{R}st} \end{cases} \tag{1}$$

where $\nabla^4 = \frac{\partial^4}{\partial x^4} + 2\frac{\partial^4}{\partial x^2 \partial y^2} + \frac{\partial^4}{\partial y^4}$, $D = \frac{Eh^3}{12(1-\nu^2)}$ is the bending stiffness of the host plate, and $w(x,y)$ is the displacement of the host plate at coordinate (x,y). The terms h, E, ρ and ν denote the plate's thickness, Young's modulus, density and Poisson's ratio, respectively. The term C_p is the piezoelectric inherent capacity, $V_{\mathrm{R}st}$ is the difference in electric potential between the piezoelectric electrodes, $q_{\mathrm{R}st}$ and Z ($Z = R + \mathrm{j}\omega L$) are the electric charge and impedance of the shunt circuit, k_{me} is the modal piezoelectric coupling stiffness, and $\delta(x-x_s)\delta(y-y_t)$ is the Dirac delta function.

In Equation (1), the effects of piezo-elements are equivalent to 1-DOF dynamic vibration absorbers and lumped masses with location (x_s, y_t), where $x_s = sa_1$ and $y_t = ta_2$ ($s,t = 1,2,3\cdots$). The above terms a_1 and a_2 are lattice constant in the x and y directions.

The bending vibration displacement of an infinite local resonant plate can be expressed in the form of Fourier series superposition

Journal of Physics: Conference Series **2458** (2023) 012018 doi:10.1088/1742-6596/2458/1/012018

$$w(x,y)=\sum_{m=-\infty}^{\infty}\sum_{n=-\infty}^{\infty}W_{mn}\mathrm{e}^{-\mathrm{j}k_m x}\mathrm{e}^{-\mathrm{j}k_n y} \tag{2}$$

where $k_m=G_{xm}+\alpha$, $k_n=G_{yn}+\beta$, (α,β) is Bloch wave vector. $G_{xm}=2m\pi/a_1$, $G_{xn}=2n\pi/a_2$ ($m,n=-N,\cdots,N$, N is an integer), $\mathrm{G}_{mn}=(G_{xm},G_{yn})$ is the inverted lattice vector of a plane wave.

Following Bloch's theorem, the displacement response of the plate can be written as

$$\begin{cases} w(x_s,y_t)=w_0\mathrm{e}^{-\mathrm{j}\alpha x_s}\mathrm{e}^{-\mathrm{j}\beta y_t} \\ q_{\mathrm{R}st}=q_{\mathrm{R}}\mathrm{e}^{-\mathrm{j}\alpha x_s}\mathrm{e}^{-\mathrm{j}\beta y_t} \end{cases} \tag{3}$$

where w_0 and q_R denote the displacement of the plate and the electric charge at coordinate $(0,0)$, respectively. Based on the properties of Dirac delta functions, it can be obtained that

$$\sum_{s=-\infty}^{\infty}\sum_{t=-\infty}^{\infty}\mathrm{e}^{-\mathrm{j}\alpha x_s}\mathrm{e}^{-\mathrm{j}\beta y_t}\delta(x-x_s)\delta(y-y_s)=\mathrm{e}^{-\mathrm{j}\alpha x}\mathrm{e}^{-\mathrm{j}\beta y}\sum_{s=-\infty}^{\infty}\sum_{t=-\infty}^{\infty}\delta(x-x_s)\delta(y-y_t) \tag{4}$$

Substituting Equation (3) and Equation (4) in Equation (1) gives

$$\begin{cases} D\nabla^4 w(x,y)-\omega^2\rho h w(x,y) \\ =\left\{\left[-\frac{1}{C_p}(k_{me}w_0-q_{\mathrm{R}})+m\omega^2 w_0\right]\mathrm{e}^{-\mathrm{j}\alpha x}\mathrm{e}^{-\mathrm{j}\beta y}\right\}\sum_{s=-\infty}^{\infty}\sum_{t=-\infty}^{\infty}\left[\delta(x-x_s)\delta(y-y_t)\right] \\ L\ddot{q}_{\mathrm{R}st}+R\dot{q}_{\mathrm{R}st}+\frac{1}{C_p}\left[q_{\mathrm{R}st}-k_{me}w(x,y)\right]=0 \end{cases} \tag{5}$$

The equation $f(x,y)=\sum_{s=-\infty}^{\infty}\sum_{t=-\infty}^{\infty}\delta(x-x_s)\delta(y-y_t)$ is a periodic function in a continuous period, so it can be expressed as a Fourier series

$$f(x,y)=\sum_{m=-\infty}^{\infty}\sum_{n=-\infty}^{\infty}F_{mn}\mathrm{e}^{-\mathrm{j}G_{xm}x}\mathrm{e}^{-\mathrm{j}G_{yn}y} \tag{6}$$

By multiplying both sides of this equation by $\mathrm{e}^{\mathrm{j}G_{x\bar{m}}x}\mathrm{e}^{\mathrm{j}G_{y\bar{n}}y}$ and integrating over the interval $[-a_1/2,-a_1/2]\times[-a_2/2,-a_2/2]$, the equation $F_{mn}=1/S$ can be obtained. Substituting Equation (2) and (6) into (5) and truncating the infinite plane wave to a finite one, it can be obtained that

$$\begin{cases} \left[DS\left(k_m^4+2k_m^2k_n^2+k_n^4\right)W_{mn}+\left(\frac{k_{me}}{C_p}-m\omega^2\right)\sum_{o=-M}^{M}\sum_{p=-N}^{N}W_{op}-\frac{1}{C_p}q_R\right]-\omega^2\left(\rho hSW_{mn}\right)=0 \\ \left(-\frac{k_{me}}{C_p}\sum_{o=-M}^{M}\sum_{p=-N}^{N}W_{op}+\frac{1}{C_p}q_R\right)-\omega^2 Lq_R+\mathrm{j}\omega Rq_R=0 \end{cases} \tag{7}$$

The first formulation in Equation (7) can be rearranged as

$$\begin{cases} \left[\mathrm{K}(\alpha,\beta)-\omega^2\mathrm{M}(\alpha,\beta)\right]\Phi=0 \\ \Phi=\left[\mathrm{W}w_R\right]^{\mathrm{T}} \\ \mathrm{W}=\left[W_{mn}\right]_{1\times\left[(2M+1)\times(2N+1)\right]} \end{cases} \tag{8}$$

where W is a column vector with $(2M+1)\times(2N+1)$ entries. The terms M,N are integers. The characteristic frequencies can be calculated through Equation (8) as the Bloch wave vector $k=(\alpha,\beta)$ is given.

3. Synthetic impedance circuit and adaptive control method

In this work, the DSP circuit of synthetic impedance presented by Fleming[9,10] was combined with the piezoelectric locally resonant structure. A block diagram of the DSP circuit of synthetic impedance is depicted in Figure 2.

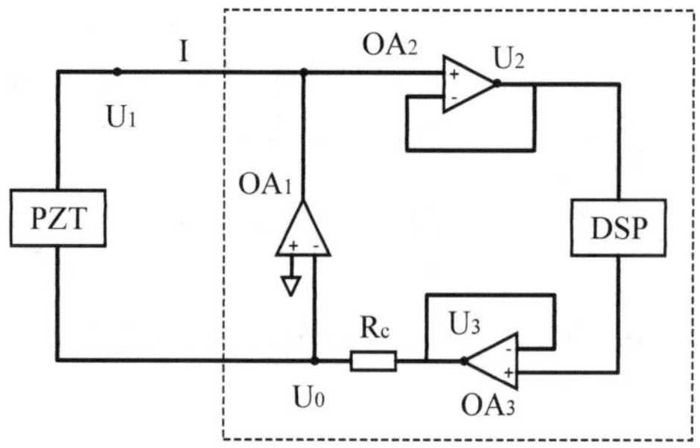

Figure 2. Synthetic impedance circuit

The OA_1 (operational amplifier) works as a voltage-controlled current source. The current I produced by such source is determined as follows under the assumption of ideal operational amplifiers: U_0 represents virtual zero voltage in the circuit, $U_2=U_1=I\times Z$, and $U_3=H(s)\times U_2$, therefore $I=(U_3/R_c)$. $H(s)$ is a frequency-dependent complex transfer function. The equation for ideally synthesized impedance is

$$Z_{ideal}(i\omega)=\frac{R_c}{H(s)} \tag{9}$$

The impedance related to the frequency of structural vibration is thus determined by the transfer function $H(s)$ prescribed in the DSP. Based on Fleming's design, an adaptive control method was presented, the key of which is to identify the peak frequency of structural vibration, and then adjust the transfer function adaptively.

In this work, the control system was intelligentized with a peak signal detection algorithm using z-scores[11]. It is based on the principle of dispersion: by using the original data, a new moving weighted sample data space is built, whose mean and standard deviation is calculated. A new datapoint is a peak if the difference between the datapoint and the moving mean is greater than T times of the moving standard deviation. The definition of z-scores is

$$Z(i)=\frac{y(i)-\left[\sum_{p=i-1-L}^{p=i-1} Y(p)\Big/L\right]}{\sigma\left[Y(i-1-L:i-1)\right]},\ (i,p=1,2,3\cdots) \tag{10}$$

where $y(i)$ is the detected datapoint. L is the length of a moving window, which determines the moving mean and standard deviation indirectly. Y is the weighted sample data, whose initial value is $[y(1),\cdots,y(L+1)]$. It will update when a new datapoint is detected. The updated weighted datapoint is

$$\begin{cases} Y(i) = I \times y(i) + (1-I) \times Y(i-1) & Z(i) \geq T \\ Y(i) = y(i) & Z(i) < T \end{cases} \tag{11}$$

With the peak signal detection algorithm, the steps of the proposed adaptive control method can be listed as

(1) Capture structural vibration and obtain the spectral characteristics by FFT,

(2) Obtain peak frequency of structural vibration with the peak signal detection algorithm,

(3) Adjust transfer function with DSP according to the impedance of the target, which can be calculated with the peak frequency through the existing theory[12],

(4) Optimize the transfer function iteratively.

4. Results and discussion

To evaluate the vibration and noise control effect of the adaptive structure, the finite element model of acoustic field and piezoelectric local resonance plate with synthetic impedance circuit was established.

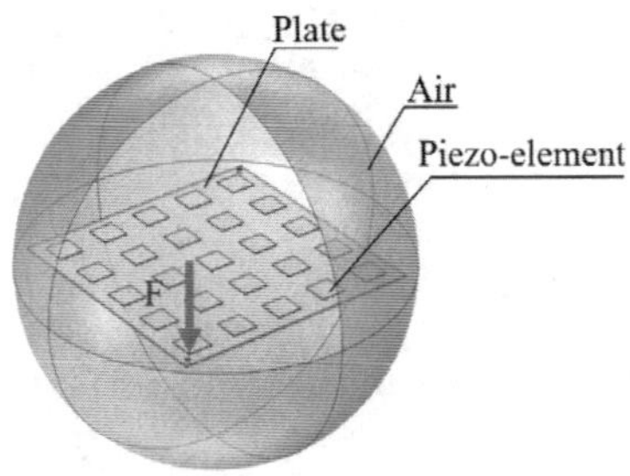

Figure 3. Piezoelectric local resonant plate

As shown in Figure 3, the average acceleration of the plate with a simply supported boundary is calculated under broadband excitation. The acoustic power is obtained by integrating on the spherical surface of air. The material parameters of piezoelectric patches and the plate are shown in Table 1.

Table 1. Material parameters

	Material	Young's module (Pa)	Poisson's ratio	Density (kg/m^3)	Size(m)	Relative permittivity
Plate	Steel	2×10^{11}	0.29	7800	$0.4\times0.4\times0.003$	/
Piezoelectric patches	PZT-5H	1.27×10^{11}	0.30	7500	$0.04\times0.04\times0.002$	1433

In this work, vibration response and acoustic radiation are calculated using the finite element method. The peak signal detection algorithm is programmed in MATLAB. The programming will update the control parameters according to the outputs of finite element software and return control parameters to the software in return. Frequency identification results and outputs of the algorithm were shown in Figure 4 and Table 2.

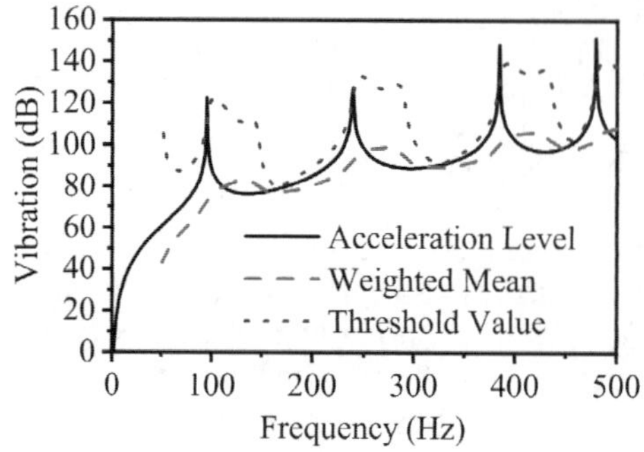

Figure 4. Frequency identification results

Table 2. Output of the programming

	1st	2nd	3rd	4th
Modal frequency (Hz)	95	239	384	479
Output (Hz)	95~96	238~240	383~385	478~480

As shown in Figure 4, the detection of peak frequencies process starts from a predetermined datapoint. The control system compares the acceleration to the Z-score. Datapoint is the peak in the condition that the acceleration is greater than the Z-score. In this research, the peak frequency refers to the modal frequency. The first four modal frequencies are 95Hz, 239Hz, 384Hz and 479Hz. In Table 2, it can be seen that the range of peak frequencies is recognizable, and the error range is acceptable.

In this research, $R_c = 10\text{k}\Omega$, $C_p = 20.5\text{nF}$, the programming is set to suppress the vibro-acoustic response at the second peak frequency, so the final transfer function is

$$H(s) = 10^4/(6300 + 35121.5j) \tag{12}$$

which has been adjusted in MATLAB iteratively. The results were shown in Figure 5 to Figure 7.

It can be seen from Figure 5 that a band gap is generated by the piezoelectric locally resonant structure, which ranges from 237.5Hz to 240.5Hz. The resonant frequency of the shunt damping circuit is close to the second modal frequency of the plate, so the corresponding acceleration level and acoustic power level are both reduced by 34.4dB, taking the first place. As shown in Figure 6 and Figure 7, the vibro-acoustic responses at the first, second and fourth peak frequencies are also suppressed. The corresponding acceleration level are reduced by 13.2dB, 33.9dB and 30.7dB respectively, and the acoustic power level are reduced by 13.2dB, 33.8dB and 28.3dB respectively.

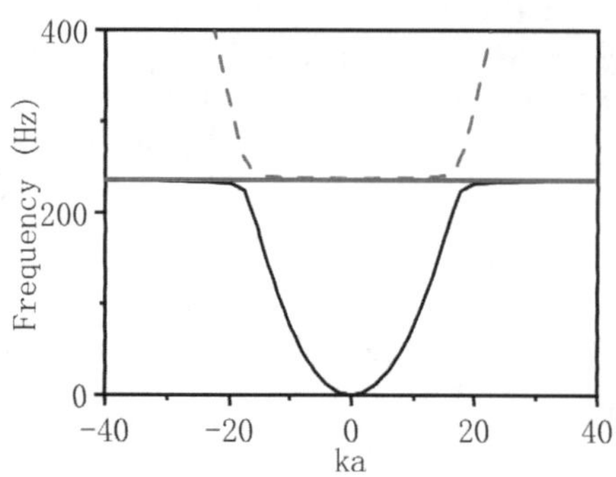

Figure 5. Band structure

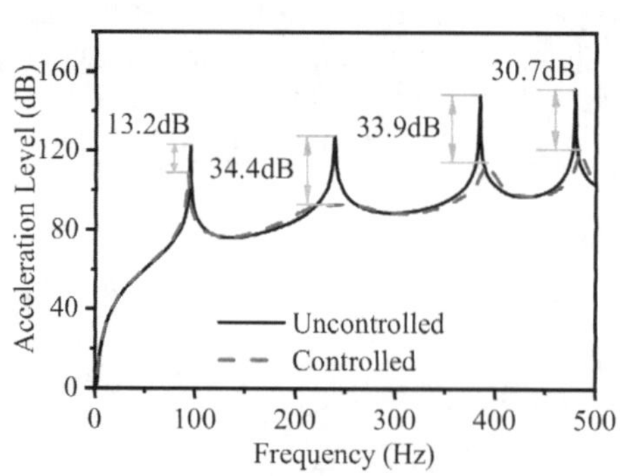

Figure 6. Vibration response

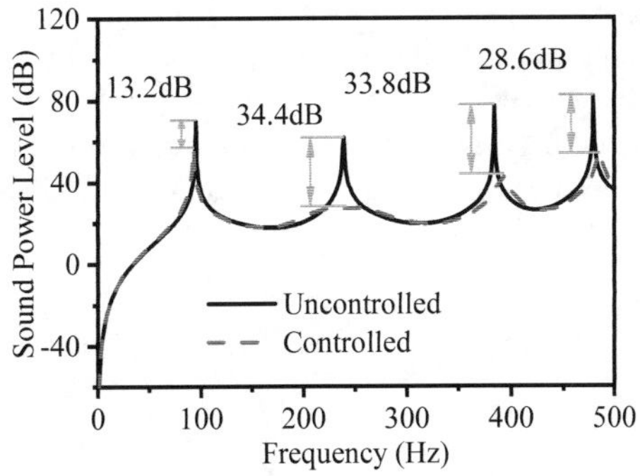

Figure 7. Acoustic radiation

5. Conclusion

A new adaptive vibration and noise control method was proposed in this work. And it was used on the piezoelectric locally resonant structure attached with a shunt damping circuit of synthetic impedance. Simulation analysis was carried out to evaluate the effect of the designed method and structure. The results show that the adaptive control method can identify peak frequencies of structural vibration, and the vibro-acoustic responses were suppressed significantly within the designed frequency range. The presented method and structure are promising to reduce vibration and noise in time-varying excitation situations. What's more, the designed structure and circuit have the potential for multimodal vibration and noise control, which remains to be studied in the future.

Acknowledgments

The authors would like to gratefully acknowledge the support of Ningbo Natural Science Foundation (2021J057).

References

[1] Lesieutre G. 1998 Vibration damping and control using shunted piezoelectric materials. *Shock and Vibration Digest.* **30(3)** 187-195

[2] De M B and Preumont A. 2008 Vibration damping with negative capacitance shunts: theory and experiment. *Smart Materials and Structures.* **17(3)** 4006-4032

[3] Fleming A J, Behrens S and Moheimani S O R. 2000 Synthetic impedance for implementation of piezoelectric shunt-damping circuits. *Electronics Letters.* **36(18)** 1525-1526

[4] Wang S, Wang M and Guo Z. 2021 Adjustable low-frequency bandgap of flexural wave in an Euler-Bernoulli meta-beam with inertial amplified resonators. *Physics Letters A.* **417** 127671

[5] Sugino C, Leadenham S, Ruzzene M and Erturk A. 2017 An investigation of electroelastic bandgap formation in locally resonant piezoelectric metastructures. *Smart Materials & Structures.* **26** 2144-2154

[6] Nečásek J, Václavík J and Marton P. 2016 Digital synthetic impedance for application in vibration damping. *Review of Scientific Instruments.* **87** 24704

[7] Yi K, Matten G, Ouisse M, Sadoulet-Reboul E, Collet M and Chevallier G. 2020 Programmable metamaterials with digital synthetic impedance circuits for vibration control. *Smart Materials & Structures.* **29** 035005

[8] Guo Z, Sheng M and Pan J. 2017 Effect of boundary conditions on the band-gap properties of flexural waves in a periodic compound plate. *Journal of Sound and Vibration.* **395** 102-126

[9] Fleming A J and Moheimani S O R, 2003 Adaptive piezoelectric shunt damping. *Smart Materials and Structures.* **12** 36-48

[10] Fleming A J, Behrens S and Moheimani S O R. 2001 Innovations in piezoelectric shunt damping. *Smart Structures and Devices.* **4235** 89-101

[11] Brakel, Van J, 2020 "Robust peak detection algorithm using z-scores". https://stackoverflow.com/questions/22583391/peak-signal-detection-in-realtime-timeseries-data/22640362 #22640362

[12] Giovanni G. 2001 A critical analysis of electric shunt circuits employed in piezoelectric passive vibration damping. *Smart Materials and Structures.* **10(5)** 1059-1068

Influence of the Spiral Angle of the Blade on the Material Loading Characteristics of the Slurry Pump

Zhuyang Chen [1a*], **Fei Yan** [1b*]

[1.] School of Mechanical Engineering, Jiangsu University of Science and Technology, Zhenjiang, Jiangsu 212003

[a*]chenzhuyang@stu.just.edu.cn, [b*]yanfei@just.edu.cn

Abstract: To study the effect of rotational speed on the internal flow characteristics and operating performance of a deep-sea mining slurry pump, a new type of slurry pump delivery system with changing vane helix angle is adopted in this paper. Based on the coupled computational fluid dynamics-discrete element method (CFD-DEM), the realizable k-εturbulence model and SIMPLE algorithm are applied to numerically simulate the solid-liquid two-phase flow field in the slurry pump to study the effect of rotational speed on the flow characteristics and distribution of particles in the slurry pump. The results show that with the increase in rotational speed, the separation of the boundary layer at the impeller runner is promoted, which weakens the particle overflow capacity and increases the hydraulic loss. The impact at the inlet of the guide vane runner is more violent. The fluid flow separation at the inlet is suppressed. The overflow capacity is increased and the hydraulic loss is weakened. With the increase in rotational speed, the more the average number of collisions between particles and vanes in each time period, the more the particles collide with the impeller and space. The collision probability and collision speed of the guide vane overflow parts become larger, thus laterally reflecting that the larger the rotational speed, the greater the wear rate of particles on the blade is.

1. Introduction

Since the beginning of the 20th century, the world has conducted extensive research on deep-sea polymetallic nodule mining technology. And it is generally believed that the slurry pump hydraulic pipeline lifting system has the most promising industrial application[1]. The return and overflow capacity of domestically developed slurry pumps are often insufficient, leading to blockage of the flow path. Therefore, we need to investigate the flow law of solid-liquid two-phase flow in the slurry pump.

Unlike centrifugal pumps that convey single-phase fluids (such as clear water), the solid-liquid two-phase flow in centrifugal slurry pumps has different trajectories, velocity characteristics, and force states. Therefore, Cai Baoyuan proposed the "two-phase flow" theory of centrifugal pumps and its design principle[2]. Walker C[3] proposed the theoretical basis of energy conversion between solid and liquid phases in slurry pumps, He Xijie put forward the principle and method of solid-liquid two-phase flow design for slurry pumps [4]. And some scholars[5] conducted an experimental study on slurry pumps with different vane parameters and analyzed the parameters such as vane profile. Some researchers conducted experimental studies on different vane parameters of slurry pumps, analyzed the influence of vane profile and other parameters on pump performance, and obtained the design principles and methods of slurry pumps. In recent years, numerical simulation of the internal flow field of slurry pumps and optimization of the design of the overflow components have been carried out to improve the pump's performance and service life. Therefore, the study of vane parameters is of great significance.

The centrifugal slurry pump's internal particle distribution and motion characteristics have a significant impact on the overflowing sections' wall wear, which in turn impacts the slurry pump's service life. Jingting Zhao conducted an in-depth study on the force, motion, and distribution pattern of particles inside the slurry pump[6]. In some studies[7], the motion trajectory of particles in the slurry pump was studied through high-speed photography technology, and the flow law of particles was obtained. The solid-liquid two-phase flow inside the mine slurry pump was studied by particle imaging velocimetry technique, and the laws of particle volume fraction distribution, velocity distribution, and motion trajectory inside the pump under different particle property conditions were obtained[8]. Liu Juan et al. used a combination of numerical simulations and experimental studies to investigate the trajectory and distribution of particle motion inside the slurry pump and its erosion and wear on the wall surface[9]. The above studies are of great significance for the design of low-wear and high-performance slurry pumps.

In this paper, a slurry pump that changes the spiral angle of the impeller blades is proposed to solve the problems of dewatering, backflow, and easy clogging of the slurry pump by changing the flow field condition of the impeller part. CFD-DEM method is used for coupling calculation, and the particle motion state of liquid-solid two-phase flow is analyzed under different speed conditions.

2. Geometric modeling and meshing

At present, the structural forms of the developed mine slurry pumps are all sectional multistage centrifugal pumps[10]. The flow field calculation domain consists of four parts: inlet extension, impeller, space guide vane, and outlet extension. The whole pump is placed in the vertical direction, with the inlet end at the bottom. And the pump lifts the ore slurry vertically. As the impeller and space guide vane are space-twisted entities, it is more difficult to model directly in the pre-processing software Workbench. So the solid modeling is first carried out with the help of the 3D modeling software Solidworks, as shown in Figure 1. The interface file (*.txt) saved in Solidworks is imported into Workbench software for computational meshing, and the mesh cells are obtained as shown in Figure 2. In this paper, the tetrahedral mesh was used for the computational model mesh, and the total number of mesh cells was 300521.

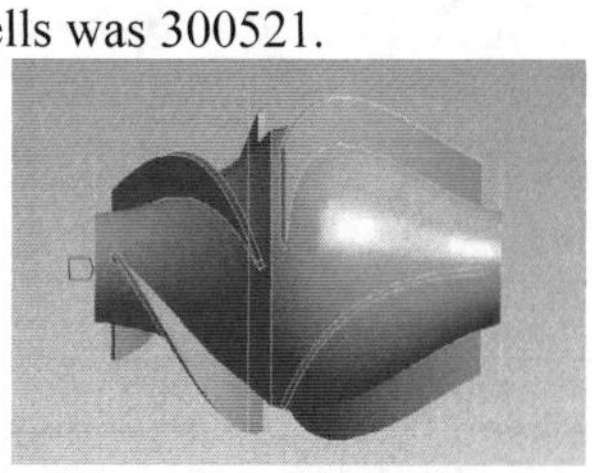

Figure 1 Solid model of impeller and space guide vane

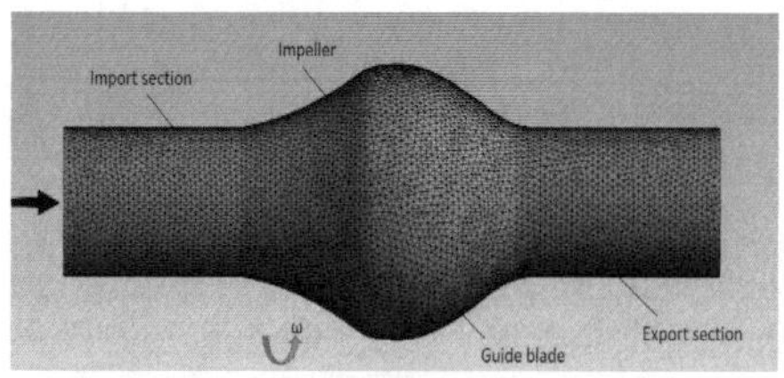

Figure 2 3D mesh model of slurry pump

3. Results and discussions

3.1 Unidirectional flow characteristics of different helix angles

To analyze the flow in the impeller internal flow field, a cross-section of ZX=-150 mm near the impeller inlet is created in CFD-Post, based on which the pressure and velocity clouds of this cross-section are generated for different impeller blade helix angles, as shown in Figure 3 and Figure 4.

Figure 3 shows the pressure cloud diagram at the cross-section of the impeller blades with different spiral angles. As can be seen from the figure, from the lower cover plate to the upper cover plate, the pressure is uniformly distributed along the direction of the blade curve in a block shape, showing a gradually rising trend. The static pressure at the pressure surface of the blade is larger than the static pressure at the suction surface. The maximum pressure is 0.219 MPa and the minimum pressure is

0.042 MPa.The low-pressure area is mainly concentrated in the inlet section of the impeller. When the water body enters the inlet area of the impeller, it is rotated by the impeller to do work, and the speed increases and the pressure decreases. The blade suction surface has a large low-pressure area. With the increase of the spiral angle, the low-pressure area shows a gradually increasing trend. The blade under the cover area pressure value is the largest, mainly concentrated in the blade near the pressure surface. With the reduction of the spiral angle, the blade under the coverage area of the high-pressure area gradually increased. It can be seen that the reduction of the impeller helix angle improves the pressure distribution inside the impeller and makes the pressure distribution on the cross-section more uniform.

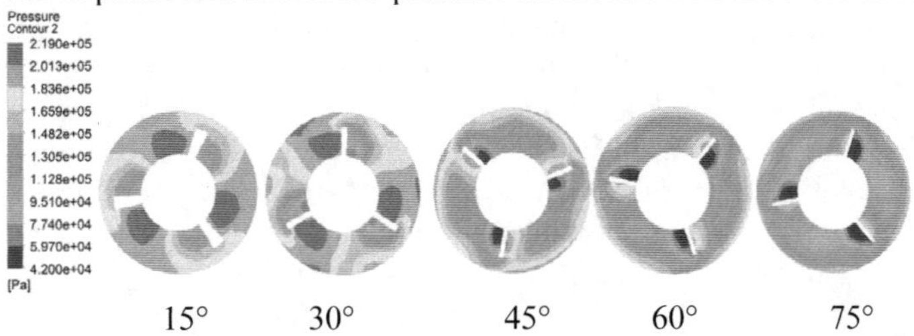

Figure 3 Pressure clouds of this section at different impeller blade helix angles

Figure 4 shows the velocity cloud at the cross-section of the impeller blades with different helix angles. As can be seen from the figure, the impeller inlet near the upper cover has a local high-speed zone, which is caused by the acceleration of fluid into the impeller through its rotation. In the impeller spiral angle of 15 degrees to 75 degrees, the inlet area of the flow rate gradually increased. This is because the spiral angle becomes larger, the width of the inlet flow channel becomes larger, the speed rises, and the maximum speed reaches 19.8 m/s. With the increase in the value of the impeller spiral angle, the blade pressure near the surface of the high-speed area is reduced. The suction surface near the low-speed area is reduced. This is because the impeller blade spiral angle is larger, and the relative length of the blade is smaller. The blade on the fluid flow restraint effect is weakened. The blade surface of the fluid is easy to flow off. The flow field also becomes relatively unstable, and fluid off-flow loss increased. Velocity distribution in the pressure surface and suction surface becomes uniform gradually. In this example, the reduction of the helix angle is more significant to improve the phenomenon of decoupling in the impeller.

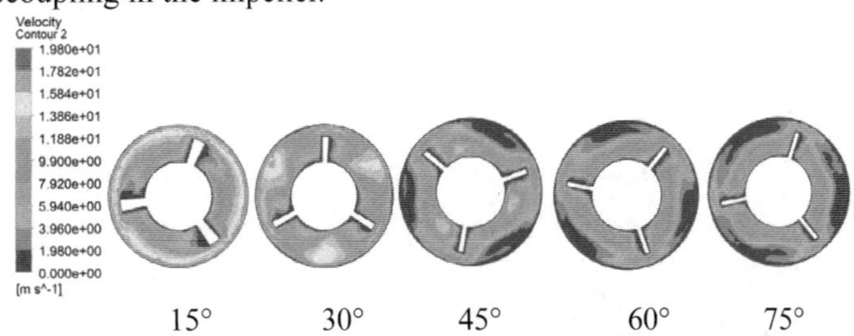

Figure 4 Velocity clouds of this section at different impeller blade helix angles

1) In summary, the impeller spiral angle selection has a crucial impact on the impeller's internal flow field. The spiral angle value, which is selected, can significantly improve the phenomenon of deliquescence. As the deep-sea mining slurry pump needs to take into account the particle flow and pump shutdown particle backflow capacity, it can not be achieved by changing the number of impeller blades and guide vane number. The impeller blade spiral angle can not be too small. Otherwise, the particles will gradually accumulate in the impeller guide vane domain, resulting in pump blockage. Therefore, the impeller blade helix angle of 30° is obtained as the best model by combining the factors of hydraulic performance and particle passage capacity.

3.2 Effect of rotational speed on the flow characteristics of particles in the slurry pump

Under the working flow rate Q_v=420 m^3/h and rotational speed n=1450 r/min, the backflow of particle cluster with particle size d=20 mm in the single-stage pump was simulated numerically. The effect of particle motion on the flow field in the impeller and guide vanes at different speed conditions is studied. Model1 is n=1200 r/min, model2 is n=1450 r/min, and model3 is n=1700 r/min.

3.2.1 Flow field characteristics inside the impeller

Figure 5(a) shows the velocity change curve in the impeller runner at different rotational speeds. The particles in the impeller runner at different rotational speeds show a similar upward trend, with the maximum velocity at the peak for model 3, the second for model 2, and the smallest for model 1. And the outlet velocity increases a lot relative to the inlet velocity. The impeller overflow area of the slurry pump is gradually increasing, while the medium flow velocity in the flow channel is gradually decreasing, causing large velocity fluctuations, local flow diffusion, and flow channel geometry. Figure 2 points out that for the vane inlet position, due to the vane crowding effect, the flow velocity increases. It is 2 points after the speed suddenly decreased because of the diffusion of the flow channel. It is 3 points after the point speed increases because of the flow separation and low-speed area.

Figure 5(b) shows the static pressure distribution of the fluid in the impeller runner under three-speed conditions. The overall pressure in the impeller region shows a “peak-shaped” trend, with the highest pressure at the peak of model 1, followed by model 2, and the smallest at model 3. The main reason for the pressure drop is the increase in flow velocity. The inflection point appears at 6 points, not only with the exit cross-sectional area diffusion but also with the vortex in the flow channel. The vortex caused by the unstable flow to a certain extent reduces the efficiency of the flow. The combined effect of slip is part of the kinetic energy into the fluid energy. The above shows that as the rotational speed increases, the separation of the boundary layer at the impeller flow path is promoted, which weakens the particle over-flow capacity and increases the hydraulic loss.

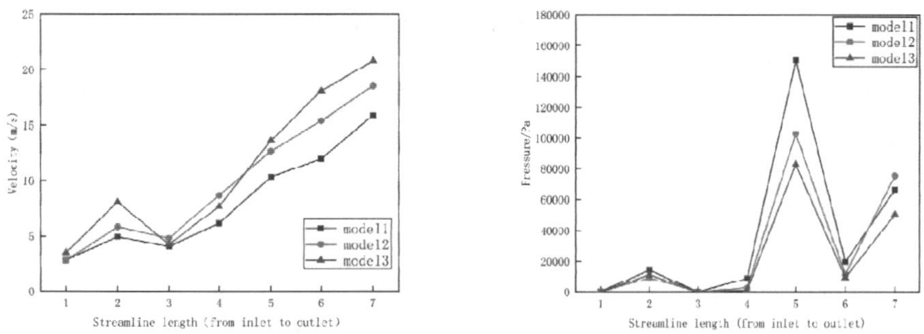

(a) Velocity variation in the impeller runner (b) Pressure variation in the impeller runner

Figure 5 Velocity and a pressure change of impeller runner

3.2.2 Flow field characteristics inside the guide vane

Figure 6(a) shows the velocity characteristics of the flow field inside the guide vane, and the overall flow velocity shows a decreasing trend. Compared with the continuous work of the blades in the impeller, the fluid in the guide vane is a process of mutual transformation of static and dynamic pressure with momentum moment conservation. The energy change characteristics of the medium in the guide vane is a process of decreasing speed and increasing pressure. The difference in rotational speed causes the change of speed drop. The particle flow field velocity of model 1 decreases more slowly. Model 2 is the second. The particle flow velocity in model 3 decreases the fastest, and the export velocity decreases a lot relative to the inlet velocity.

The pressure variation curves in the direction of the flow line inside the guide lobe are shown in Figure 6(b). The overall pressure inside the guide lobe rises slowly and then drops steeply. Model 1 has a significantly smaller trend than model 2 and model 3, and a peak occurs at point 3. The pressure of model 1 is the smallest, model 2 is larger, and model 3 is the largest. It reveals that as the speed increases, the impact at the inlet of the guide vane flow channel is more violent. The flow in this region is more chaotic. The fluid flow separation at the inlet is suppressed, the overflow capacity increases and the hydraulic loss is weakened.

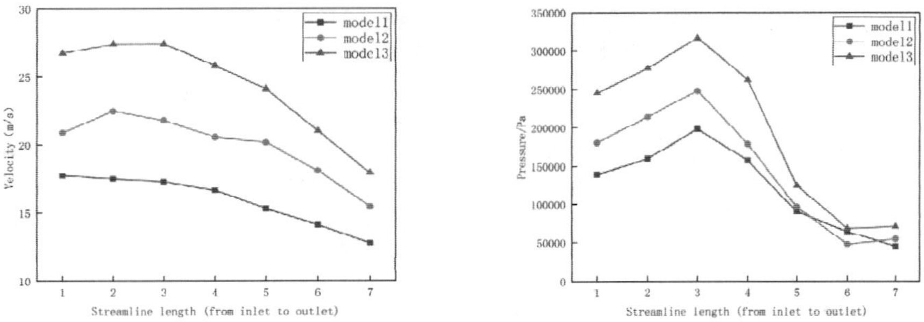

(a) Velocity variation in the guide vane flow path (b) Pressure variation in the guide vane flow path

Figure 6 Velocity and pressure changes in guide vane passage

3.3 Collision characteristics

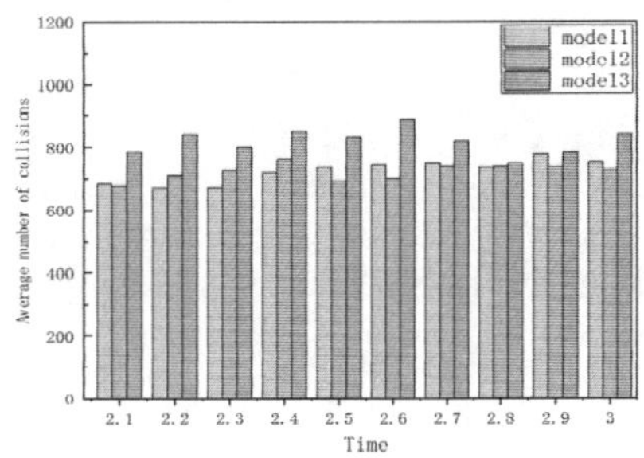

Figure 7 Average number of collisions between particles and blades at different time periods under three working conditions

Figure 7 shows the average number of collisions between particles and vanes at different periods under the three working conditions. Due to the different rotational speeds of the slurry pump, the particles move differently in the flow field, which leads to significant differences in the average number of collisions between the particles and the vanes at different times. But all three conditions show a trend of periodic fluctuations. The fluctuation range of model 3 is the largest, and the fluctuation range of model 1 and model 2 is not much different. Both are less than the fluctuation of model 3. This is because as the pump speed rises, the particles experience increased acceleration in the impeller and the likelihood and speed of impact between the particles and the impeller overflow components decrease substantially. The impeller components' collision probability and collision velocity increase, lateralizing the impeller blades' increased wear rate with higher speed. The particles flow into the space guide vane area through the impeller outlet and rotate out at a high rate. The particles flow into the space guide vane area through the impeller outlet and rotate out at a high rate. The higher the rotational speed, the higher the particle implication velocity at the impeller outlet and the higher the collision velocity with the guide vane overflow parts is. As a result, the particle wear rate on the guide vane blade also rises. The conclusion is that the impeller and the guiding vane overflow sections both experience large increases in wear rate as speed rises, which is consistent with reality. The speed can be appropriately decreased, or the impeller blade and guide vane blade should be made of higher wear-resistant materials.

4. Conclusion

In this paper, a new type of slurry pump conveying system with changing vane spiral angle is numerically simulated, and the study has the following conclusions.

1) The spiral angle value selected smaller can significantly improve the phenomenon of deliquescence. Considering the particle flow and pump shutdown particle backflow capacity, the impeller blade spiral angle can not be selected as too small. Otherwise, the pump will be blocked. Therefore, the impeller blade helix angle of 30° is obtained as the best model by combining the factors of hydraulic performance and particle passage capacity.

2) As the speed increases, the separation of the boundary layer at the impeller runner is promoted, weakening the particle overflow capacity and increasing the hydraulic loss. The impact at the inlet of the guide vane runner is more violent, the fluid flow separation at the inlet is suppressed, the overflow capacity is increased, and the hydraulic loss is weakened.

3) As the speed increases, there is a more average number of collisions between particles and blades in each time period. The collision probability and collision velocity between particles and impeller and space guide vane overflow components become larger, thus laterally reflecting that the larger the speed, the greater the wear rate of particles on blades is.

References

[1] ZOU Weisheng, HUANG Jiazhen. Deep-sea mining of oceanic manganese nodules with ore-raising technology[J]. Mining and Metallurgical Engineering, 2006, 26(3): 1-5.
[2] Cai Baoyuan. The two-phase flow theory of centrifugal pumps and its design principle[J]. Science Bulletin, 1983, 8: 56-71.
[3] Walker C I. The erect of impeller geometry on the performance of centrifugal slurry pumps[C]/4th International Conference on Bulk Materials Handling and Transportation (Symposium on Freight Pipelines). Wollongong, Australia, 1992: 97-101.
[4] He XJ. Working principle and design method of the slurry pump[J]. Fluid Machinery, 1994, 22(9): 17-22.
[5] Zangeneh M, Goto A, Harada H. On the design criteria for suppression of Secondary flows in centrifugal and mixed flow impellers[J]. Journal of turbomachinery, 1998, 120(4): 723-735.
[6] Zhao Jingting, Zhao Zhenhai. Movement of solid particles in the flow channel of centrifugal pumps[J]. Water Pump Technology,1990(1): 1-6.
[7] Dai Jiang, Wu Yulin, Sun Zixiang, etc.. Calculation of solid-liquid two-phase turbulent flow in the impeller of the centrifugal pump[J]. Journal of Engineering Thermophysics, 1996, 17(1): 46-49.
[8] Xu Hongyuan. Particle imaging velocimetry (PIV) analysis and research of solid particle velocity field in centrifugal pump flow channel[J]. Journal of Agricultural Engineering, 1998, 14(3): 112-116.
[9] Wu B, Yan H Z, Xu H L, et al. Numerical simulation of solid-phase particle erosion characteristics in slurry pumps[J]. Journal of Central South University (Natural Science Edition), 2012, 43(1):124-129.
[10] L. Yuanwen, L. Shaojun, H. Xiaozhou, Research on reflux in deep-sea mining pump based on DEM-CFD, Marine Georesources and Geotechnology. 38 (2020) 744-75.

Sound Quality Improvement of Dynamic Loudspeaker by Material Selection of Diaphragm

Gengyue Dong

Department of Materials, Imperial College London, London SW7 2AZ, UK

gengyue.dong19@imperial.ac.uk

Abstract: In terms of sound quality, mechanical stability, and cost, the moving coil speaker is exceptional to the moving iron and electrostatic speaker. The diaphragm in the speaker is crucial in reproducing the sound wave from the electrical signal input. The materials of the diaphragm should be both rigid and lightweight. However, there are few quantitative studies on the selection of diaphragm materials. Connecting the equation of mass and the deflection of a circular plate yields a performance index that maximizes Young's Modulus and minimizes the density. A line graph in logarithmic can be generated from the index to select materials from various categories in the Ashby plot provided by the Granta CES Edupack software. The chosen materials would be simulated in COMSOL and their sound pressure level plots would be analyzed based on the numerical solution. The Ashby plot and the numerical solutions by COMSOL have indicated that paper is the best material among the selected materials with the most stable frequency response in the mid-range frequency range, a high-performance index, and a low price per unit volume.

1. Introduction

The first speaker invented in 1861 was the loudspeaker of the first telephone. It has been used in various electronic devices, which need speakers for sound output daily. A speaker always comprises two main parts: the electrical circuit and the transducer. Signals are input from the terminals of the circuit and go through an amplifier [1], which ensures the signal is large enough to give a response to the transducer. Lastly, the amplified signal was transferred to the sound wave by the transducer. The first transducer is a moving iron driver, which transfers the electrical signal to a fixed coil, and moves a permanent magnet by the induced Lorentz force. Therefore, the moving magnet vibrates the diaphragm or the cone and produces the sound wave. However, the early moving-iron speaker suffers from poor low-frequency (20-300Hz) and high-frequency (8-20kHz) responses and a strong ringing effect due to the undamped oscillation [2]. Therefore, it was soon replaced by the moving coil or dynamic driver in the 1920s.

The dynamic driver comprises three main parts: a permanent magnet, a voice coil, and a diaphragm. Instead of a fixed coil as the moving-iron transducer, the voice coil oscillates under the coupling effect of a fixed permanent magnet and a time-varying current passing through the coil. The diaphragm attached to the moving coil vibrates and reproduces acoustical signals. The cone suspension also holds the diaphragm to the support and the spider connects the diaphragm to the voice coil, as shown in Figure 1.

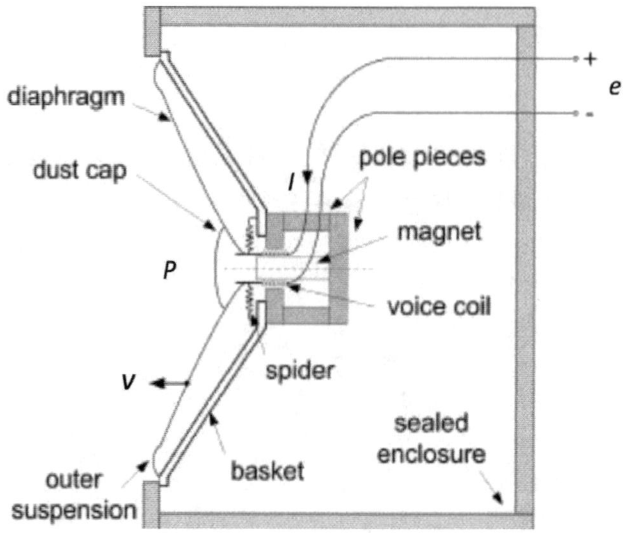

Figure 1. According to the image adapted from [3], the cross-sectional layout of a moving-coil loudspeaker can be seen: The voice coil is connected to a terminal with Voltage e and Generating Current I. The Resultant Pressure P vibrates the diaphragm and generates sound waves.

On the other hand, the electrostatic driver does not rely on magnets to generate acoustical pressures; instead, it consists of two metallic grids coupled to a current source and sandwiching a diaphragm as seen in Figure 2.

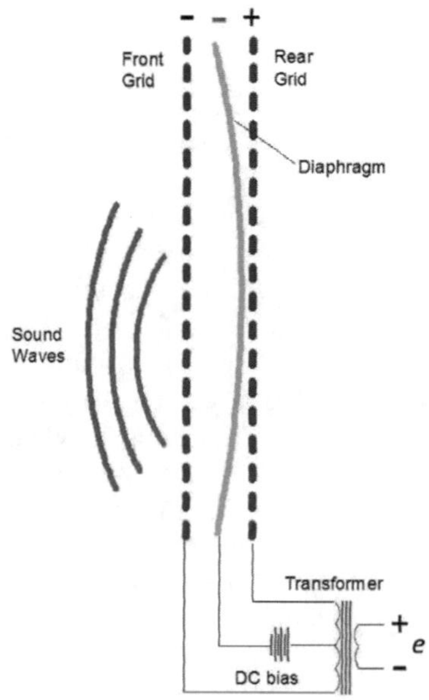

Figure 2. According to the image adapted from [4], a schematic representation of the electrostatic speaker can be seen: A Signal e was sent to both grids, and as the bias changed, the diaphragm was attracted and repelled, producing a sound wave.

When an electrical signal is delivered to the electrodes, an electric field is generated, causing the diaphragm to be attracted to one of the plates. The continual change in the electric field causes the diaphragm to vibrate, compressing the air through the perforated gaps and producing a sound wave. The diaphragm is designed to be as lightweight and flexible as possible. Due to the inherent strength of electrostatic drivers, electrostatic speakers provide superior transient response performance compared to dynamic drivers. In place of the moving-coil transducer's woofer, midrange driver, and tweeter, the electrostatic driver employs a single driver that spans the whole frequency range. It also has a steadier sound pressure level (SPL) than dynamic speakers and covers the audible frequency band (20 Hz to 2 kHz). Nonetheless, the electrostatic driver is more costly than the moving coil driver owing to more intricate production. In addition, the diaphragm of the dynamic driver is much more damage resistant than that of the electrostatic driver, which is more vulnerable.

Therefore, this paper will cover the material choices of the dynamic transducer. The emphasis would be on the cone, and its suspension or coating, while the selection criterion would be on mechanical properties and prices.

2. Principle of the Dynamic Loudspeaker

This paragraph follows a section title so it should not be indented. The moving coil loudspeaker comprises three parts: the electrical system, the mechanical system, and the acoustical system, as shown in Figure 3. The model simplifies each part of the loudspeaker. In the electrical system, the amplifier is represented as an equivalent resistorR_G while the voice coil is considered a combination of a resistor R_{EC} and inductor L_C. The Lorentz force induced by the voice coil is transferred from the input signal, in a ratio Bl according to Equation (1).

$$F = Bl \cdot I \tag{1}$$

where I is the current passing through the voice coil in Unit A, and bl is the force factor transferring current to a force in the unit, N/A.

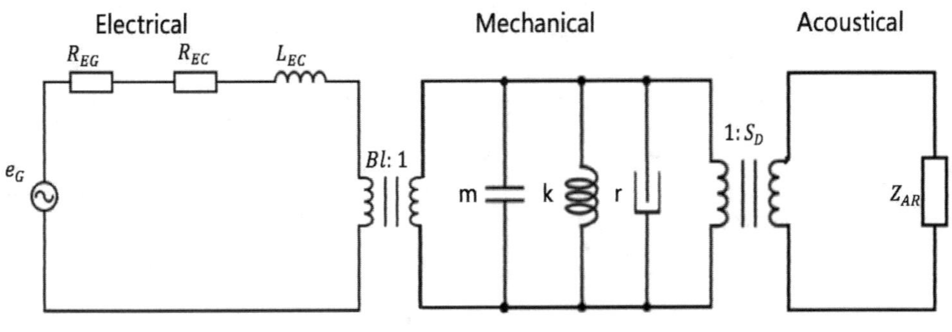

Figure 3. A schematic representation of the loudspeaker.

On the other hand, the mechanical system simplifies the transducer into the moving mass of the diaphragm and the voice coil, m, Hooke's constant, k, and the damping coefficient, r, of the diaphragm suspension. The force exerted on the diaphragm transfers to the variation of air pressure, which can be evaluated by force divided by the area of the diaphragm,S_D.

Lastly, the acoustical system contains the loudspeaker mounted on the baffle which radiates the sound wave to the air. The system is simplified to an acoustical impedance Z_{AR}, which is depicted by Equation (2) relating to the velocity of the air particles [5].

$$p = Z_{AR} v \tag{2}$$

where p is the pressure applied to the air particles, and v is the velocity of air particles.

2.1. Voltage driving speaker vs. current driving speaker

The input signal can be driven by controlling the applied voltage or applied current [6]. The voltage driving speaker would result in the magnitude of the sound wave depending on the electrical and mechanical parameters, $R_{EG}, R_{EC}, L_{EC}, m, k, and\ r$.

On the other hand, the current driving type creates a sound wave that only depends on the mechanical parameters $m, k, and\ r$.

2.1.1. Voltage driving mode. For the simplest case, a sinusoidal voltage signal with mono frequency, $e_G(t)$, is assumed to be delivered to the voice coil. The force can be written in terms of $R_{EG}, R_{EC}, and\ L_{EC}$ by Solving (3).

$$\begin{cases}(R_{EG}+R_{EC})+L_{EC}\dfrac{dI}{dt}=e_G(t)=V_0e^{i\omega t}\\ t=0\,s, I=0\,A\end{cases} \tag{3}$$

where $e_G(t) = V_0e^{i\omega t}$, V_0 is the amplitude of voltage, and ω is the angular freqeuncy of the applied voltage.

After the analytical solution of I (t) is found, the force and displacement on the diaphragm can be expressed as Equations (4) and (5) respectively.

$$F = BlI = Re\left\{\frac{BlV_0}{R_{EG}+R_{EC}+i\omega L_{EC}}\left(e^{i\omega t} - e^{-\frac{(R_{EG}+R_{EC})t}{L_{EC}}}\right)\right\} \tag{4}$$

$$u = Re\{e^{\frac{-r}{2m}}\left(Ae^{i\frac{\sqrt{r^2-4mk}}{2m}t}+Be^{-i\frac{\sqrt{r^2-4mk}}{2m}t}\right)+ \\ \frac{Bl}{k-m\omega^2+ir\omega}\cdot\frac{V_0(R_{EG}+R_{EC})}{(R_{EG}+R_{EC})^2+(\omega L_{EC})^2}e^{i\omega t} \\ -\frac{V_0Bl}{\sqrt{(R_{EG}+R_{EC})^2+(\omega L_{EC})^2}}exp(-i\tan^{-1}\frac{\omega L_{EC}}{R_{EG}+R_{EC}}) \\ \cdot\frac{L_{EC}^2}{m(R_{EG}+R_{EC})^2-r(R_{EG}+R_{EC})L_{EC}+kL_{EC}}e^{-\frac{R_{EG}+R_{EC}}{L_{EC}}t}\} \tag{5}$$

where u is the displacement of air particles from the equilibrium position.

2.1.2. Current driving mode. The current driving mode controls the applied current, resulting in a zero phase difference between the signal and force as Equation (8).

$$I = I_0e^{i\omega t} \tag{6}$$

$$F = Re\{BlI_0e^{i\omega t}\} \tag{7}$$

$$u = Re\{e^{\frac{-r}{2m}}\left(Ae^{i\frac{\sqrt{r^2-4mk}}{2m}t}+Be^{-i\frac{\sqrt{r^2-4mk}}{2m}t}\right)+\frac{BlI_0}{k-m\omega^2+ir\omega}e^{i\omega t}\} \tag{8}$$

Therefore, the Force (5) and the Displacement (6) of the transducer would only depend on the applied current signal and the parameters in the mechanical system. This circumstance makes it easier to examine the influence of diaphragm materials on sound quality. Therefore, it is assumed that the speaker is operated by the current driving mode.

3. The standard of material selections of diaphragms

First of all, the diaphragm should be light as possible. When the signal current goes through the voice coil, an oscillating force acts on the diaphragm. Then, the displacement of the diaphragm is proportional to $\frac{BlI_0e^{i\omega t}}{k-m\omega^2+ir\omega}$ which means it is inversely proportional to the mass. Therefore, as the mass is reduced, the time delay between the moments of force applied and the sound wave radiated is reduced, which makes the input signal more resolved.

The diaphragm should also eliminate the ringing effect due to the resonance response. The ringing effect gives the 'hum/hiss' like noise to ruin the experience of clients [7]. This happens at the resonance frequency, ω_{res}, which makes the amplitude of the displacement completely imaginary, and the equation of the frequency is shown in Equation (9).

$$\omega_{res} = \sqrt{\frac{k}{m}} \sim \frac{Young's\ modulus}{density} \times geometric\ factor \tag{9}$$

where Hooke's constant k can be related to Young's modulus times a geometric factor. For instance, k of a beam equal to Young's modulus times area is divided by the length.

The first step in removing the ringing effect is to select the diaphragm material with a great damping coefficient. In an ideal case, the damping ratio equals 1, where the ringing effect disappears most quickly because the system is in its critical damping state. Apart from the material selection of diaphragms, critical damping could be also achieved by coating the cone with viscoelastic polymers. The damping coefficient might also be determined by selecting an appropriate cone suspension and spider. However, the selection of coating would not be addressed in this study.

The resonance frequency also limits the bass response of the driver. A previous study of the dynamic driver's sound pressure level (SPL) [6] has pointed out that the pressure level drops significantly as the frequency is lowered below the resonance frequency as shown in Figure 4. The resonance behavior results in a low volume within the bass range. This may be resolved by decreasing the resonance frequency to the lower bound of the audible frequency, around 20 Hz. This may be accomplished by decreasing Young's modulus or increasing the material's density. However, it has been noted that the mass should be as small as possible, thus it is preferred to merely reduce the stiffness to lower the resonance frequency.

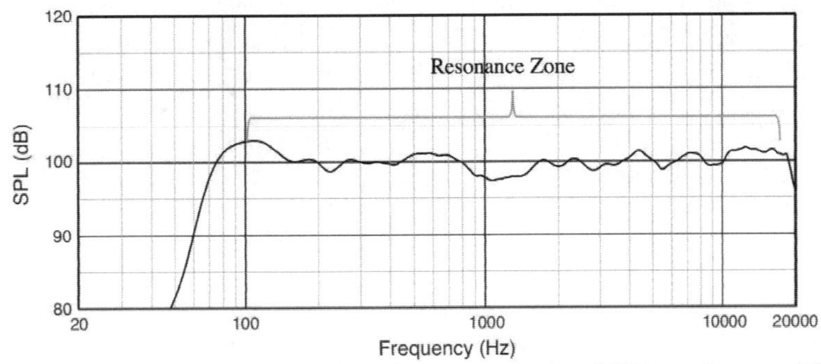

Figure 4. According to the image adapted from [6], an example of SPL can be seen: The resonance zone was labeled with the green bracket and the resonance frequency locates on the left end of the region.

The vibration modes of the cone also affect the sound quality in terms of the SPL. The diaphragm is assumed to be a rigid piston in the previous calculation. The real diaphragm material could be deflected under a high frequency of vibration. Therefore, the finite stiffness vibrates the diaphragm into various modes. If the diaphragm could be thought of as a 2D-vibrating membrane, vibration modes can be predicted by the wave Equation (10).

$$\frac{\partial^2 u(t,r,\theta)}{\partial t^2} = c^2 \nabla^2 u(t,r,\theta) \tag{10}$$

e general solution of (10) can be obtained by separating variables $u(t,r,\theta) = G(t)P(r)Q(\theta)$ to give Solution (11).

$$u_{nm}^{c} = \left(A_{nm}\cos\left(\frac{\beta_{nm}ct}{R}\right) + B_{nm}\sin\left(\frac{\beta_{nm}ct}{R}\right)\right) J_\beta\left(\frac{\beta_{nm}r}{R}\right)\cos(n\theta)$$

$$u_{nm}^{s} = \left(A_{nm}\cos\left(\frac{\beta_{nm}ct}{R}\right) + B_{nm}\sin\left(\frac{\beta_{nm}ct}{R}\right)\right) J_\beta\left(\frac{\beta_{nm}r}{R}\right)\sin(n\theta) \tag{11}$$

The solution in (12) shows quantized modes under different frequencies, f_{nm}.

$$f_{nm} = \frac{\beta_{nm}c}{2\pi R} = \frac{\beta_{nm}}{2\pi R}\sqrt{\frac{E}{\rho}} \tag{12}$$

Each vibrational mode of the membrane can be represented by the suffix (n, m) from Equation (12). Different forms of variation and the cone breakdown would result in the radiation of different regions on the membrane out of phase, hence resulting in an uneven frequency response on SPL plots. Therefore, the $(0,1)$ vibration mode is required to keep the vibration as a rigid piston to radiate a more uniform volume of sound. Therefore, the f_{nm} should be as high as possible, requiring the stiffness to be maximized and the density to be minimized.

Additionally, the price of the moving driver should be as low as possible, as the materials chosen should be for commercial use to compete with other loudspeakers.

Therefore, under the constraint of lower costs, a good diaphragm should always be lighter and stiffer to improve the signal resolution and have a uniform sensitivity of output signals. However, the material in nature can hardly have low mass and great stiffness simultaneously, which means there should be some trade-offs when selecting the material.

Meanwhile, a greater stiffness would give a weak response to the bass regime. Therefore, separate speakers are needed in terms of low, mid-range, and high frequency respectively to cover the whole range of audio frequency.

4. Selecting diaphragm materials and simulation

4.1. Maximize the material index

The material indices can be derived by minimizing the ratio between the deflection of the diaphragm and the force exerted on it. The diaphragm deflection could be approximated to a circular membrane with Radius R and Thickness h. The deflection of it can be expressed as (13) [8].

$$W = \frac{3F\,(r^2 - R^2)^2\,(1 - v^2)}{16Eh^3} \tag{13}$$

Since the deflection of each point on the diaphragm varies continuously, the maximum value of the deflection, W_{max} can be written in (14):

$$W_{max} = \frac{3(1 - v^2)FR^4}{16Eh^2} \tag{14}$$

Meanwhile, the Thickness h of the membrane could be written as the function of deflection, W_{max} shown in (15).

$$h = \frac{1}{E^{\frac{1}{3}}}\left(\frac{3(1 - v^2)FR^4}{16EW_{max}}\right) \tag{15}$$

The mass of the membrane shown can be correlated to the deflection by a substitute (15) into (16).

$$m = \rho V = \rho\pi R^2 h \tag{16}$$

Therefore, the mass could be written as (11) which consists of two parts, the geometrical and functional factors in (12), and the material index or performance index in (17).

$$m = \left(\frac{3(1 - v^2)FR^4}{16W_{max}}\right)^{\frac{1}{3}} \cdot \left(\frac{\rho}{E^{\frac{1}{3}}}\right) \tag{17}$$

$$\left(\frac{3(1 - v^2)FR^4}{16W_{max}}\right)^{\frac{1}{3}} \ldots geometrical\ and\ functional\ factors \tag{18}$$

$$\left(\frac{\rho}{E^{\frac{1}{3}}}\right) \ldots material\ factors \tag{19}$$

The material index term or performance index P shown in (20) should be maximized to find the optimal diaphragm material with the strongest stiffness and lowest mass.

$$P = \frac{E^{\frac{1}{3}}}{\rho} \tag{20}$$

Therefore, the Ashby plot should be utilized to determine the optimal material selection in Granta CES Edu Pack. To filter materials, the software generates an Ashby plot for various materials, including metal, polymer, and composites. On a logarithmic scale, an Ashby plot of Young's modulus versus density could identify the materials with the greatest P. A straight line of the material index P on the logarithmic scale shown in Equation (21) should also be plotted to impose a constraint for filtering the materials.

$$logE = 3log\rho + 3logP \tag{21}$$

A price constraint should also be imposed on the selected material to filter out materials below $10GBP/Kg$, which makes the speaker cheap enough to be commercialized.

4.2. COMSOL

The COMSOL dynamic driver model downloaded from the COMSOL program gallery was used for testing the diaphragm material performance. All the crucial parameters of geometry keep constant throughout the whole study, and the cross-section of the model is shown in Figure 5.

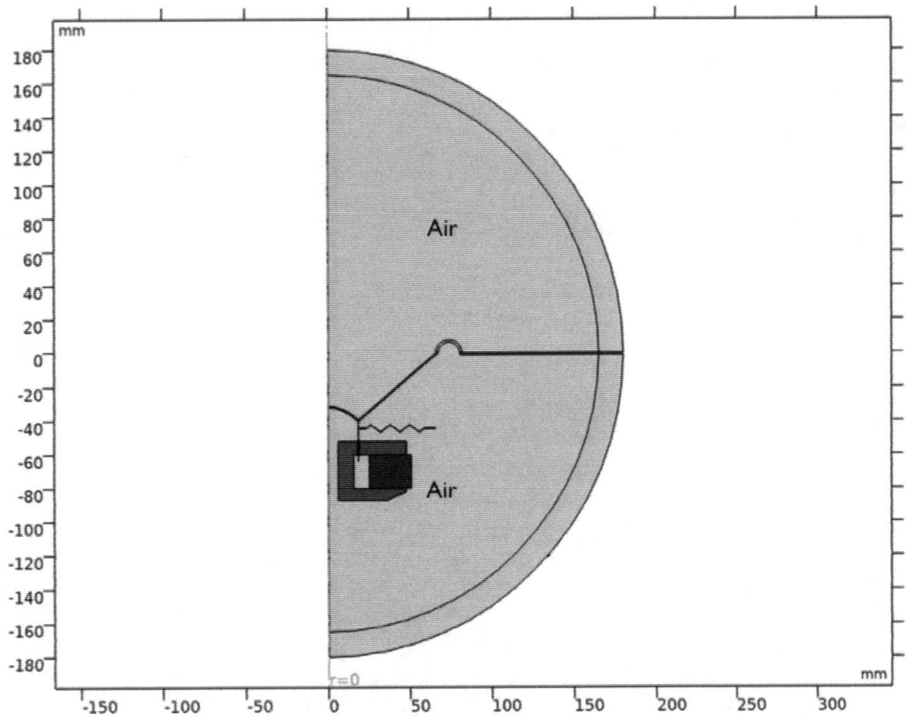

Figure 5. Cross-section of the model and the surrounding atmosphere.

The diaphragm material was replaced with the Granta-selected materials. All materials here are simplified to the object with only four parameters: density, stiffness, Poisson ratio, and the damping coefficient.

From 20Hz to 20kHz, COMSOL would compute the mechanical response of the cone in the frequency domain, and the result would generate SPL plots. The value of SPL in COMSOL follows the definition in (22).

$$SPL = 10\, lg\left(\frac{P_{rms}^2}{P_{ref}^2}\right) \tag{22}$$

In addition, all dimension parameters were held constant throughout the entire simulation series. Therefore, the change of SPL with the component size is not considered in this work.

5. Result and Discussion

From the three categories of materials—metal, polymer, and composites—the top two of each category are chosen. Ceramics are not considered in this study because no price of ceramic falls within the commercial constraints.

5.1. Metals

As shown in Figure 6, aluminum and magnesium alloys have the highest material index below £5/kg. Since density and stiffness vary among a range of their alloys, Table 1 only lists their average material properties. As shown in Table 1, Mg has a higher performance index than Al, and its cost per unit volume is lower. Therefore, magnesium is the material of choice for metallic diaphragms, which are both lightweight and vibrate in a piston mode over a broad frequency range. To verify the conclusion, both are tested in the model to produce a frequency response curve.

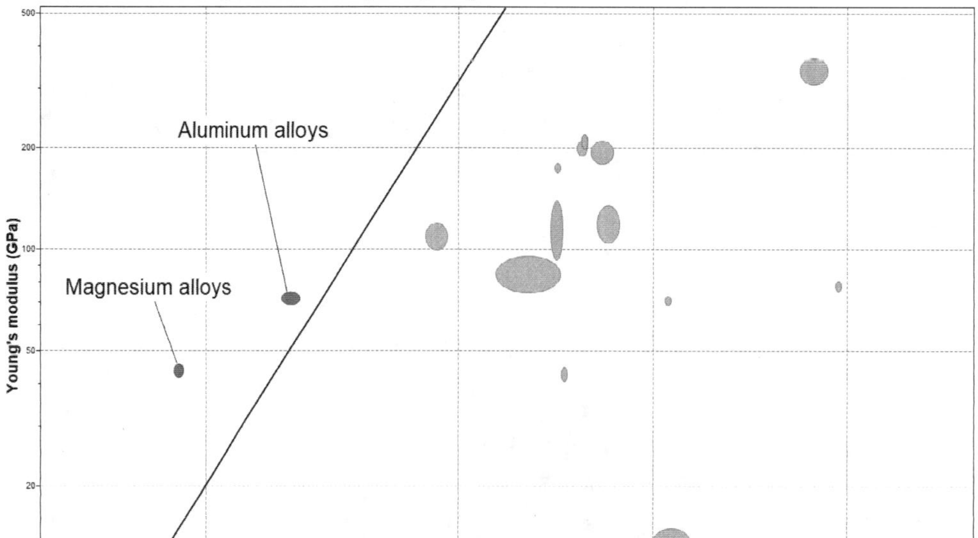

Figure 6. The Ashy plot includes all metals in the database of Granta: The red bubbles above the line represent the metals with the highest performance indices.

Table 1. Parameters of selected alloys.

	Density(kg/m^3)	Young's Modulus(GPa)	Performance Index	Poisson Ratio	Loss Coefficient	Price (GBP/cm^3)
Al	2725	72	1.53	0.33	0.0235	0.005
Mg	1810	44	2.00	0.33	0.0155	0.004

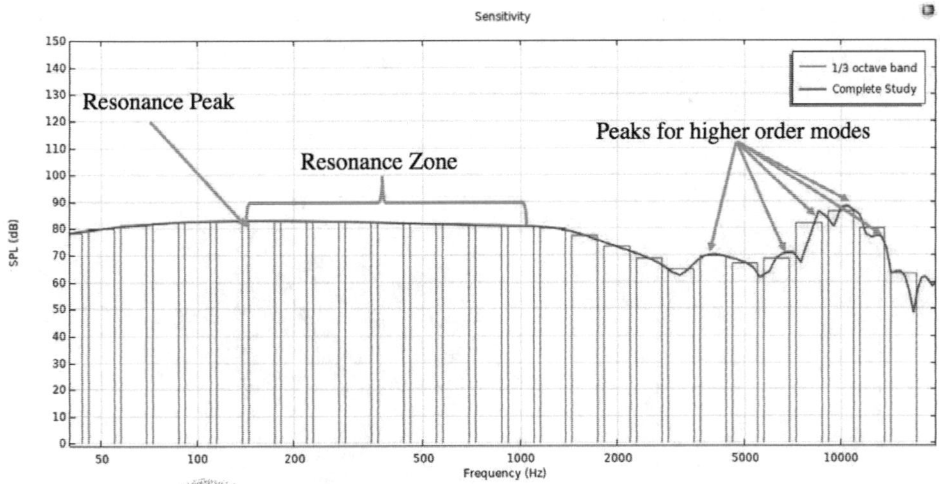

Figure 7. SPL plot of Aluminum.

The response of both alloys for low and mid-range frequencies from 100Hz to 1500Hz stays relatively stable. This region is called the resonance zone [9], which should have been between two resonance spikes. The frequency of spikes labeled in Figure 7 and Figure 8 is proportional to the performance index. The peaks reflecting the vibrational modes of the cone are also labeled in Figures 7 and 8. Meanwhile, the pressure level of Mg is slightly greater than Al over the resonance zone due to the lower dampening ratio, which results in a higher volume of sound for the Mg diaphragm.

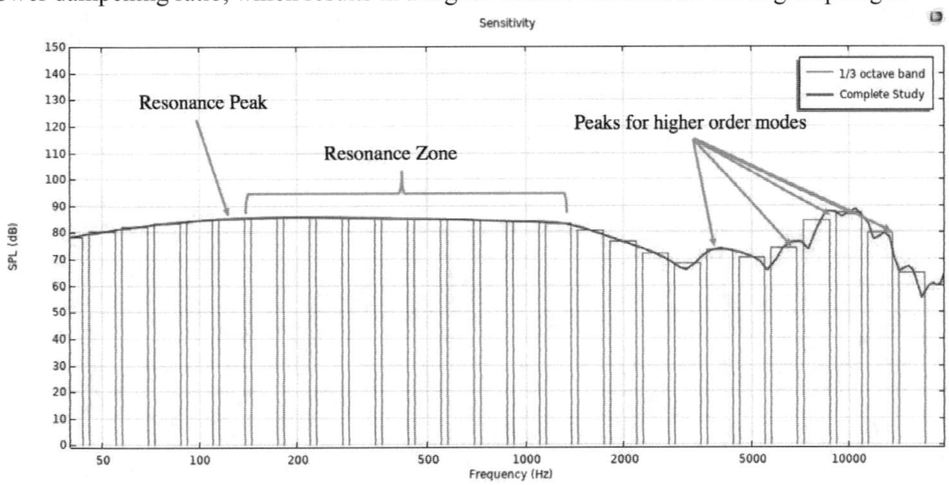

Figure 8. SPL of magnesium.

5.2. *Polymers*

Table 2. Parameters of selected polymers.

	Density (kg/m^3)	Young's Modulus (Gpa)	Performance Index ($Pa^{1/3}/kgm^{-3}$)	Price (GBP/cm^3)	Poisson Ratio	Loss Coefficient
Polyester	1220	3.24	1.21	0.002	0.392	0.0101035
Polystyrene	1045	3.25	1.42	0.003	0.395	0.01485

Polyester and polystyrene are the top two polymers extracted with the greatest performance indices from the Ashby plot in Figure 9. Although polystyrene has a greater performance index, the price of a polyester diaphragm would be 30% lower than the polystyrene one according to Table 2.

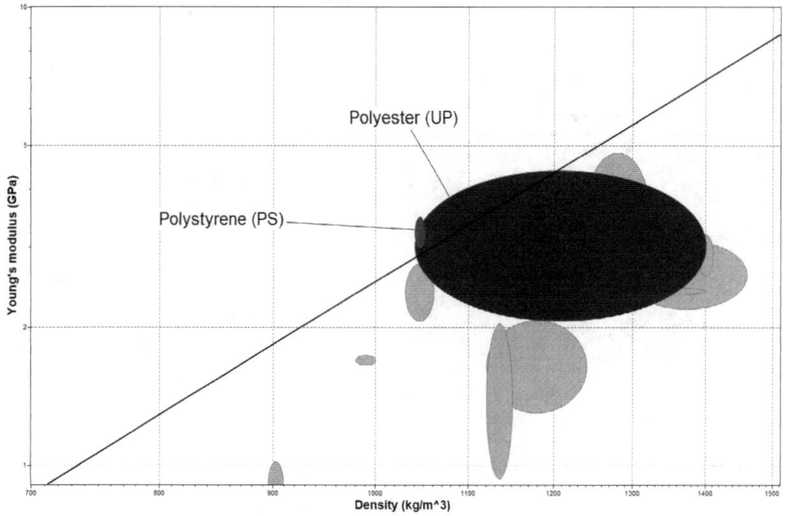

Figure 9. Ashby plot of polymers: The small blue bubble represents polystyrene with a higher performance index and smaller fluctuations in stiffness and density. On the other hand, both density and Young's modulus vary widely for polyester, the larger bubble.

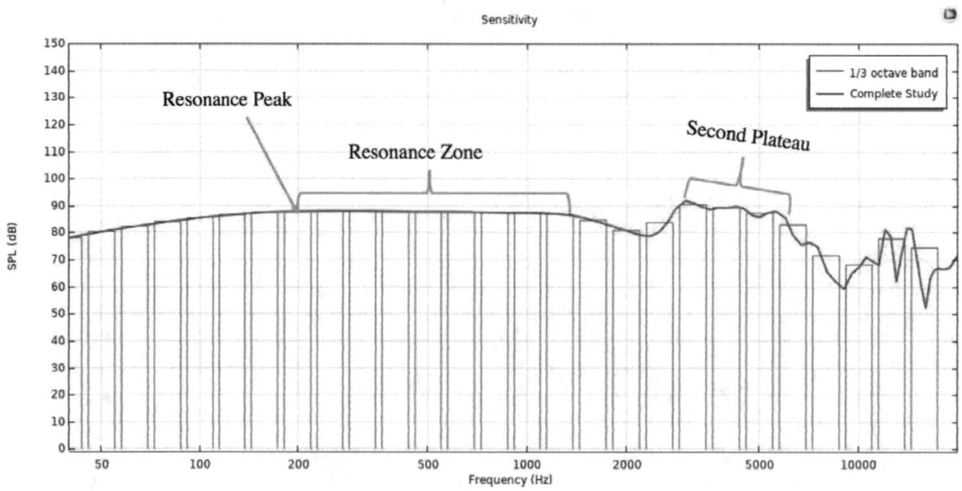

Figure 10. SPL of polyester.

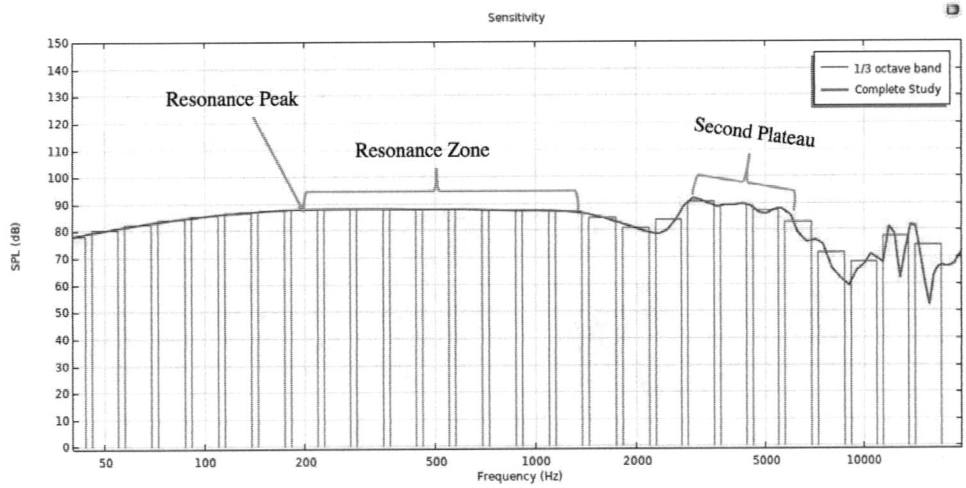

Figure 11. SPL of polystyrene.

Polyester and polystyrene are the top two polymers extracted with the greatest performance indices from the Ashby plot in Figure 9. Although polystyrene has a greater performance index, the price of a polyester diaphragm would be 30% lower than the one in Table 2.

The resonance frequency of both polymers is greater than the metal. The resonance peak of the polyester labeled in Figure 10 is at 200 Hz, while that of polystyrene locates at 250 Hz in Figure 11. After a small dip from 1500 to 2500 Hz, a second plateau would be observed in the SPL curves for both polymers between 2200 and 6000 Hz. However, the second plateau is not as flat as the resonance zone due to higher-order vibrational modes.

5.3. *Composite*

Natural materials with low density and high rigidity are difficult to accommodate with a single material type. Therefore, composites, combining materials with high rigidity, and low density, are chosen as diaphragms. In Figure 12, the Ashby plots select two types of composites: paper and Al/SiC.

Table 3. Parameters of selected composites.

	Density (kg/m^3)	Young's Modulus (GPA)	Performance Index ((Pa$^{1/3}$/kgm^{-3})	Price (GBP/cm^3)	Poisson Ratio	Loss Coefficient
Al/SiC	2780	90.5	1.61	0.013	0.3	0.005
Paper	925	5.95	1.96	0.001	0.375	0.125

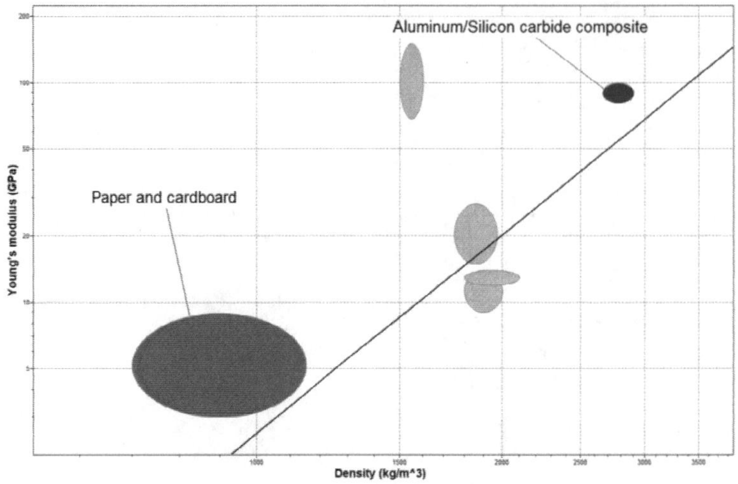

Figure 12. Ashby plot of composites: The large green bubbles represent paper and cardboard, which are low stiffness and lightweight. The brown one, in contrast, represents an Al/SiC composite with a high density and Young's modulus.

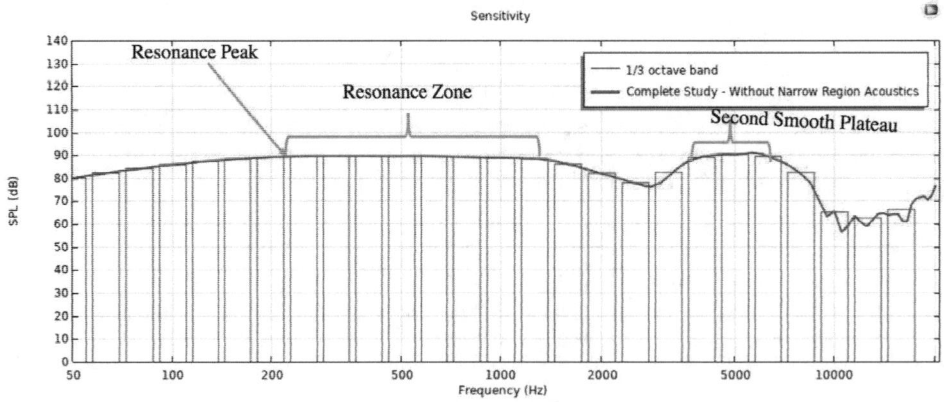

Figure 13. SPL of paper

Paper is produced by pressing multiple cellulose fiber sheets. The performance index displayed in Table 3 is the second highest after magnesium, and the cost is the lowest at 0.001 GBP/cm3. Despite a little dip to 77dB between 1000Hz and 3500Hz, the frequency response stabilizes at 90dB in the midrange from 100 to 7000Hz. The frequency response of a paper cone is very similar to that of a polymer cone, except for a smoother and flatter second plateau, as labeled in Figure 13. The smoother curve compared to the polymer may result from the 10 times damping coefficient of alloys.

Meanwhile, Al/SiC composites with lower performance index are also considered, which comprise the high-stiffness-ceramic particles in the aluminum metal matrix with low density. The frequency response is also nearly identical to that of paper in the woofer and mid-range frequency, but the second smooth plateau is absent from the curve, which makes the paper more preferred to be a composite diaphragm. The SPL curve is not displayed for simplicity in this case.

5.4. *Carbon Nanotubes (CNT)*

Table 4. Properties of CNT.

	Density (kg/m^3)	Young's Modulus (Gpa)	Performance Index	Price (GBP/cm^3)	Poisson Ratio	Loss coefficient
CNT	1385	1315	7.91028524	49.86	0.2445	N.A.

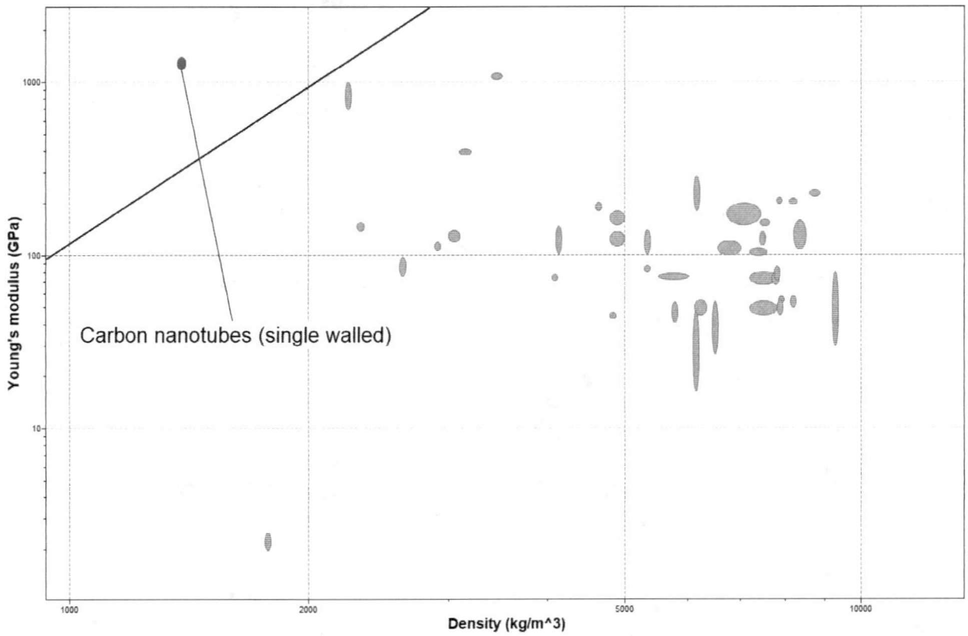

Figure 14. Ashby plot of CNT: Carbon nanotube stands out from all other materials in the database with its low density of 1385kg/m3 and exceptionally high stiffness of 1315 GPa.

Carbon nanotubes (CNT) would surpass all other materials in the database if the cost were no longer considered. The performance line selects CNT in the database including all categories of material shown in Figure 14. The database does not include the resistance ratio, thus the loss coefficient was assumed to be 0.04. The performance index in Table 4 is approximately 8, which is a three to four-fold performance index of the other selected materials owing to its extremely high stiffness and low density.

Theoretically, the SPL should be flatter than the previously listed materials, since the CNT could sustain piston vibration before the frequency of the first vibration mode. Nonetheless, the SPL curve reveals a similar performance as selected materials previously mentioned materials between 100 and 1500 Hz, but an unexpectedly poor performance above 1500 Hz, as depicted in Figure 15.

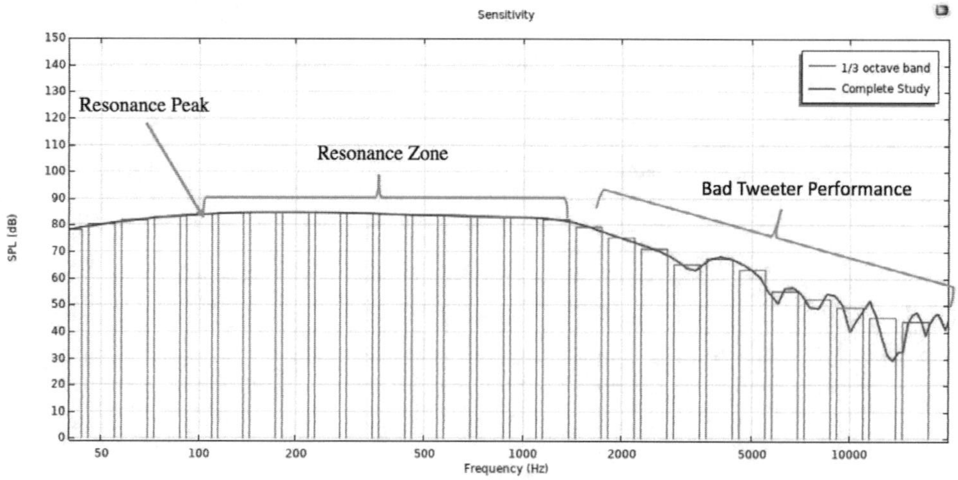

Figure 15. SPL of CNT

6. Conclusion and Further Works

The moving-coil loudspeaker has been the most successful commercialized speaker, and the frequency response is one of the most important indices for determining sound quality to provide clients with an enjoyable listening experience. The material of the diaphragm would have a significant impact on the frequency response. Therefore, this article seeks to identify the material with the broadest and most uniform frequency response in the audible range. Connecting the deflection equation and the circular plate's mass yields the initial performance index. The plot of the index with a limitation of price selects the best materials from all categories.

After the selection on Granta and simulations on COMSOL, the paper is found to be the best material resulting in the most stable frequency response and lowest price. However, none of the materials with the highest material index are particularly sensitive in the tweeter ranges. This may be attributable to the speaker's dimensional effect. However, all simulations in this study are conducted by fixing all dimensions of the modeling. Consequently, a future experiment could be conducted on cone materials of various sizes and shapes to determine the optimal material for a dynamic speaker under the geometrical constraint.

In the meantime, the deflection of the cones is not a circular plate, so it is preferable to derive the analytical form of the cone's deflection to calculate a more precise material index. However, it would be time-consuming to derive one; the numerical method could be used instead.

Reference

[1] Stereo Amplifiers Explained in a Nutshell [Internet]. [cited 2022 Aug 22]. Available from: https://www.lifewire.com/stereo-amps-3135080
[2] The Importance of a Speaker's Resonant Frequency | Pro-Audio-Land Musician News [Internet]. [cited 2022 Aug 29]. Available from: https://proaudioland.com/news/the-importance-of-a-speakers-resonant-frequency/
[3] Boulandet R, Lissek H. Acoustic impedance synthesis at the diaphragm of moving coil loudspeakers using output feedback control. 2011;
[4] Definition of electrostatic speaker | PCMag [Internet]. [cited 2022 Aug 29]. Available from: https://www.pcmag.com/encyclopedia/term/electrostatic-speaker
[5] Norton MP, Drew SJ. VIBRATION GENERATED SOUND | Fundamentals. Encyclopedia of Vibration. 2001 Jan 1;1443–55.

[6] Stutzer E, Pillonnet G, Lemarquand G, Abouchi N. Comparison between a micro speaker's voltage and current driving methods. Applied Acoustics. 2012 Nov 1;73(11):1087–98.
[7] What Causes Speaker Hum & Hiss (How to Eliminate Them Both) – My New Microphone [Internet]. [cited 2022 Aug 17]. Available from: https://mynewmicrophone.com/what-causes-speakers-hum-hiss-how-to-eliminate-them-both/
[8] Elastic Analysis of Circular Plates - Struct Ville [Internet]. [cited 2022 Aug 18]. Available from: https://structville.com/2020/11/elastic-analysis-of-circular-plates.html
[9] Audio System Resonance and Resonant Frequency | Digi Key [Internet]. [cited 2022 Aug 27]. Available from: https://www.digikey.com/en/articles/the-importance-of-resonance-and-resonant-frequency-in-audio-systems

Dual Mesh Technique for Hydroacoustic Prediction of Cavitating Flows around a Three-Dimensional Bullet

Lianjie Yu [1], Weiwen Zhao [1] and Decheng Wan *[1]

[1]Computational Marine Hydrodynamic Lab (CMHL), School of Naval Architecture, Ocean and Civil Engineering, Shanghai Jiao Tong University, Shanghai, China

dcwan@sjtu.edu.cn

Abstract. Nowadays, hydroacoustics is of great importance. A major source of underwater noise is cavitation noise. One of the difficulties in cavitation noise prediction is its complex formation mechanism. In this paper, a cavitating bullet is taken as the research object, and detailed numerical simulation research of the local cavitation for this three-dimensional model is carried out. Ffowcs Williams and Hawkings equation (FW-H) is adopted. The quadrupole term is obtained by the direct volume integration method. Furthermore, dual mesh technique is adopted to achieve fast calculation. The results show that the dual mesh technology reduces the computational complexity of the nonlinear sound pressure significantly. With the appropriate acoustic grid size, the dual mesh method can ensure the same accuracy as the CFD grid. This provides a reliable way for the calculation of nonlinear sound pressure in the cavitation phenomenon.

1. Introduction

Hydrodynamic noise is a type of underwater noise caused by turbulent fluctuation in the flow field. Hydroacoustics has not only become an important indicator for the construction of “green ships” but also affects the survivability and concealment of military vessels. First of all, for “green ships”, hydrodynamic noise interferes with marine life and seriously damage the ecological nature of the marine environment. To this end, International Maritime Organization (IMO) proposed a ship noise specification to ensure that passing ships do not cause damage to marine habitats[1]. Secondly, for the concealment of military ships, sound waves are the most effective signals transmitted by underwater signals. Excessive hydrodynamic noise will reveal the position of the ship, and accurate hydrodynamic noise detection can effectively detect the presence of the enemy.

As a unique phenomenon of the underwater flow fields, cavitation has always been an important source of hydrodynamic noise[2]. There are various taxonomies of cavitation. According to the dynamic characteristics, it is split into wandering cavity, fixed cavity, vortex cavity and vibration cavitation. According to the shape, it is seperated into sheet cavity, mist cavity, tip vortex cavity, and hub vortex cavity. According to the development stage, it is split into critical cavity, local cavity, and super cavity[3]. The acoustic analogy is usually used for computing the cavitation noise. Its basic idea is to divide the noise problem into two parts, the source one and the far-field one. The former is simulated with computational fluid dynamics (CFD) and constructs monopole, dipole, and quadrupole source. The far field is calculated by the Green function. Among various acoustic analogy methods, FW-H formulation is widely used due to its simplicity and accuracy.

Large eddy simulation (LES) and Schnerr-Sauer cavitation model is used by Lidtke[4] to simulate the sheet cavitation generated by the NACA0009 hydrofoil. He found out the association between the periodic cavity and acoustic source by FW-H formulation. This demonstrates that the method can accurately predict monopole noise caused by cavitation of the foil. However, if the cavity bubble collapses to cloud, it cannot be resolved by the numerical method well. Scholars studied the blade channel noise in the Potsdam propeller test case using FW-H acoustic analogy[5]. The results are coincided with the experiments results. However, for unsteady Reynolds averaged numerical simulation (URANS), the instability of the fluid is almost zero. In lateratures[6] [7], different formal solutions of FW-H equation were used to compute the rotation noise in open steady water. It is demonstrated that the linear term dominates in the near field, while the quadropole term dominates in the far-field region.

Although there are many examples of applying the FW-H equation to calculate cavitation noise, these studies rarely consider the quadrupole term. For the cavitation phenomenon, the role of the quadrupole should not be ignored. However, the procedure of volume integration is very expensive. The dual mesh technology that appeared in recent years can solve this problem well.

The following is the structure design arrangement of this paper. The math basic formulas are introduced in section 2, including the turbulence model, acoustic formulations, and dual mesh technology. The geometric model and numerical settings are introduced in section 3. The hydrodynamic calculations is shown in section 4. Section 5 introduces the results of the sound field predicted by the dual mesh technique. And the conclusion of this paper is given in Chapter 6.

2. Mathematical basis

The mathematical foundation of this paper will be introduced into four parts: DES equation, Schnerr-Sauer cavitation model, FW-H quadrupole volume integration, and dual mesh technique.

2.1. DES equation

The cavitation problem has high requirements for the accuracy of CFD. The acoustic calculation needs to ensure the accuracy of flow field fluctuation. The feasibility of the amount of calculation is taken into account by an intermediate method of RANS and LES-DES (Detached Eddy Simulation).

The improved Spalart-Allmaras DES is used as the turbulence model in this research, namely SA-DDES[8]. Its governing equation is as follows.

$$\frac{\partial\tilde{v}}{\partial t} + div(\tilde{v}u) = \frac{1}{C_\sigma}\{div[(v+\tilde{v})grad\tilde{v}] + C_{b2}\frac{\partial\tilde{v}}{\partial x_i}\frac{\partial\tilde{v}}{\partial x_j}\} + C_{b1}\tilde{S}\tilde{v} - C_{w1}f_w(\frac{\tilde{v}}{\tilde{d}})^2 \tag{1}$$

Terms on the left side are the transient term and convection term respectively. The first expression on the right side is diffusion term, the second expression is the source term, and the third expression is dissipation term. DDES equation modifies $\tilde{d}$ from formula.

$$\tilde{d} = d_w - f_d \cdot \max(d_w - C_{DES}\Delta, 0) \tag{2}$$

$$f_d = 1 - \tanh[(8r_d)^3] \tag{3}$$

$$r_d = \frac{v_t + v}{\sqrt{U_{ij}U_{ij}(\kappa d_w)^2}} \tag{4}$$

2.2. Schnerr-Sauer cavitation model

The cavitation model, or mass transport model, originates from the Rayleigh-Plesset formulation. Adding source terms can describe the process of condensation and evaporation in cavitation. In this research, Schnerr-Sauer mass transport model[9] is selected to use. The mass conversion equation of the model is as follows.

$$\frac{\partial(\rho_v\alpha_v)}{\partial t} + \frac{\partial(\rho_v\alpha_v u_j)}{\partial x_j} = \dot{m}^+ - \dot{m}^- \tag{5}$$

Where, α_v represents the vapor volume ratio, and $\dot{m}^+$ and $\dot{m}^-$ stands for the evaporation and condensation. The phase variation arises as follows.

$$\dot{m}^{+} = \frac{\rho_v \rho_l}{\rho} \alpha_v (1 - \alpha_v) \frac{3}{R_b} \sqrt{\frac{2}{3} \frac{\max(p_v - p, 0)}{\rho_l}} \tag{6}$$

$$\dot{m}^{-} = \frac{\rho_v \rho_l}{\rho} \alpha_v (1 - \alpha_v) \frac{3}{R_b} \sqrt{\frac{2}{3} \frac{\max(p - p_v, 0)}{\rho_l}} \tag{7}$$

Among them, R_b represents the cavity radius, and p_v represents the saturation critical pressure. The radius of cavity is calculated as follows.

$$\mathrm{R_b} = \left(\frac{\alpha_v}{(1-\alpha_v)} \frac{3}{4\pi} \frac{1}{N_b}\right)^{\frac{1}{3}} \tag{8}$$

$N_b = 10^{13} m^{-3}$ means cavity number density.

2.3. FW-H equation and quadrupole volume integration

The basic equation of acoustic analogy-Lighthill equation, is derived on the basis of the N-S equation. There are various types of Lighthill equations. Among them, Farassat Formulation 1A equation[10] is adopted, whose expression is as follows.

$$4\pi p'_T(x,t) = \int_{f=0} \left[\frac{\rho_0 \dot{v}_n}{r(1-M_r)^2} + \frac{\rho_0 v_n \hat{r}_i \dot{M}_i}{r(1-M_r)^3}\right]_{ret} dS + \int_{f=0} \left[\frac{\rho_0 c v_n (M_r - M^2)}{r^2 (1-M_r)^3}\right]_{ret} dS \tag{9}$$

$$4\pi p'_L(x,t) = \int_{f=0} \left[\frac{\dot{p}\cos\theta}{cr(1-M_r)^2} + \frac{\hat{r}_i \dot{M}_i p\cos\theta}{cr(1-M_r)^3}\right]_{ret} dS + \int_{f=0} \left[\frac{p(\cos\theta - M_i n_i)}{r^2(1-M_r)^2} + \frac{(M_r - M^2) p\cos\theta}{r^2(1-M_r)^3}\right]_{ret} dS \tag{10}$$

Here, p'_T represents thickness component. p'_L represents loading component, and (x,t) (y,t) mean space and time variables for hydrophones and sources respectively. $r = |x - y|$ represents the vector from acoustic source to probes. $\hat{r_i} = \frac{(x-y)}{r}$ is the norm of r. $\mathrm{M_r} = \frac{\hat{r_i} v_i}{c}$ means Mach number. $1 - M_r$ is nemed as Doppler factor. $\dot{v_n} = \frac{\partial}{\partial \tau}(v \cdot n)$ is the derivative of velocity.

To obtain the analytic expression of the quadrupole term, it is necessary to deduce the second-order derivative of $\left(\frac{\partial^2}{\partial x_i \partial x_j}\left[\frac{\delta(g)}{r^*}\right]\right)$. According to the chain derivative rule, the calculation is as follows.

$$\frac{\partial}{\partial x_i}\left[\frac{\delta(g)}{r^*}\right] = -\frac{\hat{r_i}}{c_0 r^*} \frac{\partial}{\partial t}[\delta(\mathrm{g})] - \frac{\hat{r_i}^*}{r^{*2}} \delta(g) \tag{11}$$

With the combination of the derivative term simplification and the formula, the analytic expression of the quadrupole term under a low Mach number is obtained.

$$\frac{\partial^2}{\partial x_i \partial x_j}\left[\frac{\delta(g)}{r}\right] = \frac{1}{c_0^2} \frac{\hat{r_i}\hat{r_j}}{r} \frac{\partial^2}{\partial t^2}[\delta(\mathrm{g})] + \frac{1}{c_0}\left[\frac{3\hat{r_i}\hat{r_j} - \delta_{ij}}{r^2}\right] \frac{\partial}{\partial t}[\delta(\mathrm{g})] + \left[\frac{3\hat{r_i}\hat{r_j} - \delta_{ij}}{r^3}\right] \delta(\mathrm{g}) \tag{12}$$

2.4. Dual mesh technology

The main computational effort in FW-H formulation is focused on quadrupole (volume integration) term. If we use N_{cell} to represent the number of quadrupole sound source grids to be calculated, N_{obs} will be used to represent the number of sound pressure test points, and N_t to represent the time step required for each grid calculation. For the "calculate-as-you-report" mode, the time complexity is in the order of $O(N_{cell} N_{obs} N_t)$. And the amount of disk storage in the "calculate before report" mode is in the order of $O(N_{cell} N_t)$. In actual engineering, this corresponds to the CFD simulation results of several terabytes. The loading of the disk becomes an important obstacle limiting the application of quadrupole volume fraction engineering.

To this end, we use a dual mesh technology to solve the trouble. The sound mesh is different from CFD mesh and is thicker, thus reducing storage and computation by reducing N_{cell}, as shown in Figure

1. For the calculation efficiency and applicability, a simple summation method is used for integration. The physical quantities on the acoustic mesh are calculated by CFD mesh weighted volume summation[11].

$$\phi_i^{acoustic} = \sum_{j=0}^{N_i} \phi_{i,j}^{CFD} \cdot V_{i,j}^{CFD} \tag{13}$$

ϕ represents any physical quantity. $\phi_i^{acoustic} = (\int \phi dV)_i^{acoustic}$ is the corresponding volume integration. Subscript i means ith acoustic element, while (i,j) means ith acoustic element in jth CFD grid. V^{CFD} represents the volume of the CFD grid.

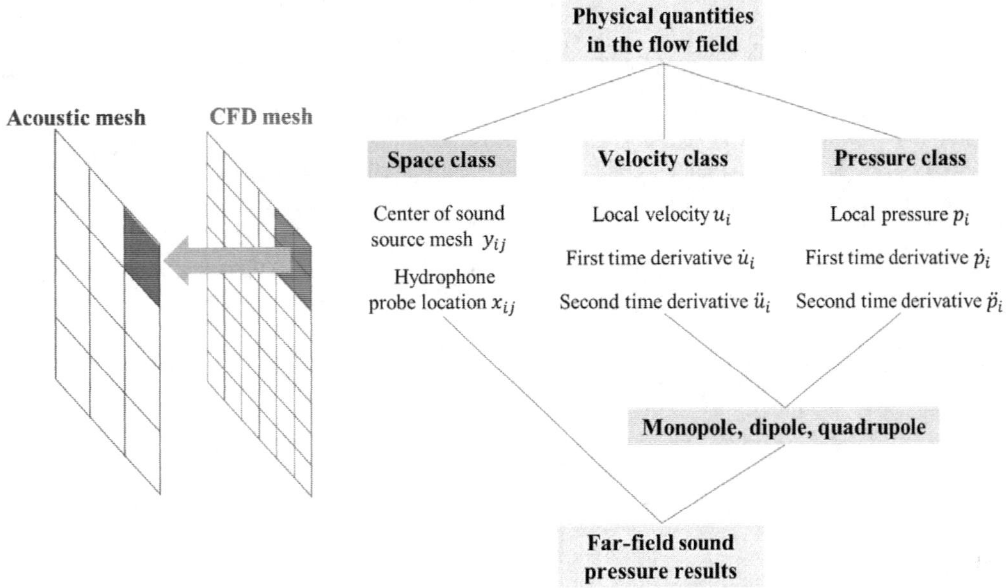

(a) The diagram of the dual mesh technique (b) The process of acoustic quantities extracted from the flow field

Figure 1 The sketch of the dual mesh technique and the flow chart of quantities

3. Geometry and numerical settings

3.1. Bullet geometry

The research object used in this paper is a three-dimensional bullet, which is a rotating geometry. The *y-z* section of the geometry is shown in Figure 2. The specific geometric parameters are shown in Table 1.

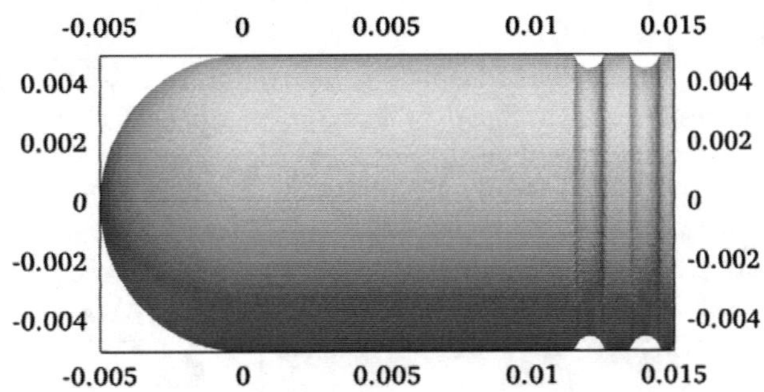

Figure 2 Schematic of the position of the bullet's y-z profile in the Cartesian coordinate system.

3.2. Numerical setup

The platform OpenFOAM is adopted to predict the fluid and acoustic field. PIMPLE decoupling algorithm is used in this paper. The working condition is to simulate the bullet being shot into the water at high speed, as is shown in Figure 3. The incoming flow is along the *z* direction whose velocity is 20 m/s. The background pressure is set at 1 atm. The fluid density is $1000\ \mathrm{kg/m^3}$. The saturation pressure is 2300 Pa, and the corresponding cavitation number is 0.4885.

Table 1 Geometry parameters of the bullet

Parameter	Value
Length of the whole bullet	0.015 m
Length of the head part	0.002 m
Diameter of the head part	0.005 m

The structured grids are adopted. The sum of grids is 3.73 million, and the y+ value is less than 20. The Courant number is kept at less than 2 in the whole calculation process. The simulation duration is 0.02 s, and the time step is set to 1e-5s. Velocity inlet boundary condition and pressure outlet condition are adopted in this case. The spatial discretization is set as the linear upwind format, and the Euler format is used for the time discretization.

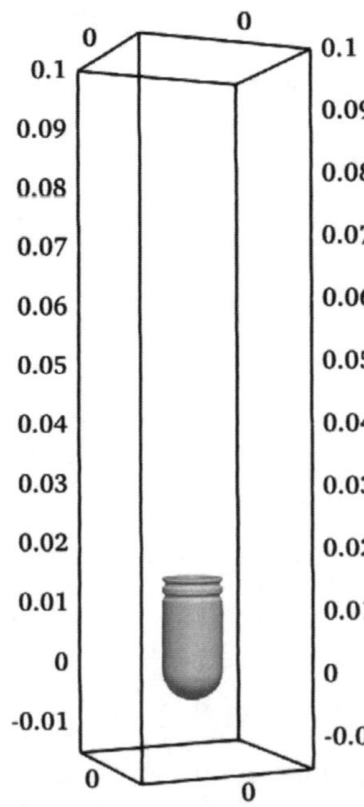

Figure 3 Schematic diagram of the computational domain for the cavitation bullet

4. Results and discussion

The hydrodynamic and acoustic results of cavitation bullets are discussed in this chapter. Accurate hydrodynamic prediction results are the basis of acoustic prediction. Therefore, the hydrodynamic results are introduced first, and then the acoustic prediction results are introduced.

4.1. Hydrodynamic results

Figure 4 shows the evolution of cavitation around the bullet over time. With the increase of time, the range of cavitation gradually expands and reaches a stable range of cavitation after 0.01 s. Therefore, the sampling interval of acoustic prediction is in the range of 0.01 s-0.02 s.

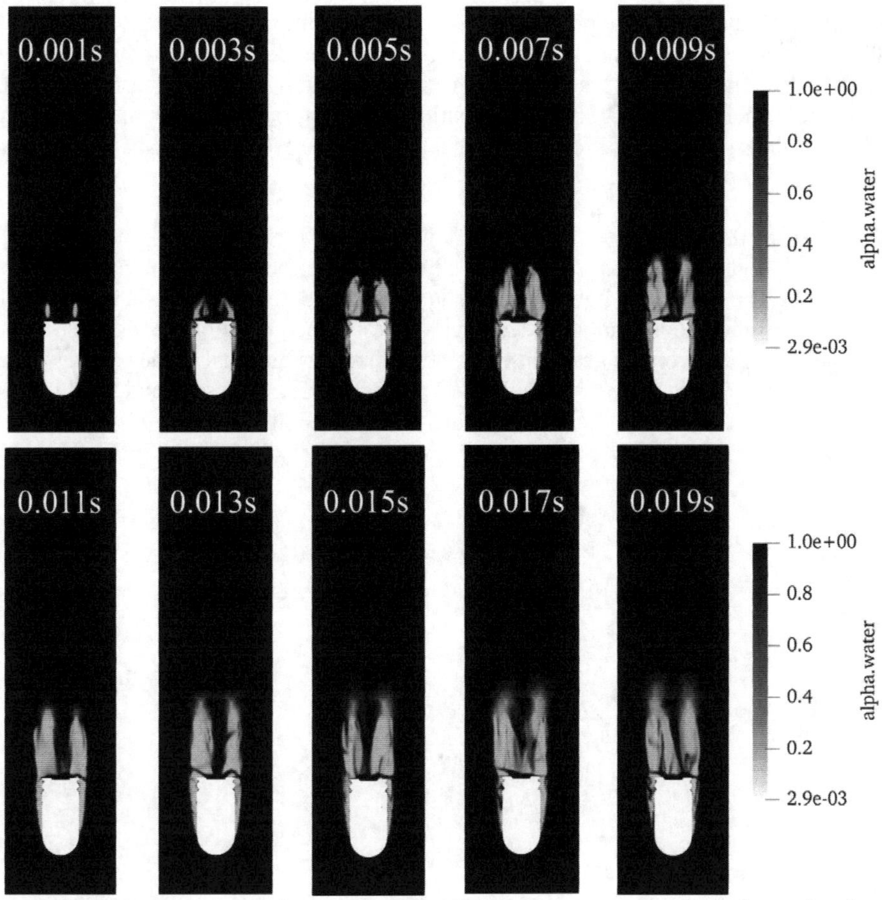

Figure 4 Bullet cavitation time evolution: Alpha water represents the volume fraction

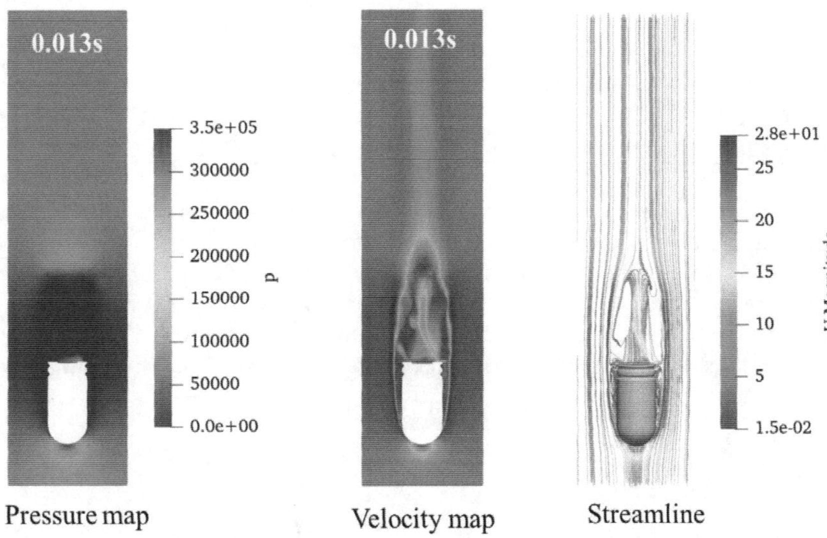

Figure 5 Instantaneous pressure map, velocity map, and streamline in slice x=0.

Figure 5 shows the distribution of pressure, velocity, and streamlines around the bullet. The selected time is 0.013 s, which reaches the stable state of cavitation. It can be seen that the pressure at the head of the bullet is higher, and there is a pressure trough at the tail part, which is the reason for the formation of cavitation. The velocity and streamlined distribution also conform to the law of flow around. It can be concluded that the hydrodynamic forecast results are accurate and noise predictions can be carried out.

4.2. Hydroacoustic results

To study the sound field characteristics, four near-field test points are set near the bullet (x=0 plane), as shown in Figure 6. The volume integration range is a cuboid, both 0.02 m in length and width, and 0.04m in height from the bullet head to outlet patch. To verify computational efficiency and accuracy of dual mesh method, three-volume grid densities are set. The first is equal to the CFD grid size (0.001 m, regarded as 'CFD'). The second grid size is 0.002 m (regarded as 'coarse'), and the third grid size is 0.004 m (regarded as 'coarser'). The total computing time and storage capacity of the three meshes are shown in Table 2.

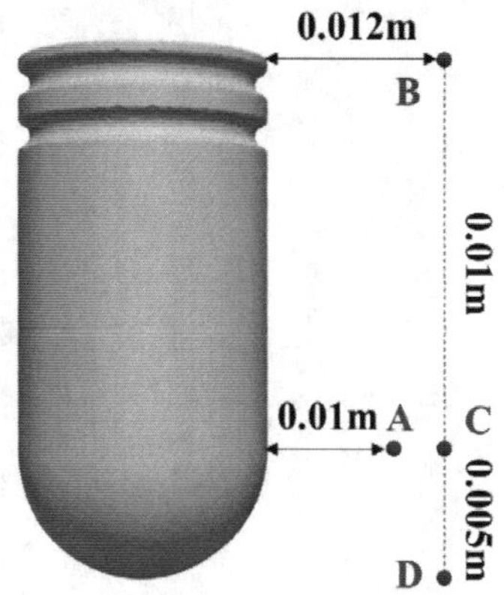

Figure 6 The locations of acoustic probes on the plane x=0

Table 2 The computation time and storage of three-volume integration grids

Grid	Size	Computation time	Storage
CFD	0.001 m	51.2 hour	1.3 G
Coarse	0.002 m	8.8 hour	254 M
Coarser	0.004 m	6.4 hour	191 M

Figure 7 presents sound pressure prediction time history for points 'A' and 'C'. It is found that sound pressure does not have a distinct period due to the cavitation phenomenon. The sound pressure results of the three-volume mesh density predictions are similar. However, the calculation time and storage amount are greatly reduced after the double mesh technology is used.

Figure 8 presents the frequency-range distribution of sound pressure level (SPL) for points 'A' and 'C'. Similarly, the three-volume mesh density prediction results are very similar. Therefore, the feasibility and accuracy of the dual mesh technique are demonstrated.

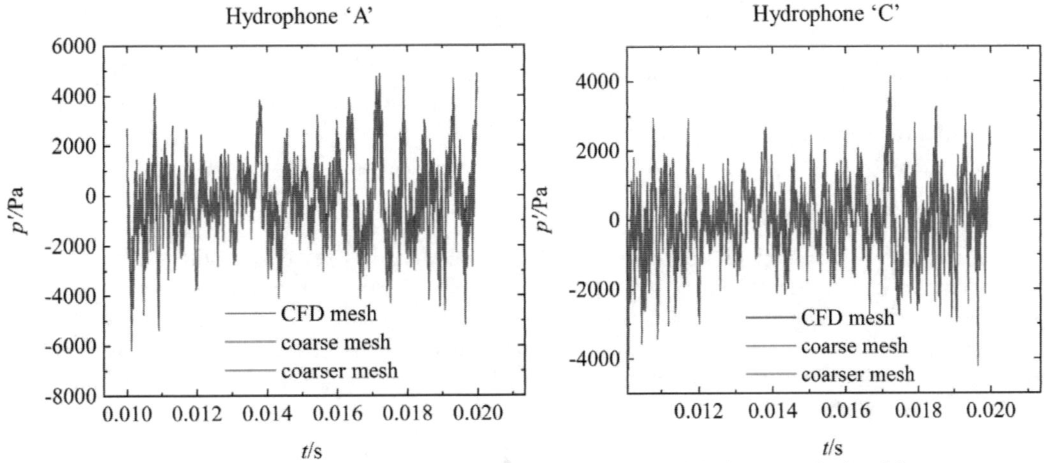

Figure 7 The sound pressure time history predicted with three mesh densities

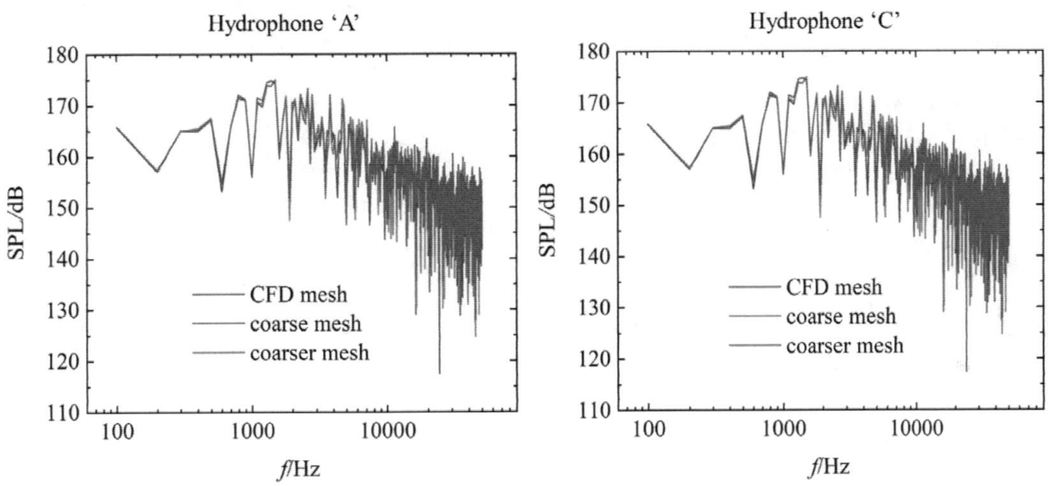

Figure 8 The SPL predicted wtih three mesh densities

5. Conclusion

In this research, OpenFOAM is adopted to study the problem of cavitation noise around the bullet. The dual mesh technology is used to calculate the quadrupole term by volume integration, and the calculation amount is reduced by coarser grid densities.Conclusions are summarized as follows.

(1) DES turbulence model can simulate the fluid field for the bullet in the cavitation state effectively.

(2) The dual mesh method can predict the flow noise accurately, which is not much different from the prediction results directly using the CFD grid.

(3) The use of dual mesh technology can reduce the calculation time and requires effecitve storage space. Compared to CFD grids, the use of coarser grids can reduce the calculation time by 87% and save storage space by 86%.

References

[1] Marine environment protection committee 2014 *Noise from Commercial Shipping and its Adverse Impacts on Marine life in: IMO (Ed.)* **MEPC** 66/17

[2] Lianjie Yu, Weiwen Zhao, Decheng Wan, Jianwei Wu 2022 Nonlinear noise of hydrofoil cavitation considering sound velocity variation and phase transitions *Ocean Engineering* **264** 112506

[3] Lianjie Yu, Jianwei Wu, Decheng Wan 2022 Correlation analysis between underwater noise and Liutex for DTMB4119 propeller *Journal of Hydrodynamics* **34**(4) 585-595

[4] Lidtke Artur K, Stephen R, Turnock and Victor F Humphrey 2016 Characterisation of sheet cavity noise of a hydrofoil using the Ffowcs Williams–Hawkings acoustic analogy *Computers & Fluids* **130** 8-23

[5] Lidtke Artur K, Victor F Humphrey and Stephen R Turnock 2016 Feasibility study into a computational approach for marine propeller noise and cavitation modelling *Ocean Engineering* **120** 152-159

[6] Ianniello S, R Muscari, and A Di Mascio 2013 Ship underwater noise assessment by the acoustic analogy. Part I: nonlinear analysis of a marine propeller in a uniform flow *Journal of marine Science and technology* **18.4** 547-570

[7] Bensow R and M. Liefvendahl 2016 An acoustic analogy and scale-resolving flow simulation methodology for the prediction of propeller radiated noise *31st Symposium on Naval Hydrodynamics*

[8] Spalart P R, Allmaras S R 1992 A One-Equation Turbulence Model for Aerodynamics Flows *In 30th aerospace sciences meeting and exhibit* p 439

[9] Schnerr G H and Sauer J 2001 Physical and numerical modeling of unsteady cavitation dynamics. *In Fourth international conference on multiphase flow (Vol. 1).* ICMF New Orleans

[10] Ffowcs Williams J E and Hawkings D L 1969 Sound generation by turbulence and surfaces in arbitrary motion. *Philosophical Transactions of the Royal Society of London. Series A, Mathematical and Physical Sciences* **264** 321-342

[11] Wang Youjiang, Tommi Mikkola and Spyros Hirdaris 2022 Fast and storage-saving method for direct volumetric integration of FWH acoustic analogy *Ocean Engineering* **261** 112087

Dynamic Response Analysis of Suspended Tunnel under Wave-current Combined Action

Songsong YAN, Peilin DOU, Baojin JING, Xiaoliu ZHANG

(College of Ship and Marine Engineering, Jiangsu University of Science and Technology, Zhenjiang 202206)

(China Institute of Ship and Ocean Engineering Design, Shanghai 200011)

Yan1996090946@163.com

Abstract: The environment in the ocean is complex, and waves and currents produce complex coupling effects on the structures in the ocean. The flow-solid coupling theory and calculation method are applied to establish a three-dimensional model to study the changes in the flow field under the joint action of waves and currents through the comparative analysis of water velocity at the upper and lower near-wall surfaces of the suspension tunnel tube body. The dynamic response of the tension-leg suspension tunnel is calculated using unidirectional fluid-structure coupling. The relationship between different wave heights, different anchor angles, different flow velocities, and different depths and the deformation of the suspension tunnel in the presence of waves is investigated. The results show that the wave height and velocity have a great influence on the floating tunnel, and the displacement changes in both horizontal and vertical directions are large. With the increase of construction depth, the influence of waves on the floating tunnel becomes smaller. When the angle of the anchor cable is changed, the horizontal direction and the vertical direction change in opposite laws, so it is very important to select a suitable angle for the anchor cable. The research results provide a reference for the design of similar floating tunnels.

1. Introduction

At present, the development of marine transportation in China is flourishing. Compared with traditional cross-sea bridges and subsea tunnels, the construction of submerged floating tunnels is not affected by the water depth in the ocean, the span is small, and its cost will not increase with the depth of the ocean. Selecting the appropriate construction depth will not only affect the navigation and operation of ships on the sea surface but also the development of landscape tourism and other functions; the reclamation of less land in the maintenance of the seabed ecological environment has a significant advantage that is more environmentally friendly[1].

Wave is the most basic environmental load on solid structures in the ocean. The wave height, wavelength, and period of waves have obvious effects on structures in the water. To study the specific numerical analysis of the submerged tunnel under the action of waves, the basin and the submerged tunnel should be modeled, and the waves should be simulated in the basin. FLUENT software provides different wave-making methods, including the momentum source wave-making method, the velocity boundary method, the push plate wave-making method, and the rocking plate wave-making method. This chapter uses the method of velocity boundary wave making to calculate the fluid. Then the structure is imported into the structural analysis module for structural analysis, focusing on the displacement response of the submerged floating tunnel in different environments, and summarizing the role of waves.

There are three kinds of support modes for submerged floating tunnels: buoy structure, tension leg type, and rigid pile column type. Through the current research of scholars, the tension-leg submerged floating tunnel has more advantages in realization and safety. To ensure the safety performance of the submerged floating tunnel, a variety of environmental conditions should be selected to study the performance of the submerged floating tunnel, especially the displacement response leg setting position, setting form, setting angle, and other factors are studied.

Huang et al. [2] studied the effects of wave period, wave height, and inclination on the dynamic response of the system. The results show that the coupling effect between the tunnel and drilling rig is not only affected by wave height, but also by wave period. Tian [3] established a nonlinear vibration mathematical object model of tunnel fluid interaction and studied the displacement response of submerged floating tunnel under the combined action of internal wave and ocean current. Zou Wei et al. [4] started from the design line shape of the submerged floating tunnel and compared the force of different incident angles and pipe lengths through the numerical calculation to provide suggestions for the design of the submerged floating tunnel. Wang Guangdi [5] et al. established a submerged floating tunnel model with a circular cross-section and used the Morison equation to calculate wave and current loads. The variation rules of wave force, seawater depth, suspension depth, current, and tunnel cross-section form were summarized. It is concluded that in the transition zone, the wave force decreases with the increase of water depth, and the change rate decreases continuously. In the deep-water area, the seawater depth will not affect the wave force, and the suspension depth can be adjusted accordingly. Zou Pengxu et al. [6] calculated the dynamic response of the submerged floating tunnel, the wave force on the pipe body, the relative variation amplitude of the anchor cable force, and the oblique anchor cable force according to different anchor cable arrangement modes and different pre-tension distribution modes, and analyzed the influence of IMA (initial angle between oblique anchor cable and vertical direction), anchor cable arrangement mode and anchor cable, and tension distribution on dynamic response.

2. Finite element method and tunnel model

2.1. Wave force calculation

For larger structures, the viscous force can be ignored and the dynamic pressure distribution on the wet surface of the structure can be calculated by the potential flow theory, and then the wave force on the structure can be calculated by integration [9]. In [10], it was proposed that according to the Bernoulli equation, the pressure p at any point on a wet surface can be derived by the following formula:

$$p=-\rho\frac{\partial\phi}{\partial t}\ (1)$$

In the formula: ρ for fluid density; t is time;

ϕ total velocity potential of the structure.

The horizontal (x direction) wave force F_h and vertical (y direction) wave force of the structure can be expressed as: F_v

$$F_h(t)=\iint_S p dS_x\ (2)$$

$$F_v(t)=\iint_S p dS_y\ (3)$$

In the formula: S is the wet surface of the structure; x and y are the horizontal and vertical coordinates of the wet surface grid points, respectively. Therefore, this paper decided to use the potential flow theory algorithm to calculate wave loads.

The wave force and constraint are important factors affecting the displacement of the submerged floating tunnel. The magnitude of the wave force is related to the velocity and structural coordinates, and the constraint includes the setting of the anchor cable. Therefore, it is necessary to explore the influence of flow velocity, wave height, construction depth, and anchor cable setting on the maximum displacement of the submerged floating tunnel.

2.2 Slurry tunnel model

In this section, it is necessary to consider the effect of both the water flow and the wave on the suspended tunnel pipeline. The VOF two-phase flow model is used in FLUENT. According to the lane speed and width specified in JTG.D60-2015 ' General Specification for Design of Highway Bridges and Culverts ', the tunnel structure in this paper adopts a circular section pipe with two-way eight lanes. Each pipeline is 48 m in diameter and 150 m in length. A watertight connection is used between pipelines, and the material used in the pipeline is steel.

When establishing the flow field area model, it is not necessary to model the fluid part and the air part separately. In the VOF model, the air-water separation is completed by establishing the free liquid surface, and the free liquid surface is at 110 m. The size of the flow field is calculated to be 200m * 150m * 150m. The size suspended tunnel pipe is a circular pipe with a diameter of 20m and a length of 150m, and the wall thickness is 2m. Y-axis is vertically upward, X-axis represents the downstream direction of water flow, and Z-axis represents the cross-flow direction of water flow; the left side of the basin is set as the velocity inlet, the right side is the pressure outlet, the upper and lower sides, and the left and right sides are set as walls, and the contact surface between the fluid and the pipeline is the body, as shown in Figure 1.

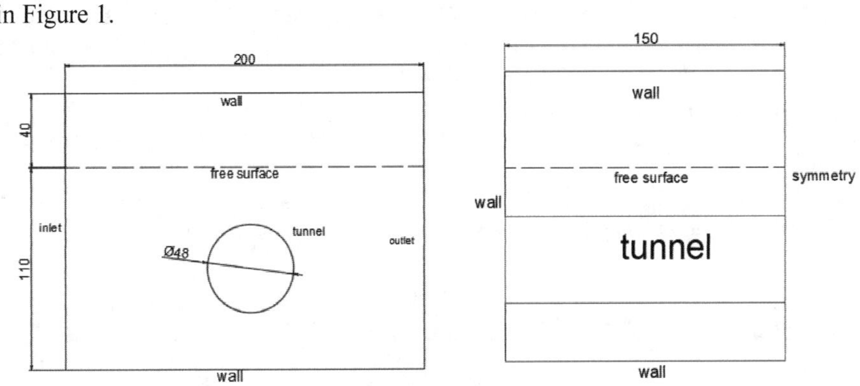

Fig. 1 Plan of submerged floating tunnel mode

The anchor cable is set on the waistline of the suspended tunnel. In [7-9], the ultimate spacing of the anchor cable arrangement of the suspended tunnel with a diameter of 23 m is 494.32 m, and the arrangement spacing selected in [10] is 50 m and 100 m. The diameter of the model used in this paper is large, and the final anchor cable spacing is symmetrically arranged at an interval of 50m, with a total of four pairs of eight anchor cables. The tunnel pipeline is connected by the MPC method without considering the gravity of the anchor cable. The angle of the anchor cable is the angle with the negative direction of the y-axis, as shown in Figure 2.

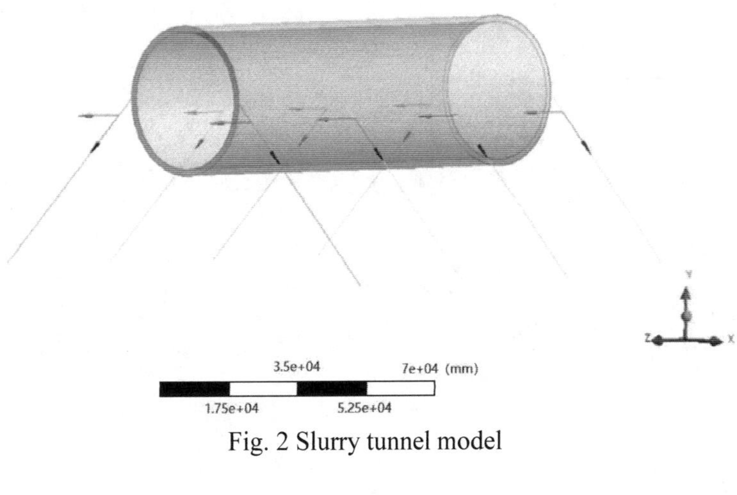

Fig. 2 Slurry tunnel model

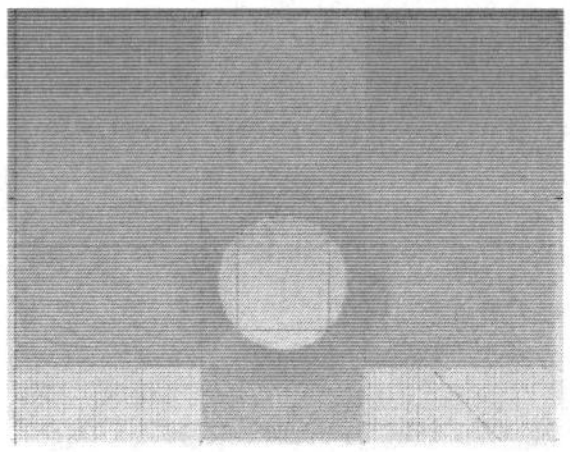

Fig. 3 Gridding

Fig. 4 Gas-water two-phase distribution

ICEM is used for grid division, and the grids around the submerged tunnel pipeline and the upper, lower, left, and right flow fields should be encrypted. In the numerical simulation calculation, the quality of the grid affects the calculation results. Firstly, the grid independence verification is carried out in the calculation straight line. By dividing the grids of different sizes and comparing the results, the maximum grid division method is taken when the results tend to be stable. The ICEM is divided into 9 parts. The number of grid layers in the near wall area of the pipeline is 33, the minimum size is 0.3, and the magnification is 1.2. The number of layers of the flow field grid increases to four times of the original grid. The final meshing result is shown in Figure 3. The initialized model and the phase display are shown in Figure 4 below.

To calculate the influence of wave and current interaction, the water area is given speed in the simulation calculation, and the calculation condition information is shown in Table 1.

Table 1 Condition information

Wave height (m)	Water depth (m)	Anchor angle (°)	Flow velocity (m/s)
1	30	45	1.2
2	30	45	1.2
3	30	45	1.2
4	30	45	1.2
5	30	45	1.2
2	20	45	1.2
2	40	45	1.2
2	20	45	0.2
2	20	45	0.6
2	20	45	1.8
2	30	30	1.2
2	30	40	1.2
2	30	50	1.2

3. Analysis of Slurry Tunnel under 2 Wave - Current Interaction

3.1 Pressure of tunnel section

3.1.1 Pipe pressure at different flow rates

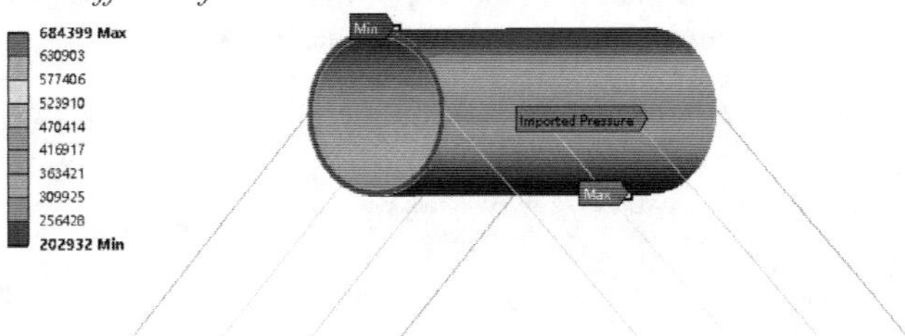

(a) flow velocity 0.6m/s

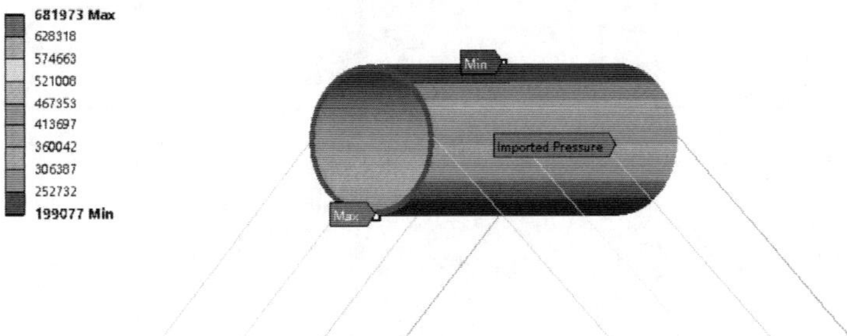

(b) flow velocity 1.2m/s

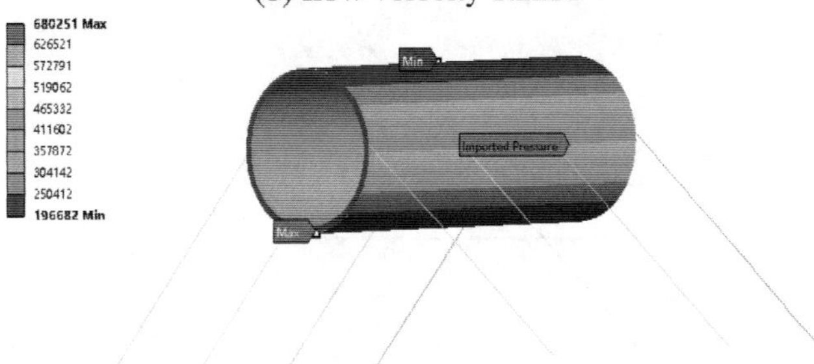

(c) flow velocity 1.8m/s

Fig. 5 Pipe pressure at different flow rates

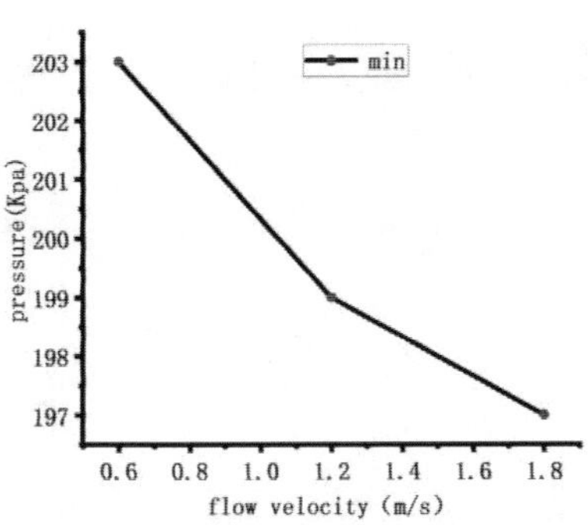

(a) Minimum pipeline pressure

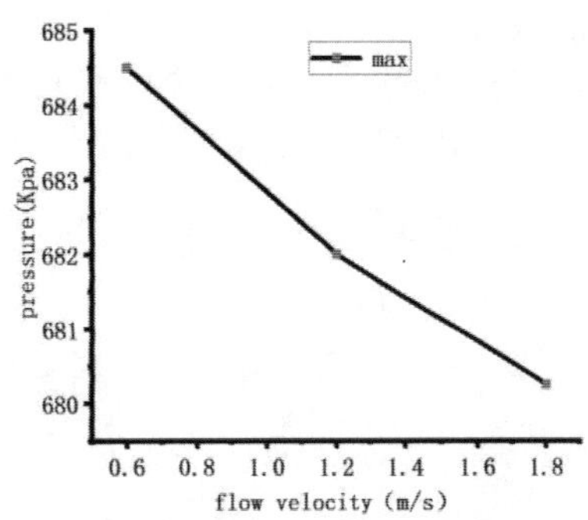

(b) Maximum pipeline pressure

Fig. 6 Pipe pressure line chart

Through the above pressure cloud diagram and numerical comparison results, it can be seen that under the combined action of waves and water flow, the pressure at the bottom of the suspended tunnel pipeline is the largest, and the pressure at the top is the smallest. With the increase in flow velocity, the maximum and minimum pressure of the suspended tunnel pipeline decrease. This situation conforms to the Bernoulli principle, and its essence is the conservation of the mechanical energy of the fluid. When the kinetic energy increases, the pressure potential energy decreases.

3.1.2 Pipe pressure at different wave heights

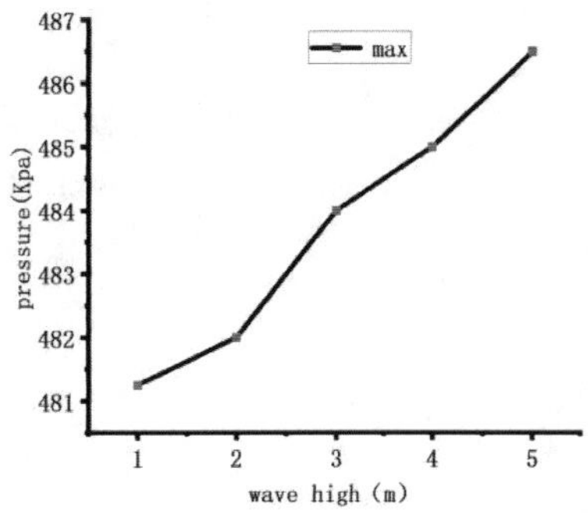

(a) Maximum pressure of the pipeline with different wave heights

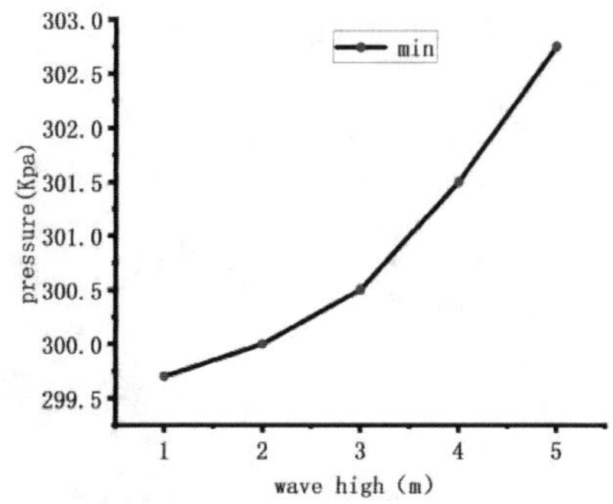

(b) Minimum pressure of the pipeline with different wave heights

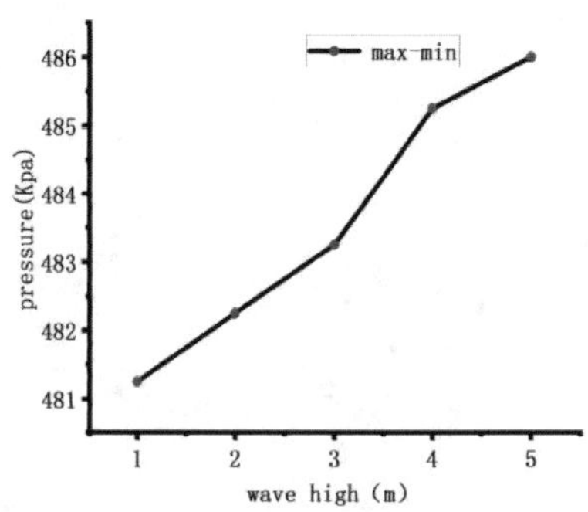

(c) Pressure difference of pipeline with different wave heights

Fig. 7 Piping pressure line chart

The numerical comparison results of the line chart show that with the increase of the wave height, the maximum and minimum values of the pipeline pressure increase, and the difference between the maximum and minimum values of the pipeline pressure also increases, which indicates that the higher the wave height, the greater the pressure change on the pipe.

3.1.3 Pipe pressure at different water depths

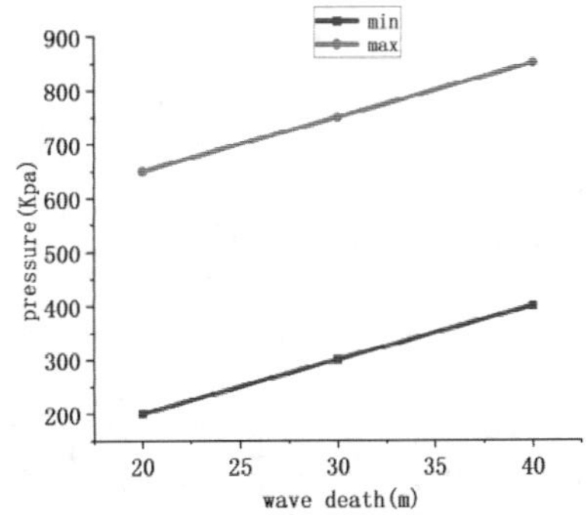

Fig. 8 Piping pressure line chart

The numerical comparison results of the line chart show that the maximum and minimum values of the pipeline pressure increase with the deepening of the construction depth of the submerged floating tunnel. The reason is that the submerged floating tunnel bears not only the wave force but also the hydrostatic pressure in the water. When the submerged floating tunnel is built deeper, the hydrostatic pressure is greater.

In general, the pressure changes of SFT under three different factors of flow velocity, wave height, and water depth are consistent with the change of test data in [11], indicating that the calculation results are credible.

3.2 Displacement response analysis

The submerged floating tunnel is built in the sea, and the wave, water velocity, and construction depth have an impact on the stress and dynamic response of the tunnel. The wave height, anchor cable angle, water flow velocity, and construction depth are taken as variables to analyze their influence on the X and Y direction displacement of the submerged floating tunnel, which is beneficial to solve the safety problem of the submerged floating tunnel and provide reasonable suggestions for the construction of the submerged floating tunnel.

3.2.1 Displacement Response Analysis under Different Flow Velocity

When the submerged depth of the submerged floating tunnel is 30m, the wave height is 2m, and the angle of the anchor cable is 45 °, the four working conditions of the submerged floating tunnel at the flow velocity of 0.2m / s, 0.6m / s, 1.2m / s, and 1.8m / s are discussed respectively.

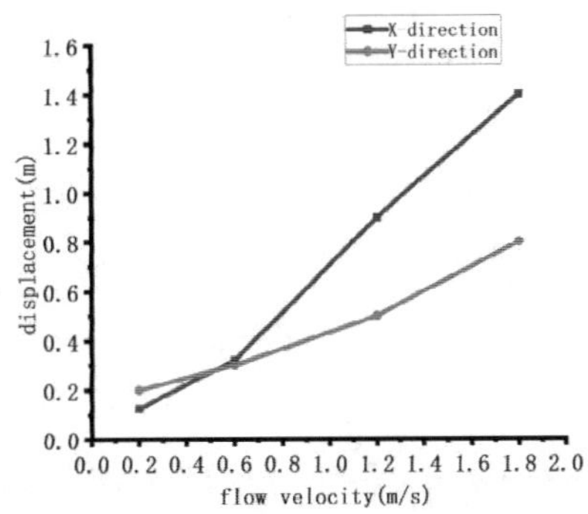

Fig. 9 The extreme values of displacement in X and Y directions at different flow rates

Table 2 Displacement extreme value of the submerged floating tunnel

Water flow velocity (m/s)	0.2	0.6	1.2	1.8
X (cm)	13.60	37.60	92.21	132.14
Y (cm)	19.50	31.60	58.00	77.10

According to the analysis of the above figure and table 2 :

On the whole, the maximum displacement of the submerged floating tunnel pipeline in the horizontal and vertical directions increases with the increase of the flow velocity. When the flow velocity is very small, for example, the horizontal and vertical displacement extremes of the submerged floating tunnel are close to each other when the flow velocity is less than 0.4m / s, and the difference is very small. When the flow velocity is greater than 0.4m / s and increases continuously, the horizontal and vertical displacement extremes of the submerged floating tunnel increase. The horizontal displacement extreme is greater than the vertical displacement extreme, and the horizontal growth rate is faster. The reason for this phenomenon is: in the fourth chapter, it is verified that the wave mainly affects the vertical displacement of the submerged tunnel. When the water flow and the wave act together, the water flow has a horizontal impact force on the submerged tunnel. Therefore, the horizontal displacement of the submerged tunnel is larger in the case of high flow velocity.

3.2.2 Displacement response analysis under different wave heights

When the underwater depth of the suspended tunnel construction is 30 m, the water flow velocity is 1.2m / s, and the anchor cable angle is 45 °, the five working conditions of the suspended tunnel at the wave height of 1m, 2m, 3m, 4m, and 5m are discussed respectively.

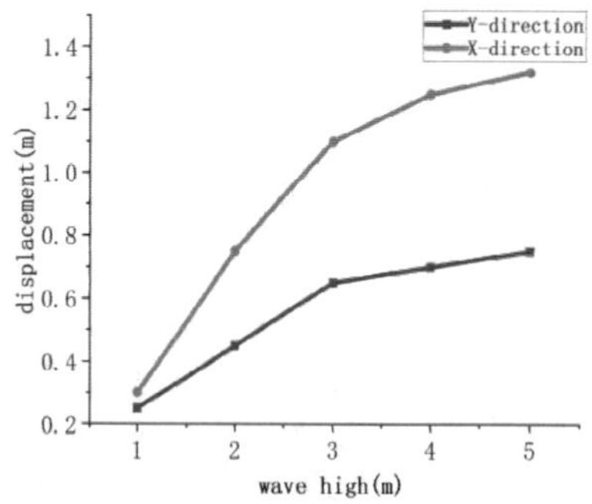

Fig. 10 Extreme values of X-and Y-direction displacements at different wave heights

Table 3 Displacement extreme value of the submerged floating tunnel

Wave height（m）	1	2	3	4	5
X (cm)	30.80	76.70	111.03	126.01	135.70
Y (cm)	26.20	48.60	64.80	72.10	76.80

According to the analysis of the above figure and table 3:

On the whole, the displacement extremum of the submerged floating tunnel pipeline in the horizontal direction and the vertical direction increases with the increase of the wave height. When the wave height is 1- 3m, it increases rapidly, and then the growth rate slows down. Under the condition that the construction depth and flow velocity of the submerged floating tunnel is constant, this calculation result fully proves that the influence of wave height on the submerged floating tunnel pipeline is huge. No matter which waves height, the displacement extreme value is greater in the horizontal direction than in the vertical direction, which reflects that the force in the horizontal direction is greater when the submerged floating tunnel is superimposed by waves and currents.

3.2.3 Displacement response analysis under different anchor cable angles

When the submerged depth of the submerged floating tunnel is 30 m, the flow velocity is 1.2m / s, and the wave height is 2m, due to the limited waters on both sides of the tunnel pipe section during modeling, the maximum inclination angle between the cable line and the vertical direction is 50 °. Therefore, the four working conditions of the cable angle of the submerged floating tunnel are 30 °, 40 °, 45 °, and 50 °.

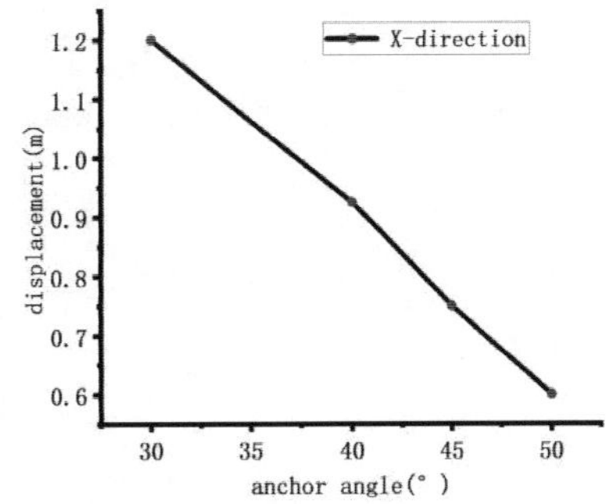

Figure 11. Extreme value of X-direction displacement under different angles of anchor cable Figure

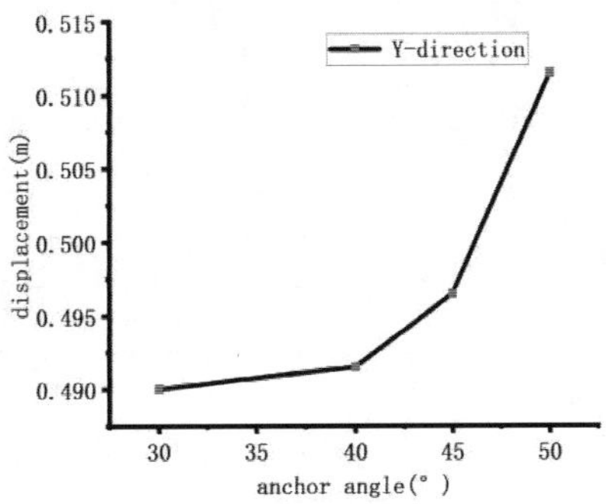

Figure 12. Extreme value of Y-direction displacement under different angles of anchor cable

Table 4 Displacement extreme value of the submerged floating tunnel

Anchor angle (°)	30	40	45	50
X (cm)	118.90	92.20	76.70	61.50
Y (cm)	49.00	49.10	49.60	51.00

According to the analysis of the above figure, it can be seen that with the increase of the angle between the submerged floating tunnel and the cable rope, the maximum displacement of the tunnel along the horizontal direction gradually decreases under the condition that other conditions remain unchanged. This is because when the angle between the submerged floating tunnel and the cable rope increases, the length of the cable rope becomes longer and longer, and the mapping length in the horizontal direction becomes longer and stronger. With the increase of the angle between the submerged tunnel and the cable rope, the maximum displacement of the tunnel along the vertical direction increases gradually, and the vertical displacement increases rapidly from 45 ° to 50 °. The reason for this phenomenon is that when the angle of the cable rope is 45 ° with the vertical direction, the mapping length in the horizontal direction and the vertical direction is equal, and when the angle of the cable rope increases, the distance between the fixed end of a pair of cable ropes and the seabed increases. Although the length of the cable rope increases, the increase in the distance weakens the stability in the vertical direction.

From Table 4, it can be concluded that although the extreme value of horizontal displacement is decreasing, it is still larger than the extreme value of vertical displacement in numerical value. Therefore, when considering the setting of anchor cables, a hybrid arrangement can be adopted, not only with anchor cables in the cable-stayed direction but also with vertical downward anchor cables to maintain the safety of the submerged floating tunnel.

3.2.4 Displacement response analysis under different water depths

When the flow velocity is 1.2m / s, the wave height is 2m, and the angle of the anchor cable is 45 °, the three working conditions of the submerged floating tunnel underwater 20m, 30m, and 40m are discussed respectively.

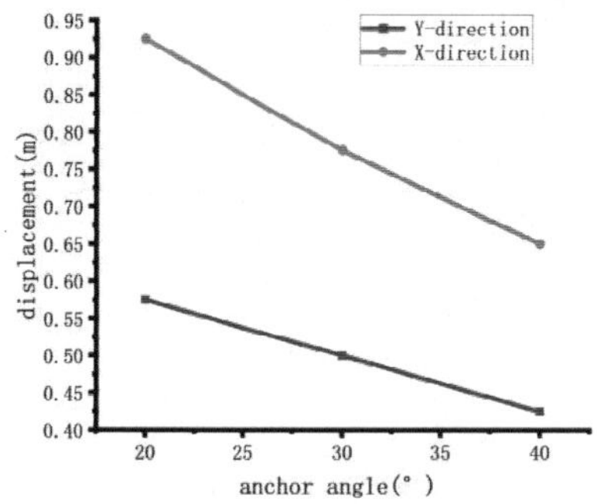

Fig. 13 The extreme values of displacement in X and Y directions under different water depths

Table 5 Displacement extreme value of the submerged floating tunnel

Water depth (m)	20	30	40
X(cm)	92.20	76.70	65.90
Y(cm)	58.00	48.60	42.10

According to the analysis of the above figure and table 5:

On the whole, the displacement extremum of the submerged floating tunnel pipeline in the horizontal and vertical directions decreases with the increase of water depth. In each water depth, the displacement in the horizontal direction is greater than that in the vertical direction, which further indicates that the influence of waves on the submerged floating tunnel decreases with the increase of construction depth. Compared with the effect of waves, the effect of water flow is more obvious.

4. Conclusion

The environment in the ocean is complex, and waves and currents have complex coupling effects on structures in the ocean. Using the fluid-solid coupling theory and calculation method, a three-dimensional model is established, and the dynamic response of the tension leg suspension tunnel is calculated by one-way fluid-solid coupling. The relationship between different wave heights, different anchor cable angles, different flow rates, different depths, and the deformation of the submerged floating tunnel is studied.

The numerical value in the X direction, that is, the horizontal direction, shows that the X-direction displacement extremum increases with the increase of wave height; increases with the increase of flow velocity; decreases with the increase of anchor cable angle; it decreases with the increase of construction depth.

The numerical value in the Y direction, that is, the vertical direction, shows that the extreme value of Y-direction displacement increases with the increase of wave height. increases with the increase of flow velocity; increases with the increase of anchor cable angle; it decreases with the increase of construction depth.

It shows that wave height and velocity have a great impact on the floating tunnel, and the displacement changes in both horizontal and vertical directions are large. However, with the increase of construction depth, the influence of waves on the floating tunnel becomes smaller, and the displacement in both horizontal and numerical directions decreases. When the angle of the anchor cable is changed, the horizontal direction and vertical direction have opposite change rules, which is the only group of opposite changes. Therefore, it is very important to select an appropriate angle for the anchor cable.

References

[1] Tengchao Ye. Research Value and Application Prospect of Suspension Tunnel [J]. Sichuan Building Materials, 2018, 44 (01): 171+173
[2] Huang G, Song Y, Sun Z, et al. Experimental investigation on the coupling between immersion rig and tunnel element during freeboard elimination[J]. Ocean Engineering,2019,186(Aug.15).
[3] XUE F T, MAM S D, HUAN P P, et al. Dynamic Responses of the Submerged Floating Tunnel Under Combined Internal Wave and Sea Current Effect[J]. Applied Mathematics and Mechanics, 2014, 35(1):71-80.
[4] Wei Zou, Wei Lin, Xiao Liu. Doing a study on wave forces acting on suspension tunnels with different linear shapes [J]. China Harbour Construction, 2020, 40 (02): 45-48+67
[5] Guang di Wang, Xiao jun Zhou, Bo Gao. Study on wave current load analysis of underwater suspension tunnel [J]. Railway Architecture, 2007 (10): 48-51
[6] Peng xu Zou, Meng yuan Liu, Liang zhi Chen. Study on hydrodynamic characteristics of pipe anchor cable coupling system of suspension tunnel under wave action [J]. Modern Tunnel Technology, 2021,58 (03): 2021.03.020
[7] Yin gang Qin, Sheng Guo Zhou, Xiao jun Zhou. Analysis on reasonable support spacing of underwater suspension tunnel [J]. Hydrodynamics Research and Progress A, 2008 (03): 309-313
[8] Ao Xiang Liu, Wei Lin, Hai qing Yin, etc Comprehensive Analysis and Evaluation Study on Structure of Multiple Anchorage Schemes for Suspension Tunnel [J]. China Harbour Construction, 2020, 40 (02): 128-133
[9] Garrison C J, Field J B, MD May. DRAG AND INERTIA FORCES ON A CYLINDER IN PERIODIC FLOW[J]. Journal of the Waterway, 1977, 103(2):193-204.
[10] Wei Zou, Xiaodong Liu, Wei Lin, et al Study on Discharge Characteristics 2: Wave Loads [J]. China Harbour Construction, 2020, 40 (02): 63-67
[11] Lantao Zhao Study on reasonable section type of suspension tunnel for wave-current coupling [D]. Chongqing Jiaotong University, 2020

Analysis of 5 MW Blade Two-Way Fluid-Structure Interaction Characteristics

Xihan Lin[1], Jiang Liu[1, 2, *], Weiting Jiang[1] and Zaiguo Fu[1]

[1] Shanghai University of Electric Power, Shanghai 201306, China

[2] Shanghai Engineering Research Center of Energy-Saving in Heat Exchange Systems, Shanghai 200090, China

*Corresponding author's e-mail: jiangliuxjtu@shiep.edu.cn

Abstract: A CFD-CSD coupled method was employed to establish an aeroelastic model of the NREL 5 MW wind turbine. The two-way fluid-structure interaction (FSI) simulation was carried out for this aeroelastic model. The response curves of blade tip flapwise, edgewise, and torsional deformation under normal operating conditions was obtained. The results show that the blade deformation shows a trend of increasing and then decreasing from the cut-in wind speed to the cut-out wind speed with the rise in wind speed. The maximum deformation appears in the vicinity of rated working conditions. Furthermore, the difference in aerodynamic performance between two-way FSI and one-way FSI has been studied to reveal the influence of the aeroelastic effect on performance. The two-way FSI that considers the blade deformation causes significant periodic fluctuations in the torque and thrust curves, with a maximum deviation of 3.75% in the average torque at 20 m/s. Therefore, the two-way FSI that introduces the aeroelastic effect can predict the aerodynamic performance more accurately.

1. Introduction

Energy is at the heart of modern society and economic development. The widespread use of oil, coal, and natural gas to meet the enormous energy demand has resulted in rising carbon dioxide (CO2) and other toxic gas emissions. To address environmental degradation, mineral resource depletion, and other issues, as well as the traditional energy crisis, renewable energy has emerged as an essential means for the global energy industry. According to Stanford University's Department of Civil and Environmental Engineering, the global wind turbine power generation capacity in 2000 was 72 TW. Even if only 20% of the energy were used, it would be enough to meet all the world's energy consumption at the time (1.6 - 1.8 TW) [1]. As a result, wind energy has enormous development and utilization potential. The size of wind turbines has tended to be large-scale as wind turbine technology has advanced in recent years, with rotor diameters ranging from a few meters to a hundred meters. Nowadays, wind turbines with rotor diameters of more than 200 meters will inevitably lead to blades becoming slenderer and more flexible. Large and flexible blades cause us to think of the blade as an elastomer that undergoes non-linear deformation, and the longer the blade, the greater the elastic effect. The flow field surrounding the blade will be significantly affected by the deformation of the blade. Hence, the aeroelastic problem of the blade should be paid attention to in wind turbine research.

The wind energy community has long been concerned about the safe and stable operation of wind turbines, and the aeroelastic problem is very significant for the operation of large wind turbines. In 2011, Y. Bazilevs et al. [2] first performed the two-way fluid-structure interaction three-dimensional simulation on a full-size large offshore wind turbine. Dong OK Yu et al. [3] used a coupled CFD-CSD approach on the NREL 5 MW reference wind turbine. The effects of two different configurations, gravity, and pitch angle, on the blade aeroelastic response and aerodynamic loads are investigated. Lin Wang et al. [4] performed a one-way FSI simulation of the WindPACT 1.5 MW wind turbine to study the static structural response of the blade as well as the pressure distribution. B. Dose et al. [5] used an open-source CFD code OpenFOAM, with a built-in structural solver BeamFOAM to develop a fluid-structure coupling tool for high-precision wind turbine simulation. The effects of various constructions, including blade cone angle, hub tilt angle, and yaw angle, on blade aeroelastic deformation and aerodynamic performance were investigated. Zhanwei Li et al. [6] developed an aeroelastic model of the blade using BEM and geometrically exact beam theory (GEBT). They investigated the effects of unsteady inflow conditions such as tower shadow (TS), wind shear (WS), and yaw inflow, as well as the impact of gravity on the aerodynamic performance and aeroelastic properties of the wind turbine. Currently, the research on the aeroelastic characteristics of 5 MW wind turbine blades is not comprehensive enough, mainly focusing on studying low wind speed conditions before the pitch-controlled and lacking research on high wind speed conditions after the pitch-controlled. Therefore, this work investigates the aeroelastic deformation characteristics and aerodynamic performance of the NREL 5 MW wind turbine under eight wind speeds in normal operating conditions based on the two-way FSI calculations.

2. Wind Turbine Geometry

This work uses a 5 MW wind turbine designed by the National Renewable Energy Laboratory. Table 1 lists the detailed parameters of the wind turbine, which has a rated output of 5 MW at rated working conditions of 11.4 m/s wind speed and 12.1 rpm rotor speed. Since this work emphasizes the blades, just the rotor-alone is modelled, and tilt and cone angles are neglected.

Table 1 5 MW wind turbine essential parameters

Parameter	Value
Rated power	5 MW
Rotor/hub diameter	126 m/3 m
Number of rotor blades	3
Airfoil section	DU and NACA airfoils
Rated wind speed	11.4 m/s
Rated rotor speed	12.1 rpm
Cut-in/cut-out wind speed	3/25 m/s
Cut-in/cut-out rotor speed	6.9/12.1 rpm

3. Mesh Generation, Computational Method, and Model Validation

3.1. Computational zone, boundary conditions, and mesh generation

Figure 1 presents the computational region of the flow field, which includes the external stationary region and the internal rotating region. The stationary region is a cylindrical region with a length of 3.5 D and a diameter of 2.4 D, the rotating region has a length of 0.16 D and a diameter of 1.2 D, where D denotes the rotor diameter, and the hub center is located at the computational region's coordinate origin. The inlet boundary condition of the calculation region is set to the velocity inlet. In contrast, the pressure outlet is set as the outlet boundary condition, the sidewall boundary condition of the stationary region is set to symmetry, and the blade surface is set to no-slip wall. The part where the external stationary region and the internal rotating region contact is set to the interface. The Fluent Meshing software is used to mesh both the computational domain's internal and external flow fields as unstructured tetrahedral mesh owing to the blade's complicated construction. The flow field has 4.97 million meshes, with 3.03 million meshes in the internal rotating region and 1.94 million in the external stationary region.

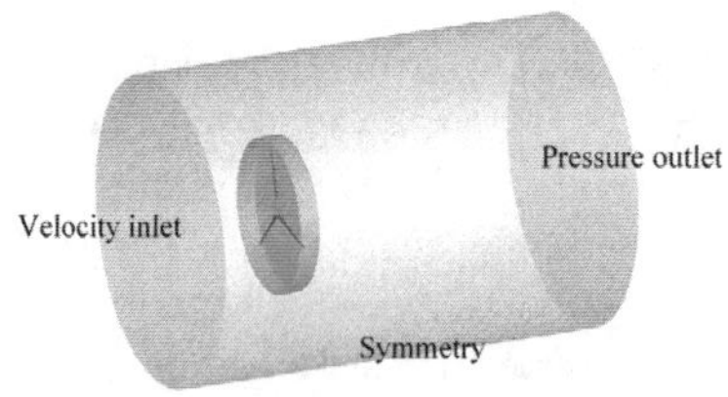

Figure 1 The flow field computational region

This structure model employs the SHELL181 element on the blade for multiple thickness assignments, resulting in a simplified blade hollow model with a thickness ranging from 0.078 m at the blade root to 0.018 m at the blade tip. An orthotropic E-glass/epoxy composite [7] was employed to lay up the blade to strengthen the blade's stiffness in the flapwise, and edgewise directions, and $[\pm45/0/90_2/0_3]_s$ lay-up was applied. The mechanical properties of the orthotropic composite for each layer are given in Table 2, where E_1, E_2, G_{12}, and V_{12} represent the elastic modulus, shear modulus, and Poisson's ratio in the vertical fiber direction and along the fiber direction, respectively. The blade and hub meshing are adopted with Triangles mesh, with a total of 88229 cells and 44033 nodes.

Table 2 Mechanical properties of an orthotropic E-glass/epoxy composite

E_1 (GPa)	E_2 (GPa)	G_{12} (GPa)	V_{12} (-)	ρ (g/cm^3)
39	8.6	3.8	0.28	2.1

3.2. Computational method

The aerodynamic load of the blade is calculated using a CFD solver based on Navier-Stokes equations, and the aeroelastic response of the blade is calculated using a FEM-based CSD solver. The aerodynamic loads of the blades were obtained by solving the Unsteady Reynolds-Averaged Navier-Stokes (URANS) equations, where the URANS equations are defined as follows:

$$\frac{\partial}{\partial t}(\rho u_i)+\frac{\partial}{\partial x_j}(\rho u_i u_j)=-\frac{\partial p}{\partial x_i}+\frac{\partial}{\partial x_j}\left(\mu\frac{\partial u_i}{\partial x_j}-\rho\overline{u_i'u_j'}\right)+S_i \tag{1}$$

Where ρ is the air density, t is the time, u_i and u_j represent the Reynolds mean velocity component of fluid, p is the pressure, μ indicates the coefficient of dynamic viscosity, S_i is the generalized source term ($i, j = 1,\ 2,\ 3$).

The structural dynamic response of the blade is calculated using the control equations listed below:

$$Mu''+Cu'+Ku=F(t) \tag{2}$$

Where M , C , and K represent the mass matrix, damping matrix, and stiffness matrix, respectively; u'' , u' , and u represent the acceleration, velocity, and displacement of the blade structure, respectively; and $F(t)$ represents the force acting on the blade.

The research employs a loosely coupled method to solve the CFD solver and the CSD solver by solving the fluid control equations and the solid control equations sequentially. This means that the transient flow field is solved first, followed by the dynamic response of the structure. The same time step is applied to the transient simulation of the flow and structure fields.

3.3. Model validation

Five different mesh schemes were selected for mesh-independent verification to ensure the accuracy of the simulation. As seen from Table 3, the error of the output torque decreases continuously as the number of meshes increases. The scheme with a mesh number of 4.97 million is selected for further research considering accuracy and computational efficiency.

Table 3 Output torque for different mesh numbers

Number (million)	2.15	3.21	4.10	4.97	6.13	Reference
Torque (kN·m)	3541	3719	3758	4012	4207	4180
Error (%)	15.29	11.03	10.10	4.02	0.65	

The aerodynamic and structural models are verified separately to confirm the validity of the aeroelastic model. To begin, the aerodynamic model is validated by performing sufficient steady calculations. Figure 2 compares the rotor torque at various wind speeds to the results calculated by FAST [8]. The simulation results are very close to the FAST results up to the rated wind speed of 11.4 m/s. The torque predicted by CFD is marginally lower than the torque predicted by FAST, and the error grows as the wind speed rises, reaching a peak of 9.80% at 20 m/s, indicating that the aerodynamic model of this work meets the accuracy requirements and can be used for further two-way FSI calculations.

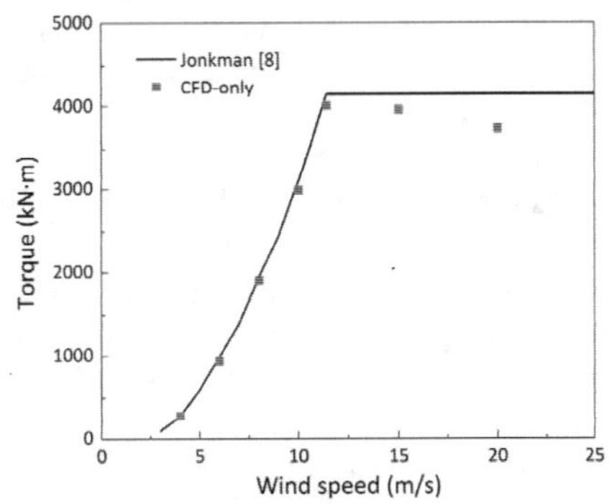

Figure 2 Torque comparison of various wind speeds

The modal analysis is used to confirm the rationality of the structure model. A single blade without load and rotation is employed in the modal calculations, and the blade root is set to be fixed support. Table 4 compares the results of Jeong [9] with the first three modes of natural frequencies obtained from modal calculations. Differences in blade material and lay-up parameters (number of layers, thickness, orientation) resulted in different natural frequencies.

Table 4 The first three natural frequencies of the blade

Model number	Jeong (Hz)	Present (Hz)
1	0.673	0.559
2	1.106	1.297
3	1.926	1.686

4. Results and Discussions

In this study, the two-way FSI calculation is employed to reveal the aeroelastic deformation behavior of the blade in the normal operating range. Between the cut-in and cut-out wind speeds, 4 m/s, 6 m/s, 8 m/s, 10 m/s, 11.4 m/s, 13 m/s, 15 m/s, and 20 m/s were chosen to calculate. The results of the converged steady calculations are used as the initial flow field in the two-way FSI calculations to ensure the accuracy of the unsteady calculations. The two-way FSI simulations for all operating conditions go through the same eight rotational periods, and the time step is the time it takes to rotate the blade by one degree. Figure 3 (a) presents the definition of blade labeling and rotor rotational direction. Figure 3 (b) depicts the positive flapwise and edgewise deformation directions with arrows. The positive torsional deformation translates into a nose-up rotation.

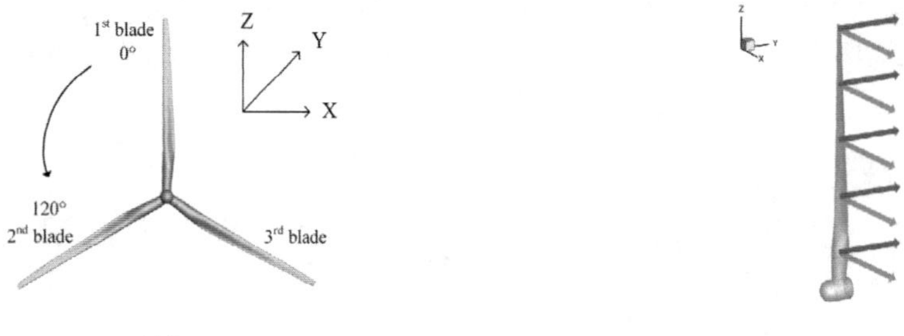

(a) (b)

Figure 3 Definition of wind turbine direction
(a) Definition of blade labeling and rotor rotation direction; (b) The direction of flapwise deformation and edgewise deformation. Blue arrows indicate a positive flapwise direction, whereas green arrows indicate a positive edgewise direction.

4.1. Blade deformation characteristics at various wind velocity

Figure 4 depicts the variation of aeroelastic deformation with azimuth at the blade tip under various velocities. As seen in Figure 4, the flapwise, edgewise, and torsional deformations increase and decrease as velocity increases. In the rated wind speed range, flapwise, edgewise, and torsional deformation increase with increasing velocity and amplitude. This is because as the wind speed increases in the rated wind speed range, the inflow angle and angle of attack increase, causing the aerodynamic loads to increase gradually, resulting in aeroelastic deformation to increase gradually. However, the aeroelastic deformation of the blade does not always increase with increasing wind speed. From 11.4 m/s increased to 13 m/s, due to exceeding the rated wind speed, the wind turbine stalled, and the blade began to pitch, leading to the blade angle of attack is reduced so that the flapwise and torsional deformation appeared to a significant reduction, torsional deformation even appeared with the opposite torsional angle. The edgewise deformation is not the same as the other two directions' deformation, from 11.4 m/s to 13 m/s edgewise deformation but also increased because the edgewise direction and blade aerodynamic tangential force direction are consistent. Due to inertial force, edgewise deformation at the beginning of the blade becomes pitch still appeared to slightly increase. In the pitch-controlled operational range, when the velocity continues to rise, the flow angle remains increased, the pitch angle will increase, and the angle of attack decreases gradually so that the flapwise, edgewise, and torsional deformation show a trend of reduction. However, the amplitude does not decrease and still maintains a significant fluctuation. In conclusion, the wind turbine operating in high-velocity condition is more susceptible to the fatigue fracture phenomenon. The wind turbine design should pay more attention to the blade in the high-velocity aeroelastic response and the load situation.

Figure 4 (d) illustrates the blade tip aeroelastic deformation law with wind speed. As seen in Figure 4 (d), the flapwise deformation increases with increasing wind speed toward the flapwise positive direction and reaches its maximum value at 11.4 m/s. The edgewise and torsional deformation increased in the negative direction as the wind speed increased. However, when the wind speed exceeds 11.4 m/s, the blade pitch angle gradually increases and the aeroelastic deformation begins to decrease, which is consistent with the previous studies [4, 8, 9]. In the normal operating range, as the wind speed gradually grows, the blade's aeroelastic deformation first increases and then decreases, with the maximum deformation occurring in the vicinity of rated operating conditions. Furthermore, the maximum blade tip flapwise deformation of 2.647 m is much smaller than the horizontal distance from the hub center to the yaw axis of 5.019 m [8], indicating that the wind turbine model adopted in this study will not have blade-to-tower collisions during regular operation.

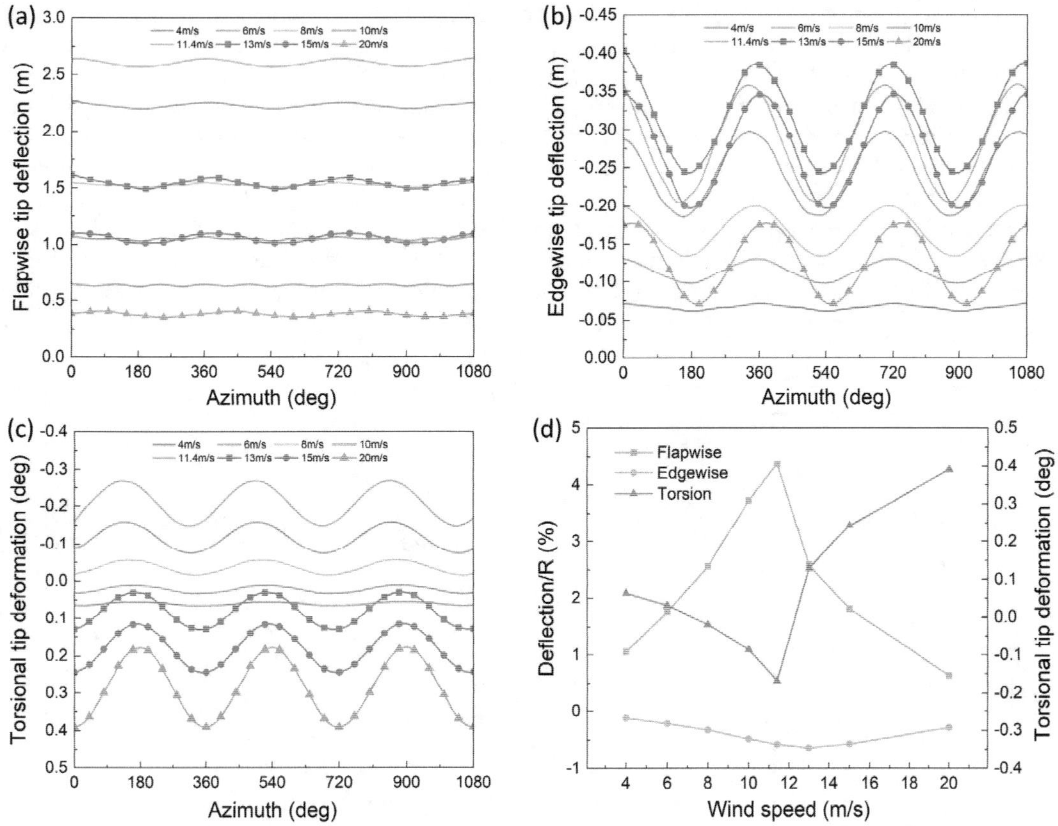

Figure 4 Variations in the azimuth of blade tip deformations at various velocities
(a) Flapwise deflection; (b) Edgewise deflection; (c) Torsional deformation; (d) Blade tip deformations at various wind speeds

4.2. Overall aerodynamic performance under two-way FSI

Two-way FSI considers the influence of structural deformation on the fluid so that it significantly impacts the aerodynamic performance. To highlight the ability of two-way FSI to capture fluctuations in aerodynamic loads, one-way FSI is introduced to contrast with it. The impact of two-way FSI and one-way FSI on the overall aerodynamic performance is discussed in this section. Figure 5 depicts the variation of torque and thrust with azimuth for six wind speeds (4 m/s, 8 m/s, 11.4 m/s, 13 m/s, 15 m/s, and 20 m/s), comparing the aerodynamic performance of two-way FSI and one-way FSI. Figure 5 shows that the one-way FSI performance curve is relatively flat and less fluctuating, whereas the two-way FSI performance curve shows a harmonic motion trend. This is because the one-way FSI ignores the influence of structural deformation on the flow field, resulting in minor aerodynamic load variation.

The blade undergoes periodic deformation, which changes the surrounding flow field, resulting in periodic fluctuations in the overall performance curve. In the rated wind speed range, the average values of two-way FSI and one-way FSI torque at 4 m/s, 8 m/s, and 11.4 m/s differed by 0.521%, 0.533%, and 0.562%, respectively; the average values of thrust differed by 0.153%, 0.867%, and 1.704%, respectively. In the low-velocity region, as the increase of wind speed, the blade angle of attack increases, as does the blade deformation and amplitude, resulting in a gradual rise in the aeroelastic effect. In the pitch-controlled operational range, the pitch angle gradually increases. In contrast, the blade angle of attack suddenly drops, and thus the aeroelastic effect sharply decreases. Still, the high-velocity region response amplitude did not decrease as the pitch angle increased, so the aeroelastic effect appeared to increase the trend. The average values of torque at 13 m/s, 15 m/s, and 20 m/s differed by 0.350%, 0.876%, and 3.748%, respectively; the average values of thrust differed by 0.429%, 0.645%, and 2.693%, respectively. The aeroelastic effect gradually increases as the wind speed increases. When the wind speed exceeds the rated wind speed of 11.4 m/s, the blade begins to pitch, the aeroelastic effect is sharply reduced, and then increases as the wind speed grows.

Furthermore, the magnitude of the variation in torque error is significantly greater than that of the variation in thrust, indicating that the aeroelastic effect is more significant for torque. It can be derived that the higher the wind speed during wind turbine operation, the more significant the impact of the aeroelastic effect on the aerodynamic performance. As a result, for wind turbine simulation in high wind speed conditions, the aeroelastic effect should be considered to ensure an accurate forecast of the large wind turbine's aerodynamic performance.

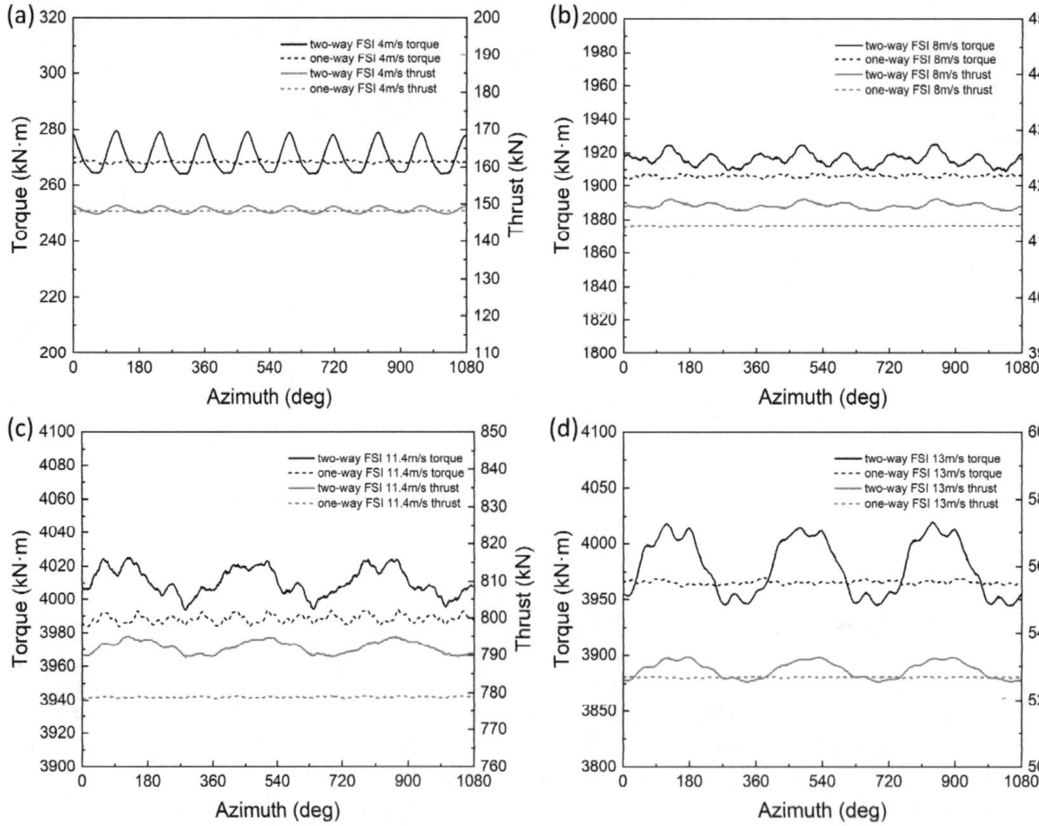

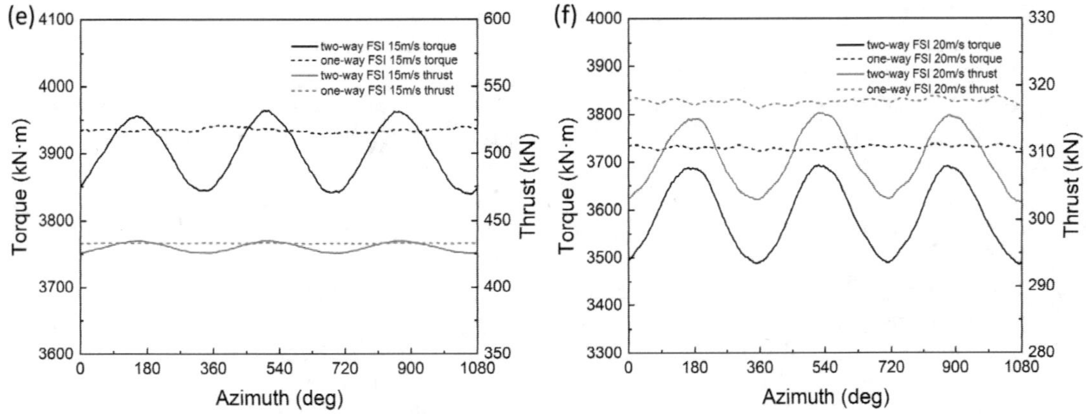

Figure 5 The azimuthal variations of torque and thrust
(a) 4 m/s; (b) 8 m/s; (c) 11.4 m/s; (d) 13 m/s; (e) 15 m/s; (f) 20 m/s.

5. Conclusions

In this study, an aeroelastic model of a large horizontal-axis wind turbine is developed by combining the CFD and CSD methods. The aeroelastic response and aerodynamic performance of two-way FSI were analyzed, and the following conclusions were drawn:

(1) Blade tip flapwise, edgewise, and torsional deformation with azimuthal variations show the periodic variation trend. Still, flapwise direction deformation is much larger than edgewise and torsional direction deformation, indicating that flapwise direction deformation dominates the aeroelastic deformation of the blade. The amplitude of flapwise deformation compared to its deformation is significantly smaller than that of the other two directions, indicating that flapwise deformation is more stable.

(2) In the rated wind speed range, as the velocity grows, the inflow angle increases and followed by the angle of attack increases, therefore, the blade deformation gradually increases, and the deformation amplitude also gradually increases. In the pitch-controlled operational range, with further growth of wind speed, the blade begins to pitch, and the inflow angle continuously increases. Still, the angle of attack gradually decreases, resulting in the aeroelastic deformation of the blade in the high-velocity region decreasing instead.

(3) Two-way FSI causes the overall torque and thrust to fluctuate periodically because it accounts for the influence of structural deformation on the fluid. The impact of the aeroelastic effect on aerodynamic performance will gradually increase as the wind speed grows, with a sudden drop when the blade begins to pitch, followed by a trend of increase.

Acknowledgments

This study was supported by research funds from Shanghai 2020 Annual Science and Technology Innovation Action Plan: Social development and Science & Technology Project (No. 20dz1205302).

References

[1] Archer, C. L., and Jacobson, M. Z., 2005 *Evaluation of global wind power.* Journal of Geophysical Research: Atmospheres, 110 (D12).

[2] Bazilevs, Y., Hsu, M. C., Kiendl, J., Wüchner, R., and Bletzinger, K. U., 2011 *3D simulation of wind turbine rotors at full scale. Part II: Fluid-structure interaction modeling with composite blades.* International Journal for Numerical Methods in Fluids, 65 (1-3), pp. 236-253.

[3] Yu, D. O., and Kwon, O. J., 2014 *Predicting wind turbine blade loads and aeroelastic response using a coupled CFD–CSD method.* Renewable Energy, 70, pp. 184-196.

[4] Wang, L., Quant, R., and Kolios, A., 2016 *Fluid structure interaction modelling of horizontal-axis wind turbine blades based on CFD and FEA*. Journal of Wind Engineering and Industrial Aerodynamics, 158, pp. 11-25.

[5] Dose, B., Rahimi, H., Herráez, I., Stoevesandt, B., and Peinke, J., 2018 *Fluid-structure coupled computations of the NREL 5 MW wind turbine by means of CFD*. Renewable Energy, 129, pp. 591-605.

[6] Li, Z., Wen, B., Dong, X., Peng, Z., Qu, Y., and Zhang, W., 2020 *Aerodynamic and aeroelastic characteristics of flexible wind turbine blades under periodic unsteady inflows*. Journal of Wind Engineering and Industrial Aerodynamics, 197.

[7] Daniel, I. M., Ishai, O., Daniel, I. M., and Daniel, I., 1994, Engineering mechanics of composite materials, Oxford university press New York.

[8] Jonkman, J., Butterfield, S., Musial, W., and Scott, G., 2009, "Definition of a 5-MW Reference Wind Turbine for Offshore System Development," United States.

[9] Jeong, M.-S., Cha, M.-C., Kim, S.-W., Lee, I., and Kim, T., 2014 *Effects of torsional degree of freedom, geometric nonlinearity, and gravity on aeroelastic behavior of large-scale horizontal axis wind turbine blades under varying wind speed conditions*. Journal of Renewable and Sustainable Energy, 6 (2).

The Correlation between the Numerical Analysis of Wake Dynamics and the General Theory of Aerodynamic Instability of a Flapping Foil

Liping Sun[1*], Peijian Zhou[2], and Yixiang Bao[3]

[1] China Energy Technology and Economics Research Institute, China Energy Investment Corporation Ltd., Beijing 102211, China

[2] College of Metrology and Measurement Engineering, China Jiliang University, Hangzhou 310018, China

[3] National Institute of Clean and Low Carbon Energy, China Energy Investment Corporation Ltd., Beijing 102209, China

*sunliping@zju.edu.cn

Abstract: We study the correlation between the wake dynamics analyzed by numerical simulation and the general theory of aerodynamic instability of a flapping foil. The pure pitching motion and the pure heaving motion are considered in this paper. The main parameter is the Reynolds number, here we fixed the Reynolds number at Re = 1700. We found an interesting phenomenon in our research, which is that the lift amplitude C_Y of the numerical simulation is consistent with the the results of the general theory of aerodynamic instability, even at low frequencies where the additional mass effect is less important. And the average thrust C_T(mean) increases exponentially with the Strouhal number $S_{RA} = fA/U$, which is also an interesting phenomenon. By making the lift amplitude C_Y dimensionless with the amplitude θ_0 and h_0, we find that the dimensionless lift amplitudes C_Y/θ_0 or C_Y/h_0 are distributed along the same trend under different amplitudes. Through comparison and analysis of the pure pitching motion and the pure heaving motion, We clearly found that the lift on the heaving foil is slightly lower and the change of the lift is slightly more dramatic when the frequency is small.

1. Introduction

Aerodynamic researchers have realized early the importance of airfoil rotation in the study of fluttering wings. As proposed by Munkt [1], when two-dimensional airfoils are translated and at the same time simultaneously rotated with small amplitudes, additional circulation to maintain the Kutta condition at the trailing edge of the airfoil is required. Research related to this idea has aroused the interest of many scientists. Theodorsen [2] later proposed the general theory of aerodynamic instability. He got the aerodynamic forces of an oscillating airfoil which has three independent degrees of freedom, and he also pointed out the main theory, which is mainly based on potential flow and the Kutta condition. In recent years, many researchers have used experimental or numerical simulation methods to analyze aerodynamics and wake dynamics. Deng [3,4,5,6] used the finite volume method to calculate the lift and thrust forces of a flapping foil. Bohl and Koochesfahani's [7] experiments performed the dynamics and analyzed the vortical structure in the wake behind an oscillating airfoil, especially under the working condition with high frequency.

In conclusion, the dynamics in the wale behind a flapping foil and the general theory of aerodynamic instability have been extensively studied. However, there is little research on the correlation between the numerical analysis of wake dynamics and the general theory of aerodynamic instability of a flapping foil. In this paper, we study the general theory of aerodynamic instability of a flapping foil, we also consider the pure pitching motion and the pure heaving motion. Based on the general theory of Theodorsen [2], we derive the lift and moment equations of the flapping foil, and then studied the correlation between the numerical analysis of wake dynamics and the general theory. Then, we make the lift amplitude dimensionless with the amplitude to explore the evolution and development trends. Finally, we make a direct comparison of the characteristics between the pure pitching foil and the pure heaving foil.

2. Definition of the studied

In this paper, the motion of a flapping foil usually includes vertical heaving motion and rotation around the center of the foil (which is called pitching motion), as we have previously used [3]. In our research, as shown in figure.1, we choose a NACA0015 airfoil as the research object, and apply a specified motion to the foil, and make the NACA0015 foil sinusoidal oscillation, pure pitching motion or pure heaving motion. Here, A is the distance between the highest point and the lowest point reached by the tail tip during the flapping motion of the foil, c is the chord length of the NACA0015 foil, and U is the uniform flow velocity at infinity. And in our research, we defined the amplitudes of pitching motion and heaving motion as θ_0 and h_0, respectively, and their typical motion equations are as follows:

$$\theta(t) = \theta_0 \, sin(2\pi f t)$$
$$h(t) = h_0 \, sin(2\pi f t + \varphi)$$

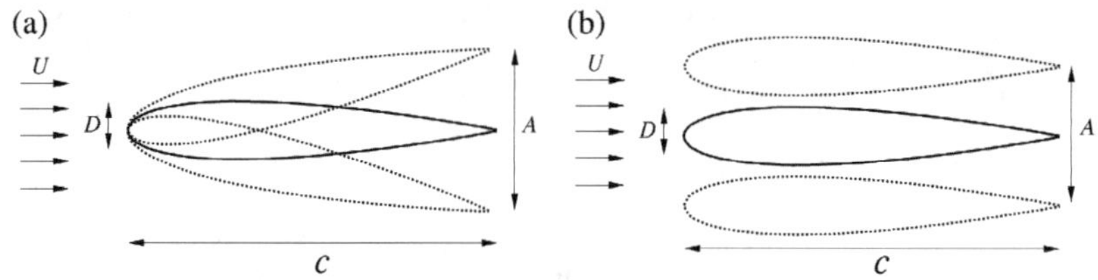

Figure 1. Schematic diagram of the foil profiles. (a) the pure pitching foil (b) the pure heaving foil.

Here, f is the flapping frequency, φ is the phase difference between rotational motion and translational motion. For a flapping wing with a specified motion, the phase difference is generally taken as $\pi/2$. In our research, we define the Reynolds number $Re = Uc/\upsilon$, where υ is the kinematic viscosity, c is the chord length, and U is the uniform flow velocity at infinity. We choose $Re = 1700$ for our simulations. And $Re = 1700$ is equivalent to $Re_D = 255$, where we use the thickness D of an airfoil as feature length to define the Reynolds number, $Re_D = UD/\upsilon$. The amplitude and frequency that characterize the wake dynamics are usually dimensionless to obtain the dimensionless amplitude A_D and frequency S_r, where Sr can also be equivalently called the Strouhal number, which is defined as:

$$A_D = A/D \text{ and } S_r = fD/U$$

In the study of biomimetic dynamics, many researchers are widely used to using an exception parameter in the study of the wake characteristics, which is usually defined as:

$$S_{rA} = fA/U = A_D \times S_r$$

The propulsive efficiency of a flapping foil is usually defined as:

$$\eta = \frac{U\langle T\rangle}{\langle P\rangle}$$

where T is the thrust of the foil, here, $\langle T\rangle$ and $\langle P\rangle$ represent the average thrust and average input power respectively, which is time average.Therefore, we can calculate the instantaneous input power by the method as follows:

$$P(t) = -Y(t)\frac{dh}{dt} - M(t)\frac{d\theta}{dt}$$

Where $Y(t)$ is the vertical force, which means the lift force, $M(t)$ is the moment especially for a pitching foil ,which means the moment at the pitching axis (the center of rotation in this paper is at the head of the foil). The $\frac{dh}{dt}$ and $\frac{d\theta}{dt}$ are the velocity of the heaving motion and the angular velocity of the pitching motion, respectively. In this study, the first term on the right of the above equation is usually used as the input power, especially for the pure heaving foil, while the second term on the right of the above equation is used to calculate the input power, mainly used for the pure pitching foil.

The motion of flapping foil mainly includes two degrees of freedom of rotation and translation. According to the motion of its two degrees of freedom, the motion of flapping foil usually can be divided into pure pitching motion, pure heaving motion, and pitching and heaving coupled motion.

3. Methods

3.1. The finite volume method

In our research, by using the finite-volume method, we numerical simulate the unsteady flow around the airfoil. We assume that the fluid is Newtonian fluid, and the flow is also incompressible flow. Due to the low Reynolds number, we think that no turbulence model is needed for the flow. For the pure pitching foil, the circle inside the resign of numerical simulation has a moving grid, the radius of the circle is 5.0c, and the center is fixed at the leading edge of the airfoil, where the grid of the calculated area outside the circle is fixed, just as we have previously used [4,5]. While for a pure heaving foil, the whole calculated area moves vertically up and down with the heaving foil, as we have previously used [6]. As shown in the following equation, here, the integral form of the equations, which usually defined in an arbitrary moving volume V, and assumed the volume which bounded by a closed surface S states

$$\frac{d}{dt}\int_V \rho U dV + \oint_S ds.\rho(U - U_b)U = \oint_S ds.(-pI + \nu\nabla U) \tag{1}$$

Where, U is the fluid velocity at infinity, U_b means the boundary velocity of a cell, ρ is the density of the fluid, p is the pressure, and ν is the kinematic viscosity of the fluid, which is defined as $\nu = \mu/\rho$. If you want to know more details about the discretization and implementation of the computational boundary conditions, please refer to [8].

In our study, for two-dimensional simulations, the computational area is a rectangle with dimensions of $20c \times 18c$. The whole mesh of the area is about 209492 cells for the simulations. We used the same solver as in the previous article [3,4,5], in which the solver has been validated.

3.2. The general theory of the aerodynamic instability

The linear theory is generally believed to be based on the theory of potential flow theory and the Kutta condition. As shown in Figure 2, the airfoil is represented by a straight line. The chord length is $2b$, and b is the unit length. Suppose the chord extends along the x axis, and for the wing, the leading edge corresponds to $x=-1$ and the trailing edge corresponds to $x=+1$, the coordinate $x=a$ represents the rotation axis of the wing, and $x=c$ represents the coordinate of the aileron hinge. Other related parameters are shown in Figure 2.

We assume that the motion of each degree of freedom of the wing is sinusoidal oscillation, and can be expressed as follows:

$$\alpha = \alpha_0 e^{i((pt+\varphi_0)}$$

$$\beta = \beta_0 e^{i((pt+\varphi_1)} \qquad (2)$$
$$h = h_0 e^{i((pt+\varphi_2)}$$

The α_0, β_0, and h_0 are constants and represent the maximum amplitudes in different degrees of freedom, respectively. φ_0, φ_1 and φ_2 are the phase angle sin different degrees. And the parameter p determines the frequency of oscillation. The relationship between p and frequency is:

$$p = \frac{kv}{b} \qquad (3)$$

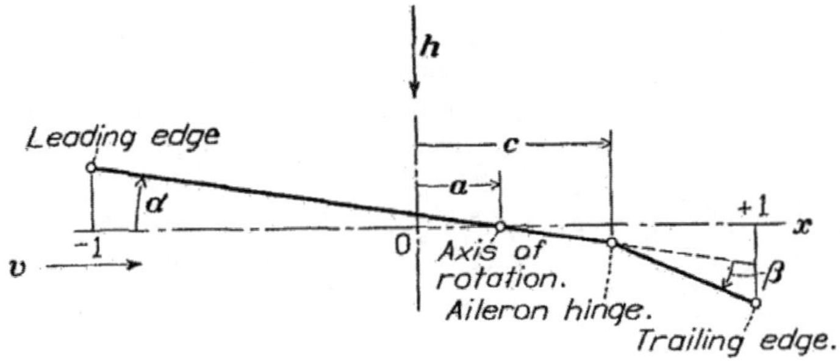

Figure 2. Parameters of the wing and aileron combinations

Therefore, the parameter k is defined as $k = pb/v$. It can be seen that $2\pi/k$ is the wavelength between the continuous waves in the wake with the half-chord length $b = c/2$as the reference length.

According to the research of Theodorsen [2], the calculation formulas of the lift and moment of the three-degree-of-freedom airfoil can be expressed as:

$$P = -\rho b^2(v\pi\dot{\alpha} + \pi\ddot{h} - \pi ba\ddot{\alpha} - vT_4\dot{\beta} - T_1\ddot{\beta}) - 2\pi\rho vbC(K)Q \qquad (4)$$

$$M_\alpha = -\rho b^2\left[\pi(\frac{1}{2} - a)vb\dot{\alpha} + \pi b^2(\frac{1}{8} + a^2)\ddot{\alpha} + T_{15}v^2\beta + T_{16}v\dot{\beta} + 2T_{13}b^2\ddot{\beta} - a\pi b\ddot{h}\right] + 2\rho vb^2\pi(a + \frac{1}{2})C(k)Q \qquad (5)$$

$$M_\beta = -\rho b^2\left[T_{17}vb\dot{\alpha} + 2T_{13}b^2\ddot{\alpha} + \frac{1}{\pi}v^2T_{18}\beta - \frac{1}{2\pi}v^2T_{19}\dot{\beta} - \frac{1}{\pi}T_3b^2\ddot{\beta} - T_1b\ddot{h}\right] - \rho vb^2T_{12}C(k)Q \qquad (6)$$

In the above equation, the definition of Q is:

$$Q = v\alpha + \dot{h} + b(\frac{1}{2} - a)\dot{\alpha} + \frac{1}{\pi}T_{10}v\beta + \frac{b}{2\pi}T_{11}\dot{\beta} \qquad (7)$$

In those equations, the real part of P is related to the real part of the motion equation, just as $\alpha = \alpha_0\,cos(pt + \varphi_0)$, and the imaginary part of P is related to the imaginary part of the motion equation, like $\alpha = \alpha_0\,sin(pt + \varphi_0)$, the same way for M_α and M_β. For more details, you can refer to the research of Theodorsen [2].

In our study of the analysis of the motion equation of the flapping foil, we only need to decompose one of the real parts or the imaginary part of the equation. Here we choose to analyze the imaginary part, so the motion equation of the flapping foil can be expressed as:

$$\alpha = \alpha_0\,sin(pt + \varphi_0)$$
$$\beta = \beta_0\,sin(pt + \varphi_1) \qquad (8)$$
$$h = h_0\,sin(pt + \varphi_2)$$

The α_0, β_0 and h_0 are constants and represent the maximum amplitudes in different degrees of freedom, respectively, φ_0, φ_1, and φ_2 are the phase angles.

The lift and moment of the flapping foil can be expressed as follows:

$$P = -\rho b^2[v\pi\alpha_0p\,cos(pt + \varphi_0) - \pi h_0p^2\,sin(pt + \varphi_2) + \pi ba\,\alpha_0p^2\,sin(pt + \varphi_0) - vT_4\beta_0p\,cos(pt + \varphi_1) + T_1b\,\beta_0p^2\,sin(pt + \varphi_1)] - 2\pi\rho vbF\left[v\alpha_0\,sin(pt + \varphi_0) + h_0p\,cos(pt + \varphi_2) + b(\frac{1}{2} - a)\alpha_0p\,cos(pt + \varphi_0) + \frac{T_{10}}{\pi}v\beta_0\,sin(pt + \varphi_1) + \frac{T_{11}}{2\pi}b\beta_0p\,cos(pt + \varphi_1)\right] -$$

$$2\pi\rho v b G\left[v\alpha_0 \cos(pt+\varphi_0) - h_0 p \sin(pt+\varphi_2) - b(\tfrac{1}{2}-a)\alpha_0 p \sin(pt+\varphi_0) + \tfrac{T_{10}}{\pi} v\beta_0 \cos(pt+\varphi_1) - \tfrac{T_{11}}{2\pi} b\beta_0 p \sin(pt+\varphi_1)\right] \quad (9)$$

$$M_\alpha = -\rho b^2\left[\pi(\tfrac{1}{2}-a) v b\alpha_0 p \cos(pt+\varphi_0) - \pi b^2(\tfrac{1}{8}+a^2)\alpha_0 p^2 \sin(pt+\varphi_0) + T_{15} v^2 \beta_0 \sin(pt+\varphi_1) + T_{16} v b\beta_0 p \cos(pt+\varphi_1) - 2T_{13} b^2 \beta_0 p^2 \sin(pt+\varphi_1) + a\pi b\, h_0 p^2 \sin(pt+\varphi_2)\right] + 2\rho v b^2 \pi(\tfrac{1}{2}+a)F\left[v\alpha_0 \sin(pt+\varphi_0) + h_0 p \cos(pt+\varphi_2) + b(\tfrac{1}{2}-a)\alpha_0 p \cos(pt+\varphi_0) + \tfrac{T_{10}}{\pi} v\beta_0 \sin(pt+\varphi_1) + \tfrac{T_{11}}{2\pi} b\beta_0 p \cos(pt+\varphi_1)\right] + 2\rho v b^2 \pi(\tfrac{1}{2}+a)G\left[v\alpha_0 \cos(pt+\varphi_0) - h_0 p \sin(pt+\varphi_2) - b(\tfrac{1}{2}-a)\alpha_0 p \sin(pt+\varphi_0) + \tfrac{T_{10}}{\pi} v\beta_0 \cos(pt+\varphi_1) - \tfrac{T_{11}}{2\pi} b\beta_0 p \sin(pt+\varphi_1)\right] \quad (10)$$

$$M_\beta = -\rho b^2\left[T_{17} v b\alpha_0 p \cos(pt+\varphi_0) - 2T_{13} b^2 \alpha_0 p^2 \sin(pt+\varphi_0) + \tfrac{1}{\pi} v^2 T_{18}\beta_0 \sin(pt+\varphi_1) - \tfrac{1}{2\pi} v b T_{19}\beta_0 p \cos(pt+\varphi_1) + \tfrac{1}{\pi} T_3 b^2 \beta_0 p^2 \sin(pt+\varphi_1) + T_1 b h_0 p^2 \sin(pt+\varphi_2)\right] - \rho v b^2 T_{12} F\left[v\alpha_0 \sin(pt+\varphi_0) + h_0 p \cos(pt+\varphi_2) + b(\tfrac{1}{2}-a)\alpha_0 p \cos(pt+\varphi_0) + \tfrac{T_{10}}{\pi} v\beta_0 \sin(pt+\varphi_1) + \tfrac{T_{11}}{2\pi} b\beta_0 p \cos(pt+\varphi_1)\right] - \rho v b^2 T_{12} G\left[v\alpha_0 \cos(pt+\varphi_0) - h_0 p \sin(pt+\varphi_2) - b(\tfrac{1}{2}-a)\alpha_0 p \sin(pt+\varphi_0) + \tfrac{T_{10}}{\pi} v\beta_0 \cos(pt+\varphi_1) - \tfrac{T_{11}}{2\pi} b\beta_0 p \sin(pt+\varphi_1)\right] \quad (11)$$

In our research, the chord length of the foil is c, and $c = 1$, the center of rotation is at the head of the foil, so $a = 0$, the motion equation of the pitching is as follows:

$$\theta(t) = \theta_0 \sin(2\pi f t) \quad (12)$$

The lift force of the pitching foil is:

$$P = -\rho b^2[v\pi\alpha_0 p \cos(pt+\varphi_0)] - 2\pi\rho v b F\left[v\alpha_0 \sin(pt+\varphi_0) + b(\tfrac{1}{2}-a)\alpha_0 p \cos(pt+\varphi_0)\right] - 2\pi\rho v b G\left[v\alpha_0 \cos(pt+\varphi_0) - b(\tfrac{1}{2}-a)\alpha_0 p \sin(pt+\varphi_0)\right] \quad (13)$$

For the heaving foil in our research, the motion equation is:

$$h(t) = h_0 \sin(2\pi f t + \tfrac{\pi}{2}) \quad (14)$$

The lift force of the heaving foil is:

$$P = -\rho b^2[-\pi h_0 p^2 \sin(pt+\varphi_2)] - 2\pi\rho v b F[h_0 p \cos(pt+\varphi_2)] - 2\pi\rho v b G[-h_0 p \sin(pt+\varphi_2)] \quad (15)$$

4. Results

We employed several non-dimensional parameters, the Strouhal number $S_r = fD/U$, $k = \pi f c/U$, $S_{rA} = fA/U = A_D \times S_r$, and the non-dimensional flapping amplitude $A_D = A/D$, which characterize the dynamical features of the flow wakes. It is possible to make a direct comparison between the two methods and we also make a horizontal comparison between the pure pitching foil and the pure heaving foil.

In Figure 3, we show the lift amplitude C_Y of a pitching foil as a function of the Strouhal number $k = \pi f c/U$ at different pitch angles when $Re = Uc/v = 1700$, which is compared with the results derived from the general theory of aerodynamic instability based on Theodorsen [2]. It can be found that the numerical simulation results agree very well with the results of the general theory of aerodynamic instability, even at low frequencies where the additional mass effect is less important.

As shown in Figure 3, we also dimensionless the lift amplitude C_Y with the flapping amplitude θ_0, and analyze the variation of C_Y/θ_0 as a function of the Strouhal number k at different pitch angles when $Re = 1700$. It should be noted that if we dimensionless the moment of the flapping foil, it is more reasonable to use the square of the amplitude θ_0 to dimensionless. And we get the result of the general theory of aerodynamic instability. We can find the results is very effective over the entire pitch angle range. At different pitch angles, all dimensionless lift amplitudes are distributed along the same trend.

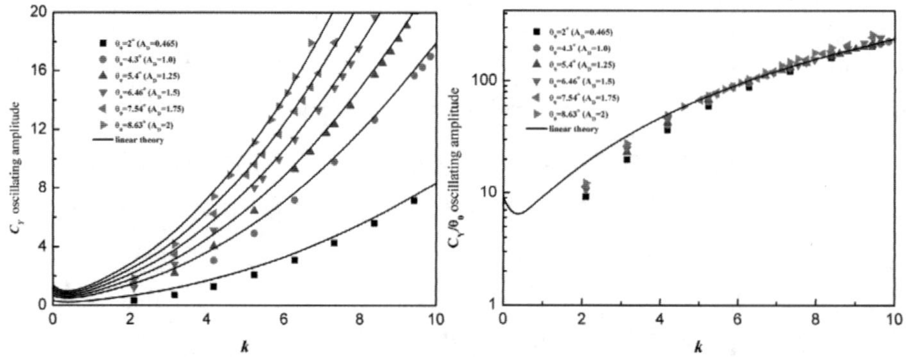

Figure 3. $Re = 1700$, the lift amplitude C_Y of a pitching foil as a function of the Strouhal number $k = \pi fc/U$ at different pitch angles. And the dimensionless lift amplitude C_Y/θ_0 as a function of the Strouhal number k.

In addition, we also studied the relationship between the average thrust C_T(mean) of the pitching foil and the Strouhal number $S_{rA} = fA/U = A_D \times S_r$, as shown in figure 4. The average thrust C_T at all amplitudes showed the same trend with S_{rA}. Through fitting, it is found that the average thrust C_T(mean) increases exponentially with the Strouhal number S_{rA}, which is an interesting phenomenon.

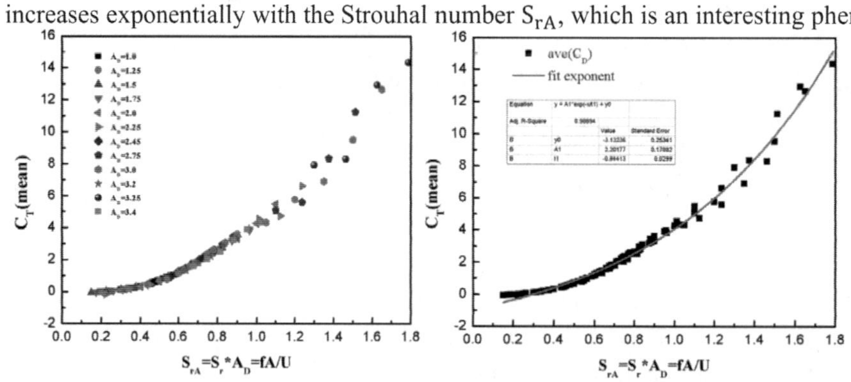

Figure 4. $Re = 1700$, the average thrust C_T(mean) of the pitching foil as a function of the Strouhal number $S_{rA} = fA/U$.

Similarly, in Figure 5, we show the lift amplitude C_Y of a heaving foil as a function of the Strouhal number $k = \pi fc/U$ under different heaving amplitudes when $Re = 1700$, which is also compared with the results derived from the general theory of aerodynamic instability based on Theodorsen [2]. The results of the numerical simulation are in good agreement with the derivation of the general theory. We also dimensionless the lift amplitude C_Y with the heaving amplitude h_0, and analyze the variation of C_Y/h_0 as a function of the Strouhal number k. It is clear that all dimensionless lift amplitudes C_Y/h_0 are distributed along the same trend under different heaving amplitudes, as the results of a pitching foil.

However, compared with the results of a pitching foil, the lift on the heaving foil is slightly lower and the change of the lift is slightly more dramatic when the frequency is small.

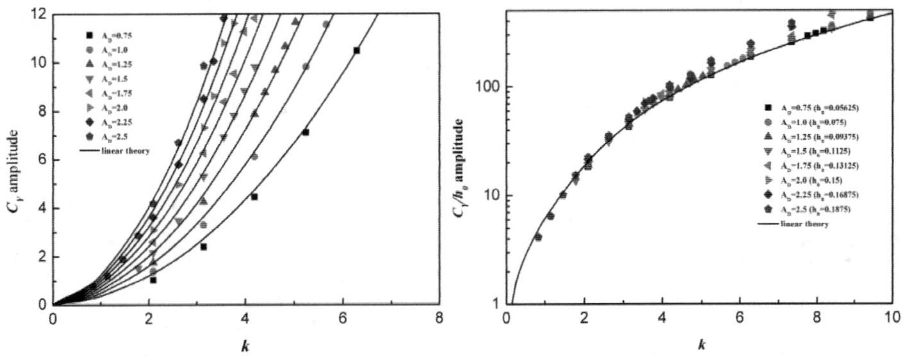

Figure 5. $Re = 1700$, the lift amplitude C_Y of a heaving foil as a function of the Strouhal number $k = \pi fc/U$ at different pitch angles. And the dimensionless lift amplitude C_Y/θ_0 as a function of the Strouhal number k.

5. Conclusions

In this paper, we have studied the correlation between the numerical analysis of wake dynamics and the general theory of aerodynamic instability of a flapping foil. Both a pure pitching motion and a pure heaving motion are considered respectively. After deriving the lift equation of the flapping foil from the general theory of aerodynamic instability, we found an interesting phenomenon that the lift amplitude C_Y of the numerical simulation are in good agreement with the results of the general theory of aerodynamic instability, and the average thrust C_T(mean) increases exponentially with the Strouhal number $S_{rA} = fA/U$. By making the lift amplitude C_Y dimensionless with the amplitude θ_0 and h_0, we find that the dimensionless lift amplitudes C_Y/θ_0 or C_Y/h_0 are distributed along the same trend under different amplitudes. It is noted that if we dimensionless the moment of the flapping foil, it is more reasonable to use the square of the amplitude θ_0 to dimensionless, which is still an open question.

Second, we make a direct comparison between the pure pitching motion and the pure heaving motion, we found that compared with the results of a pitching foil, the lift on the heaving foil is slightly lower and the change of the lift is slightly more dramatic when the frequency is small.

Acknowledgments

We warmly thank Postdoc Jinghang Lin at Yale University for fruitful discussions. This work was supported by the Science and Technology projects of China Energy Investment (Grant No. GJNY-22-151).

References

[1] Munk M M. Note on the air forces on a wing caused by pitching[R]. 1925.
[2] Theodorsen T. General theory of aerodynamic instability and the mechanism of flutter[R]. 1949.
[3] Deng J, Sun L, Teng L, et al. The correlation between wake transition and propulsive efficiency of a flapping foil: A numerical study[J]. Physics of Fluids, 2016, 28(9): 094101.
[4] Deng J, Sun L, Shao X. Dynamical features of the wake behind a pitching foil[J]. Physical Review E, 2015, 92(6): 063013.
[5] Deng J, Caulfield C P. Three-dimensional transition after wake deflection behind a flapping foil[J]. Physical Review E, 2015, 91(4): 043017.
[6] Sun L, Deng J, Shao X. Three-dimensional instabilities for the flow around a heaving foil[J]. Physical Review E, 2018, 97(1): 013110.
[7] Bohl D G, Koochesfahani M M. MTV measurements of the vortical field in the wake of an airfoil oscillating at high reduced frequency[J]. Journal of Fluid Mechanics, 2009, 620: 63-88.
[8] Ferziger J H, Perić M, Street R L. Computational methods for fluid dynamics[M]. Berlin: springer, 2002.

Technologies and Applications of Pipeline-type Oil-gas-water Separation

Jun Zhang[1], Lide Su[2], Jipu Han[2], Qier Sa[2], Juntao Wang[2], Weichen Chai[2], Bingyu Li[2], Xingfu Zhong[1,*]

[1] Institute of Mechanics, Chinese Academy of Sciences, Beijing 100190, China

[2] Petro China Erlian oilfield company, Xilinhot 02600, China

[*] Corresponding author, E-mail: xfzhong@imech.ac.cn

Abstract: The separation of oil, gas, and water mixture is an important process in the production of crude oil. Traditional separation tanks are large, heavy, long in separation time, and high cost to manufacture. This paper introduces a separation technology with the main characteristics of the pipeline structure. This technology uses the principles of centrifugation, expansion, air flotation, and gravity separation. Pipes and other equipment are the core equipment, which has the advantages of flexible process composition, small footprint, and high separation efficiency. This technology can be used for rapid oil, gas, and water separation, solving the problems of expansion, upgrading old oilfield stations, realizing the purpose of re-injection of water in-situ, energy saving, and reducing emissions.

1. Introduction

The produced liquid from the well is usually a mixture of crude oil, natural gas, water, and other impurities. From extraction to the refinery to user use, oil-gas-water separation is the most important process in the petroleum industry. At present, most of China's onshore oilfields have entered the middle and late stage of exploitation, the moisture content of most oil wells in Daqing, Shengli, Liaohe, and other oilfields is as high as 90%, and a large part of the production cost is used for oil-gas-water separation and oily sewage treatment. To improve oil recovery, the oilfield adopts active water flooding, polymorphic oil flooding, foam flooding, ternary composite flooding, and other technologies, which makes the composition of the produced liquid more complex, and the oil-water emulsification is becoming more and more serious, which further increases the difficulty of multiphase separation. On the other hand, China's offshore oil exploitation has achieved remarkable results, with offshore oil and gas production approaching 100 million tons in 2021, accounting for more than 25% of the country's oil production. However, due to the particularity of the marine environment, there are huge differences between offshore oil exploitation and onshore oilfield output liquid treatment technology. Offshore oil exploitation requires higher and new technologies to solve the problems of large production liquid treatment, high separation index, and small space faced by offshore platform operations.

Since 2000, the scientific research institutions headed by the Institute of Mechanics of the Chinese Academy of Sciences have carried out the innovation of high-efficiency oil-gas-water separation technology through in-depth research on the multiphase flow mechanism of oil, gas, and water, put forward the concept of pipeline separation, and developed into a series of pipeline separation devices and technologies with spiral pipe, T-shaped pipe, swirl pipe, air floatation pipe, and coalescing pipe as the core, breaking through the traditional separation method based on gravity settlement of large tanks in oilfields [1,2,3].

2. Mathematical model

2.1 Two-fluid model

Both phases of oil and water can be regarded as incompressible fluids, and assuming that the two phases are in thermodynamic equilibrium, the equation for conservation of mass and the equation for conservation of momentum in the two-fluid model can be written as:

$$\frac{\partial \alpha_k \rho_k}{\partial t} + \nabla \cdot \left(\alpha_k \rho_k \vec{u}_k\right) = 0 \qquad (k = w, o) \tag{1}$$

$$\frac{\partial\left(\alpha_k \rho_k \vec{u}_k\right)}{\partial t} + \nabla \cdot \left(\alpha_k \rho_k \vec{u}_k \vec{u}_k\right) = -\alpha_k \nabla p_k + \nabla \cdot \alpha_k \tau_k + \nabla \cdot \alpha_k \tau_k^t$$
$$+ \alpha_k \rho_k \vec{g} - \alpha_k \rho_k \left[\vec{\Omega} \times \left(\vec{\Omega} \times \vec{r}\right) + 2\vec{\Omega} \times \vec{u}_k\right] + \vec{M}_k \tag{2}$$

In the formula, the subscript k represents the oil phase (o) and the water phase (w). If the effect of surface tension is ignored, the pressure of the phases at the interface is equal, in which case there are:

$$\vec{M}_w = -\vec{M}_o \tag{3}$$

If the discrete phase exists in the form of droplets, the interphase forces include drag, additional mass, and lift, expressed as:

$$\vec{M}_o = \vec{M}_o^d + \vec{M}_o^{vm} + \vec{M}_o^l \tag{4}$$

$$\vec{M}_o^l = C_L \alpha_o \rho_w \left(\vec{u}_w - \vec{u}_o\right) \times \left(\nabla \times \vec{u}_w\right) \tag{5}$$

$$\vec{M}_o^{vm} = C_{vm} \alpha_o \rho_w \left(\frac{D_w \vec{u}_w}{Dt} - \frac{D_o \vec{u}_o}{Dt}\right) \tag{6}$$

$$\vec{M}_o^d = \frac{3}{4} C_D \frac{\alpha_o \rho_w}{d_o} \left|\vec{u}_w - \vec{u}_o\right| \left(\vec{u}_w - \vec{u}_o\right) \tag{7}$$

$$C_D = \begin{cases} \dfrac{24\left(1 + 0.15\,\mathrm{Re}_o^{0.687}\right)}{\mathrm{Re}} & ,\mathrm{Re}_o \le 1000 \\ 0.44 & ,\mathrm{Re}_o > 1000 \end{cases} \tag{8}$$

In this paper, d_o is the diameter of the oil droplet and Re_o is the Reynolds number of the oil droplet.

2.2 Turbulence model

Turbulence models in multiphase flows are much more complex than single-phase flows, and the $k - \varepsilon$ model is the most widely used. For cases where the layered flow and phase density ratio are close to 1, the turbulence characteristics can be well captured by using the mixed turbulence model [2]. In the hybrid turbulence model, the equations for k and ε are:

$$\frac{\partial}{\partial t}(\rho_m k) + \nabla \cdot (\rho_m \vec{u}_m k) = \nabla \cdot \left(\frac{\mu_{t,m}}{\sigma_k} \nabla k\right) + G_{k,m} - \rho_m \varepsilon \tag{9}$$

$$\frac{\partial}{\partial t}(\rho_m \varepsilon) + \nabla \cdot (\rho_m \vec{u}_m \varepsilon) = \nabla \cdot \left(\frac{\mu_{t,m}}{\sigma_\varepsilon} \nabla \varepsilon\right) + \frac{\varepsilon}{k}(C_{1\varepsilon} G_{k,m} - C_{2\varepsilon} \rho_m \varepsilon) \tag{10}$$

In the above formula, the definition of mixing density ρ_m and mixing speed $\vec{u}_m$ can be written as:

$$\rho_m = \sum \alpha_k \rho_k \tag{11}$$

$$\vec{u}_m = \frac{\sum \alpha_k \rho_k \vec{u}_k}{\sum \alpha_k \rho_k} \tag{12}$$

The turbulent viscosity μ_t and turbulent flow energy $G_{k,m}$ can be expressed as:

$$\mu_{t,m} = \rho_m C_\mu \frac{k^2}{\varepsilon} \tag{13}$$

$$G_{k,m} = \mu_{t,m}\left[\nabla \vec{u}_m + \left(\nabla \vec{u}_m\right)^T\right] : \nabla \vec{u}_m \tag{14}$$

The values of each constant are the same as those of single-phase flow, i.e., Cμ=0.09, σk=1.0, σε=1.3, C1=1.44, and C2=1.92.

3. Structural model validation

3.1 Oil-water gravity sedimentation model

A 2-dimensional model is used to simulate the separation process of oil-water phases, the height of the model is 0.41 m, and the width is 1.0 m, according to the actual dimensions of the pipeline or storage tank. The physical parameters of oil and water are shown in Table 1:

Table 1. Physical properties and inlet conditions of the medium.

Medium	Density ($kg\cdot m^{-3}$)	Viscosity (mPa·s)	Inlet flow rate (m/s)	Phase content (%)
LP-14 White oil	836.0	31.0	1.0	0.305
Tap water	998.0	1.003	1.0	0.695

Figure 1 shows the growth curve of the oil-water mixture of oil droplets with different particle sizes over time under gravity sedimentation. The oil-water phases are gradually separated, and the thickness of the mixed layer is reduced, which is consistent with the experimental observations. The oil-water separation process at droplet diameters of 0.1 mm, 0.15 mm, 0.2 mm, and 0.25 mm was numerically simulated. Figure 1 shows the change in water layer height with time, and the results show that the water layer height increases linearly.

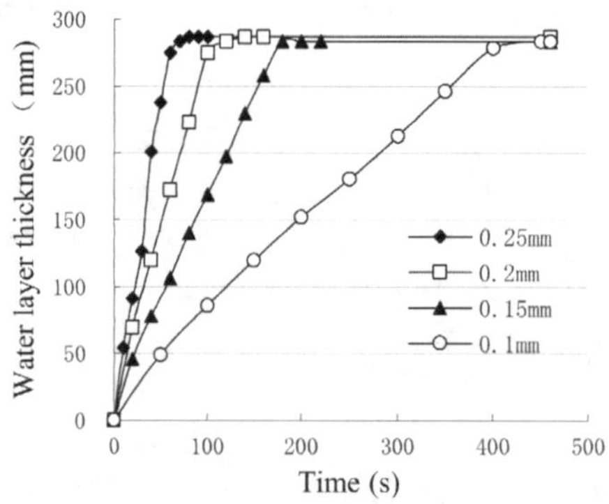

Fig.1 Calculation results of oil-water two-phase Gravity sedimentation model

3.2 T-shaped piper separation model

The T-shaped pipe consists of 2 to 3 horizontal tubes and multiple vertical tubes, which are separated by gravity and stratification as a pre-separation device for gas-liquid and liquid-liquid. When the oil-water or gas-liquid two-phase mixture flows in the horizontal pipe, the heavy phase is subjected to gravity and descends to the bottom horizontal pipe through the T-shaped bifurcation structure, and the vertical pipe converges. The light phase with less density floats up to the top horizontal pipe through the T-shaped structure and the vertical pipe, forming a two-phase layered flow. The T-shaped bifurcation structure plays the role of two-phase flow path selection, the vertical pipe serves as the channel for the dynamic exchange of flow between the two phases, and the horizontal pipe achieves the purpose of transportation and aggregation [4,5].

The separation effect is mainly determined by the design of the T-shaped pipe bifurcation structure, the diameter of horizontal, and vertical pipes, the height and number of vertical pipes, and the diversion ratio of upper and lower horizontal pipes. The early T-shaped pipe design is relatively simple, and the current T-shaped pipe has evolved a variety of structural forms according to the separation needs, such as the horizontal pipe at the top of the T-shaped pipe. The gas-liquid separation is changed to an inclined design, the vertical pipe is changed to an inclined design scheme to reduce the disturbance at the outlet, the rectifier and diversion device are added to the inlet and outlet, and the functional structure is added inside the pipeline. Figure 2 shows the color map of the distribution of the volumetric phase content of the oil phase in the three-layer T-pipe when the oil density is 900kg/m3 and the inlet oil content is 15%.

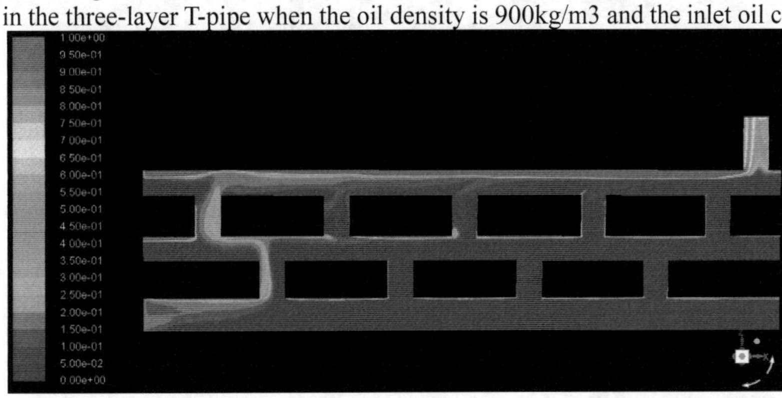

Fig. 2 T-tube separation numerical simulation of oil phase content distribution

3.3 Cyclone pipe separation model

The separation of cyclone pipe takes advantage of the density difference of the miscible medium to achieve efficient separation of different media through the motion paths generated by different centrifugal forces of the rotating flow field. According to the different rotation methods and inlet and outlet structures, the swirl tube separator can be divided into columnar cyclone separator and guide vane type cyclone separator [6,7,8].

The cylindrical cyclone rotates tangentially through the horizontal inlet, the oil-water two phases rotate and flow at high speed in the cylindrical separation section, the medium of each phase produces different centrifugal forces under the action of density difference, and the heavy phase (such as water or sand) gathers towards the pipe wall under the action of gravity and centrifugal force and spirals downward, and finally discharges from the bottom outlet; Light phases (such as oil or gas) gather in the center of the cyclone, in the opposite direction of axial motion to the heavy phase, and are finally discharged from the top outlet. The deflector-type cyclone rotates axially through the fluid, and the light phase gathers to form an oil core or gas core, which is led out through the axial outlet; The heavy mass phase is thrown towards the side wall of the pipe and is led out through the interface pipe section at the side wall. Compared with the columnar cyclone, the internal flow field of the deflector-type cyclone is symmetrical and stable, and there is no axial reverse flow; Radial dimensions can be installed horizontally, vertically or obliquely, or downscale, giving it a good advantage in downhole or size-constrained scenarios.

The results of the oil-water separation experiment of the column-type cyclone show that by controlling the mixing ratio of oil and water, the oil and water entering the cyclone can form an oil core under the action of centrifugal force. The valve that controls the top light phase outlet and the bottom-heavy phase outlet adjusts the shunt ratio (the ratio of the top outlet flow to the inlet flow), and as the split ratio increases from 0 to 0.95, the oil core can form shapes of different thicknesses and lengths.

The process and effect of oil-water separation were further studied by numerical simulation (as shown in Figure 3: red represents the oil phase, blue represents the water phase), in which the model setting conditions are: pipeline diameter 75 mm, inlet flow 10 $m^3 \cdot h^{-1}$, inlet oil content of 8%, oil viscosity 245 mPa·s, oil outlet flow of 20% of the total inlet flow, and oil phase density from 850 $kg \cdot m^{-3}$. With the increase of the density of the oil phase, the oil content of the outlet gradually increases, and the oil phase content in the formed oil core gradually decreases. Separation of oil-water phases with similar densities is more difficult.

Fig.3 Cyclone pipe structure model and calculation results of oil-water two-phase separation

4. Engineering cases

After years of continuous research and improvement, pipeline separation technology has become increasingly mature, and the different pipe separation structures in this technology can be applied separately or in combination.

4.1 Treatment of oily sewage

Adopted the same structured cyclone, the separation and oil removal efficiency can be improved through a multi-stage series connection, and the equipment is connected in parallel to solve the problem of processing capacity. Swirl sewage treatment plants are already in operation with capacities from 1 to 800 m3/h per unit. There are 5 sets of equipment in parallel to treat 18000 m^3 of polymer-bearing wastewater per day in a joint station of Dagang Oilfield, the oil concentration of the inlet sewage of the equipment is 100-500 mg/l, the oil content of the outlet can be reduced to 60-80 mg/l after treatment, and the single-stage cyclone oil removal rate is greater than 75%. The experimental data is shown in Figure 4.

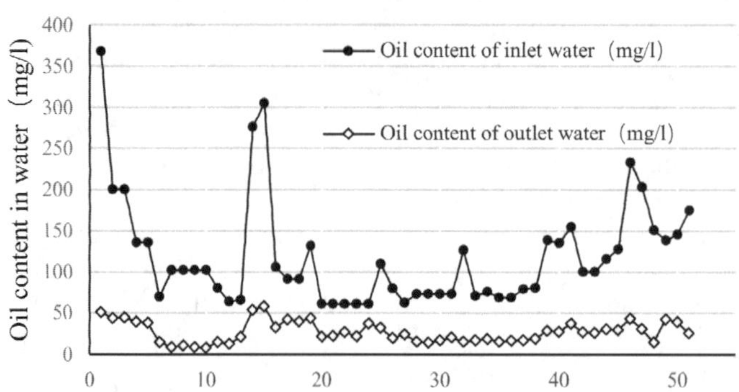

Fig. 4 Oil removal by cyclone experimental

4.2 Water separation treatment of offshore-produced liquid
Given the rapid water separation and treatment needs of offshore produced liquid in an oilfield, a three-stage cyclone treatment test device (as shown in Figure 5) is designed, the first-stage cyclone separates most of the oil and gas in the produced liquid, and the second and third stages treat the oily wastewater separated by the first stage cyclone. According to the different separation tasks, each stage cyclone adopts different specifications and parameter structures, the oil content of the sewage separated by the first stage cyclone is less than 1000mg/l, the oil content of the second stage cyclone is no more than 500mg/l, and the oil content of the third stage cyclone is below 50mg/l.

Fig. 5 Offshore platform separation equipment

4.3 Oil, gas, and water three-phase separation treatment
To realize the purpose of local water distribution and on-site reinjection of a metering and transfer station in a northwest oilfield, through the combination of multi-stage pipeline cyclone, T-pipe, pipeline coalescer, S-type pipe separator, and other structures, the rapid separation of oil, gas, and water of the oil well-produced liquid is realized. The gas is used for heating boilers, water is injected back into the ground, and oil is transported abroad. The equipment is designed to handle a capacity of 400 m^3/d and achieve a water separation of 200 m^3/d. The effluent contains $\leq$ 50 mg/l, and then after the two-stage sedimentation and oil removal and suspended solids removal by the water injection skid, it can be directly injected back into the ground. After removing water, the pipeline separation equipment reduces the external transmission pressure by 50%, alleviates the scaling and corrosion problems of the pipeline, and prolongs the service life of the pipeline.

5. Conclusion
Compared with traditional separation technology and equipment, pipeline separation technology has the following relative advantages in oil, gas, water, and sand separation:

1) Physical separation method based on the principle of mechanical composite, fast separation speed, high efficiency, no chemical agents, no risk of secondary pollution;
2) Compared with the traditional gravity settling tank, it can reduce the land occupation and volume by more than 40%, and the operating weight of the equipment is reduced by more than 50%, which is especially suitable for engineering construction or transformation under the condition of limited space and bearing capacity;
3) Modular process design, easy to integrate with existing processes, flexible layout, easy to move;
4) Simple and fast manufacturing and installation, can be assembled on-site, no heavy lifting equipment is required;
5) Easy operation and management, convenient maintenance, and maintenance.

Given the separation needs of two or more mixed media such as oil, gas, and water in oil and gas fields, the pipeline separation technology developed based on hydrodynamic process and simulation breaks through the traditional separation method based on gravity settlement to achieve rapid and efficient dynamic separation under pipeline flow. The pipeline separator includes a series of separation devices with the pipeline as the main structure, such as a T-shaped pipe, spiral pipe, swirl tube, and swirl air flotation separation tube, which comprehensively adopts the mechanical principles of gravity, buoyancy, centrifugal, air flotation, expansion and so on, which greatly improves the heterogeneous separation efficiency. Different separation devices can be used alone or in combination, and the combination mode is fast and flexible, which adapts to the current development needs of separation technology in the oil and gas industry and has broad application prospects. Engineering applications show that pipeline separation technology can greatly reduce the cost of oilfield production fluid treatment, show good technical and economic advantages, and have good economic and social benefits and environmental benefits.

References

[1] SHI S Y, XU J Y, SUN H Q, et al. Experimental study of a vane-type pipe separator for oil-water separation[J]. *Chemical Engineering Research and Design*, **2012, 90(10)**: 1652-1659.
[2] YANG, LELE, WANG, JING, MA, YONG, et al. Oil-Water-Gas three-phase separation in multitube T-Junction separators[J]. *Water*,**2019,11(12)**:2655.
[3] ZHONG X F, WU Y X, LI S M, et al. Investigation of pipe separation technology in the oilfield[J]. *Advanced Materials Research*, **2013, 616-618**: 833-836
[4] YANG L, AZZOPARDI B J. Phase split of liquid-liquid two-phase flow at a horizontal T-junction[J]. *International Journal of Multiphase Flow*, **2007, 33(2)**: 207-216.
[5] PANDEY S, GUPTA A, CHAKRABARTI D P, et al. Liquid-liquid two-phase flow through a horizontal T-Junction[J]. *Chemical Engineering Research & Design*, **2006, 84(10)**: 895-904.
[6] Zhang J, He YT, Liu S, et al. Oil-water separation in a cylindrical cyclone with vortex finder[J]. *PHYSICS OF FLUIDS*, **2022,34(3)**:11.
[7] WEI P K, WANG D, NIU P M, et al. A novel centrifugal gas-liquid pipe separator for high-velocity wet gas separation[J]. *International Journal of Multiphase Flow*, **2020, 124**: 103190.
[8] Liu S, Zhang J, Xu JY. An investigation of a gas-liquid swirling flow with shear-thinning power-law liquids[J]. *PHYSICS OF FLUIDS*, **2022,34(7)**:17.

Analysis of Aircraft Parking Stability Based on Atmosphere Boundary Theory

J Liu[1,2], J L Ge[3], H Yang[2], W S Chen[2], J H Wu[1] and Q L QU[3]

[1] School of Transportation Science and Engineering, Beihang University, Beijing 100191, China

[2]Airport Research Institute, China Academy of Civil Aviation Science and Technology, Beijing, 100028, China

[3] Key Laboratory of Fluid Mechanics of Ministry of Education, Beihang University, Beijing, 100191, China

liujia_zhdz@126.com

Abstract: Based on numerical simulation technology, the aerodynamic load research of civil aircraft parked on the ground is conducted. The CRM model parked on the ground is established by CFD software. Based on the atmospheric boundary layer theory, the seven incoming flow conditions for numerical calculation are designed. And the aerodynamic loads' change of the aircraft model is calculated when the -90° sideslip angle. The result shows that the mooring load on the nose landing gear mainly comes from the longitudinal load caused by the upwards pitching moment on the flat tail and the lateral load caused by the yaw moment; since the rolling moment coefficient is less than the pitching moment coefficient by two orders of magnitude, which hardly affects the mooring load. The study provides a reference for reducing the mooring load of civil aircraft under heavy wind conditions.

1. Introduction

1.1. Importance of Ground Stability

Aircraft ground stability refers to the parked aircraft which is influenced by the strong crosswind showing serious results such as sideslipping or tilting. At present, the airport generally adopts the methods of placing wheel blocks or mooring to reinforce the aircraft. However, the bearing load of each mode is based on the aerodynamic load of the aircraft. In the stage of aircraft design, the aerodynamic load data is mainly provided by wind tunnel experiments. In recent years, with the development of CFD (Computational Fluid Dynamics) and computer technology, the numerical simulation methods of the flow field are becoming more and more important in aerodynamic prediction. Through the numerical simulation of CFD, the calculation results can not only predict the force of the model accurately but also provide more abundant spatial flow field information. It helps to explain the generation mechanism of the aerodynamic load of the parked aircraft, as well as the characteristics of flow field change.

1.2. Research Status

Gong C [1] studied and analyzed the stability of a certain civil aircraft parked in a strong crosswind based on the results of wind tunnel experiments. The conclusion showed the aircraft had a large pitching moment under the vertical cross wind and the nose had the tendency to tilt up. Zhiqiang W [2] analyzed the changes in the parameters such as the sink motion of the wake vortex, the vorticity attenuation, the lateral motion of the wake vortex, and the vortex center velocity under the influence of different crosswinds. The results showed that the vorticity of the wake vortex and the rolling moment increased rapidly under the crosswind. The wake vortex center distance decreases and then rebounds with separation quickly. Weibin G [3] adopted the finite element method to establish the mooring load model of large fixed-wing civil aircraft and analyzed the influence factors, such as direction and aircraft weight of mooring load under wind load.

1.3. Existing Deficiencies

Although some scholars have used wind tunnel experiment data [4-7] to study the stability of parked aircraft under crosswind, there are few studies on the aerodynamic load calculation of aircraft with the wind field structure under the atmospheric boundary layer. For parked aircraft, the aerodynamic load was related to the wind direction, velocity, and aerodynamic layout and the flow field was complex. The wind tunnel method could not obtain specific flow field information, but CFD could provide more abundant flow field information as well as a better explanation of the causes of force changes. According to the atmospheric boundary layer theory, the structure of the near-surface wind field is a gradient. The wind velocity and turbulence intensity are all functions of the ground friction coefficient. Therefore, the different geomorphic features and regional meteorological conditions had a great impact on the aerodynamic load of aircraft parked on the ground, which needed detailed analysis further.

This paper used the CFD method to simulate the near-ground flow field surrounding the CRM (Common Research Model) with RANS (Reynolds Averaged Navier Stokes) and SST k-ω turbulence model. Then, the gradient wind field function of atmospheric boundary layer theory was constructed to calculate the aerodynamic load of the CRM model with different wind velocities under vertical crosswind wind. Through the analysis of component loads and spatial flow field, the changes of pitching moment, rolling moment, and yaw moment with wind velocities had obtained. The conclusion provides the necessary theoretical basis for further mooring load calculation.

2. Calculation method of aircraft aerodynamic load

Using the CFD method, the aerodynamic load distribution and pitching moment could be calculated under different attack angles and velocities. Analysis of the typical transport aircraft aerodynamic load with the configuration of wing body/flat tail/ vertical tail could provide the basis for aircraft design.

2.1. Aircraft Model

The CRM model was designed and developed by NASA's SFW (Subsonic Fixed Wing) aerodynamic technology research group and the DPW (Drag Prediction Workshop) committee. The main purpose was to provide a reference shape for CFD verification and validation. The CRM model was a typical modern transport aircraft configuration, with a Ma number of 0.85 and a lift coefficient of 0.50. The model is divided into various configurations, including wing body, wing/body/flat tail, and wing-body /flat tail/pylon/nacelle assembly.

In this paper, the CRM model adopted a wing body/flat tail/vertical tail configuration. The CRM model geometric profile was shown in Figure 1.

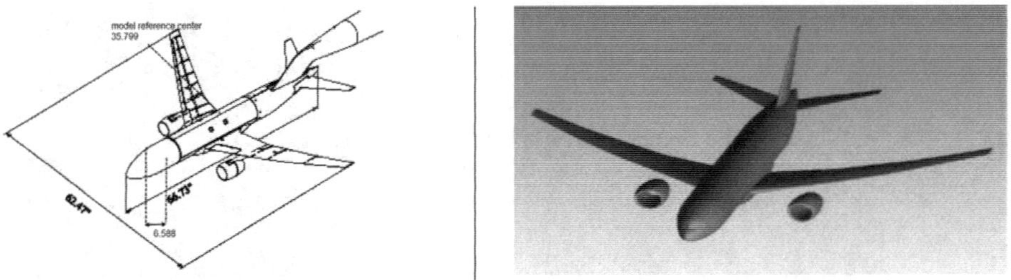

(a) Main scale length	(b) CRM numerical model

Figure 1. CRM model.

The CRM experimental model was a scaled model with a ratio of 0.027. The aircraft model in this paper is compared with the actual CRM aircraft drawing. Therefore, the shape parameters of the CFD model could be obtained through scale transformation, as shown in the following table.

Table 1. Main parameter of CRM model.

Parameter	Wing Span (m^{-1})	Aspect Ratio	Reference Area (m2)	Reference Chord Length (m^{-1})
Value	58.763	9	191.8	7.005

2.2. Calculation Domain and Grid System

To simplify the research and improve the grid quality, the numerical model removed the pylon, nacelle, and other components in the original model WBTN, which had a small impact on the overall aerodynamic layout of the aircraft. Meanwhile, the CRM vertical tail provided by NASA was added. The calculation domain took the nose as the coordinate origin, 600m width from the front, left, and right, 400m above, and 1000m behind. The X direction of the coordinate system was the flow direction. The Y direction was the span direction. And the Z direction was the vertical direction. The pointwise software was used to generate the unstructured grid of CRM for the whole aircraft. The aircraft surface was local refinement. The boundary layer flow on the solid wall surface was divided by a prism grid. The height of the first layer of the grid on the aircraft surface was set as 3×10^{-5}m satisfying that y+ equals 1 (y+ was dimensionless distance, which was used to estimate the height of the boundary layer grid near the wall. y was the distance between the first layer center and wall. μ was the dynamic viscosity, ρ was the density, and was the wall shear stress). The overall calculation domain and grid distribution were shown in Figure 2 and Figure 3 with a grid number of about 16 million.

(a) Whole domain (b) Whole CRM model grid

Figure 2. CRM model grid.

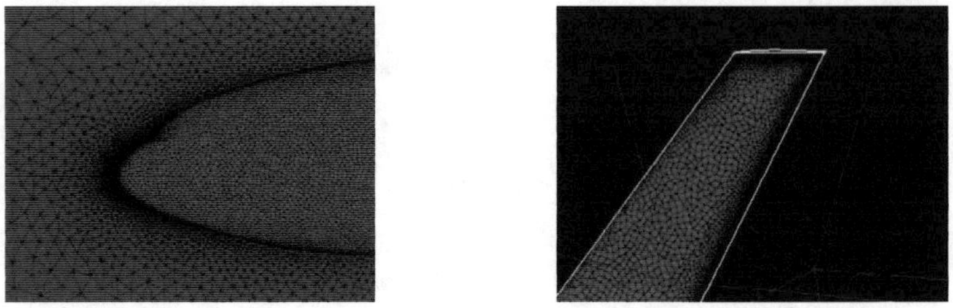

(a) Nose details (b) Wing details

Figure 3. Detail grid.

2.3. Governing Equations and Boundary Conditions

2.3.1. Governing Equations. This paper belongs to the low-speed flow problem, so the energy equation is not considered in the solution of the governing equ010a0tion. Viscous incompressible N-S equation, continuity equation, and momentum equation based on Reynolds mean are respectively as followed.

$$\frac{\partial u_i}{\partial x_i} = 0 \tag{1}$$

$$\frac{\partial(\rho \bar{u}_i)}{\partial t} + \frac{\partial(\rho \bar{u}_i \bar{u}_j)}{\partial x_j} = -\frac{\partial \bar{p}}{\partial x_i} + \frac{\partial}{\partial x_j}(\mu \frac{\partial \bar{u}_i}{\partial x_j} - \rho \overline{u_i' u_j'}) \tag{2}$$

The steady flow based on the pressure model was chosen to discretize with the finite volume method. The pressure, momentum, energy, and turbulent diffusion were calculated in the second-order upwind scheme.

2.3.2. Turbulence model: Compared with the standard k-ω model, when the SST k-ω model closes the N-S equation, the transverse dissipation derivative term is added, and the transport process of turbulent shear stress is considered in the definition of turbulent viscosity. The turbulence constants used in the model are also different. Therefore, the SST k-ω turbulence model has higher accuracy in calculating the flow field around the aircraft.

$$\frac{\partial(\rho k)}{\partial t} + \frac{\partial}{\partial x_i}(\rho k u_i) = \frac{\partial}{\partial x_j}(\Gamma_k \frac{\partial k}{\partial x_j}) + \widetilde{G_k} - Y_k + S_k \tag{3}$$

$$\frac{\partial(\rho \omega)}{\partial t} + \frac{\partial}{\partial x_i}(\rho \omega u_i) = \frac{\partial}{\partial x_j}(\Gamma_\omega \frac{\partial \omega}{\partial x_j}) + G_\omega - Y_\omega + D_\omega + S_\omega \tag{4}$$

2.3.3. Boundary Condition: The incoming flow boundary was the velocity inlet. The outlet boundary condition was free outflow. The top boundary condition was symmetric. The aircraft surface and the ground adopted a non-slip wall. The other surrounding boundaries were set as the outlet boundary according to the incoming flow direction.

2.4. Model Verification

In this paper, a half model was adopted during the verification and the comparison data referred to the CFD calculation results on the NASA website. The grid of the half-model verification was 11 million unstructured grids as shown in Figure 4. Using Fluent software, the CFD result showed the aerodynamic coefficient of CRM WBT0 with the 0°, 1°, and 2°attack angle. When the attack angle was 0°, the pressure coefficient of the upper wing and spatial streamline distribution was shown in Figure 5. The specific comparison results were shown in Table 2.

Journal of Physics: Conference Series **2458** (2023) 012026 doi:10.1088/1742-6596/2458/1/012026

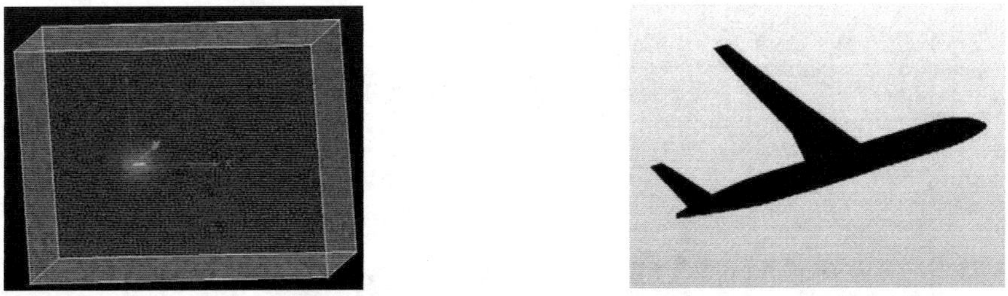

(a) Whole domain of the half model (b) Detail of half model

Figure 4. CRM half model.

(a) Pressure coefficient, α=0° (b) Streamline, α=0°

Figure 5. Pressure coefficient and spatial streamline distribution.

Table 2. Comparison of aerodynamic coefficients.

Case Name	Ma	Alpha	C_L	C_D
NASA value	0.85	0	0.1186	0.0181
CFD value	0.85	0	0.1243	0.0220
NASA value	0.85	1	0.2600	0.0199
CFD value	0.85	1	0.2700	0.0253
NASA value	0.85	2	0.4026	0.0237
CFD value	0.85	2	0.3768	0.0320

It was shown in Table 2 that the error of the lift coefficient was 0.0057 and the error of the drag coefficient was 0.0039 with the condition of 0° attack angle. It verified the accuracy and reliability of the model.

2.5. Working Conditions on Ground

The previous research mainly considered the cruise state with the uniform incoming flow, as well as the height being higher than the atmospheric boundary layer. When the wind blows over the ground surface, it often forms a thin layer, which is called the wind field flow force phenomenon. It satisfies the wind field distribution characteristic below 500m height, and the boundary layer flow is shown in Figure 6.

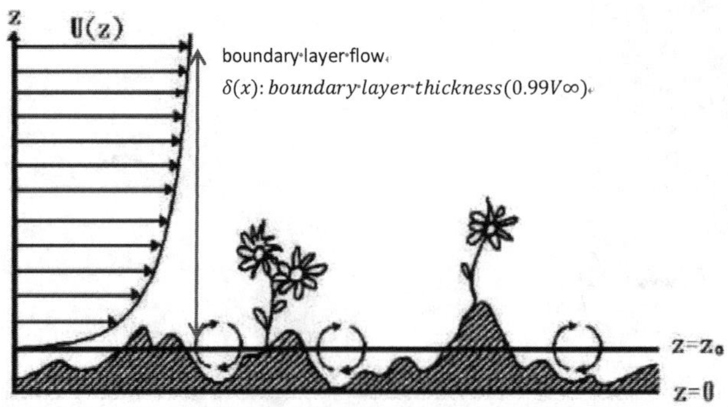

Figure 6. Boundary layer flow.

Since the flow field structure near the ground is greatly affected by the surface roughness, the average wind speed profile is a function of the surface roughness length and the friction velocity. It's usually defined as the logarithmic law or exponential law. In this paper, the exponential law function is used to simulate the wind speed profile of the atmospheric boundary layer, the formula is shown below.

$$U_Z = U_{10}(\frac{Z}{Z_{10}})^{\alpha} \tag{5}$$

Where Z_{10} was the reference height of 10m, U_{10} was the wind velocity at the reference height, and alpha was the ground roughness index.

$$C_p = \frac{2(p-p_0)}{\rho v_H^2} \tag{6}$$

Where p was the average pressure of the aircraft surface node, p_0 and v_H are the static pressure and average pressure at the reference height (H = 10m), respectively. pwas the air density, taken as 1.225kg/m^3.

3. Aircraft stability analysis

In this paper, the aerodynamic load of aircraft parked on the ground under different working conditions can be obtained by CFD simulation of the whole CRM flow field. According to previous studies, when the sideslip angle was -90°, the pitching moment coefficient could reach the maximum value. Therefore, this paper calculated X kinds of vertical wind, as shown in Table 3. Therefore, this paper calculated 7 kinds of vertical wind, as shown in Table 3. Ansys Fluent was used to simulate the flow field of the whole aircraft and Tec plot software was used for the flow field structure post-process after numerical simulation.

Table 3. Seven working conditions

Case Name	**Wind Velocity at Different Heights (ms^{-1})**			
	H=5m	**H=10m (Body center height)**	**H=15m**	**H=20m (Vertical tail height)**
Case 1	4.48	5.00	5.34	5.59
Case 2	8.95	10.00	10.67	11.17
Case 3	13.43	15.00	16.01	16.76
Case 4	17.9	20.00	21.34	22.35
Case 5	22.38	25.00	26.68	27.93
Case 6	26.85	30.00	32.01	33.52
Case 7	31.33	35.00	37.35	39.11

The calculation results of aerodynamic coefficients with the wind axis, under the 7 working conditions, were shown in Figure 7. The main factors were C_L (lift coefficient), C_D (drag coefficient), and C_X (side force coefficient). At -90° sideslip angle, with the increase of wind velocities, C_L increases fastest, while C_D and C_X decreased slowly and changed little.

Meanwhile, the results of aerodynamic coefficients with the body axis, under the 7 working conditions, were shown in Figure 8. The main factors were C_m (pitching moment coefficient), C_l (rolling moment), and C_n (yaw moment coefficient). At -90° sideslip angle, with the increase of wind velocities, Cm increases fastest, while C_n and C_l changed little.

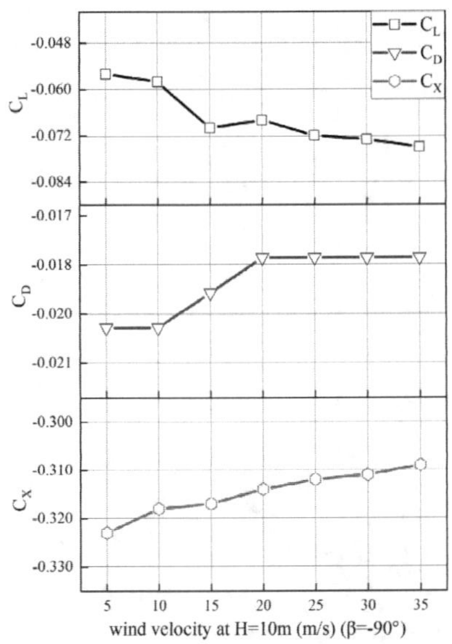

Figure 7. CFD results of Wind Axis Aerodynamic Coefficient (=-90°)

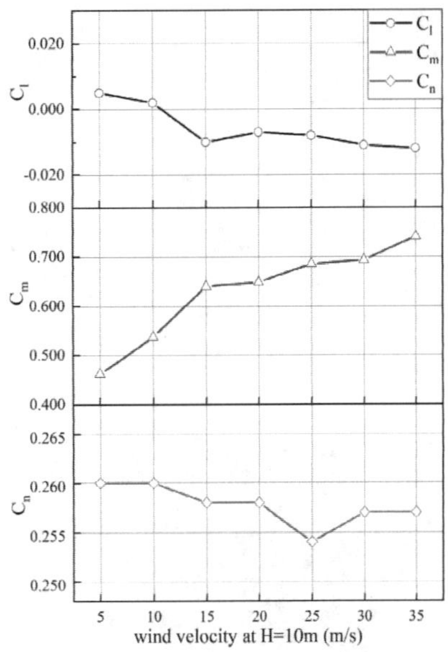

Figure 8. CFD results of Body Axis Aerodynamic Coefficient (=-90°)

To study the contribution of the flat tail and the vertical tail to the aerodynamic moment, the simulation of three different configurations was conducted under the wind speed of case 7 in this paper. The aerodynamic coefficients of the body axis were shown in Table 4. It showed that the flat tail and vertical tail had little effect on the rolling moment. But the increase in pitching moment caused by the flat tail was significant. And the vertical tail gave a great contribution to the yaw moment.

Table 4. Body Axis Aerodynamic Coefficient of different configuration in Case 7

Case 7 Aerodynamic Coefficient	**Body Axis Aerodynamic Coefficient of Different Configurations**		
	Wing Body	**Wing Body/ flat tail**	**Wing Body/ flat tail/ vertical tail**
C_l	-0.016	-0.021	-0.012
C_m	0.580	0.725	0.740
C_n	0.159	0.161	0.256

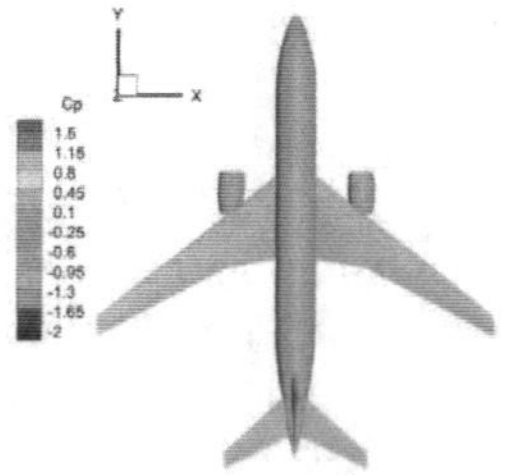

(a) Case 1

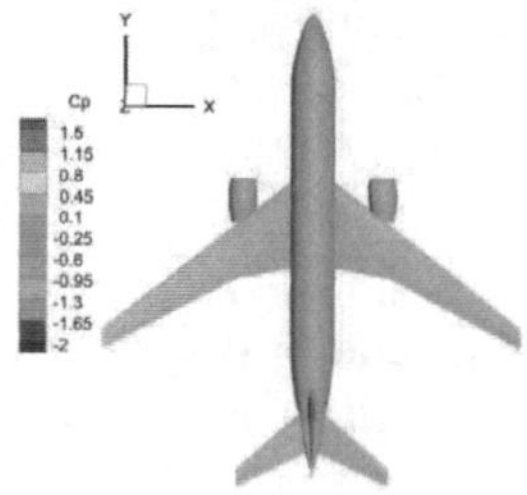

(b) Case 3

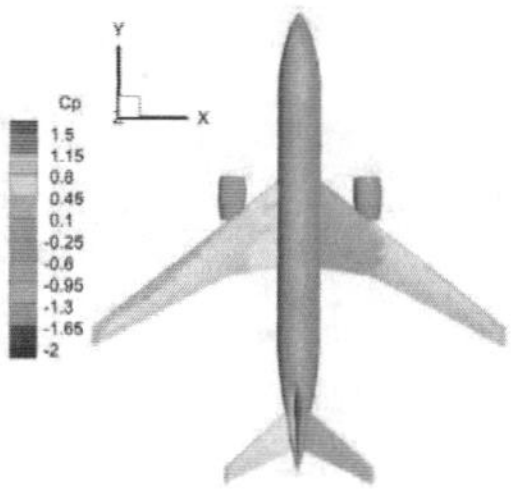

(c) Case 5

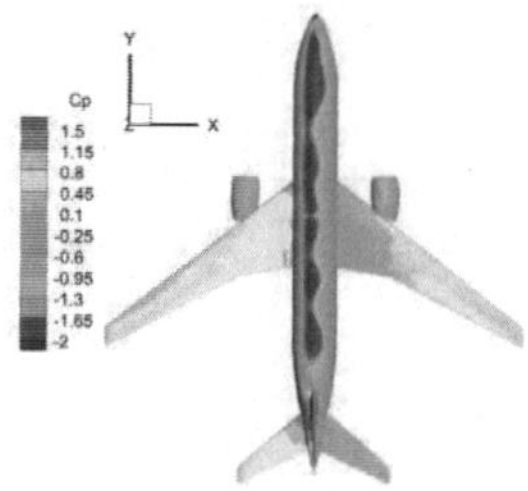

(d) Case 7

Figure 9. Pressure contour of CRM (β=-90°).

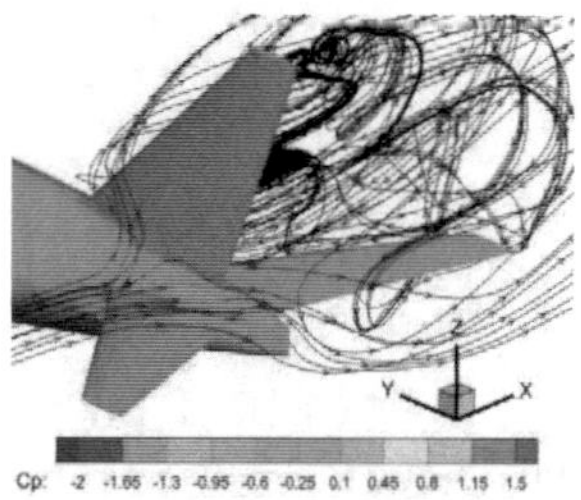

(a) Case 1

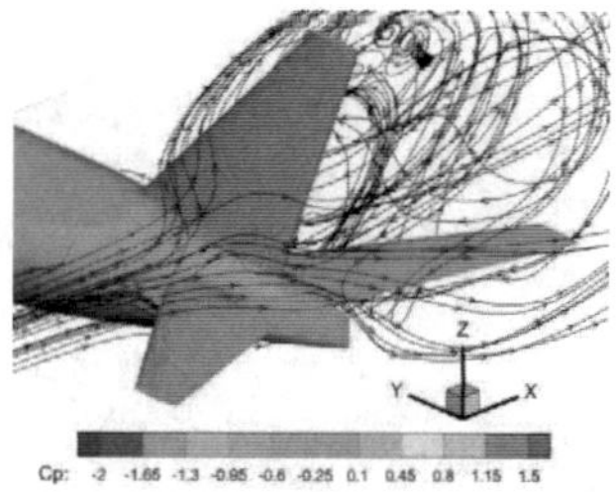

(b) Case 3

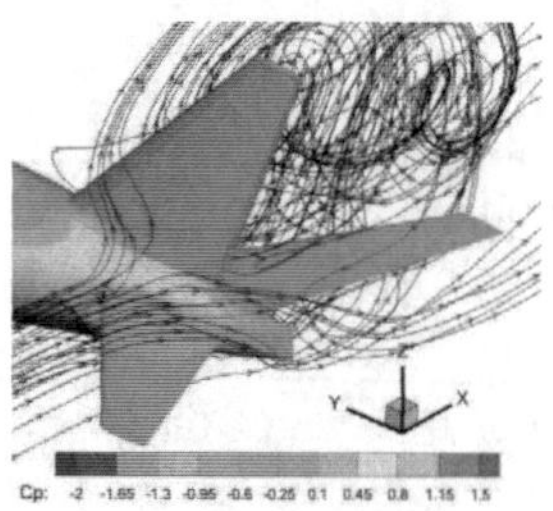

(c) Case 5

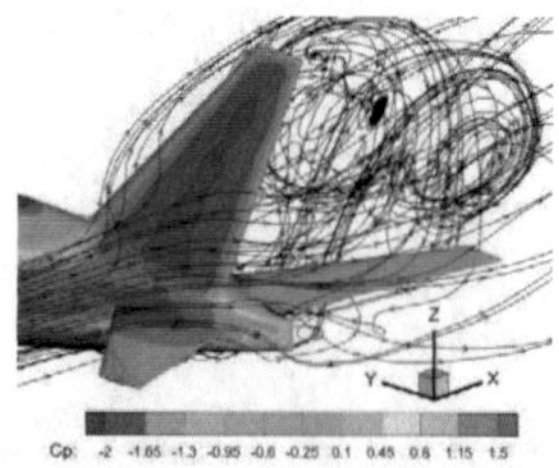

(d) Case 7

Figure 10. Flow field at the flat tail (β=-90°).

Figure 9 showed the pressure on the body of the aircraft. The change of Cm was mainly determined by the aerodynamic force on the flat tail. By the windward side, the pressures on the flat tail and the rear fuselage were heavy, which caused the large lift moment. In this situation, the aircraft tended to tilt up. This pressure was mainly caused by the stagnation and splitting effect of the vertical tail to the transverse airflow. This stagnation resulted in a high-pressure region near the upper surface of the flat tail and the rear fuselage is shown in Figure 10. At the same time, the impact of the splitting flow on the flat tail resulted in a downward aerodynamic force. Cn was quite smaller than Cm. But CRM had a large aspect ratio (8-12), so the magnitude of the yaw moment and pitching moment were equivalent. Its impact on the mooring load could not be ignored. C_l mainly affected the mooring load of the two main landing gears. But C_l was two magnitudes less than C_m, so its influence on mooring load was not dominant.

4. Conclusion and PROSPECT

In this paper, the aerodynamic loads of aircraft parked on the ground based on the atmospheric boundary layer theory were obtained by simulating the flow field of CRM. According to the flow field boundary conditions of parking on the ground, the working conditions of different wind velocities were designed with a -90° sideslip angle. The changes in aerodynamic force and aerodynamic moment were compared to explain the flow mechanism. The main conclusions were as follows:

- The cause of the pitching moment was that the vertical tail split the airflow, which made the flat tail and the rear fuselage receive downward aerodynamic force.
- The separation at the vertical tail increases the lateral resistance of the vertical tail, which caused the large yaw moment. At this time, the front landing gear bore a large lateral load.
- C_l is less than C_m by two orders of magnitude and its influence on mooring load was insignificant.

According to the body pressure distribution results under heavy wind conditions, it is known that, when the aircraft was parked on the ground, if the elevator and the flat tail are deflected downward, the aerodynamic force on the flat tail can be reduced. At the same time, if the rudder can be pre-bias set along the wind, the upward pitching moment can be further reduced, as well as reducing the mooring load.

Acknowledgments

The work of this paper has been supported by the 2020 Research on Airport Operation Security Related Regulations and Standards and New Technology Application Verification, which is sponsored by the Civil Aviation Administration of China.

References

[1] Gong C and Yifei L, "Analysis of Aircraft Parking Stability in the Situation of Crosswind by Wind Tunnel Test," Civil Aircraft Design & Research, 2015.
[2] Zhiqian W, Zhiyuan L and Wei L, "Research on Aircraft Wake Vortex Strength Dissipation and Vortex Motion under Crosswind Impact", Journal of Air Force Engineering University (Natural Science Edition), 2017.
[3] Weibin G, Xiufen J, and Jian M, "Non-linear analysis of mooring load for large fixed-wing civil aircraft", Chinese Journal of Applied Mechanics, 2014.
[4] Yipeng L, Gong C, Chuanliang G, and Yongyi S, "Calculation Research on Tie-down Aerodynamic Load for Civil Aircraft Based on Numerical Wind Tunnel Technique," Civil Aircraft Design & Research, 2016.
[5] Wang Y T, Sun Y, Meng D H, et a1, "Numerical simulation of aerodynamic characteristics of CRM-WB configuration with a support system and wing deformation", Acta Aeronautica et Astronautica Sinica, 2017.
[6] Wei T, Li C, Hui W, and Pingtao Z, "Experimental study on the control of fixed floor boundary layer in the low-speed wind tunnel," Journal of Experiments in Fluid Mechanics, 2017.
[7] Jinping H, Jian Y, Jianyin L, Shanxun L and Jun D, "Study on method of Aircraft Mooring Load Calculation", Trainer, 2016.

Study on the Tribological Properties of Polymer Interface and the Mapping Mechanism of Friction Noise in a Wide Temperature Range

Ganlin Cheng[1], Fei Guo[1, *], Xiaohong Jia[1]

[1]State Key Laboratory of Tribology in Advanced Equipment, Tsinghua University, Beijing 100084, China

*Corresponding author: Fei Guo.

cgl19@mails.tsinghua.edu.cn; guof2014@mail.tsinghua.edu.cn; jiaxh@tsinghua.edu.cn

Abstract. In order to study the tribological properties and frictional noise mapping mechanism of polymer interface at a wide temperature range, tribological and frictional noise acquisition tests were carried out on a Ball-on-Three-Plates Tribology Accessory. The friction coefficient and noise signal data sets were collected for 15 minutes from the friction surfaces of polytetrafluoroethylene and 0Cr18Ni9 austenitic steel pairs at six temperatures and three constant loading-speed contact points in a wide temperature range of -150°C-25°C. The wear rate of the polymer was measured and the wear state of the polymer interface was studied by field emission environmental scanning electron microscopy (SEM). Based on the analysis of the time-frequency domain characteristics of friction noise and the qualitative mapping relationship between the wide-temperature domain tribological properties and friction noise, the theoretical and experimental basis is provided for establishing online early warning monitoring of the tribological performance of polymer-metal friction pairs based on the time-frequency domain characteristics of friction noise.

1. Introduction

Polymers have been widely used in sliding friction systems in wide temperature range applications due to their excellent friction and wear properties in the absence of external lubrication. However, the wide temperature range and the limited application of liquid lubricants have resulted in extremely harsh tribological applications between friction pairs[1, 2]. The friction noise generation of polymer-metal pairs is usually caused by the interaction between the friction region and the dynamic process of the whole mechanical system, which causes friction instability at the macro level. It is usually accompanied by severe vibration and/ or high-intensity noise in the friction connection[3, 4]. Jibiki *et al.*[5] found that a certain fretting period is required to generate friction noise by studying the fretting wear between carbon steels. When the friction coefficient was sharply reduced, the greater the friction noise was, the better correlation between the decibel level and the amount of wear will be. Mo *et al.*[6] found that during the sliding process, the influence of the roughness peak between the metal disk and the ceramic ball causes vibration, which generated noise. The size proportion of the groove texture on the friction surface was related to the friction noise. In general, most of the friction noise research focuses on the influence of material properties and working conditions on friction noise, but research on friction noise from the perspective of tribology is rarely reported. And the research on the mapping relationship between frictional interface noise and tribological evolution is not deeply studied.

Therefore, it is of great practical significance to systematically study the mapping relationship between frictional noise and tribological properties of polymer-metal pairs in a wide temperature range, reveal their mechanism, and establish a qualitative mapping relationship between them. In this paper, the influence of different temperatures and working conditions on friction noise and its mechanism of action are systematically studied by conducting tribological noise test collection, time-frequency domain collection analysis, and qualitative friction-wear mapping analysis. All these provide a theoretical and experimental basis for establishing online early warning monitoring of tribological properties of polymer-metal friction pairs based on the characteristics of friction noise in the time-frequency domain.

2. Tribological experimental device and the conditions and methods for collecting friction noise

To study the tribological properties of different polymer–metal pairs and working conditions in a wide temperature range, a tribological experiment was conducted on a Ball-on-Three-Plates Tribology Accessory (T-BTP, Anton Paar, Austria) of a rheometer and the specific contact friction mode, as shown in Figure 1. The polymer plate was made from one plastic material, namely polytetrafluoroethylene (PCTFE). The size of the plate was (14.92 ± 0.05) mm in length* (5.88 ± 0.02) mm in width* (3.06 ± 0.05) mm in height. One alloy steel (304) was used for the metal ball test materials. The diameter of the ball was 12.7 mm. The seven temperature conditions were 55, 25, 0, -20, -70, -120, and -150 °C. The low temperatures were achieved by pressurizing and cooling a self-pressurized liquid-nitrogen tank. The three load/ speed conditions were 5 N/ 750 rpm, 10 N/ 1000 rpm, and 15 N/1250 rpm. The friction experiment was conducted for 1200 s. The friction contact mode was point contact between the metal ball and three metal plates, and the lubrication mode was dry friction. A field-emission environmental scanning electron microscope (FEI Quanta 200 FEG, United States) was used to characterize the polymer surface after rubbing. The wear amount was measured by weighing, and the corresponding wear rate was calculated. A sound-pressure microphone sensor (378c1, PCB Piezotronics Inc., America) was installed 40 mm outside the three-ball device to collect the friction noise signal with a sensitivity of 1.0 mV/Pa and a frequency range of 0-70 kHz. The dynamic measurement range was 173 db per 20 μPa. An acoustic acquisition and analysis system (INV3068WP-4, Hangzhou Yiheng Technology Co., Ltd., China) was used to record and analyze the friction noise signal. The sound was sampled at a frequency of 51200, 25 frames per second, 40 ms per frame, and 2048 points were collected. The structure is used to predict and recognize the friction coefficient of the polymer interface through machine learning from the friction noise in a wide temperature domain.

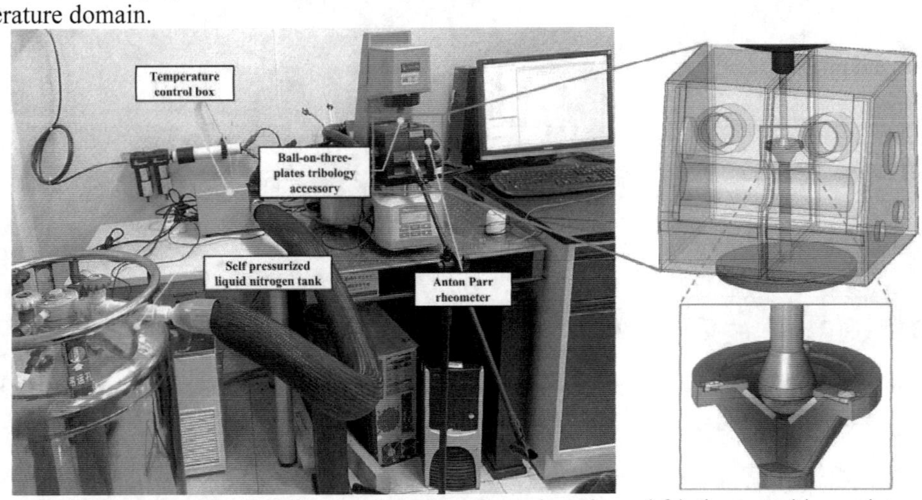

Figure 1 Tribology test module for a wide temperature domain and friction matching point contacts

3. Analysis of tribological test results

3.1. Comparison of friction noise characteristics and wear performance under different working conditions
Figure 2 shows the amplitude, time-frequency analysis diagram, and scanning electron microscope diagram of friction noise at different temperature points of a working condition for 15N-1250 rad/min. It can be seen that at the temperature point of 25°C, the temperature of the friction interface between the polymer and the metal gradually increases in the early stage of the test. And the melting of the polymer plays a certain role in lubrication, without producing friction noise. With the increase in interface temperature, the polymer interface appeared serious delamination and abrasive wear due to the harsh working environment of high temperature and pressure. The friction coefficient and sound pressure are further increased, and the friction noise frequency mainly ranges from 0-15 kHz, with no obvious distribution rule in the frequency range. At the temperature points of 0°C, -20°C, and -70°C, the wear is characterized by the combination of adhesive wear and abrasive wear. At this time, the friction noise always exists, and the frequency distribution range is wide, covering 0-25 kHz. At -20°C, the frequency distribution is regular. It can be inferred that the wear particles in the friction interface are relatively concentrated. A part of the debris removed from the friction area is embedded in the polymer, and no new particles are generated. At two temperature points of -120 °C and -150 °C, the friction interface is mainly characterized by adhesive wear. And the frequency distribution of friction noise is relatively sparse, with stripe distribution at individual frequency bands and low sound pressure values.

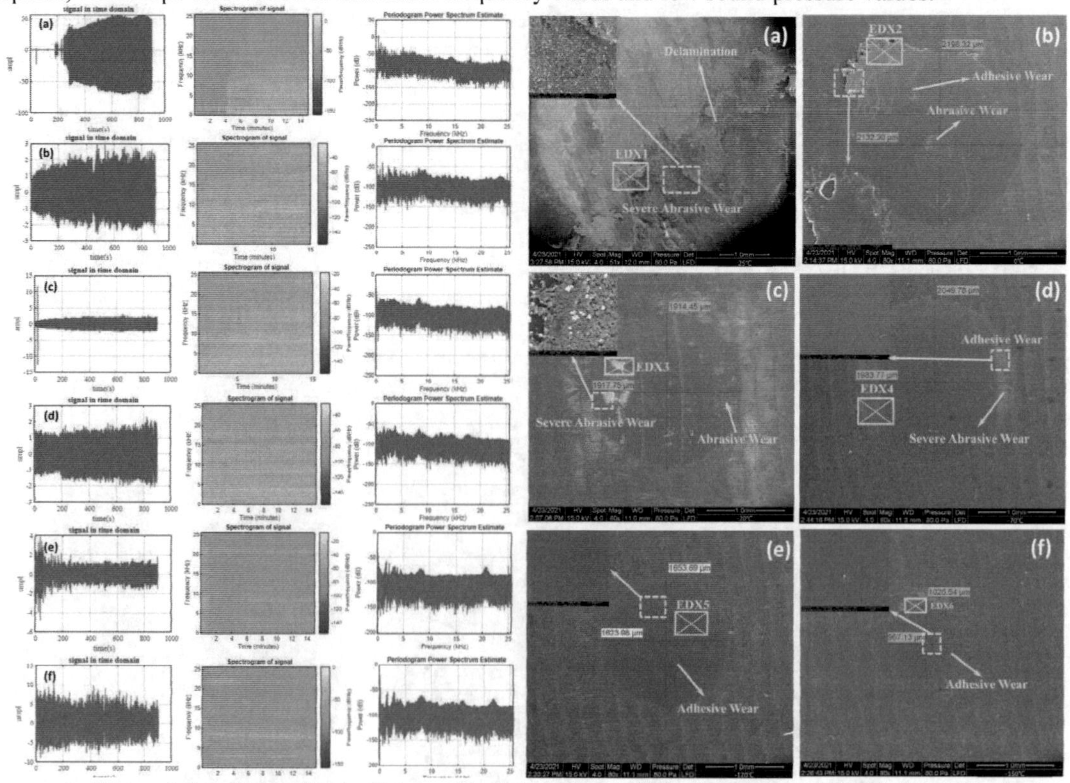

Figure 2 Sound amplitude diagram, time domain analysis diagram, and SEM diagram (from left to right). (a). 25°C, (b). 0°C, (c). -20°C, (d). -70°C, (e). - 120°C, (f). -150°C.

In the analysis of the reasons why the friction coefficient, wear rate, and friction noise of the polymer interface are all higher under the conditions of 25°C and 15N-1250 rad/min, surface element analysis is carried out, as shown in Table 1. The results of energy spectrum analysis show that in addition to the original C, F and Cl elements on the surface of 25°C compared with 0°C, Ni, Cr, Fe, and other metal elements are also added. And the proportion of O element increases. It can be seen that under the high PV working environment, the polymer and metal undergo mechanochemical reactions with the increase of interface temperature. And the metal elements on the metal surface are grafted and oxidized to adhere to the polymer surface, resulting in the phenomenon of "soft wear hard". In the actual application scenario, the working environment with high temperature and high PV should be avoided as far as possible to avoid serious or permanent failure of the polymer-metal interface.

Table 1 Analysis and comparison of main elements on PCTFE worn surface

	EDX1		EDX2		EDX3		EDX4		EDX5		EDX6	
Element	Wt%	At%	Wt%	At%	Wt%	At%	Wt%	At%	Wt%	At%	Wt%	At%
C	8.26	16.8	9.70	17.5	13.1	22.9	14.8	25.0	13.7	23.9	12.2	21.4
O	4.13	6.03	1.37	1.86	2.47	3.24	5.18	6.57	3.24	4.22	2.88	3.80
F	39.8	49.0	49.2	56.3	46.2	51.1	45.6	48.8	45.2	46.7	46.8	52.0
Cl	36.8	24.3	39.7	24.3	38.3	22.7	34.4	19.7	37.8	22.2	38.2	22.8
Fe	5.93	2.48	-	-	-	-			-	-	-	-
Ni	3.15	1.25	-	-	-	-			-	-	-	-
Cr	1.83	0.82	-	-	-	-			-	-	-	-

3.2. Research on mapping mechanism between frictional noise characteristics and tribological properties

The detected noise sound pressure is generally a random process with time as a parameter, and the characteristics are usually extracted in the time domain and frequency domain. In order to obtain the relationship between the time domain characteristics of friction noise and the friction coefficient at the corresponding time of each second in the time series, the root mean square value of the sound amplitude at each second is extracted through time domain analysis, and the sound pressure values within the time range are averaged. Compared with the mean friction coefficient and the final wear rate, the results shown in Figure 3 show a decrease in temperature. It is found that there is a good response relationship between them. Under the three working conditions, when the friction coefficient increases or decreases with the temperature, the wear rate presents the same changing rule. At 25 °C, the polymer friction interface generates a higher temperature, and the polymer and metal will have a mechanochemical reaction, leading to an increase in adhesion and resulting in greater wear and friction coefficient. With the temperature decreasing, the wear rate generally decreases with the increase in material strength. But at -150°C, there will be a small amount of mechanical loss[9], which increases the response of friction coefficient and wear rate. The relationship between sound pressure and wear rate was further studied. Under 5N-750Rpm of low PV working conditions, the number of abrasive particles produced was small. The main wear form is adhesive wear, and the polymer interface is mainly stuck slip motion. When the friction coefficient decreases, the sound pressure shows an increasing trend. Under the working condition of 10-1000Rpm, the worn form at the temperature points of 25°C and 0°C is mainly abrasive wear. The generation of abrasive is consistent with the changing trend of sound pressure, so the wear rate is consistent with the change law of sound. With the temperature decreasing, the shear strength of PCTFE material increases, and adhesive wear gradually becomes the main wear form. The change in sound pressure is opposite to the change in wear rate again. Under the high PV condition of 15-1250Rpm, more wear particles are produced at the temperature points of 25°C, 0°C, and -20°C. The wear rate is in direct proportion to the change rule of sound pressure. The results of the following temperature are the same as those of the previous two working conditions.

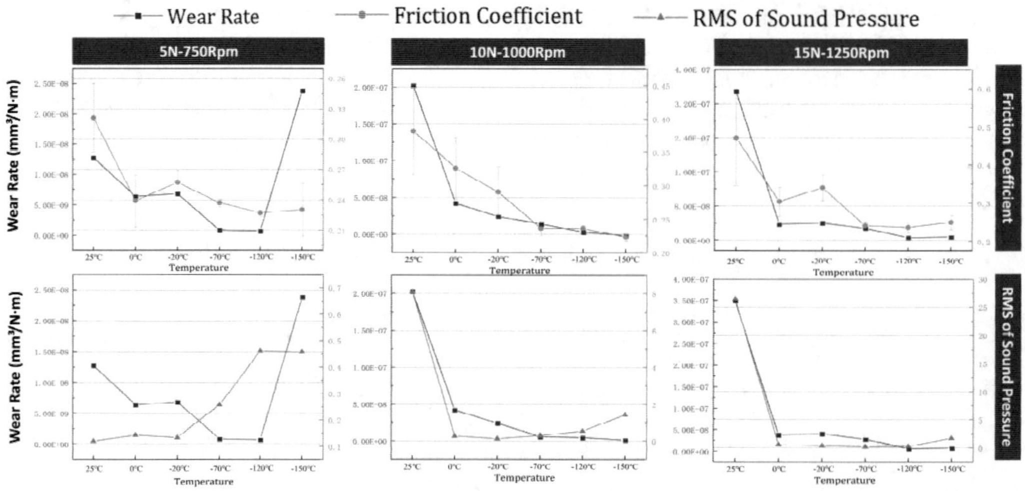

Figure 3 Comparison of wear rate, friction coefficient, and sound pressure change under different temperature conditions

As for the further exploration of the influence of polymer wear on the change of sound pressure, Figure 4 shows the relationship between the tribological properties of 2A12 and PCTFE on the time evolution of friction and sound pressure during the working condition test at 15N-1250Rpm in a wide temperature range.

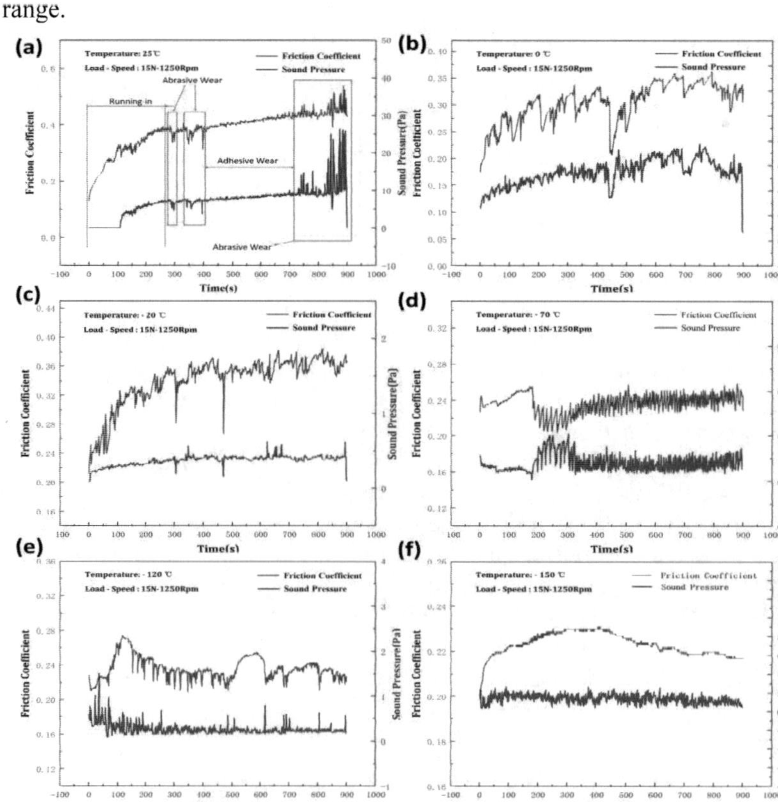

Figure 4 Relation between the evolution of tribological properties and time sound pressure at a wide temperature range. (a) 25°C, (b) 0°C, (c) -20°C, (d) -70°C, (e) -120°C, (f) -150°C

According to the temperature point analysis at 25°C, it can be seen that the two materials enter the run-in stage at the initial stage of polymer and metal friction. After a period of running in, it will enter the stick-slip stage. When wear particles appear on the friction interface with two or three body wear occurs, there will be a sudden change of sound pressure amplitude corresponding to abrasive wear. When the abrasive is embedded in the friction interface or polished, it will enter a new stick-slip stage. It can be found that the tribological properties are closely related to the tribological noise characteristics and variation rules, which makes it possible to predict the tribological properties with the same characteristics by using random nonlinear tribological noise characteristics. Furthermore, the mapping law between the change of sound pressure and the change of stick-slip and wear form with the change of temperature from 0°C to -150°C was further analyzed. It can be seen that during the temperature change from 25°C to -20°C, the changing trend of sound pressure and friction coefficient caused by abrasive particles is completely consistent and different from the adhesive wear stage. The change degree of sound pressure and the friction coefficient is larger than that during the stick-slip process. When the temperature changes from -70°C to -150°C, the friction noise is mainly caused by the stick-slip effect. And the correlation between the response of the friction coefficient and the stick-slip phenomenon increases significantly. When the friction coefficient decreases in the stick-slip process, the response of the sound pressure increases. When the friction coefficient increases, the response of the sound pressure decreases, which is contrary to the positive trend of abrupt change caused by abrasive wear, showing a negative correlation trend.

4. Conclusion

In this paper, the friction noise test collection, time-frequency domain analysis, and tribological performance mapping analysis are conducted, and the main results are as follows.

- The frequency distribution of noise caused by abrasive wear ranges from 0 to 25 kHz, and a part of the debris removed from the friction area is embedded in the polymer with no new particles generated. The frequency distribution of noise caused by adhesive wear is sparse, and it is striped in some frequency segments. The text should be set to single-line spacing.
- The wear forms of 304-PCTFE pairs are different in the wide temperature range from -150°C to 25°C. The main wear form is adhesive wear at low temperatures. With the increase in temperature, abrasive wear gradually occurs at the polymer interface. When the PV value is high at 25°C, mechanical and chemical reactions will occur between the polymer and metal, which should be paid attention to in the practical application environment.
- When the friction coefficient of 304-PCTFE pairs increases or decreases with temperature in a wide temperature range, the changing trend of sound pressure caused by abrasive particles in abrasive wear is completely consistent with that of the friction coefficient. When the friction coefficient decreases in the stick-slip process of adhesive wear, the friction coefficient decreases. The sound pressure will respond to an increase. When the friction coefficient increases, the sound pressure will respond to a decrease. This is contrary to the abrupt positive change trend caused by abrasive particles in abrasive wear, showing a negative correlation trend.

Acknowledgment

This work was supported by the National Natural Science Foundation of China under Grant U1937602, the Major National R&D Projects of China under Grant J2019-IV-0020-0088, and the Independent Research Project of State Key Laboratory of Tribology in Advanced Equipment under Grant SKLT2022B10.

References

[1] Gradt T, Theiler G. 2009 *ESMATS.* **13**
[2] Theiler G, Gradt T. 2010 *Wear.* vol 269 pp 278-284
[3] Pikovskii A, Rozenblyum M, Kurths J. 2003 *Tekhnosfera Moscow Synchronization*
[4] Sergienko V, Bukharov S. 2008 *Tribologia.* vol 129 p 137
[5] Jibiki T, Shima M, Akita H, et al. 2001 *Wear.* vol 251 pp 1492-1503
[6] Dong C, Mo J, Yuan C, Bai X, Tian Y. 2019 *Tribology Letters.* vol 67 pp 1-12

Analysis of Intake Performance of Forebody and Inlet on Pre-deformation

Siyi Li[1], Lei Liu[1], Shenshen Liu[1*], Xiaofeng Yang[1], Ziyi Wang[1], Yanxia Du[1]

[1]State Key Laboratory of Aerodynamics, China Aerodynamics Research and Development Center, Mianyang 621000, China

Shenshen Liu email: liushenshen@cardc.cn

Abstract: Under long cruise conditions, the forebody and inlet of a hypersonic vehicle were sensitive to deform under the aerodynamic thermal load, which hurt aerodynamic characteristics, intake performance, and aircraft safety. A pre-deformation method was proposed by using the numerical method of fluid-thermal-structural coupling based on the in-house software, aiming at the static deformation of a typical forebody-inlet model. Three pre-deformed models were combined by the forebody/inlet deformation and then compared the intake performance between the pre-deformed models and the ideal rigid model. The results show that the pre-deformation of the forebody can not only ensure the intake performance partly but also avoid the risk of excessive pressure on the intake wall. The pre-deformation of the inlet can reduce the flow separation near the ramp wall and reduce the pressure of intake, but it also reduces the total pressure recovery coefficient and affects intake performance negatively.

1. Introduction

Modern hypersonic vehicle pursues high speed and almost adopts thin body shape in flight vehicle design. In some specific and complex parts of the vehicle (wings, rudders, etc.), the design requires high mobility and makes these structures greatly flexible. The forebody-inlet of the hypersonic vehicle is characterized by its special sharp and thin shape and exposure to the complex inflow environment [1]. The aerodynamic/thermal/structural coupling impact is significant which will endanger flight safety.

To overcome the negative effect of structural flexibility on aerodynamic characteristics and intake performance, the passive method was generally adopted by increasing the structural rigidity in the previous design. But there is a problem that adds the structural weight and violates the design requirement of increasing the thrust-weight ratio of the aero engine. With the continuous development of aeronautical materials, the strong performance of the material has been significantly improved, however, the improvement in material rigidity is limited. Therefore, in the process of flight vehicle design, if the strength of aeronautical materials is further improved in requirements, the rigidity of the material must be reduced, and the structure also tends to be flexible. To meet the requirements of structural rigidity and minimize the weight of the structure at the same time, a method that causes flexible pre-deformation on the structure can be adopted and meets the ideal requirements of the aerodynamic design.

In previous aircraft designs, structural pre-deformation was first implemented in wings design. Palacios R, etc. [2] came up with a technology of active flexible slender wings. Wang Yun, etc. [3] adopted the pre-deformation method on the flexibility of high-ratio-aspect composite wings by using the finite element method and the fluid-structure coupling method to analyze aerodynamic characteristics.

In this paper, the coupled analysis platform for thermal environment and structure response (FL-CAPTER [4]) which was developed by China Aerodynamics Research and Development Center is adopted to study the effect of pre-deformation of forebody-inlet on intake performance and compare with the ideal rigid model at the same time.

2. Numerical calculation method and principle

A numerical coupling method (CFD-CTSD) is adopted to calculate. With the hexahedral structured grid in a fluid field, the finite volume method is adopted to calculate the Navier-Stokes equations. The implicit time-stepping method of LU-SGS is used on the elements. The Navier-Stokes equation with dimensionless form in a rectangular coordinate system is shown below.

$$\frac{\partial Q}{\partial t}+\frac{\partial E}{\partial x}+\frac{\partial F}{\partial y}+\frac{\partial G}{\partial z}=\frac{1}{\mathrm{Re}}\left(\frac{\partial E_{\mathrm{V}}}{\partial x}+\frac{\partial F_{\mathrm{V}}}{\partial y}+\frac{\partial G_{\mathrm{V}}}{\partial z}\right) \tag{1}$$

Where Q is the conserved state vector, E, F, and G are the inviscid flux vector, E_{v}, F_{v}, and G_{v} are the viscous flux vector, and Re is the dimensionless coefficient.

With the unstructured grid in a structural thermal field, the finite volume method is adopted to calculate the heat conduction equation. Explicit time stepping method of second order Runge-Kutta is used on the elements. The heat conduction equation with no inner heat source in a rectangular coordinate system is shown below.

$$\rho C_p\frac{\partial T}{\partial t}=\frac{\partial}{\partial x}\left(k\frac{\partial T}{\partial x}\right)+\frac{\partial}{\partial y}\left(k\frac{\partial T}{\partial y}\right)+\frac{\partial}{\partial z}\left(k\frac{\partial T}{\partial z}\right) \tag{2}$$

With the unstructured grid in the structural stress field, the finite element method is adopted to calculate the elasticity equations. Aiming the thermal stress, it can be considered as only linear strain or none shear strain. The thermal deformation is the initial strain ε_0:

$$\varepsilon_0=\alpha(\phi-\phi_0) \tag{3}$$

The stress and strain relation are shown below.

$$\sigma=D(\varepsilon-\varepsilon_0) \tag{4}$$

The materials involved in this paper are isotropous, so the linear expansion coefficient α is the same in all directions. The ϕ_0 is the initial structural temperature and the ϕ is the transient structural temperature which can be calculated by interpolation.

The tightly coupled method is used in this paper and the coupling mode of each field is shown in Figure 1. The time scale of heat conduction is much larger than that of the flow field which is regarded as steady in coupling calculation so that it can improve computational efficiency. Similarly, structural stress and strain calculations can be regarded as steady.

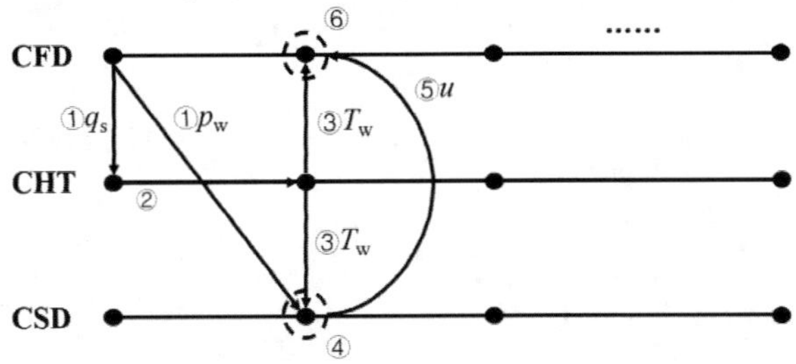

Figure 1. Fluid-thermal-structural coupling data transmission.

3. Modelling and inflow conditions

According to the rigidity assumption, the ideal aerodynamic shape and structure model of the forebody-inlet are established. The GK-01 [5] shape is used in this paper (Figure 2) and the freestream condition is shown in Table 1. Fluid and structural grids are shown in Figure 3. The flow grid is encrypted at the boundary layer and the first layer mesh thickness is 1×10^{-6}m.

Table 1. Freestream condition

Ma	T_∞(K)	ρ_∞(kg/m^3)	P_∞(Pa)	T_{w0}(K)
7.0	220.1	5.08×10^{-3}	3210.2	300

Aiming to the structural materials, MoCu is used in the head of solid, GH49 is used in the thin-wall structure, wall thickness is 10mm, and material parameters are shown in [6].

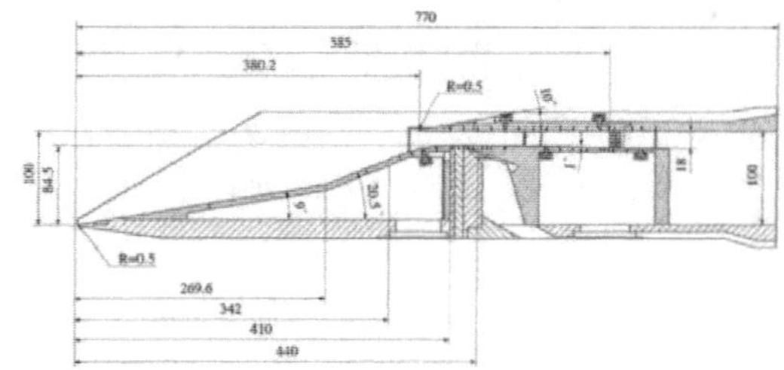

Figure 2. GK-01 model size(mm).

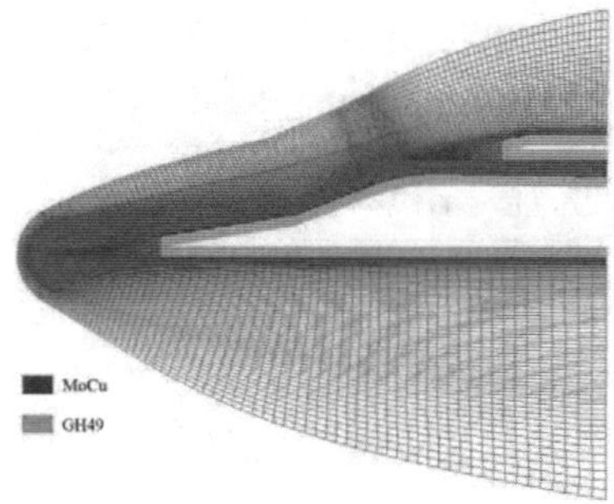

Figure 3. Grid of ideal model.

4. Pre-deformation of model

The fluid-thermal-structural coupling is calculated and the ideal model (none pre-deformed model) is used. The coupling time is the 50s. Out contour of the structure before and after deformation is shown in Figure 4 (a), where the maximum deformation is upward to 15.5mm at the forehead of the forebody, and the second is the cowl lip of the inlet, which moves down 9.8mm. So, under long cruise conditions, structure deforms in aerodynamic and thermal environments. In [7], it showed that thermal deformation can affect aerodynamic characteristics and inlet parameters.

Therefore, the pre-deformation method is used. According to the design requirements, the structure is pre-deformed to make the shape of the deformed structure conform to the design shape [8]. Three pre-deformed models are designed including the forebody pre-deformed model, inlet pre-deformed model, and forebody-inlet pre-deformed model. The deformation is used for modeling. Three models' out contour compared with the ideal rigid model (original shape) is shown in Figure 4 (b-d).

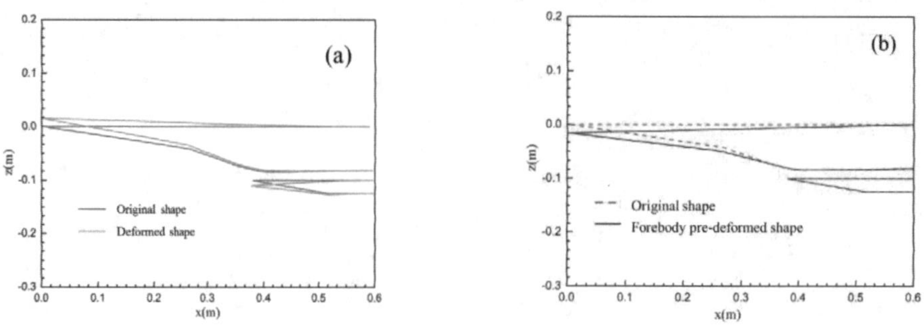

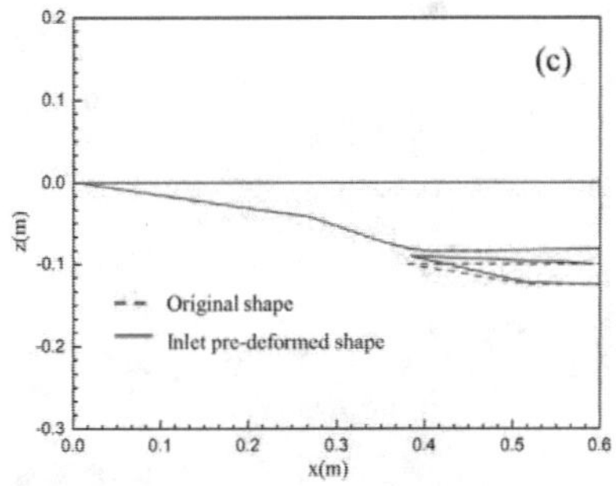

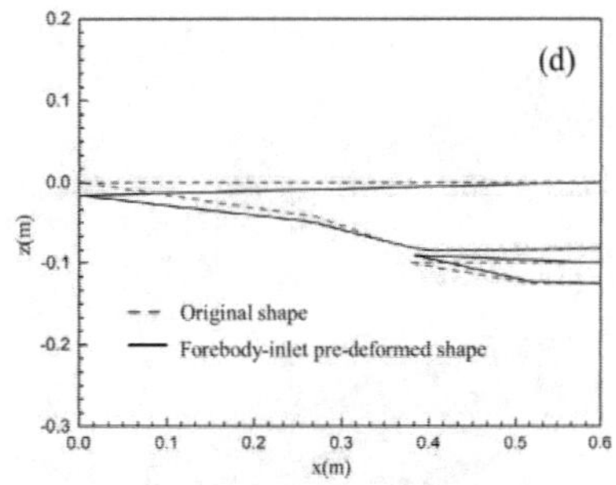

Figure 4. Out contour of structures. (a) None pre-deformed models before and after deformation; (b) Forebody pre-deformed model and ideal rigid model; (c) Inlet pre-deformed model and ideal rigid model; (d) Forebody-inlet pre-deformed model and ideal rigid model.

5. Analysis of calculation results

Three models are calculated in the same method as the none pre-deformed model. The inlet pressure contours of the three models at the initial time compared with the ideal rigid model are shown in Figure 5. It is shown that the typical flow characteristic is the reflection of multiple shock waves in the inlet, and the shock reflection of the ideal rigid model is the most obvious. The inlet pressure of the three models is lower than that of the ideal rigid model, especially the inlet pre-deformed model and the forebody-inlet deformed model are more significant on lower pressure, which is caused by the smaller entrance of the inlet of the two models at the initial time and less mass flow enter the inlet.

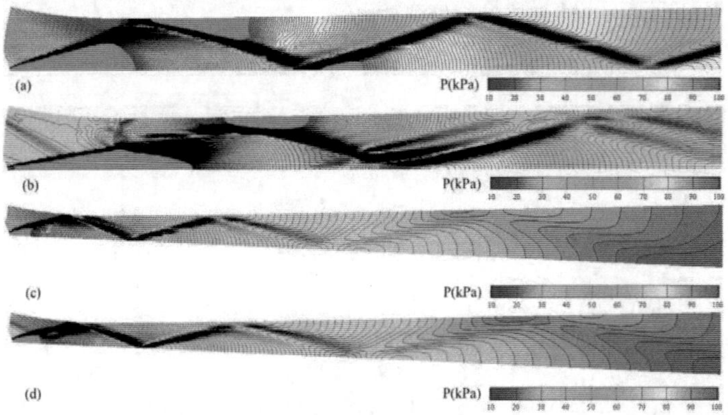

Figure 5. Pressure contours of the inlet for different models. (a) Ideal rigid model; (b) Forebody pre-deformed model; (c) Inlet pre-deformed model; (d) Forebody-inlet pre-deformed model.

Mach number contours of the three models at 50s and the ideal rigid model are shown in Figure 6. It is shown that the oblique shock wave over the ramp surface enters the cowl lip in the ideal design condition and causes multiple shock reflections inside the inlet.

Aiming at the forebody pre-deformed model, although the shape of the deformed forebody tends to be the ideal state under aerodynamic heating, however, the size of the inlet's entrance area determines whether the wave system inside the inlet is in the design stage. So only forebody pre-deformation can not change the deviation of the inlet from the design state after deformation. Aiming at the inlet pre-deformed model, it is shown that the inlet tends to the design state after deformation, and it can reduce the problem of increasing flow separation caused by the deviation of the inlet from the design stage. Aiming at the forebody-inlet pre-deformed model can also reduce the problem of flow separation near the ramp surface. In short, the pre-deformation design can make the structure tend to the ideal design state during flight and obtain the ideal aerodynamic characteristics.

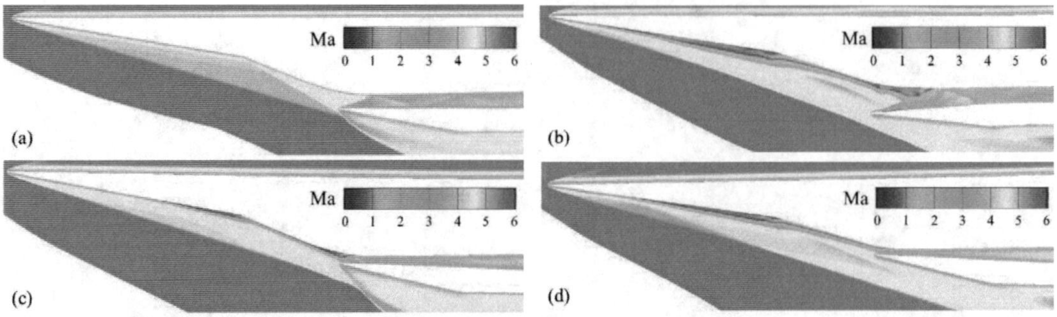

Figure 6. Mach number contours for different models. (a) Ideal rigid model; (b) Forebody pre-deformed model after deformation; (c) Inlet pre-deformed model after deformation; (d) Forebody-inlet pre-deformed model after deformation.

The shapes of the inlet pre-deformed model and the forebody-inlet pre-deformed model can be observed in Figure 6, and the two models' inlets fail to restore the position of the ideal rigid model after deformation. Comparison of the out contour of structures between the three models at 50s (after deformation) and the ideal rigid model are shown in Figure 7.

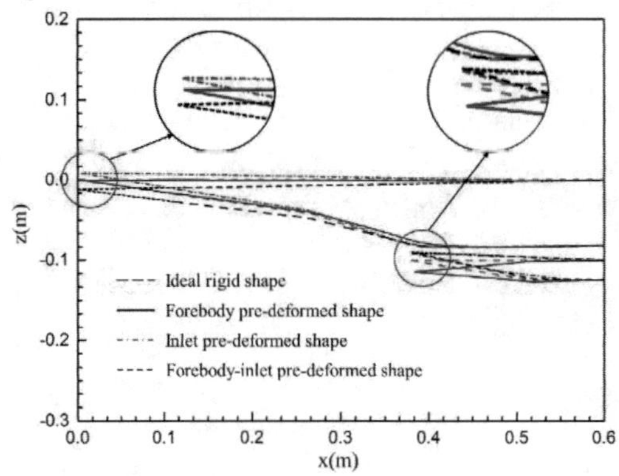

Figure 7. Out contour of structures for different models.

After deformation, the forebody restores to 98% of the ideal state in the forebody pre-deformed model, the inlet restores to 13.3% of the ideal state in the inlet pre-deformed model, the forebody restores to 25.8%, and the inlet only restores to 3% of the ideal state in the forebody-inlet pre-deformed model. Because the pre-deformation design changes the flow field state near the wall, especially for the inlet pre-deformed model and the forebody-inlet pre-deformed model, the pressure inside the inlet at the initial time is lower (Figure 5). The insufficient aerodynamic force makes it difficult for the pre-deformed inlet to restore to the ideal state.

In respect of the issues above, the optimized model of the forebody-inlet pre-deformed model has been used which the pre-deformation of the forebody is 7.75mm and the pre-deformation of the inlet is 4.9mm. The same method is used to calculate until the 50s. Mach number contours of the optimized model at 50s are shown in Figure 8. It is shown that the flow separation near the ramp wall is significantly reduced compared with Figure 6 (b). Combined with Figure 6, it can be seen that flow separation near the ramp wall is mainly determined by the position of the inlet. The out-contour comparison between the optimized model after deformation and the ideal rigid model is shown in Figure 9. It is shown that the inlet restores to 102% of the ideal state in the optimized model, which almost achieves the ideal design state of the inlet.

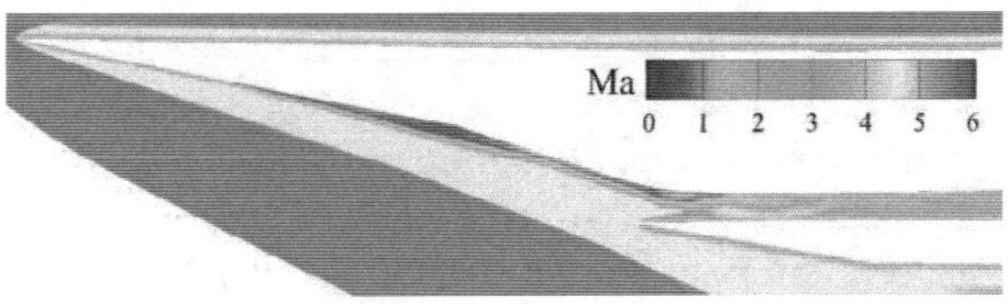

Figure 8. Mach number contours of the optimized model at 50s.

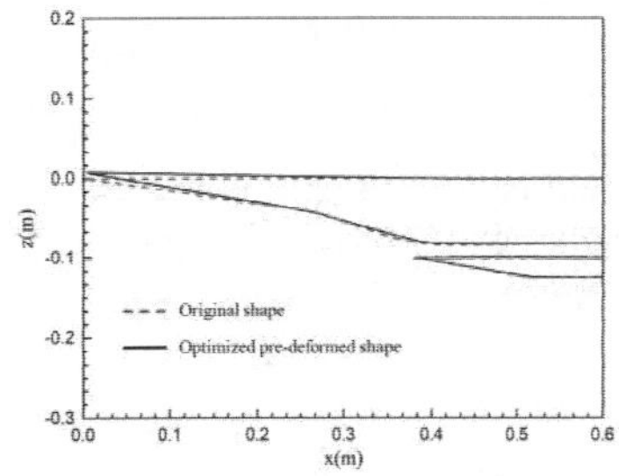

Figure 9. Out contour of the optimized model after deformation and the ideal rigid model.

Intake performance can be measured by the inlet mass flow coefficient φ, the total pressure recovery coefficient δ, and the static pressure ratio R_p. The intake parameters of the three models are at the initial and deformed state while comparing with the ideal rigid state and deformed state of none pre-deformed model are shown in Table 2.

Table 2. Intake parameters for different models.

	None pre-deformed model		Forebody pre-deformed model		Inlet pre-deformed model		Forebody-inlet pre-deformed model	
	Ideal rigid (Initial)	Deformed	Initial	Deformed	Initial	Deformed	Initial	Deformed
φ	0.395	0.578	0.514	0.389	0.11	0.096	0.23	0.083
δ	0.326	0.261	0.394	0.3	0.188	0.073	0.232	0.066
R_p	51.027	88.107	32.93	30.52	15.65	15.88	15.379	11.482

It is shown that aiming to the forebody pre-deformed model after deformation, the total pressure recovery coefficient is only 7.9% lower than the ideal rigid state, however, the static pressure ratio is 40.18% lower than the ideal rigid one. Compare with the none pre-deformed model after deformation, the total pressure recovery coefficient and the static pressure ratio of the forebody pre-deformed model after deformation are 14.9% higher and 65.36% lower, respectively. It shows that the forebody pre-deformed model can ensure the intake performance partly, and avoid the risk of excessive pressure on the intake wall. Although the inlet pre-deformation can reduce the inlet pressure and the flow separation near the ramp wall in Figure 5, the low total pressure recovery coefficient indicates that the inlet pre-deformation affects intake performance negatively, which is detrimental to engine start.

Table 3. Intake parameters for optimized model

	φ	δ	R_p
Initial	0.3	0.303	19.064
Deformed	0.49	0.262	44.13

The intake parameters of the optimized model are shown in Table 3. It is shown that aiming to the optimized model after deformation, the total pressure recovery coefficient is 0.4% higher than the none pre-deformed model after deformation, and the static pressure ratio is 13.5% lower than the ideal rigid one and 49.9% lower than the none pre-deformed model after deformation. Compare with the forebody pre-deformed model after deformation, although the total pressure recovery coefficient of the optimized model after deformation is 12.6% lower, the inlet mass flow coefficient is 25.96% higher which is good for aero engine work. Although the intake performance of the optimized model after deformation is similar to the none pre-deformed model after deformation, the pressure of intake can be reduced. And there is an advantage in that mass flow is boosted compared with the forebody pre-deformed model after deformation.

6. Conclusion

In this paper, a tightly coupled method is used to calculate three models in pre-deformation design. And the conclusions are as follows:

(1) The pre-deformation design of the forebody and inlet can make the structure tend to the desired shape during cruise flight. The forebody pre-deformation can make the forebody restore to the ideal rigid position after deformation. Because the inlet entrance area is reduced and the pressure of intake is lower for the inlet pre-deformation, the aerodynamic force is not enough to restore the pre-deformed inlet to the ideal rigid position. A reasonable amount of pre-deformation can be conducive to aerodynamic characteristics and intake performance.

(2) The pre-deformation of the forebody can ensure the intake performance partly, and avoid the risk of excessive pressure on the intake wall. However, it cannot solve the problem of flow separation near the ramp wall. The pre-deformation of the inlet can reduce the flow separation near the ramp wall and reduce the pressure of intake, but it also affects the inlet performance negatively, which hurts the engine start.

(3) By setting the pre-deformation of the inlet to 50% of the original inlet deformation, the inlet can be restored to the ideal rigid position after deformation and the flow separation near the ramp wall is reduced. However, the intake performance cannot be effectively improved and is similar to the none pre-deformed model after deformation. The advantage is that the pressure inside the inlet can be reduced compared with the none pre-deformed model after deformation and mass flow is boosted compared with the forebody pre-deformed model after deformation. So, the inlet pre-deformation should be carefully considered in the design of the inlet.

Acknowledgments

This research work is supported by the National Natural Science Foundation of China (Grant No. 11972359).

References

[1] Reinartz B U, Herrmann C D, Ballmann J. 2003. Aerodynamic performance analysis of a hypersonic inlet isolator using computation and experiment. *Journal of Propulsion and Power*, **19**: 868-875.

[2] Palacios R, Cesnik C. 2005. Static nonlinear aeroelasticity of flexible slender wings incompressible flow. In: *46th AIAA/ASME/ASCE/AHS/ASC Structures, Structural Dynamics, and Materials Conference*. Austin, Texas: 1-10.

[3] Wang Y, Xu J F. 2013. Analysis of aerodynamics characteristic of the flexible wing on pre-deformation. *Journal of Nanchang Hangkong University: Natural Sciences*, **27**: 47-51.

[4] Gui Y W, Liu L, Dail G Y, Zhang L T. 2017. Research status on hypersonic vehicle fluid-thermal-structural coupling and software development. *Acta Aeronautica et Astronautica Sinica*, **38**: 1-19.

[5] Hohn O, Gulhan A. 2010. Experimental investigation on the influence of yaw angle on the inlet performance at Mach 7. In: *48th AIAA Aerospace Sciences Meeting Including the New Horizons Forum and Aerospace Exposition*. Orlando, Florida: 1-16.

[6] <China Aeronautical Materials Handbook> editorial committee. 2002. *China Aeronautical Materials Handbook. Deformation superalloy casting superalloy*. vol 2. Standards Press of China, Beijing.

[7] Dai G Y, Jia H Y, Zeng L, Liu L, Qiu B. 2018. Effects of Fluid-Thermal-Structural coupling on inlet parameters of hypersonic intake. *Journal of Propulsion Technology*, **39**: 1267-1274.

[8] Zhang H, Ma D L, Ma T L. 2008. Analysis of aerodynamics characteristic of the flexible wing caused by deflection. *Journal of Beijing University of Aeronautics and Astronautics*, **34**: 487-490.

Performance Study of External Scaffolding under Wind-Induced Action

Qian Cao[1], Yiyang Huang[2], Bo Chen[1*], Pengjin Wu[2], Sheng Zhang[1], Liang Ma[1], Lilong Wei[2], Lepeng Huang[2]

[1]Chengdu First Construction Engineering of CDCEG, No. 123 Yusha Street, Chengdu, Sichuan, 610017, China.

[2]School of Civil Engineering, Chongqing University, Chongqing, 400045, China.

[*]Corresponding author's e-mail: cqu2022@126.com

Abstract. Scaffolding is an essential temporary measure in engineering construction, and its safety management is critical in the construction process. In this paper, the overall scaffolding model is simulated and analyzed by SAP2000 finite element software. The difference in force between the scaffolding under static wind load and time-range wind load is obtained so that the wind vibration coefficient of the scaffolding is suggested. The results show that the most influential factor in the stability of the scaffold is the way in which the wall elements are arranged. At the same time, the most favorable arrangement is a two-step, two-span arrangement of the wall elements, a 50° angle of scissor bracing, and a wind protection factor of 0.8 for the safety net. Also, the wind vibration factor β_z should be taken as 2.5 instead of 1 in the scaffolding code for close-meshed scaffolding at a height of 50 m.

1. Introduction

As an accessory structure during the construction of a building, external scaffolding has an independent structural system. The safety performance of scaffolding is an essential guarantee for construction safety and safeguards the prerequisite for fast and efficient construction activities. In previous studies on scaffold stability, Rubio-Romero et al.[1] emphasized the importance of diagonal scaffold bracing. And Peng et al.[2] proposed to determine the layout form of scaffolds according to the ultimate bearing capacity of the bars. Jia et al.[3] studied the fastener scaffold model and proposed a calculation model and algorithm which simplified the calculation workload.

In addition, to determine the wind load applied on the scaffold structure, the research usually uses methods such as field actual measurement and wind tunnel test. However, these two methods are complex in terms of testing process and method and require a lot of human and material resources. Therefore, numerical simulation is widely used in scaffold wind vibration analysis. Some researchers have used finite element software analysis to study the extent and law of the effect of defects in the setting of connecting wall members on the maximum bending moment of scaffold uprights under wind loads to provide a basis for the reasonable design of scaffolds[4]. Besides, some studies have also presented the dynamic characteristics of attached scaffolds under the action of pulsating winds[5]. At the same time, some projects have analyzed the maximum bending moment of scaffolding under wind loads on the bars and summarised the laws of variation of the maximum bending moment, providing data to support the safe design of scaffolding construction in super tall buildings[6].

Therefore, to derive the variability of the force conditions of the scaffold under static and time-range wind loads and to propose the optimum wind vibration coefficient for the scaffold, this study simulated and analyzed the overall scaffold model by SAP2000 finite element software. And the scaffold performance is experimented with and analyzed under the action of several influencing factors with the "uniformly dispersed and comparable" characteristics of the orthogonal test taken into consideration to obtain the test results.

2. Principle of wind load action

The wind resistance of a structure needs to be studied to ensure that it is safe from wind-induced effects. Wind speed is generally divided into two components: mean wind and pulsating wind. The mean wind can be thought of as a period of time in which the wind speed remains more or less constant from time to time and increases as the height increases. The period of the mean wind is long and differs significantly from the self-oscillation period of the structure. Therefore, the mean wind action can be regarded as static on the structure. The pulsating wind is a wind speed with a minor period change from the mean wind, usually only a few seconds. It is more similar to the self-oscillation period of the structure and can have a vibrational effect on the structure. The wind speed $V(x, y, z, t)$ acting at any point (x, y, z) on the structure is therefore the sum of the average wind speed $\bar{v}(z)$ and the pulsating wind speed $v(x, y, z, t)$, as shown in Equation 1.

$$V(x, y, z, t) = \bar{v}(z) + v(x, y, z, t) \qquad \text{Equation 1}$$

2.1. Static wind loads

When the period of the wind is more significant than 10 minutes or more, the nature of the force of the static wind can be considered static. Structural calculations are usually simplified by the essential wind speed and the primary wind pressure. The basic wind speed is the maximum average wind speed in the area obtained by statistical analysis of the data under certain specification requirement conditions through many records and observations of the wind speed by meteorological observation stations in different areas. The primary wind pressure is obtained by converting the essential wind speed. According to the Code for Structural Loads of Buildings (GB50009-2012), which stipulates the conditions for selecting the essential wind speed, and with a combination of the air density and aerodynamic studies, the conversion can be made by the following equations.

$$\omega_0 = \frac{1}{2}\rho v_0^2 \qquad \text{Equation 2}$$

OR

$$\omega = \frac{1}{2}\frac{\gamma}{g}v^2 \qquad \text{Equation 3}$$

OR

$$\omega_0 = \frac{1}{1600}v_0^2 \qquad \text{Equation 4}$$

Where, ω means wind pressure per unit area;
γ means air volume weight;
g is the acceleration of gravity;
ρ means air density;
v means wind speed;
and the basic wind pressure value must not be less than 0.3 kN/m^2.

2.2. Pulsating wind loads

The short-period winds in wind speed are pulsating winds. The pulsating winds produce a short-period variation in wind pressure on the structure. And the wind pressure has a certain randomness, which leads to random vibration of the structure. This is the most significant factor affecting the life of the structure and the comfort of human occupants. And in more severe cases, it can cause resonance of the structure, leading to rapid structural damage.

The main parameters that characterize pulsating winds are turbulence intensity, turbulence integration scale, gust coefficient, pulsating wind speed spectrum, and coherence function. The degree of turbulence reflects the variation in the overall intensity of wind fluctuations. The degree of turbulence is defined as the ratio of the mean square of the pulsating wind speed over a mean time distance at a point to the mean wind speed at that point. At the same time, the gust coefficient reflects the variation in the intensity of local fluctuations in the wind. The gust coefficient is the ratio of the maximum value of the average wind speed throughout the gust to the average wind speed over the average time distance. The turbulence integration scale reflects the spatial distribution of pulsating winds in response to the interaction. The pulsating wind speed spectrum reflects the relative influence of the vortex action on the turbulent energy at different turbulence integration scales and analyses the distribution of pulsating momentum at different "frequencies" or "scales" of vortex distribution.

2.3. Wind load values

According to the standard value of the horizontal wind load acting on the scaffold and the Code for Structural Loads of Buildings (GB50009-2012) 8.1, the wind load is calculated by the following equation.

$$\omega_k = \beta_Z . \mu_Z . \mu_S . \omega_0 \qquad \text{Equation 5}$$

ω_k means the standard value of wind load (kN/m^2).

β_Z means the wind vibration coefficient at height Z. The scaffold is attached to the main structure, and β_Z=1.0.

μ_Z means the wind pressure height variation coefficient. This paper assumes that the scaffold is located in an environment of class C ground roughness at a height of 50 m. The height of the overhanging scaffold and the height of scaffold is 15 m. Therefore, μ_Z is used to obtain μ_{50}=1.10 and μ_{65}=1.24 respectively;

μ_S is the wind load body type coefficient. In the fully enclosed state of the scaffold, μ_S=1.0φ. In the semi-enclosed state, μ_S=1.3φ. Some studies used the image analysis method to determine the porosity of the dense mesh safety net, then calculated the wind blocking coefficient, and finally obtained the new net $\varphi = 0.83$, and the old net $\varphi = 0.81$;

ω_0, the basic wind pressure (kN/m^2), according to the Code for Structural Loads of Buildings (GB50009-2012) Appendix E, can be obtained through the wind pressure table of various coastal cities in China. The larger value of the basic wind pressure for a 50-year event is 0.45 kN/m^2.

3. Analysis of the overall scaffolding model

3.1. Basic assumptions

In order to facilitate the analysis of scaffolding modeling and simplify the model theory, the following assumptions are first made. (1) The longitudinal horizontal bars are connected to the uprights by right-angle fasteners, and this connection node is regarded as a rigid connection node. (2) All corresponding bar axes are considered to be in the same plane, irrespective of the influence of the position deviation between two adjacent bars due to the fastener connection. (3) All scissor braces are considered tension units, and scissor braces are connected to horizontal bars and uprights by using swivel fasteners and articulated nodes. (4) The transverse horizontal bar and the longitudinal horizontal bar are considered semi-rigid connections. (5) The connection at the bottom end of the scaffold is an articulated system with no uneven settlement. (6) The effect of construction loads is not considered. (7) The connecting wall elements are considered to be rigidly restrained. (8) The effect of the torsional deformation of the bars on the overall stability of the scaffold is not considered. (9) Initial defects and non-linear effects of the structure are not considered. (10) The effect of dense safety risers on the structure's stiffness is ignored. (11) No consideration is given to the effect of scaffold slabs on the structure.

3.2. Overall model based on orthogonal test method design

According to the characteristics of the double-row fastener type overhanging scaffolding, a sweeping bar was set up at 20 cm from the bottom, and a rigid connection to the building to replace the wall members. The material properties and section parameters of the steel pipes are shown in Tables 1 and 2 respectively.

Table 1 Material properties of steel pipes

Elastic modulus $(E, N/m^2)$	Poisson ratio(μ)	Yield strength $(f, N/m^2)$
2.06×10^{11}	0.3	2.35×10^{8}

Table 2. Steel pipe cross-section parameter characteristics (unit: kg, m)

Outer diameter (d)	Wall thickness (t)	Cross-sectional area (A)	Moment of inertia (I)	Cross-sectional modulus (W)	Turning radius (i)	Density
4.8×10^{-2}	3.5×10^{-3}	4.89×10^{-4}	1.219×10^{-7}	5.08×10^{-6}	1.58×10^{-2}	7850

The horizontal bars (large crossbars), longitudinal horizontal bars (small crossbars), uprights, and scissor braces have ideal elastoplastic material principal relationships. They have a pre-yield modulus of elasticity E of 2.06 x 10^{11} Pa and a standard value of yield strength of 235 MPa. The SAP2000 model simulates the semi-rigid action of the bars by end release to and stiffness domain length, with the connecting wall members providing rigid lateral restraint to the frame. Gravity loads are applied to the scaffold steel tube members, and surface wind loads are applied to the scaffold.

In this paper, the finite element numerical simulation uses a fully enclosed suspended double-row scaffold with a dense safety net. It is 50 m above ground level, ten steps high, 1.5 m between steps, 15 spans, 1.5 m longitudinal spacing, and 1.05 m horizontal spacing, with a length, width, and height of 22.5 m, 1.05 m, and 15 m respectively. The wall-mounted arrangement is arranged in two steps and two spans, three steps and two spans, and two steps and three spans, as well as the scissor support angle is set at 45°, 50°, and 60°, as shown in Table 3 and 4.

Table 3 Orthogonal experiment

Working condition number	A (Wall-mounted arrangement)	B (Scissor support angle)	C (Safety net wind protection factor)
A	1	1	1
B	1	2	2
C	1	3	3
D	2	1	2
E	2	2	3
F	2	3	1
G	3	1	3
H	3	2	1
I	3	3	2

Table 4 Level factor

Level	A (Wall-mounted arrangement)	B (Scissor support angle)	C (Safety net wind protection factor)
1	Two steps and two spans	45°	0.8

2	Two steps and three spans	50°	0.9
3	Three steps and two spans	60°	1.0

In order to avoid the test results being affected by the scaffold ends not being restrained, the same restraint of connecting wall elements was added to both sides of the scaffold. The scaffold design is shown in Figure 1.

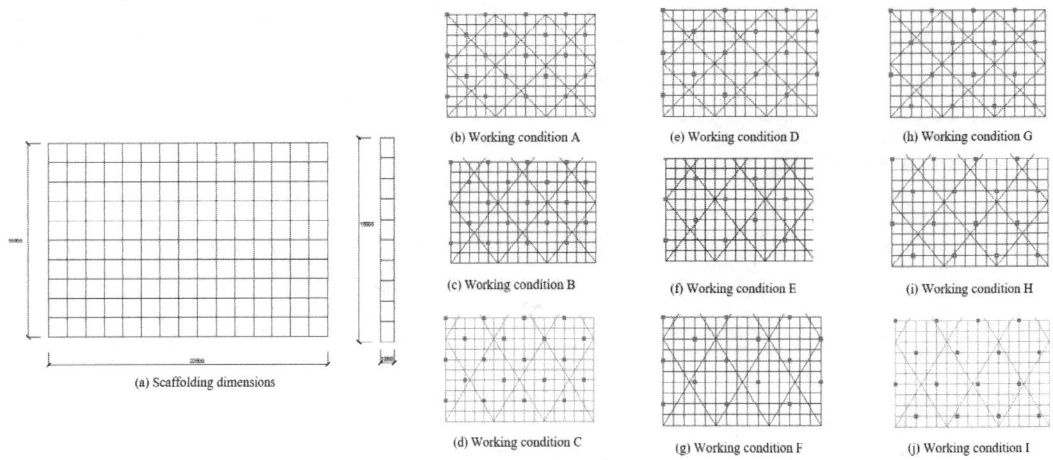

Figure 1 Scaffold design

The SAP2000 model simulates the semi-rigid action of the bars by end release to and rigid domain length. Connecting wall members provides a rigid lateral restraint to the frame. Gravity loads are applied to the scaffold steel tube members, and surface wind loads are applied to the whole scaffold. The complete model of the scaffold is shown in Figure 2.

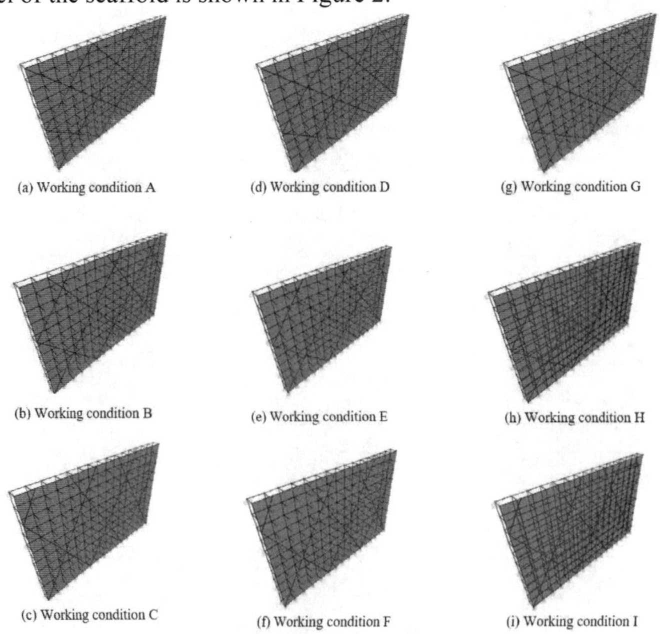

Figure 2 Scaffold model

3.3. Static analysis of the overall scaffolding model

Wind loads of 0.07 kN/m, 0.26 kN/m, and 0.45 kN/m were applied to the scaffold structures under various working conditions to obtain the scaffolding forces under each load for each working condition, as shown in Table 5.

Table 5 Overall static calculation results of scaffolding

	Rods at 0.07 kN/m				Rods at 0.26 kN/m				Rods at 0.45 kN/m			
Working condition number	Max. axial force/ kN	Max. bending moment/ kN·m	Max. stress / MPa	Max. displace ment/ mm	Max. axial force / kN	Max. bending moment / kN·m	Max. stress / MPa	Max. displace ment / mm	Max. axial force / kN	Max. bending moment / kN·m	Max. stress / MPa	Max. displace ment / mm
A	1.26	0.15	26.99	3.6	4.79	0.55	102.3	13.6	8.383	0.97	180.3	23.8
B	1.39	0.15	27.44	3.1	5.2	0.56	102.7	11.3	9.019	0.95	176.5	19.6
C	1.57	0.19	34.33	4.8	6.01	0.7	130.2	18.1	10.388	1.23	228.1	31.8
D	2.15	0.28	51.33	7.9	8.15	1.05	194.8	30.4	14.261	1.91	354	53.2
E	2.45	0.25	54.34	6.3	8.4	0.92	169.4	23.6	14.631	1.61	284.9	41.2
F	1.84	0.24	45.17	7.3	6.99	0.92	171.4	27.9	12.204	1.65	307.7	48.8
G	2.25	0.28	53.52	7.7	8.53	1.09	203	29.2	14.793	1.99	369.9	51.2
H	1.79	0.19	35.43	4.6	6.7	0.71	132.3	17.2	11.624	1.28	237.9	29.9
I	2.03	0.26	45.82	7.2	7.68	0.99	185.2	27.5	13.357	1.81	336.4	48.1

In the calculation of the maximum axial force, the optimal combination of the axial force of the scaffolding wall elements is two steps and two spans for the wall element layout, 50° for the scissor bracing angle, and 0.8 for the wind protection factor of the safety net. This affects the axial force of the wall elements in descending order, including the wall element layout, the wind protection factor of the safety net, and the scissor bracing angle. The optimum combination of the maximum axial force of the scaffolding bars is two steps and two spans for the wall elements, 60° for the angle of the scissor brace, and 0.8 for the wind protection factor of the safety net. This affects the maximum axial force of the bars in the following order, namely wall elements, the wind protection factor of the safety net, and angle of the scissor brace.

In the calculation of the maximum bending moment, the optimal combination of the maximum bending moment value of the scaffolding bars is two steps and two spans for the wall members, 50° for the angle of the scissor brace, and 0.8 for the wind protection factor of the safety net. The order of influence on the maximum bending moment value of the bars is from largest to smallest, namely wall member layout, wind protection factor of the safety net, and angle of the scissor brace.

In the calculation of the maximum stress, the optimal combination of the maximum stress values of the scaffold bars is two steps and two spans for the wall attachment, 50° for the angle of the scissor brace, and 0.8 for the wind protection factor of the safety net. The order of influence on the maximum stress values of the bars is wall attachment, the safety net's wind protection factor, and the scissor brace's angle.

In the calculation of the maximum displacement, the optimal combination of the maximum displacement of the scaffold rod nodes is two steps and two spans for the wall attachment arrangement, 50° for the angle of the scissor brace, and 0.8 for the wind protection factor of the safety net. The order of influence on the maximum displacement of the rod nodes is from largest to smallest, namely wall attachment arrangement, wind protection factor of the safety net, and angle of the scissor brace.

The optimal combination for the overall stability of the scaffold is two steps and two spans for the wall elements, a 50° angle of scissor bracing, and a 0.8 wind protection factor for the safety net. This affects the axial forces of the wall elements in the following order, namely the angle of scissor bracing, wall elements, and wind protection factor for the safety net.

3.4. Dynamic analysis of the overall scaffolding model

3.4.1 MATLAB-based time-series simulation of wind speed

In the early 1960s, Canadian researchers developed a hypothesis relating to horizontal pulsating wind loads through data measurements and calculations. And based on this hypothesis, pulsating wind speed spectra were obtained:

$$S_v(f) = fk\overline{v}_{10}^2 \frac{x^2}{f(1+x^2)^{\frac{4}{3}}} \quad \text{Equation 6}$$

$S_v(f)$ is the pulsating wind load power spectrum;
k is the coarse viscosity factor at ground level;
$\bar{v}_{10}$ is the average wind speed at this location at a height of 10 m above ground level;
f is pulsating wind frequency;
x is the turbulence integration scale factor. The calculation is as follows.

$$x = 1200\sqrt{v_{10}} \quad \text{Equation 7}$$

The primary wind speed spectrum currently used in China for calculations related to wind kinetic energy is the Davenport spectrum, which is also the wind speed spectrum used in this paper.

As shown in Figure 3, 27 spatial points are uniformly selected on the scaffold, and the time range wind speed at each point is generated by using MATLAB and the stochastic simulation method described above. The wind speed design value is $U(z_{10}) = 26.83m/s$. The VonKarman constant K is 0.4, and the ground roughness z_0 is 0.05. The wind speed time interval is 0.125 s. The sampling value is 2400, and the simulation duration is 300 s. The obtained time range wind speed is converted into time range wind pressure, which is uniformly loaded on the scaffold structure. As the time range wind speed was obtained through numerical simulation, it has a certain randomness. So three groups of time-range wind speeds were generated. After comparison and analysis, the maximum error of the three-time range wind speed was within 3%, and the maximum error of the time range wind pressure was within 6%. Therefore, the average time range wind speed of the three groups of time range wind speed was chosen as the time range wind speed for this simulation.

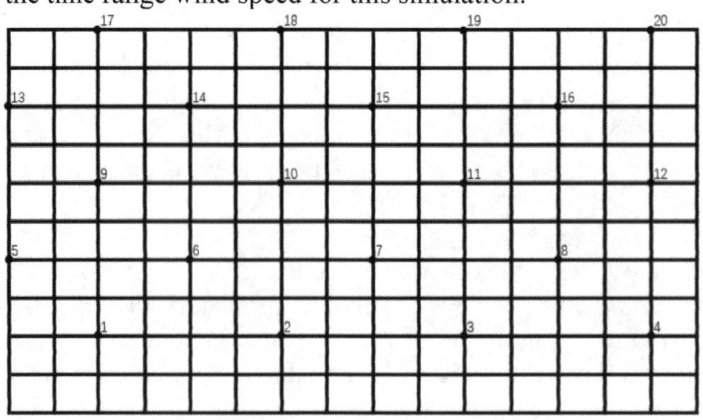

Figure 3 Simulated wind speed point

3.4.2 Scaffolding time-course wind load simulation analysis

Based on the SAP2000 software modal analysis, the first five orders of self-oscillation cycles of the scaffold structure were extracted for each working condition, as shown in Table 6.

Table 6 Self-vibration periodic table of each working condition

Working condition number	Vibration type 1	Vibration type 2	Vibration type 3	Vibration type 4	Vibration type 5
A	0.269	0.260	0.249	0.187	0.130
B	0.212	0.206	0.202	0.156	0.111
C	0.271	0.262	0.251	0.188	0.130
D	0.379	0.370	0.221	0.213	0.188
E	0.298	0.294	0.168	0.164	0.156
F	0.382	0.371	0.222	0.213	0.189
G	0.282	0.272	0.259	0.189	0.130
H	0.215	0.210	0.206	0.158	0.111
I	0.285	0.275	0.262	0.191	0.131

The pulsating wind pressure obtained in the previous section is imported into SAP2000, and only the unidirectional input is analyzed in this paper. The results are obtained by applying the pulsating wind pressure to the scaffold safety net. The pulsating wind pressure results are extracted for each working condition of the scaffold and compared with the data under static wind pressure at the points corresponding to the maximum values of displacement, bending moment, reaction force, and axial force indicators.

The maximum values of the time-course curves for each working condition are shown in Table 7. And the maximum values of the time-course loads are compared with the maximum values of the static loads to obtain the increase in each index under the time-course loads, as shown in Table 8.

Table 7 Maximum time-course curve under various working conditions

Working condition number	Max. axial force/ kN	Max. bending moment/ kN·m	Max. stress / MPa	Max. displacement/ mm
A	16.87	11.82	1.85	48.5
B	19.10	11.25	1.97	41.5
C	21.50	12.04	2.78	65.2
D	29.24	15.96	3.72	99.4
E	35.08	17.50	3.39	85.3
F	25.12	13.53	3.18	95.5
G	31.88	17.52	3.81	102.1
H	23.34	13.61	2.73	58.3
I	28.86	15.61	3.63	96.6

Table 8 Increase in each indicator under time-course loading

Working condition number	Max. axial force/ kN	Max. bending moment/ kN·m	Max. stress / MPa	Max. displacement/ mm
A	101.31%	120.11%	90.33%	103.78%
B	111.78%	108.57%	107.37%	111.73%
C	106.97%	106.24%	126.02%	105.03%
D	105.03%	80.12%	94.76%	86.84%
E	139.76%	104.92%	110.56%	108.05%
F	105.83%	102.70%	92.73%	95.70%
G	115.51%	97.54%	91.46%	99.41%
H	100.79%	98.31%	113.28%	94.98%
I	116.07%	115.64%	100.55%	100.83%

As can be seen from Table 8, the maximum values of displacement, reaction force, and axial force time-course wind loads are approximately doubled in comparison to the maximum values of static wind loads for all working conditions. Therefore, there is a significant difference in the structure's forces under time range and average wind loads. Considering the scaffolding forces under pulsating wind loads, the wind vibration coefficient Z should not be taken as 1.0, but at least 2.3 according to a conservative value.

According to the Code for Structural Loads of Buildings (GB50009-2012) 8.4 to calculate the wind vibration coefficient in the downwind direction, the wind vibration coefficient β_z at height z is calculated through the following equation.

$$\beta_Z = 1 + 2gI_{10}B_Z\sqrt{1+R^2} \qquad \text{Equation 8}$$

g is the peak factor, taken as 2.5;
I_{10} is the nominal turbulence intensity at 10 m height, taken as 0.23 for Class C ground roughness;
R is the pulsating wind load resonance factor;
B_z is the background factor for pulsating wind loads.

The pulsating wind load resonance factor and the pulsating wind load background factor are each calculated according to the following equation.

$$R = \sqrt{\frac{\pi}{6\zeta_1}\frac{{x_1}^2}{(1+{x_1}^2)^{\frac{4}{3}}}} \qquad \text{Equation 9}$$

$$x_1 = \frac{30f_1}{\sqrt{k_w\omega_0}}, x_1 > 5 \qquad \text{Equation 10}$$

f_1 is the first-order self-oscillation frequency of the structure (Hz). k_w is the ground roughness correction coefficient, 0.54 for Class C ground roughness. ζ_1 is the structural damping ratio, 0.01 for steel structures.

$$B_z = kH^{a_1}\rho_x\rho_z\frac{\phi_1(z)}{\mu_z} \qquad \text{Equation 11}$$

$\phi_1(z)$ is the 1st-order vibration coefficient of the structure, taken from the table in Appendix G. H is the total height of the structure. k *and* a_1 coefficients are obtained from the table. ρ_x *and* ρ_z are the horizontal and vertical correlation coefficients of the pulsating wind, calculated according to the following equation.

$$\rho_x = \frac{10\sqrt{B+50e^{-B/50}-50}}{B} \qquad \text{Equation 12}$$

$$\rho_z = \frac{10\sqrt{H+60e^{-H/60}-60}}{H} \qquad \text{Equation 13}$$

B, H is the width of the windward side of the structure (m) and the total height of the structure (m) respectively.

The wind vibration coefficients of the scaffold are calculated based on the above formula, and the wind vibration coefficients of the scaffold under each working condition are obtained as shown in Table 9.

Table 9 Calculation results of wind vibration coefficient under various working conditions

Working condition number	A	B	C	D	E	F	G	H	I
Wind vibration coefficient β_z	2.45	1.93	2.47	3.46	2.72	3.49	2.58	1.96	2.60

The wind vibration coefficient results calculated over the specification are compared with the simulation results, and it is found that the wind vibration coefficient varies greatly due to the different self-oscillation periods of the structure. Under most working conditions, the simulation results are in general agreement with the formula calculation results, and some other results are on the safe side. According to the above results, the wind vibration coefficient of scaffolding is taken as $\beta_z = 2.5$.

The comparison of the results of the wind vibration coefficient calculated by the specification with the simulation results found that the wind vibration coefficient varies greatly due to the different self-oscillation periods of the structure. And under most working conditions, the simulation results are in general agreement with the formula calculation results. According to the above results, the wind vibration coefficient of the scaffold is considered to be $\beta_z = 2.5$.

4. Conclusion

In this paper, an orthogonal test method is used to establish a model of a fully enclosed suspended double-row scaffold with a dense safety riser under nine working conditions. SAP2000 was used to analyze the stresses of the nine working conditions under static and pulsating wind conditions, and the maximum bending moment, displacement, and stress values of the static wind bars were obtained. The scaffold was optimally arranged by using extreme difference analysis, with two steps and two spans of connecting wall elements, 50° scissor bracing, and a safety net with a wind protection factor of 0.8. The dynamic analysis showed that the displacement and internal forces of the scaffold structure increased to varying degrees under pulsating wind loads. According to the analysis results of this paper, the wind vibration coefficient should be taken as $\beta_z = 2.5$ for a dense scaffold at a height of 50 m.

References

[1]. Rubio-Romero, J.C., M. Carmen Rubio Gámez, and J.A. Carrillo-Castrillo, *Analysis of the safety conditions of scaffolding on construction sites.* Safety Science, 2013. **55**: p. 160-164.

[2]. Peng, J.-L., et al., *Experimental and numerical studies of practical system scaffolds.* Journal of Constructional Steel Research, 2013. **91**: p. 64-75.

[3]. Jia, L., et al., *Experimental Study on Bearing Capacity of Reinforced Steel Tubular Scaffold under Uniform Loads.* Advances in Civil Engineering, 2019. **2019**: p. 4324528.

[4]. Liu, C., et al., *Experimental and numerical study on lateral stability of temporary structures.* Archives of Civil and Mechanical Engineering, 2018. **18**(4): p. 1478-1490.

[5]. Błazik-Borowa, E., et al., *Technical factors affecting safety on a scaffolding.* Life in Construction, 2017: p. 154-163.

[6]. Lipecki, T., et al., *Façade scaffolding behaviour under wind action.* Archives of Civil and Mechanical Engineering, 2020. **20**(1): p. 27.

Directivity of spherical acoustic scattering based on COMSOL

Xiaofang Sun*, Shisong Zhang, Zhongshan Yao

Department of Military Oceanography and mapping, Dalian Naval academy

*394403007@qq.com

Abstract. The acoustic scattering characteristics of spheres are the basis for studying the sound scattering law of the target. The directivity of acoustic scattering is one of the most important characteristics of the target sound scattering law. In this paper, the acoustic scattering characteristics of rigid spheres and elastic spherical shells are analyzed from the two typical basic models of the rigid sphere and elastic spherical shell. The sound scattering directivity corresponding to different ka values under different models is analyzed by COMSOL finite element simulation analysis. The change law of spheroid scattering directivity of spheres under different elastic backgrounds of the rigid sphere and elastic spherical shell is compared and studied. It is found that the scattering directivity of both rigid spheres and elastic spherical shells will change with the change of ka value. The larger the ka, the stronger the forward directivity, and the more side lobes for rigid spheres. The backward directivity is greater than the forward directivity at ka≤1, while the forward directivity of the elastic spherical shell is always greater than the backward direction. And it is not difficult to find that the directivity of rigid spheres is relatively uniform in backward scattering when ka=1. When ka>1, the directivity increases with ka, the forward directivity becomes stronger and stronger, and the side lobes gradually increase. The elastic spherical shell uses air as the internal filler, and its directivity has always been forward than backward. The forward directivity also increases with the increase of ka, and the side lobes also increase with the increase of frequency.

1. Introduction

The problem of sound scattering of spheres has always been the focus of attention. In the past target analysis, the research was often aimed at rigid spheres. The sphere does not deform under the action of incident sound waves. The sound waves do not penetrate the inside of the sphere, so it will not excite the sound field inside the sphere. Common solid metal spheres, or spherical shell spheres whose thickness and radius ratio is greater than 4%, can be approximated as rigid spheres [1-3].

However, some common sonar targets, are strictly elastomers or viscoelastic. Due to the mutual interference between the components of the echo signal, the intensity of the scattered wave changes greatly and minimally with the change in frequency. The frequency effect of this scattered wave intensity is significantly different from that of rigid objects[4]. In addition, the scattering field of elastic objects also behaves differently from the scattering field of rigid objects in other places, such as reradiation waves carrying characteristic information of the target. Therefore, the sound scattering characteristics of elastic objects are the physical basis of object detection and classification recognition[5]. In this paper, the acoustic scattering field characteristics of rigid spheres and air-filled spherical shells with different ka values will be studied based on COMSOL. The acoustic scattering direction of rigid spheres and elastic spherical shells with different ka values Characteristics are compared and analyzed.

2. Research content

2.1. Scattering sound field of a smooth rigid sphere

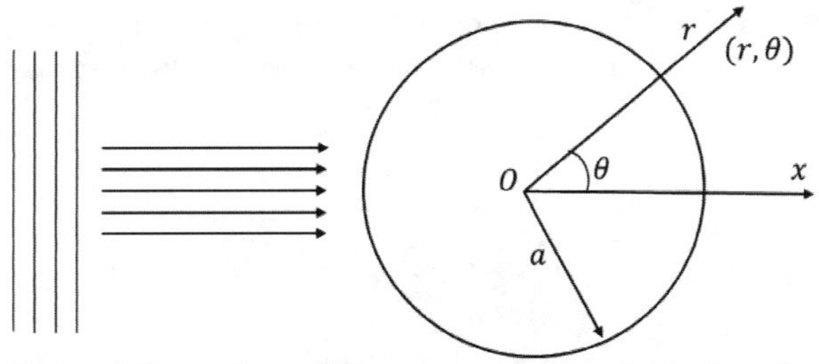

Figure.1 Scattering of plane waves on a spherical surface

As shown in Figure 1, a smooth and rigid sphere with a radius is located in an infinite fluid medium. The plane wave incident along the x-axis onto the sphere is now analyzed.

The incident plane wave can be written as the wave equation for the sound $P_i = P_0 e^{j(krcos\theta-\omega t)}$. The pressure of its scattered sound field is based on the spherical coordinate system:

$$\frac{1}{r^2}\frac{\partial}{\partial r}\left(r^2\frac{\partial p_s}{\partial r}\right)+\frac{1}{r^2 sin\theta}\frac{\partial}{\partial\theta}\left(sin\theta\frac{\partial p_s}{\partial\theta}\right)+\frac{1}{r^2 sin^2\theta}\frac{\partial^2 p_s}{\partial\varphi^2}+k^2 p_s=0 \quad (1)$$

Where r,θ,ϕ are coordinate variables in the spherical coordinate system. And k is the wavenumber, equal to the ratio of the circular frequency of the incident sound wave to the velocity in the medium around the sphere. Considering the high symmetry of the sphere, and the incident wave is symmetric to the x-axis, the scattered wave should also be symmetric to the x-axis. The above equation can be simplified to:

$$\frac{1}{r^2}\frac{\partial}{\partial r}\left(r^2\frac{\partial p_s}{\partial r}\right)+\frac{1}{r^2 sin\theta}\frac{\partial}{\partial\theta}\left(sin\theta\frac{\partial p_s}{\partial\theta}\right)+k^2 p_s=0 \quad (2)$$

This equation, using the separated variable method, can be solved :

$$p_s=\sum_{m=0}^{\infty} a_m P_m(cos\theta)h_m^{(1)}(kr) \quad (3)$$

Combined with the boundary conditions of the rigid sphere, after adding the time factor $e^{-j\omega t}$, the scattering sound pressure expression is:

$$p_s=\sum_{m=0}^{\infty}-j^m(2m+1)P_0\frac{\frac{d}{dka}j_m(ka)}{\frac{d}{dka}h_m^{(1)}(ka)}P_m(cos\theta)\,h_m^{(1)}(kr)e^{-j\omega t} \quad (4)$$

The equation is a general expression of the scattered sound field. Under far-field conditions, the ball Hankel function can be written as[6-7]:

$$\lim_{kr\to\infty} h_m^{(1)}(kr)=\frac{1}{kr}e^{j\left(kr-\frac{m+1}{2}\pi\right)} \quad (5)$$

In the meantime,

$$b_m=j^m(2m+1)\frac{\frac{d}{dka}j_m(ka)}{\frac{d}{dka}h_m^{(1)}(ka)} \quad (6)$$

$$D(\theta)=\frac{1}{ka}\sum_{m=0}^{\infty} b_m e^{-j\frac{m+1}{2}\pi}P_m(cos\theta) \quad (7)$$

The scattered wave sound pressure expression is simplified to:

$$p_s(r,\theta)=-P_0 a\frac{1}{r}D(\theta)e^{j(kr-\omega t)} \quad (8)$$

$D(\theta)$ is used to calculate the directivity function of the rigid sphere. The model immovable sphere with a radius of 0.5 m is calculated by using the COMSOL simulation software. The plane wave with a pressure amplitude of 1 Pa is selected for the incident sound wave. The sound velocity is 1500m/ s. The maximum mesh size is one-sixth of the wavelength. The perfect matching layer PML thickness is 0.5 m. The outside of the sphere is surrounded by water. The model is constructed as follows:

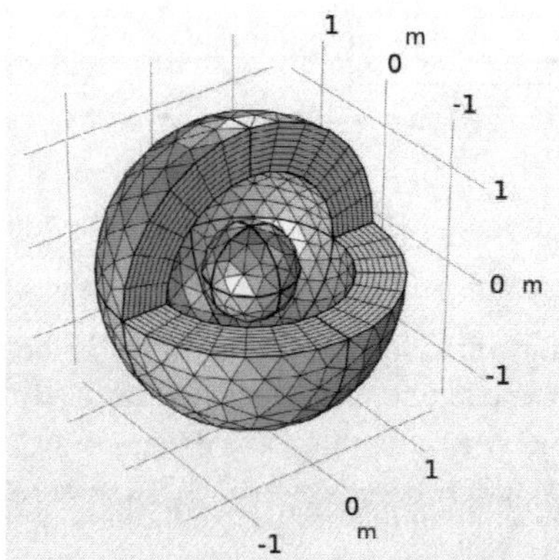

Figure.2 Mesh splitting diagram of spherical acoustic scattering

When the direction of the plane wave is along the x-axis, the external scattering sound pressure level corresponding to the XOY plane is studied. When the ka is 1, 2, 3, and 4, its directivity is analyzed. The results are as follows:

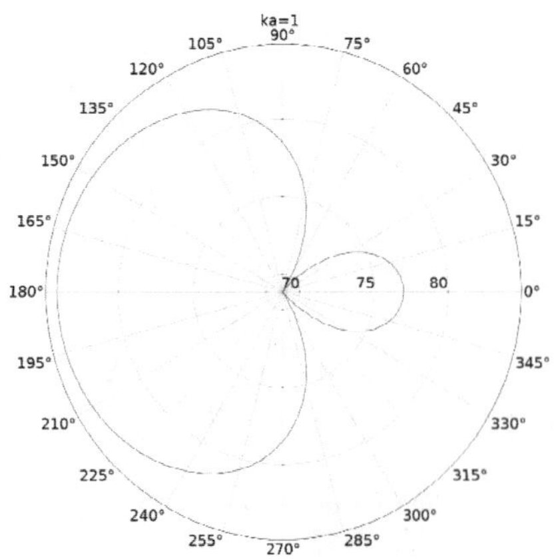

(a) Scattering directivity at ka=1

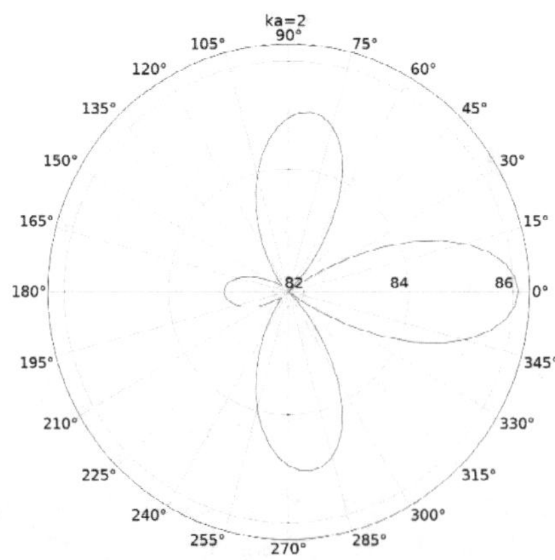

(b) Scattering directivity at ka=2

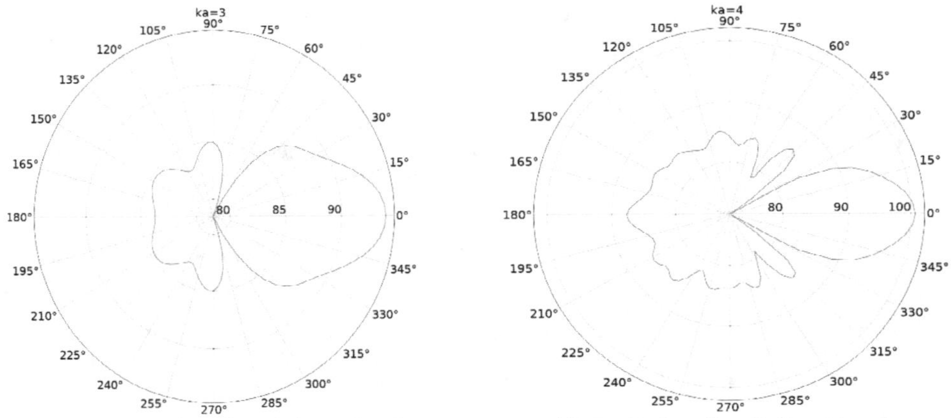

(c) Scattering directivity at ka=3 (d) Scattering directivity at ka=4

Figure.3 Acoustic scattering directivity of rigid spheres

Through the above simulation analysis, it is not difficult to find that the scattering directivity of rigid spheres is affected by the change of ka. When the ka value changes, the spatial distribution of the scattered wave also changes. When ka=1, the backscatter of the ball is relatively uniform and the forward scattering is weaker. With the increase of the ka value, the forward scatter directivity becomes stronger and stronger. The spatial distribution of the scattering field gradually becomes complex, and the directional pattern begins to appear sidelobes. The higher the frequency, the more sidelobes.

2.2. Scattering directivity of elastic spherical shells

For elastic objects, the incident sound wave can be put into the interior of the object, and excite the internal sound field. When the wavelength of the sound wave inside the object is less than the radius of the sphere, the internal fluctuation process begins to become important. At this time, the internal standing wave field will be established, and cause the object to make a simple vibration. The sound wave radiated by the object vibration is also a component of the echo signal.

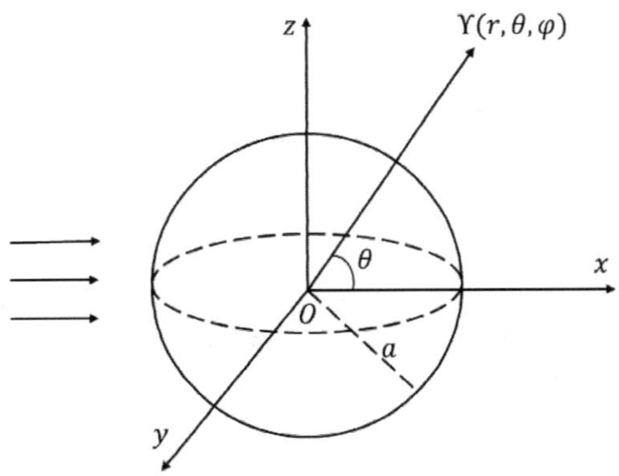

Figure.4 Schematic diagram of the incidence of elastic spherical plane waves

In the aqueous medium around the sphere, the incident sound pressure P_i and scattered sound pressure P_s in the total sound field can be expressed as:

$$P_i = e^{ikrcos\theta} = \sum_{n=0}^{\infty} i^n(2n+1)J_n(kr)P_n(cos\theta) \quad (9)$$

$$P_s = \sum_{n=0}^{\infty} i^n(2n+1)b_nH_n^{(1)}(kr)P_n(cos\theta) \quad (10)$$

Where $J_n(\cdot)$ is the nth-order Bessel function, which $P_n(\cdot)$ is the nth-order Legendre function, which $H_n^{(1)}(\cdot)$ is the first type of spherical Hankel function, and b_n is a scattering coefficient determined only by boundary conditions. Combined with the total sound pressure and normal displacement boundary conditions in water, the backscatter morphological function of the elastic sphere that can be obtained by simplifying by using the separation variable method is:

$$|f(x,\pi)| = \frac{2}{x}|\sum_{n=0}^{\infty}(2n+1)b_n(-1)^n| \quad (11)$$

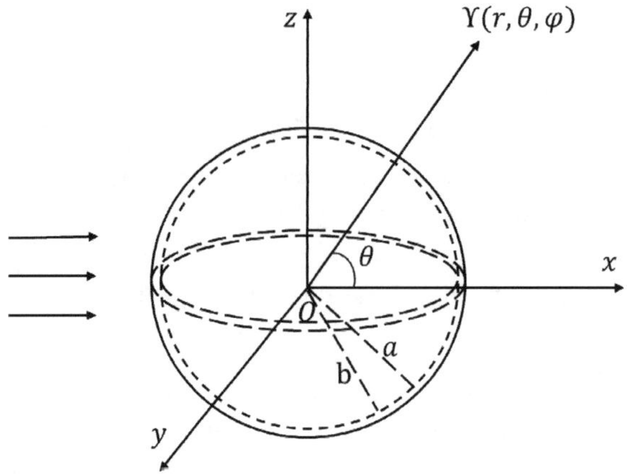

Figure.5 Schematic diagram of plane wave incidence of the elastic spherical shell

Based on the above theory, this paper takes the inner filler as fluid as an example, as is shown in figure.5. The inner radius of the spherical shell is b, the filling fluid density is $\rho_1$and the speed of sound is c_1. The sound field in the elastic spherical shell ($b \le r \le a$) is expressed as:

$$\varphi_1 = \sum_{n=0}^{\infty} i^n(2n+1)P_n(cos\theta)[c_n j_n(k_{d1}r) + d_n y_n(k_{d1}r)] \quad (12)$$

$$\psi_1 = \sum_{n=0}^{\infty} i^n(2n+1)P_n(cos\theta)[c_n j_n(k_{s1}r) + d_n y_n(k_{s1}r)] \quad (13)$$

The sound field inside the spherical shell ($r < b$) is represented by :

$$p_1 = \sum_{n=0}^{\infty} i^n(2n+1)P_n(cos\theta)g_n j_n(k_1 r) \quad (14)$$

where,$k_1 = \frac{\omega}{c_1}, k_{d1} = \frac{\omega}{c_{d1}}, k_{s1} = \frac{\omega}{c_{s1}}$. c_{d1}, c_{s1} are the longitudinal and transverse wave velocities of the elastic spherical shell respectively[8-10].

Similarly, to calculate the directivity function of the elastic spherical shell. Calculate the elastic spherical shell with the radius a=0.5 m and the thickness d=0.015m. The spherical shell is a steel body, and the internal filling is air. The incident sound wave selects a plane wave with a pressure amplitude of 1Pa and a sound velocity of 1500m/ s. The simulation is calculated as follows:

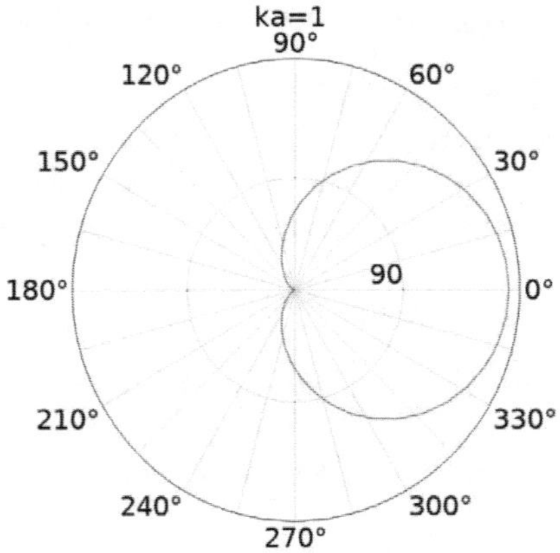

(a) Scattering directivity at ka=1

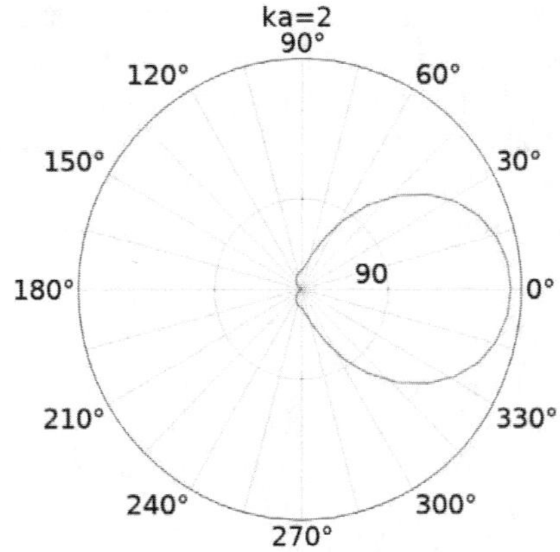

(b) Scattering directivity at ka=2

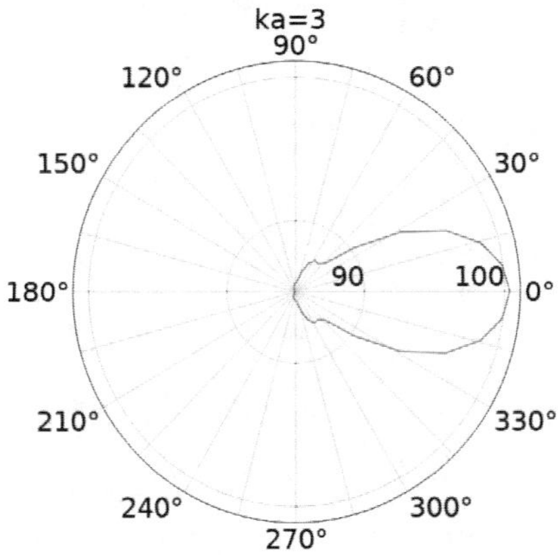

(c) Scattering directivity at ka=3

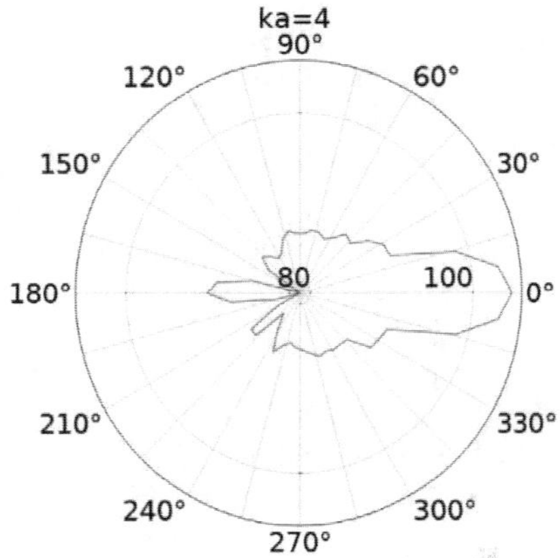

(d) Scattering directivity at ka=4

Figure.6 Acoustic scattering directivity of elastic spherical shells

Through the above simulation analysis, it is not difficult to find that the forward directivity of the spherical shell has always been greater than the backward directivity. With the increase of the ka value, the directivity is getting stronger and stronger, but the side lobes are also increasing.

3. Conclusion

The acoustic scattering echo of the target is the main information source of active sonar detection. In the acoustic countermeasure based on active sonar detection underwater, the acoustic scattering characteristics of the target are the basis of all detection work. Although the geometry of common sonar targets varies from object to object, they are close to spherical. So when doing theoretical analysis, for the convenience of mathematical processing, they can be approximated as spheres or spherical shells with other types of fillers inside. This approximation often greatly simplifies mathematical operations, and the results obtained apply to actual sonar targets.

This paper studies the scattering directivity of rigid spheres and elastic spherical shells under different ka values and uses COMSOL finite element analysis software to simulate them, which provides a certain reference for further study of the acoustic scattering characteristics of the target. There are also the following shortcomings in the research process of this paper:

First of all, when using COMSOL for simulation analysis, it is based on static seawater and does not consider the interference of seawater mobility on scattering directivity.

Secondly, the materials selected for the study of directivity are not rich enough, and there is no detailed comparison, which has certain limitations.

Because of this, the research content will be refined later. The theoretical analysis and simulation research will be combined with the actual marine environment as much as possible, to obtain more accurate simulation results. On the other hand, the research question on the directivity of spherical shell thickness on sound scattering can be considered, which can provide a basis for the design of relevant underwater acoustic structures in the future.

References

[1] Waterman P C. New Formulation of Acoustic Scattering[J]. The Journal of the Acoustical Society of America,1969, 45(6):1417-1429.
[2] Faran J J. Sound Scattering by Solid Cylinders and Spheres[J]. The Journal of the Acoustical Society of America, 1951, 23(4), 405-417.
[3] Goodman, Ralph R.Reflection and Transmission of Sound by Elastic Spherical Shells[J]. The Journal of the Acoustical Society of America, 1962, 34(3):338-344.
[4] Ugincius P, überall, Herbert. Creeping-wave analysis of acoustic scattering by elastic cylindrical shells [J]. Acoustical Society of America, 1967, 43(6):1578-1579.
[5] Zampolli M, Tesei A, Jensen F B, et al. A computationally efficient finite element model with perfectly matched layers applied to scattering from axially symmetric objects[J]. Journal of the Acoustical Society of America, 2007, 122(3):1472-1485.
[6] Kumar N, Mitra U, Narayanan S S. Robust object classification in underwater sidescan sonar images by using reliability-aware fusion of shadow features[J]. IEEE Journal of Oceanic Engineering, 2015, 40(3): 592-606.
[7] Hansen R E, Lyons A P, Sæbø T O, et al. The effect of internal wave-related features on synthetic aperture sonar[J]. IEEE Journal of Oceanic Engineering, 2015, 40(3):621-631.
[8] MD Guild, Alu A, Haberman M R.Cancellation of acoustic scattering from an elastic sphere[J]. Journal of the Acoustical Society of America, 2011, 129(3):1355-1365.
[9] Anderson S D, Sabra K G, Zakharia M E, et al. Time-frequency analysis of the bistatic acoustic scattering from a spherical elastic shell[J]. The Journal of the Acoustical Society of America, 2012, 131(1):164-173.
[10] Sabra K G, Anderson S D. Subspace array processing using spatial time-frequency distributions: Applications for denoising structural echoes of elastic targets[J]. The Journal of the Acoustical Society of America, 2014, 135(5): 2821-2835.

Optimize the Rectifier Structure to Improve the Accuracy of Gas Ultrasonic Flowmeter Under Low Flow Conditions

Bin Yang*, Ying Zhang[2], Limiao Shi[1], Xingyu Zhan[1]

1. Tianjin University of Technology and Education
2. Tianjin Sino-German University of Applied Sciences
3. 1392940805@139.com

Abstract: With the promotion of the coal-to-gas policy, the consumption of civil natural gas in China is increasing rapidly. Improving the measurement accuracy of the flowmeter is of great significance in reducing transportation costs. Computational fluid dynamics software (FLUENT) was used for simulation, the upstream choke model of a single elbow pipe was established, and the flow field in the urban pipe network's transition area was analyzed and optimized to improve the existing commercial flow Conditioner. The effects of the Reynolds number, the installation location of the ultrasonic gas flowmeter, and aperture depth on measurement accuracy were investigated. The results show that installing the flow Conditioner can improve the velocity distribution of the flow field in the pipeline. Optimizing the flow Conditioner can shorten the installation position of the flowmeter. The optimum thickness of the flow Conditioner is 0.25D. The research results provide important theoretical significance and application value for improving the industry's ultrasonic gas flowmeter measurement accuracy.

1. Introduction

Driven by the development of the global green economy, natural gas, as a high-quality and efficient low-carbon energy, is in increasing demand. Natural gas consumption is expected to exceed 4.3 trillion cubic meters in 2024. China has steadily advanced energy consumption reform recently and built a safe and clean modern energy system[1]. In 2017, the country introduced a policy of replacing coal with gas and signed a gas transfer agreement with Russia in 2019 to import 1.3 trillion cubic meters of gas annually to meet the domestic supply gap. Therefore, the natural gas pipeline network needs a high-precision flowmeter device to provide the measurement basis for natural gas trade settlement[2].

At present, there are three kinds of gas flowmeters, which are turbine, roots, and ultrasonic flowmeters. Compared with the first two flow meters, ultrasonic gas flow meters have obvious advantages in the natural gas trade measurement system due to the lack of built-in flow resistance, low-pressure loss, and high range ratio[3-4]. However, the measurement accuracy is greatly affected by the flow field distribution, the roughness of the pipeline, the structure of the upstream choke, and the degree of fluid turbulence will cause uneven velocity distribution on the channel line of the flowmeter, resulting in errors[5].

Aiming at the above problems, scholars at home and abroad have done much work with different research methods. In terms of channel structure and algorithm, Qin[6] improved the measurement accuracy of the ultrasonic flowmeter by building an accurate gas flow field model and therefore designed a multi-channel experimental device to measure the flow field distribution of gas in pipelines.

Zhao[7] used CFD software to simulate the fluid distribution in the opposite tube to study the influence of the installation position, installation direction, and pipeline length of the transducer of the dual-channel ultrasonic flowmeter on the flow measurement. Liu[8] aiming at the influence of pipe fitting connection on the flow field, established a turbulence model of the flow field for simulation based on the CFD method, obtained the flow field distribution of natural gas after passing through the pipe fitting connection, and proposed the installation requirements of the ultrasonic flowmeter. The optimization of channel structure and algorithm is generally applied to the pipeline structure under specific conditions, which has some limitations. It is an effective method to improve the detection accuracy of a flowmeter by designing a rectifier to stabilize irregular flow fields and reduce the eddy current[9].

The rectifier can be divided into blade, hole, and composite. Currently, the most widely used rectifiers at home and abroad include blade-type Etoile rectifiers, pore plate-type Laws rectifiers, and NEL rectifiers[10-11]. The current rectifier design modifies the turbulent flow field with a high Reynolds number, but the gas in the laminar and turbulent flow transition zone is rarely studied. Transition in the boundary layer within the transition zone is strongly nonlinear, accompanied by many coherent vortex structures and the formation and random disorder of many small vortex structures[12-13]. However, the study of the transition mechanism of compressible fluids is more complicated. In actual urban pipe network transportation, natural gas flow in the pipeline is mostly in transition and low turbulence zones. Therefore, this study analyzes the gas flow state and the generation mechanism, changes the boundary layer transition rule in the transition area, and optimizes the current commonly used rectifier, which can effectively improve the measurement accuracy of the ultrasonic gas flowmeter in the near turbulent scale.

2. Theoretical foundation

2.1. *Ultrasonic flowmeter*

The measurement principle diagram of the double-channel ultrasonic flowmeter is shown in Figure 1.

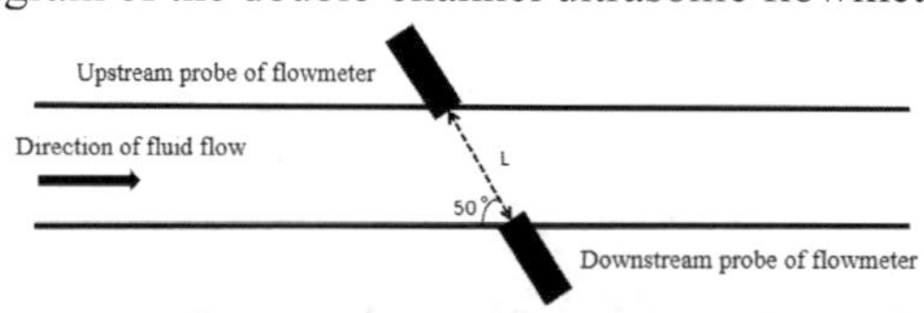

Figure 1 Schematic diagram of double channel ultrasonic flowmeter

According to the principle of time difference method, the formula of instantaneous volume flow rate is obtained by measuring the downstream and counter-current response time of ultrasonic wave in the pipeline[14]:

$$V_L = \frac{L}{\cos\theta} \times \frac{(t_2 - t_1)}{t_1 t_2} \tag{1}$$

$$Q = AV_m = \frac{V_L}{K} \cdot A \tag{2}$$

where V_m is the average flow velocity of the pipeline section fluid can be directly obtained through the integral operation of FLUENT. V_L is the average linear velocity of the fluid on the channel can be calculated by the weight integral method of FLUENT. L is the length of the channel between the probe of the flowmeter. t_1 and t_2 represent the propagation time of downstream flow and countercurrent flow of the fluid in the pipeline, respectively. θ is the Angle between the channel and the pipe axis. Q is the volume flow of fluid in a pipe. A is the cross-sectional area of a pipeline. K is the correction factor, whose function is to compensate the sound channel's average linear velocity to the pipe section's average velocity.

The measurement error calculation formula of the ultrasonic gas flowmeter is as follows[15]:

$$E = \left| \frac{V_L - V_m}{V_m} \right| \times 100\% \quad (3)$$

where E is the relative error.

2.2. Transition zone

Many coherent vortexes, such as hairpin vortexes and annular vortexes, will appear in the process of laminar flow turbulence conversion, a prominent feature of transition zones. After forming many vortex structures and the random disorder process, laminar flow is gradually transformed into fully developed turbulence[16-17]. Figs. 2 (a)-(c) show the absolute helicity (H) distribution of eddy currents in the pipeline under laminar flow (Re=1800), transition zone (Re=3500), and turbulent state (Re=20000), respectively. The eddy is mainly concentrated in the center of the pipe in the turbulent stage, while the eddy is distributed near the pipe wall in the transition zone stage.

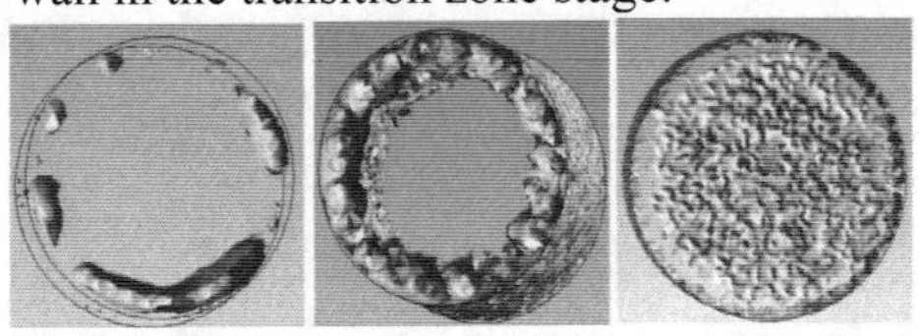

(a) H=0.49 m/s^2 (b) H=10.02 m/s^2 (c) H=175.98 m/s^2

Figure 2 Distribution diagram of vortex helix in the pipe

The N-S equation is the basic equation to control fluid motion, and its perturbation equation can be derived from it. The continuity equation and the momentum equation can be written as[18]:

$$V \bullet v = 0 \quad (4)$$

$$\frac{\partial v}{\partial t} + (v \bullet \nabla) v = -\frac{1}{\rho} \nabla p + v \nabla^2 v \quad (5)$$

where V is the velocity vector, P is the intensity of pressure, v is the coefficient of kinematic viscosity, ∇ is the differential operator, $(\frac{\partial}{\partial x}, \frac{\partial}{\partial y}, \frac{\partial}{\partial z}), x、y$ and z are the coordinates of three directions along the flow field space, respectively.

The system of parallel flow linear stability can be written as[19-20]:

$$\mathrm{i}\alpha u + \frac{dv}{dy} + \mathrm{i}\beta\omega = 0 \quad (6)$$

$$\mathrm{i}\,(\alpha U + \beta W - \omega) u + \frac{dU}{dy} v + \mathrm{i}\alpha p = \frac{1}{\mathrm{Re}} (\frac{d^2 u}{dy^2} - \lambda^2 u) \quad (7)$$

$$\mathrm{i}(\alpha U + \beta W - \omega) v + \frac{dp}{dy} = \frac{1}{\mathrm{Re}} (\frac{d^2 v}{dy^2} - \lambda^2 v) \quad (8)$$

$$\mathrm{i}\,(\alpha U + \beta W - \omega)\, W + \frac{dW}{dy} v + \mathrm{i}\beta p = \frac{1}{\mathrm{Re}} (\frac{d^2 w}{dy^2} - \lambda^2 \omega) \quad (9)$$

where $\lambda^2 = \alpha^2 + \beta^2$, and the boundary condition is $y = 0$ or $y \to \infty$, $u = v = \omega = 0$.

2.3. Design of rectifier

Zanker rectifier commonly used in the former market is a porous plate structure with 5 layers of pitch circle and 32 holes. The rectifying action of fluid through the orifice plate makes the velocity distribution of the pipe cross-section more uniform. However, the pore size at different pitch circles of Zanker positive fluid is different, the pore distribution is sparse near the pipe wall, and the eddy current elimination at the boundary layer of the transition zone is weak.

In this study, the Zanker rectifier structure was optimized. A multi-diameter circular hole with dense hole distribution near the walled pitch was designed by comprehensively considering pressure loss, processing cost, and other factors. Its structure and schematic diagram are shown in Figure 3 and Table 1.

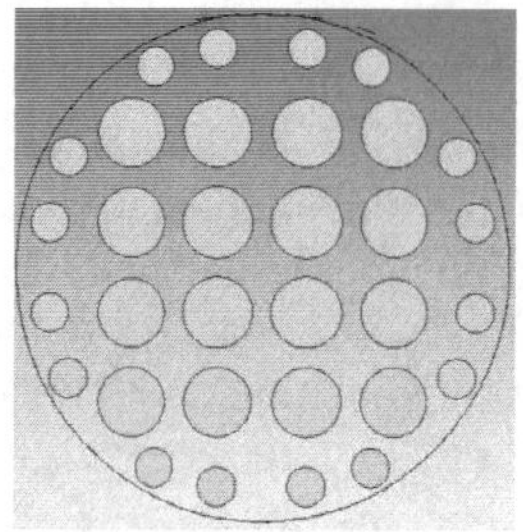

Figure 3 Schematic diagram of optimized rectifier structure

Table 1 Design of rectifier

Zanker rectifier		Optimized rectifier	
d_2=0.139D (8 holes)	D_2=0.56D	d_2=0.136D (8 holes)	D_2=0.56D
d_3=0.136D (4 holes)	D_3=0.75D	d_3=0.136D (4 holes)	D_3=0.75D
d_4=0.11D (8 holes)	D_4=0.85D	d_4=0.077D (8 holes)	D_4=0.88D
d_5=0.077D (8 holes)	D_5=0.90D	d_5=0.077D (8 holes)	D_5=0.91D

3. Fluent simulation

3.1. Physical model

To study the effect of near-wall eddy currents on the measurement accuracy of the ultrasonic flowmeter in the transition zone of circular pipe, a testing system is established, which consists of an upstream choke, straight buffer pipe, rectifier, and downstream straight pipe. The upstream choke piece selects the most common 90° single elbow in engineering, in which the elbow is 5D away from the rectifier, and the pipe diameter is D=40 mm. The physical model is drawn by DM software, and its schematic diagram is shown in Figure 4.

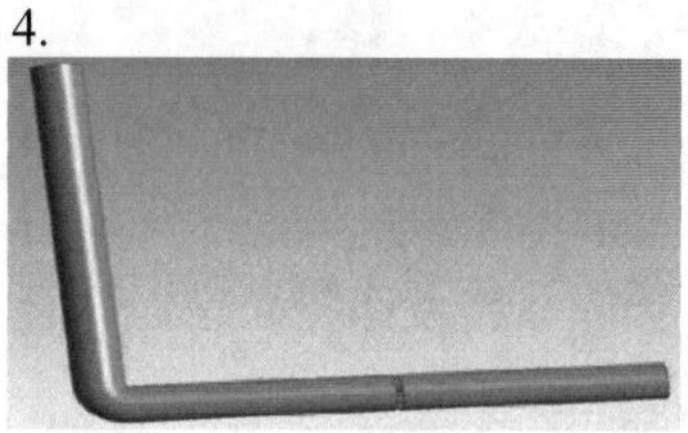

Figure 4 Schematic diagram of the pipeline model

Meshing is divided by using Meshing software in ANSYS. The single-bend pipe and rectifier adopt three-dimensional structured mesh and local encryption processing. Buffer straight pipe and downstream straight pipe segments are divided by unstructured grids, as shown in Figure 5.

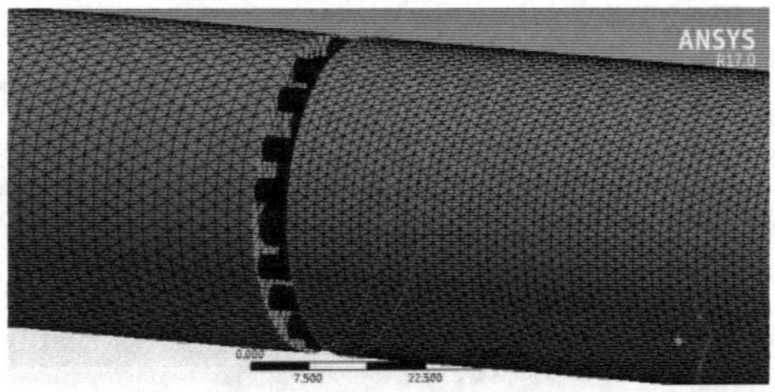

Figure 5 Schematic diagram of mesh division

Grid independence verification is carried out to ensure the quality of grid division. The same mesh division method was adopted, and the number of mesh was controlled at 0.8 million, 1.1 million, 1.3 million, and 1.6 million. After changing the flow conditions and comparing the influence of mesh size on the calculation accuracy, it is found that the simulation results are less affected by mesh density when the total number of grids is higher than 1.1 million. Considering the computing speed, the number of mesh in this study is determined to be 1.1 million.

3.2. Boundary condition setting

The pipe inlet is in velocity-inlet condition. The pipe outlet is in pressure-outlet condition. The fluid medium under atmospheric pressure is methane gas. According to the Reynolds number of the pipeline, the transition zone and turbulence zone were set as the k-Epsilon model.

3.3. Simulation parameters setting

To fully study the rectifier performance of rectifiers with different structures on the flow field in the transition region and near turbulent region, the simulation parameters are set as follows:

(1) The gas flow state covers the laminar flow area, the transition area, the near-turbulent area, and the turbulent area, and the corresponding Reynolds numbers Re are 1300, 4100, 12000, and 24500, respectively. The corresponding inlet velocity is 0.5 m/s, 1.7 m/s, 5 m/s, and 10 m/s, respectively. The velocity distribution of the fluid in the pipeline cross-section is observed.

(2) The distance between the fixed elbow and the rectifier was 5D, and the length of the buffer straight pipe was adjusted to 2D, 4D, 6D, 8D, and 10D, respectively, to analyze the optimization effect of the flow field under different rectifier structures.

(3) The thickness of the rectifier orifice was changed to 0.125d, 0.25d, 0.5d, and 0.75d, and the influence of orifice depth on the flow field was analyzed.

4. FLUENT Simulation Results

4.1. Influence of Reynolds number on the flow field

The cross-section flow field at the downstream pipeline's $4D$ position varies with the Reynolds number as shown in Figures 6-9. The thickness of the rectifier is $0.125D$. The vector of the gas flowing through the upstream elbow will change, which will cause the local eddy along the flow direction in the pipe. The velocity distribution at the pipe cross-section is symmetric, and the highest velocity point is shifted to the lower part of the cross-section. When the inlet velocity was 0.5 m/s, the flow field in the pipe was laminar, and the fluid turbulence was low. In this case, the flow velocity changes significantly from the wall surface to the center of the pipeline due to the boundary layer effect without installing a rectifier. Although the flow velocity is shifted to the maximum area, the isoline of the whole velocity is symmetrical, which can meet the integral requirements of the ultrasonic gas flowmeter. After installing the two types of rectifiers, the velocity gradient is weakened, and the flow isosurface is symmetric. This indicates that the rectifier plays a certain role in optimizing the flow field under the laminar flow state, but the effect is insignificant.

With the increase of inlet velocity, the fluid turbulent motion is enhanced. In the transition region and near the turbulent region, the detachment of the boundary layer will cause the detachment vortex, leading to the flow field near the wall being disordered and the velocity distribution disorder increasing. By comparing Figure 7 and Figure 8, it is found that the distribution uniformity of the flow field is significantly improved after the installation of the rectifier. Compared with the Zanker rectifier, the optimized rectifier's outer pitch circle is closer to the pipe wall, and its aperture is smaller, which is helpful to improve the low-speed flow field at the place far from the center of the pipe due to the detuned vortex, making the linear velocity of the flowmeter more accurate.

When the Reynolds number is 24500, the flow field is in the highly turbulent region. At this time, the velocity disturbance near the center of the pipeline is intense, and the influence of the local flow field near the wall on the overall linear velocity is relatively weakened. Due to the smaller pitch circle diameter and the denser hole in the center of the near orifice plate, the Zanker rectifier has a more obvious rectification effect in the high-turbulence region than the optimized rectifier. But generally speaking, both kinds of rectifiers can meet the test requirements on a turbulent scale.

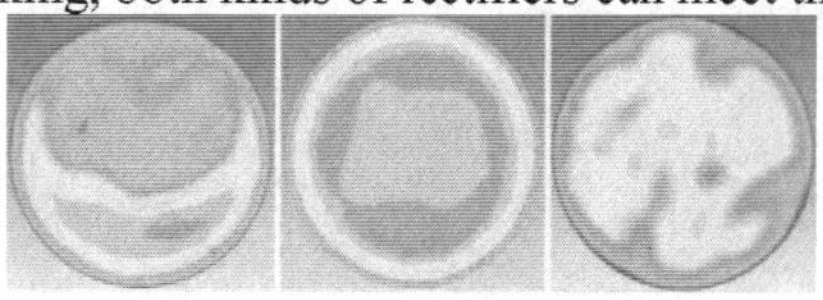

(a) (b) (c)

Figure 6 The fluid velocity distribution in the pipeline when Re=1300 ((a) no rectifier, (b) Zanker rectifier, (c) optimized rectifier)

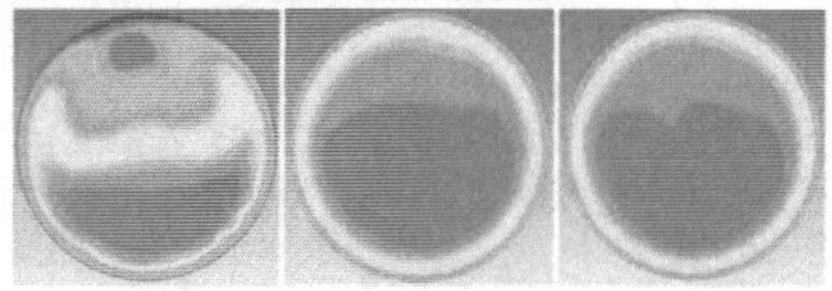

(a) (b) (c)

Figure 7 The fluid velocity distribution in the pipeline when Re=4100 ((a) no rectifier, (b) Zanker rectifier, (c) optimized rectifier)

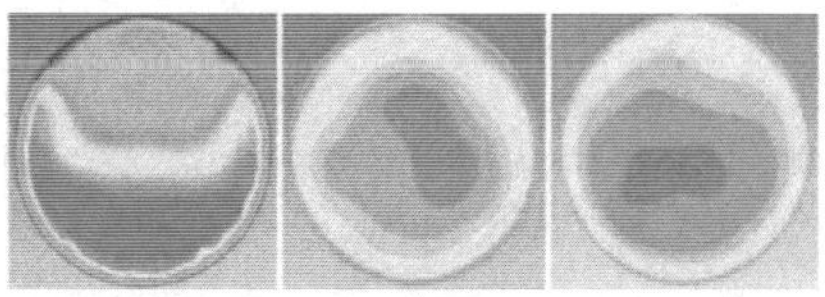

(a) (b) (c)

Figure 8 The fluid velocity distribution in the pipeline when Re=12000 ((a) no rectifier, (b) Zanker rectifier, (c) optimized rectifier)

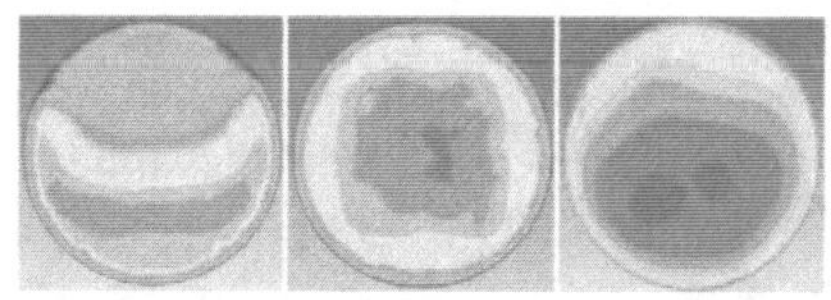

(a) (b) (c)

Figure 9 The fluid velocity distribution in the pipeline when Re=24500 ((a) no rectifier, (b) Zanker rectifier, (c) optimized rectifier)

4.2. Flow field distribution at different pipe locations

The fluid velocity distribution at different downstream pipeline locations under the transition zone (Re=4100) and near the turbulence zone (Re=12000) are shown in Figure 10-15. In the absence of a rectifier, the velocity distribution of the flow field in the pipeline near the elbow is disordered. With the forward flow of gas, the flow field gradually changes to a fully developed flow. For the flow field when Re=4100, the cross-section velocity distribution of the buffered straight pipe at $8D$ is relatively symmetric. Furthermore, even at $10D$ when Re=12000, the buffered straight pipe still has an obvious irregular distribution area. However, For Zanker and the optimized rectifier, it can be seen that the flow field is significantly optimized at the $4D$ position of the straight buffer pipe, the velocity isospheric surface is more symmetrical, and the velocity gradient changes in the near wall surface.

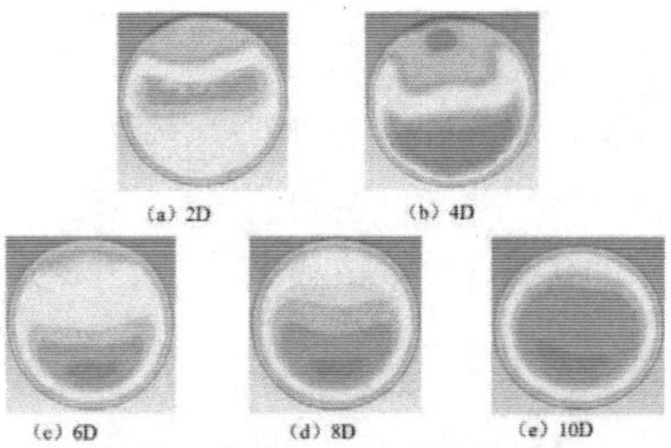

Figure 10 Fluid velocity distribution at different positions in downstream piping without rectifier when *Re*=4100

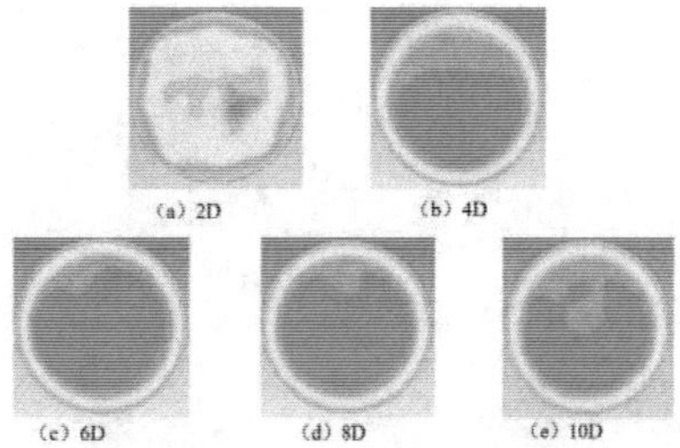

Figure 11 Fluid velocity distribution at different positions in downstream piping with Zanker rectifier when *Re*=4100

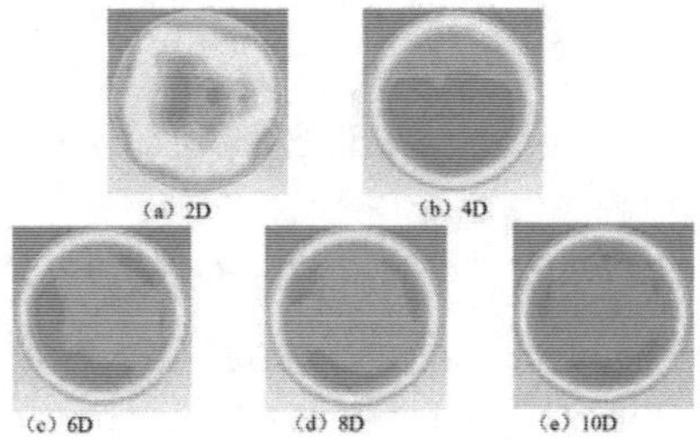

Figure 12 Fluid velocity distribution at different positions in downstream piping with optimized rectifier when *Re*=4100

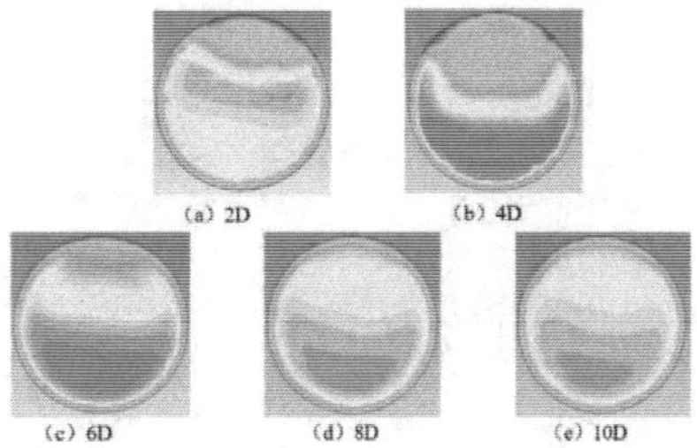

Figure 13 Fluid velocity distribution at different positions in downstream piping without rectifier when *Re*=12000

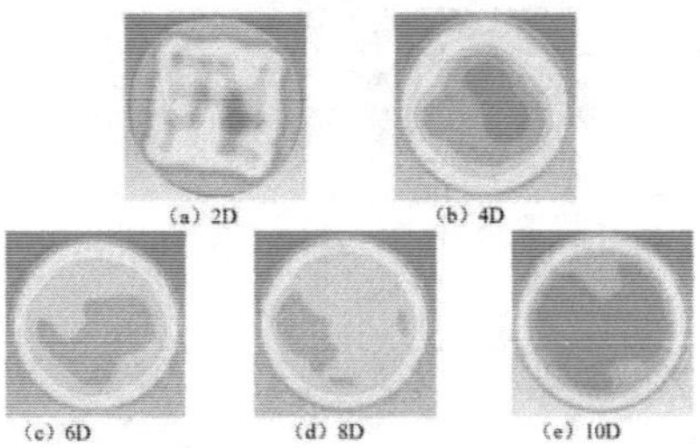

Figure 14 Fluid velocity distribution at different positions in downstream piping with Zanker rectifier when *Re*=12000

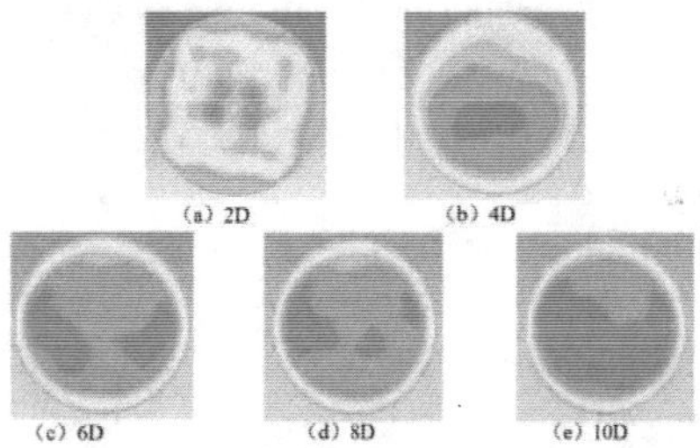

Figure 15 Fluid velocity distribution at different positions in downstream piping with optimized rectifier when *Re*=12000

The relative measurement errors of the three installation conditions at different buffered straight pipe locations are shown in Table 2. The error analysis shows that installing a rectifier is an effective method to improve measurement accuracy. The measuring conditions can be reached 4*D* after the rectifier positions in the transition zone and near the turbulence zone. The error of the optimized rectifier is controlled within ±4.5%, while the error of the optimized rectifier is controlled within ±4.0%. This indicates that the latter design structure is reasonable and more suitable for the flow field in the transition zone.

Table 2 Measurement error at different sections of pipe/%]

The type of rectifier	Cross section position	2*D*	4*D*	6*D*	8*D*	10*D*

No rectifier	Re=4100	10.36	8.42	8.76	7.76	7.84
	Re=12000	27.63	22.43	15.29	16.79	13.21
Zanker rectifier	Re=4100	9.63	4.42	3.13	3.49	3.22
	Re=12000	14.39	5.76	5.13	4.98	4.54
Optimized rectifier	Re=4100	10.41	3.69	3.82	2.97	3.14
	Re=12000	12.71	3.96	2.72	3.11	1.98

4.3. *The influence of rectifier thickness on the flow field*

The Reynolds number of the fixed pipeline is Re=12000, and the thickness of the rectifier is adjusted and optimized. The eddy current intensity of the pipeline section at the 2D position downstream is shown in Figure 16. Due to the eddy currents caused by the shedding of the lower boundary layer in the pipeline transition area and near the turbulent area, when the rectifier thickness is 0.125D, the eddy current strength gradient near the wall of the pipeline section is obvious. However, when the thickness is increased to 0.25D, the high-strength area of the section vortex becomes smaller, and the symmetry is improved. Considering the rectification effect and pipeline pressure loss, the thickness of the rectifier was optimized and set to 0.25D.

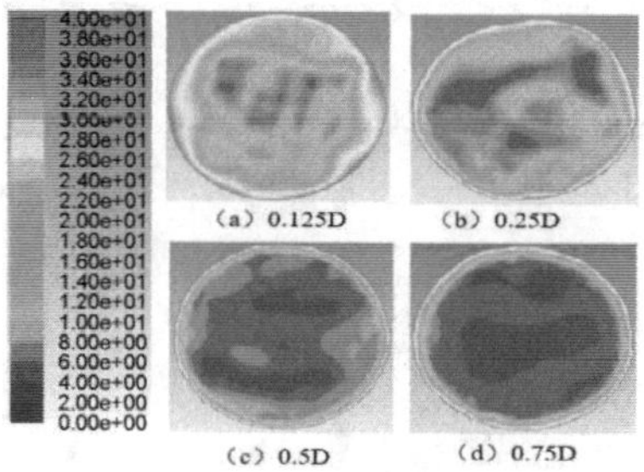

Figure 16 The eddy current intensity at the different thicknesses of the optimized rectifier

5. Experimental Results

5.1. *Experimental equipment and methods*

Figure 17 shows the experimental platform of the ultrasonic gas flowmeter. In this experiment, a small pipe diameter section was selected for testing. The gas was transported in the pipeline system, and its flow rate was regulated through the valve. The 90°C single bend serves as an upstream choke device. The distance between the rectifier and the bend is fixed, and the buffer pipe can be replaced. An ultrasonic gas flowmeter and calibration instrument measure the pipeline's real-time gas flow.

Figure 17 Ultrasonic gas flowmeter test platform

We control the change of gas velocity in the range of 0-10 m/s. Each measurement collects 80 sets of data, processes the signal to calculate the actual flow rate, and calculates the measurement error by comparing it with the standard instrument.

5.2. *Results and discussion*

The upstream elbow and rectifier position is fixed, the rectifier thickness is 0.25*D*, and the ultrasonic gas flowmeter is installed at the buffer pipe at 6*D*. The adjusted Reynolds number is the same as the simulation value, and the corresponding flow test points are 2.26 m^3/h, 7.68 m^3/h, 22.6 m^3/h, and 45.2 m^3/h, respectively. The experimental error is shown in Figure 18. In the laminar flow scale, the measurement error with or without the rectifier is within ±10%, which meets the measurement requirements. The average precision of the rectifier designed in this study is 2.53% in the transition zone and near the turbulence zone, which is better than the Zanker rectifier. The results indicate that the rectifier can shorten the measuring distance of the flowmeter after optimizing the structure, and the trend is consistent with the simulation results. Thus, the optimized rectifier is more suitable for detecting the flow field in the transition zone and near turbulence.

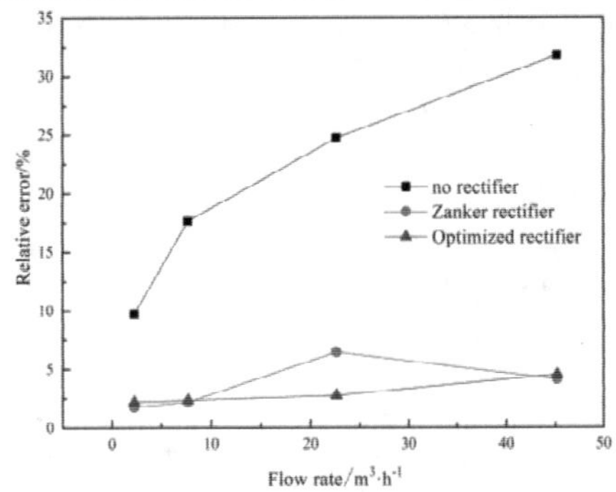

Figure 18 Experimental error analysis

6. Conclusion

In this paper, the generation mechanism and distribution of eddy currents in the transition zone are analyzed in detail using the FLUENT simulation and experiment method to optimize and improve the Zanker rectifier, which is widely used today. A new type of rectifier with a larger pitch circle area and denser equal diameter hole is designed, and the velocity distribution of the flow field at the pipe cross-section under different Reynolds Numbers with or without a rectifier is investigated. Moreover, we investigate the effect of measurement position on relative measurement error with the adjustment of the buffer length of a straight pipe. By changing the aperture depth, the optimum thickness of the rectifier under low-pressure loss is also determined. The conclusions are as follows:

(1) With the increase of Reynolds number, the velocity distribution disorder at the pipe section is enhanced, but the flow field can be improved effectively after installing the rectifier. The effect of the optimized rectifier on the eddy current near the wall is more obvious in the range of Re=4100-12000.

(2) It is effective to lengthen the buffer pipe length to reduce the measurement error. After installing the rectifier, the measurement position can be shortened to 4*D* downstream of the rectifier. The measurement error of the optimized rectifier in the transition zone and near the turbulence zone is generally less than 4%.

(3) When the optimized rectifier thickness reaches 0.25*D*, the adjustment effect of the flow field has been achieved. Taking into account the actual test to reduce the pressure loss demand, the optimal thickness of the optimized rectifier was determined to be 0.25*D*.

(4) By analyzing the effect of the rectifier and the credibility of the simulation, it is found that the relative error under different flow rates in the experiment is consistent with the overall trend of the simulation results. The average relative measurement error of the optimized rectifier in the transition and near turbulence zones is controlled within 2.53%.

(5) In the next step, more in-depth research will be carried out around the external embedded combined rectifier and standard sound velocity verification to solve the problem of the practical application of gas ultrasonic.

Declaration of conflicting interests

The author(s) declared no potential conflicts of interest concerning this article's research, authorship, and/or publication.

Acknowledgments

This work was funded by the National Natural Science Foundation of China (62073240), the Science & Technology Development Fund of Tianjin Education Commission for Higher Education (2020KJ087), Equipment and prototype of Solvetech (Tianjin)Automation Technology Co., Ltd.

References

[1] Dubey, S., Dubey, P., Rajagopalan, S., Sharma, S. (2019) Real-time implementation of Kalman filter to improve accuracy in the measurement of time of flight in an ultrasonic pulse-echo setup[J]. *Review of Scientific Instruments*., 90(2): 025105.

[2] Cai, U., Zhi, G., Dong, E., Yun, W., Jian, L., Ai, J., Shen, Y. (2018) Theory, technology and prospects of conventional and unconventional natural gas. *Petroleum Exploration and Development*. .45(4): 604-618.

[3] Zhao, N., Hu, L., Peng, X., Fang, Z., Chen, W., Fu, X. (2019) A Method Combining Measurement Tool and Numerical Simulation for Calculating Acoustic Signals of Ultrasonic Flowmeter. *IEEE Sensors Journal*. 19(24): 11805-11813.

[4] Fang, Z., Hu, L., Mao, K., Chen, W., Fu, X. (2018) Similarity Judgment-Based Double-Threshold Method for Time-of-Flight Determination in an Ultrasonic Gas Flowmeter[J]. *IEEE Transactions On Instrumentation and Measurement*. 67(1): 24-32.

[5] Zheng, D., Mei, J., Wang, M. (2016) Improvement of gas ultrasonic flowmeter measurement non-linearity based on ray tracing method. *IET Science, Measurement & Technology*. 2016, 10(6), 602-606.

[6] Qin, L., Hu, L., Mao, K., Chen, W., Fu, X. (2016) Flow profile identification with multipath transducers[J]. *Flow Measurement and Instrumentation*. 52: 148-156.

[7] Zhao, H., Peng, L., Takahashi, T., Hayashi, T., Shimizu, K., Yamamoto, T. (2015) CFD-Aided Investigation of Sound Path Position and Orientation for a Dual-Path Ultrasonic Flowmeter With Square Pipe[J]. *IEEE Sensors Journal*.15(1): 128-137.

[8] Liu, E., Tan, H., Peng, S. (2017) A CFD simulation for the ultrasonic flow meter with a header[J]. *Tehnicki Vjesnik-Technical Gazette*. 24(6): 1797.

[9] Sun, Y., Zhang, T., Zheng, D. (2018) New analysis scheme of flow-acoustic coupling for gas ultrasonic flowmeter with vortex near the transducer. *Sensors*. 18(4), 1151.

[10] Nishiguchi, H., Sawayama, T., Nagamune, K. (2016) A propagation time difference evaluation for a clamp-on ultrasonic flowmeter for low-pressure gas. *Japanese Journal of Applied Physics*. 55(7S1), 07KB05.

[11] TANG, X., Xie, X., Fan, B., Sun, Y. (2018) A fault-tolerant flow measuring method based on PSO-SVM with transit-time multipath ultrasonic gas flowmeters. *IEEE Transactions on Instrumentation and Measurement*. 67(5): 992-1005.

[12] Sircar,S., Bansal, D. 2019. Spatiotemporal linear stability of viscoelastic free shear flows: Dilute regime[J]. *Physics of Fluids*. 31(8):084104.

[13] Hallanger, A., Saetre, C., Frøysa, K. (2018) Flow profile effects due to pipe geometry in an export gas metering station-Analysis by CFD simulations[J]. *Flow Measurement and Instrumentation*. 61: 56-65.

[14] Mubarok, M., Cater, J., Zarrouk, S. (2020) Comparative CFD modelling of pressure differential flow meters for measuring two-phase geothermal fluid flow[J]. *Geothermics*. 86: 101801.

[15] Perumal, K., Ganesan, R. (2016) CFD modeling for the estimation of pressure loss coefficients of pipe fittings: An undergraduate project[J]. Computer Applications in Engineering Education. 24(2): 180-185.

[16] Willatzen, M., Kamath, H. (2008) Nonlinearities in ultrasonic flow measurement[J]. *Flow Measurement and Instrumentation*.19(2): 79-84.
[17] Zheng, D., Lv, S., Mao, Y. (2019) Effect mechanism of the non-ideal flow field on the acoustic field in the gas ultrasonic flowmeter. *IET Science, Measurement & Technology*. 13(4), 469-477.
[18] Zhang, H., Guo, C., Lin, J. (2019) Effects of velocity profiles on measuring accuracy of the transit-time ultrasonic flowmeter. *Applied Sciences*. 9(8): 1648.
[19] Zhou, H., Ji, T., Wang, R., Ge, X., Tang, X., Tang, S. (2018) Multipath ultrasonic gas flow-meter based on multiple reference waves. *Ultrasonics*. 82: 145-152.
[20] Chen, J., Zhang, K., Wang, L., Yang, M. (2020) Design of a High Precision Ultracsonic Gas Flowmeter. *Sensors*. 20(17): 4804.

Tubing Depth Optimization Study of Horizontal Well Drainage and Gas Production in WeiYuan Shale Gas Field

Mingguang CHE[1*], Jie LIU[2], Chengzhong BU[3], Kai DONG [4]

[1] Research Institute of Petroleum Exploration and Development PetroChina, Beijing

[2] Yangtze University, Wuhan Hubei

[3] CNPC Chuanqing Drilling Engineering Company Limited, Sichuan Chengdu

[4] CNPC Great Wall Drilling Engineering Company Limited, Beijing

*Corresponding author Email: chemg69@petrochina.com.cn

Abstract: Timely production with a tube in shale gas wells can effectively ensure the stable decline of gas wells. However, there is no in-depth study on which position of tubing is the most favorable for the stable production of gas wells and whether the wellbore trajectory has an impact on this position. To determine the reasonable end of tubing (EOT), firstly, through the gas water flow experiment of the whole wellbore test device of a horizontal well, the characteristics of liquid accumulation in the test section and the liquid discharge at the outlet are analyzed. The results show, when the horizontal section inclines upward and the end of the tubing reaches 1/3 of the horizontal section, the gas flow carrying liquid is the smallest. Then, OLGA software is used to simulate the wellbore effusion in the production process of gas wells with different tubing depths under different wellbore trajectories. The results show when the horizontal section inclines upward and the end of the tubing arrives at 1/3 of the horizontal section, the cumulative production of the gas well is the largest. Based on comprehensive analysis, it is most favorable for horizontal shale gas wells when the end of the tubing is about 1/3 of the horizontal section.

1. Introduction

The production rate of a shale gas well is high in the early stage of development. Most gas wells are put into production using empty casing, which can effectively maintain the production of liquid-carrying gas [1-4]. Without restriction to good production, the wellhead pressure will quickly drop to the export pressure, and then abnormal production will occur [5-8]. In this case, the normal liquid-carrying of a gas well becomes impossible, and the wellbore will have liquid accumulated inside, or even become water flooded, resulting in production shutdown [9-11]. As a result, the life cycle of a gas well decreases sharply. To ensure a stable pressure decline for gas wells, the tubing must be timely adopted for production. Therefore, it is a pressing problem demanding a solution to maintain the stable production of shale gas wells to explore the liquid accumulation pattern in the middle and late stage of shale gas wells and determine the running timing and depth of tubing, as well as other technical requirements for string selection [12]. Firstly, the characteristics of liquid accumulated in the tested pipe section and the liquid production rate at the outlet position were analyzed by indoor physical simulation. And then, the numerical simulation method was used to simulate the liquid accumulation in the wellbore during the production of the gas well for different tubing running depths and different wellbore trajectories. After that, the production situations of gas wells at different depths were compared and analyzed. Finally, a reasonable running depth of tubing was determined with the results of physical simulation and numerical simulation combined.

2. Physical Simulation for Gas-liquid Flow at Different Depths of Tubing

2.1. Testing equipment and procedures
A schematic of the testing apparatus is shown in Figure 1. This apparatus can be used in a test to simulate the gas-liquid flow pattern, the liquid-carrying by gas at different good sections, and the dewatering gas production process in a horizontal gas well. The indoor simulation test at different running depths of tubing was conducted at room temperature. The liquid production outlet was vented. The wellbore structure of a horizontal well comprises a vertical section of 90°, a deviated section with complete curvature, and a horizontal section of 0°. Given the actual field conditions, tap water and air were used as the simulation media under indoor conditions to ensure that the gas-liquid flow state was visible during the indoor test.

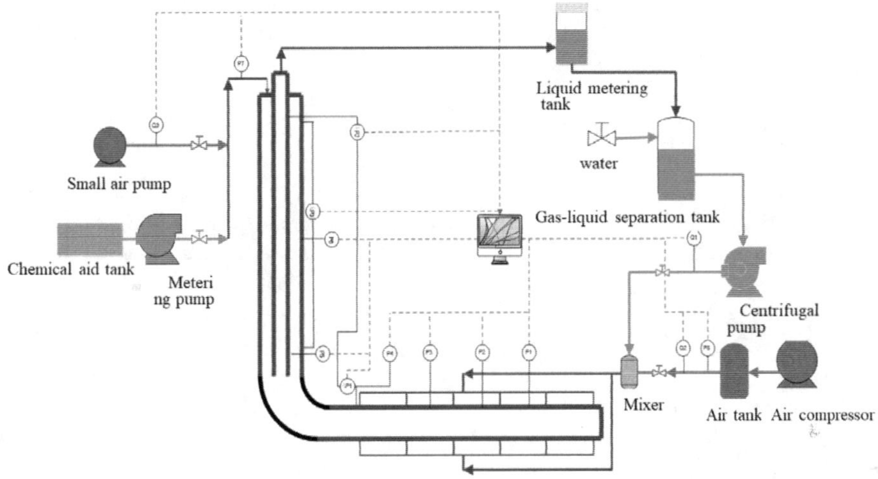

Figure 1. Schematic of testing apparatus for physical simulation of gas-liquid flow in a horizontal well.

In the production process of WeiYuan Gas Field, the average flowing bottom-hole pressure (FBHP) is 6.25MPa, the liquid production rate is 1-3 m^3/d, and the gas production rate is (1-4) $\times 10^4$ m^3/d. According to the gas state equation, the corresponding bottom hole gas flow rate is 6.7-26.7 m^3/h. In an actual gas well, the tubing has a diameter of 62 mm and the casing has a diameter of 127 mm. Therefore, 60 mm was selected as the diameter of the simulation string for a single pipe of the complete horizontal well, and the ratio of tubing to casing diameter was 1:2.

2.2. Testing steps and data processing
The testing process simulates the gas-liquid flow in the whole wellbore of a horizontal well. The gas-liquid flow test in the whole wellbore was conducted for different running depths of string when tubing was run into different positions of the horizontal section with accumulated liquid and the deviated section (in the case of liquid accumulation in the horizontal and deviated sections, respectively, tubing was run into the upper end of the deviated section, the heel end of horizontal section, 1/3 of horizontal section, 2/3 of horizontal section, and the toe end of horizontal section). During the test, the wellhead was connected to a liquid metering tank through a buffer screen. The testing tubing diameter was 60 and 30 mm. The liquid flow rate was 0.2, 0.4, 0.6, 0.8, and 1.0 m^3/h, respectively. The airflow rate was gradually reduced from 55 to 4 m^3/h (under testing conditions). During testing, the gas-water flow phenomena at different good sections, gas production rate, water production rate, and wellhead liquid production rate, as well as the pressure and differential pressure

data of pipelines in each good section were mainly recorded. 0.01 s was used as the unit to meter the wellhead liquid production rate, and 20 s was used as a stable phase during processing to sort and compare the data of the wellhead liquid production rate.

2.2.1. Analysis of wellhead liquid production rate and stability

When the liquid volume is constant, the impact of different tubing running depths on liquid carrying by the gas flow is different in a horizontal gas well. For a liquid flow rate of 0.4 m^3/h, as an example, the wellhead liquid production rate changes in response to different running depths of tubing, as shown in Figure 2.

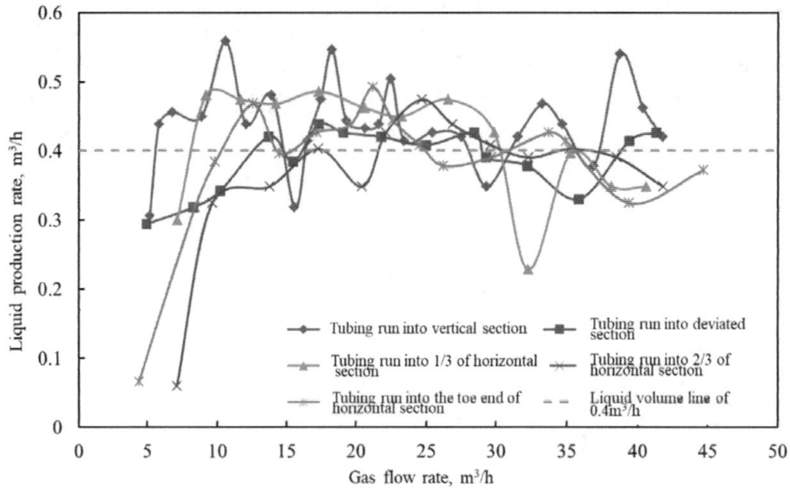

Figure 2. Gas flow rate vs. liquid production rate for different running depths of tubing.

As can be seen from Figure 2, fluctuation in liquid production rate at the outlet was the greatest when tubing was run into the vertical well section. This is mainly because it is the most difficult for gas to carry liquid in a deviated well section. Since tubing was not run into the deviated well section, the cross-section of fluid passing through the deviated section was still casing, which didn't alleviate the problem of difficult fluid-carrying in the gas well. Therefore, fluctuation in the liquid production rate was very obvious. When tubing was run into the deviated section and 1/3 of the horizontal section, the liquid production rate experienced a changing process from increasing, becoming stable, decreasing, and increasing again with the increase in gas flow rate. This process is mainly caused by the change in the gas-liquid flow regime. When the gas volume is small, the gas-liquid flow appears as a bubble flow. As the gas volume increases, a slug flow with the greatest lift efficiency begins to appear. Subsequently, the liquid production rate suddenly decreases when the gas flow rate increases to a value at which the flow regime changes from a slug flow into an annular mist flow. However, the liquid production rate will increase again when the gas flow rate further increases. When the gas production rate of the horizontal gas well allows the continuous production of liquid-carrying gas, it can be regarded that the gas well is in a relatively stable production state. Although the changing pattern of liquid production rate is similar in the two cases, the change in gas velocity in the horizontal section during flowing will result in the fluctuation in pressure because the tubing is run into 1/3 of the horizontal section. Impacted by the pressure fluctuation, gas is more likely to pass through the deviated section, and the stable gas flow rate corresponding to the wellhead liquid production rate is smaller than that when tubing is run into the deviated section.

When tubing is run into 2/3 and the toe end of the horizontal section, the liquid production rate first increases and then decreases with the increase in gas flow rate. However, the gas flow corresponding to the peak liquid production rate when tubing is run into 2/3 of the horizontal section is significantly less than that when tubing is run into the toe end of the horizontal section. This is because there is no pressure fluctuation caused by the change in gas velocity when gas flows through the horizontal section after the tubing is run into the toe end of the horizontal section.

2.2.2. Analysis of gas flow rate required for normal liquid carrying

Figure 3 shows the gas flow rates required for normal liquid carrying in the case of different wellbore trajectories, the same liquid flow rate, and different tubing running depths.

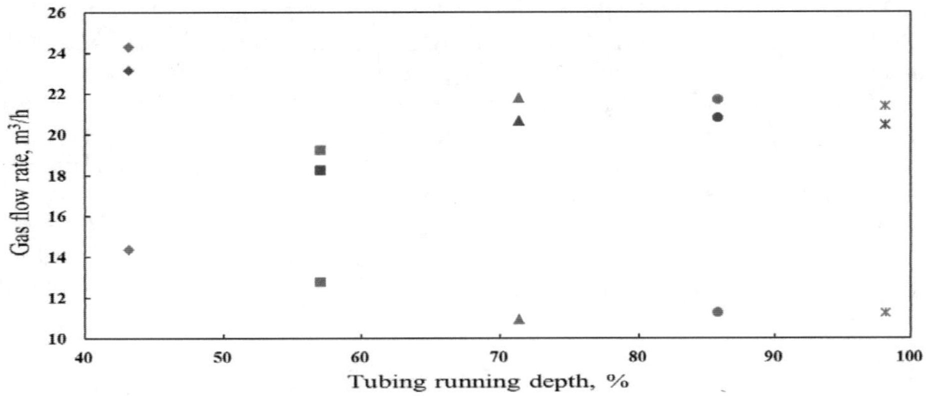

(a) Liquid flow rate = 0.2 m3/h

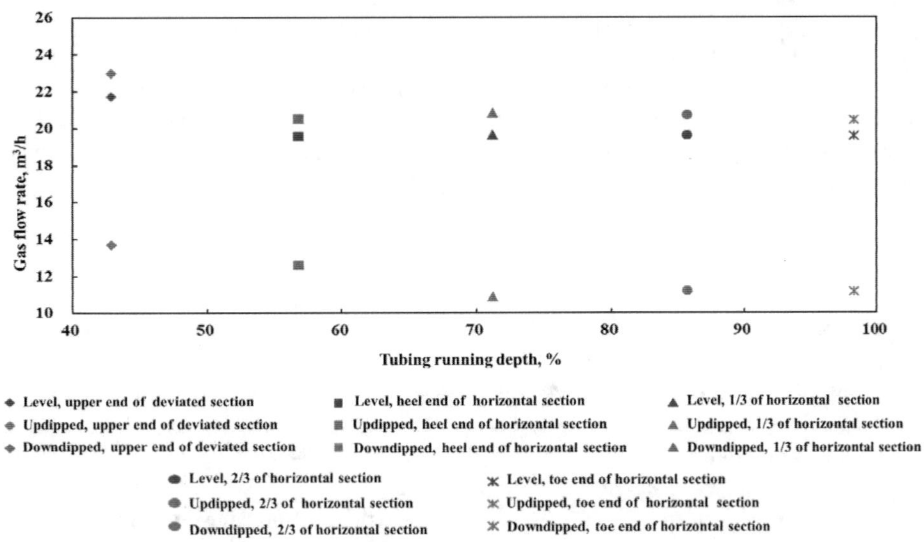

(b) Liquid flow rate = 0.8 m3/h

Figure 3. Gas flow rates are required for carrying different volumes of liquid at different running depths of tubing.

As can be seen from Figure 3 (a), in the case of a 0.2 m3/h liquid flow rate and 3 different states (Level, up-dipped, and down-dipped) of wellbore trajectories, the gas flow rate required for normal liquid carrying is the minimum when the horizontal section is up-dipped and is the maximum when the horizontal section is down-dipped. The flow rate of liquid-carrying gas for the up-dipped horizontal section is about 40% less than that when it is level, and that for the down-dipped horizontal section is about 5% more than that when it is level. The main cause is that the key for gas flow to carry liquid in a horizontal section of the wellbore lies in whether or not the flow regime in the horizontal section can change from laminar flow into the non-laminar flow. The greater the height of the liquid accumulated near the heel, the more likely it is for the liquid accumulated to be carried out of the horizontal section by the gas flow. When the horizontal section is down-dipped, all the liquid accumulates at the toe and is difficult to be carried out of the horizontal section by the gas flow.

For the same wellbore trajectory, the gas flow rate required for normal liquid carrying first decreases, and then increases, and after that decreases again with the increase in the running depth of tubing. This is mainly because it is the most difficult for gas flow to carry liquid in deviated well section. The gas flow required for normal liquid carrying can be greatly reduced by running the tubing below the deviated well section. Furthermore, when the impact of pressure fluctuation caused by the change in the velocity of gas during its flow through the horizontal and deviated sections on the deviated section is the greatest, it is more likely for gas to pass through the gap between the pipe and the liquid to carry the liquid out of the horizontal section. Therefore, the gas flow rate for normal liquid carrying is the minimum when tubing is run into the deviated section in the case that the horizontal section is in a level or down-dipped state, or when tubing is run into 1/3 of the horizontal section in the case that the section is in an up-dipped state.

As can be seen from Figure 3 (b), when the liquid flow rate is 0.8 m3/h, the variation trend of the gas flow rate for normal liquid carrying at different tubing running depths is the same as that when the liquid flow rate is 0.2 m3/h, but the minimum gas flow rates for liquid carrying are different in the two cases.

3. Numerical Simulation Study of Dewatering Gas Production Effect

OLGA is used for the simulation calculation of liquid accumulated in oil and gas pipelines. For this purpose, pipeline structure data, elevation data, and pipeline operation data are required. After the model of accumulated liquid is established, its simulation similarity needs to be validated first. Since the OLGA calculation follows the principle of reverse calculation for pressure and direct calculation for temperature, the outlet pressure and inlet temperature of the pipeline can be fixed in the simulation calculation. The inlet pressure and outlet temperature of the pipeline can be obtained through simulation calculation, and then the simulation results are compared with the actual production data to validate if the model is accurate

3.1. Basic parameters for numerical simulation

3.1.1. Productivity equation of gas well

Production data from several wells of Platform Wei-204HX were used as the basic parameters for the simulation study.

Beggs-Brill multi-phase pipeline flow calculation method was used to process the dewatering gas production testing data to calculate the FBHP corresponding to wellhead pressure. Gas production testing data is shown in Table 1.

Table 1. Dewatering gas production testing data of Well Wei-204HX-1

Item No.	Casing pressure, MPa	Water production rate, m^3/d	Choke diameter, mm	Orifice diameter, mm	Instantaneous gas flow rate, $10^4 m^3/d$	Flowing pressure of well, MPa
1	32.15	338.88	7.0	28.575	5.6288	57.1285
2	31.47	497.52	8.0	28.575	8.2872	55.8227
3	30.68	582.24	9.0	28.575	11.2272	54.3185
4	30.34	539.28	10.0	28.575	15.4896	52.6820

The above test data were processed by using the binomial productivity test data processing method and the productivity equation of the well was obtained as follows:

$$p_r^2 - p_{wf}^2 = 118.7Q_{sc} - 3.3299Q_{sc}^2$$

Where p_r and p_{wf} are the formation pressure and FBHP in MPa, respectively, and Q_{sc} is the gas production rate in 10^4 m^3/d.

Journal of Physics: Conference Series **2458** (2023) 012032 doi:10.1088/1742-6596/2458/1/012032

3.1.2. Wellbore structure of gas well

The wellbore trajectories of Wells 204HX-1 (whose horizontal section is up-warped) and 204HX-4 (whose horizontal section is down-dipped) of Platform Wei-204HX were selected as examples for analysis, and their geo-steering maps are shown in Figure 4.

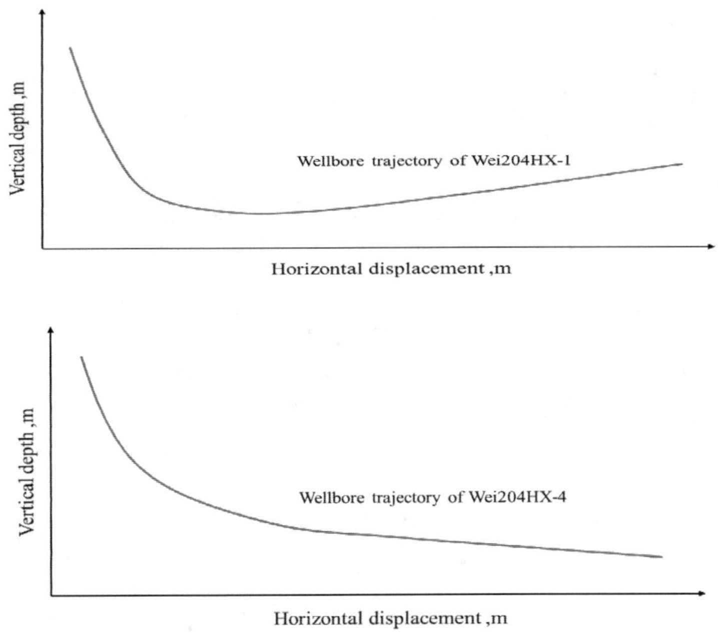

Figure 4. Geosteering-while-drilling maps of Wells Wei-204HX-1 and Wei-204HX-4

3.1.3. Formation pressure drop gradient

Since OLGA cannot describe the production decline caused by factors such as low permeability and desorption of shale gas reservoirs, formation pressure drop has to resort. These data, however, are difficult to acquire in the field. Then, the formation pressure drop gradient of the simulation well was determined by history matching. With the formation pressure drop gradient set at 0.23 MPa/d, matching results were obtained, and the results were then compared with the gas production rate of Well Wei-204HX-1, as shown in Figure 5.

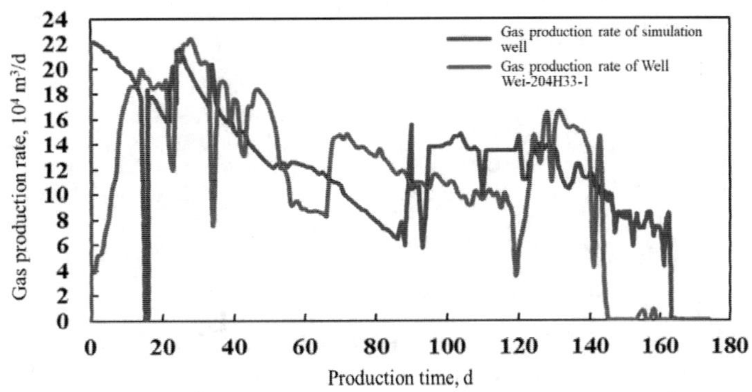

(a) Comparison of gas production rates

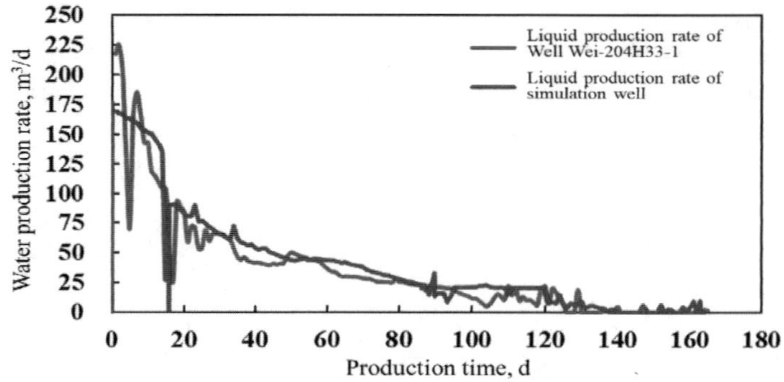

(b) Comparison of water production rates

Figure 5. Comparison between the production data of Well Wei-204HX-1 and the simulation results

The simulation results are close to the actual production data, which can satisfy the needs of the simulation. However, it takes 36 to 48 hours to finish the simulation. In a later study, the formation pressure drop gradient for simulation was increased by a factor of 60 to 0.23 MPa/(24 min) to speed up the simulation calculation time.

3.2. Numerical simulation analysis for well with an up-dipped horizontal section

The simulation results for different running depths of tubing are shown in Figure 6, which are mainly characterized as follows: 1) In the early stage of simulated production, the horizontal gas well realizes its maximum production rate when tubing is run into the vertical and deviated sections and experiences its minimum production rate when tubing is run into the heel of horizontal section. The main reason is that the gas flow velocity is small when tubing is just run into the horizontal section. In this state, liquid has the largest friction loss during its flowing from the bottom hole to the wellhead, and the FBHP is the largest when transportation pressure is the same. Therefore, the production rate is the minimum. 2) With the continuation of production, the production rate of the gas well experiences a sudden drop, and then normal production is resumed. At this point, the gas production rate when tubing is run into the toe end is the minimum due to pressure build-up in the casing-tubing annulus. Besides, due to the occurrence of critical liquid-carrying flow, the production rate of the gas well when tubing is run into the deviated section gradually becomes higher than that when tubing is run into the vertical section.

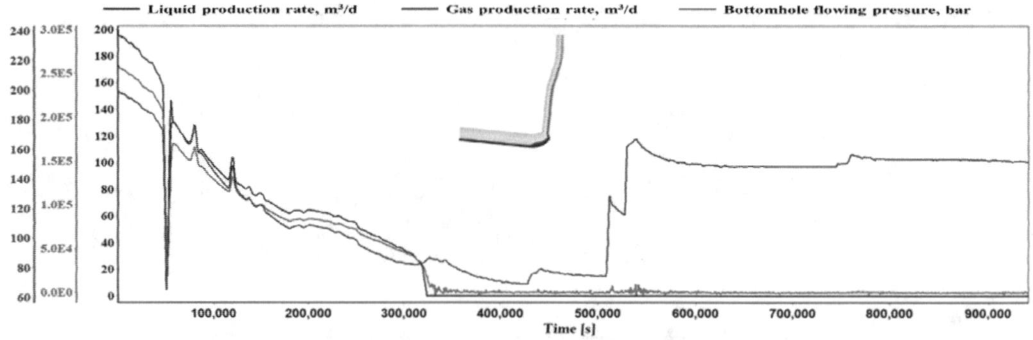

(a) Tubing runs into a vertical section

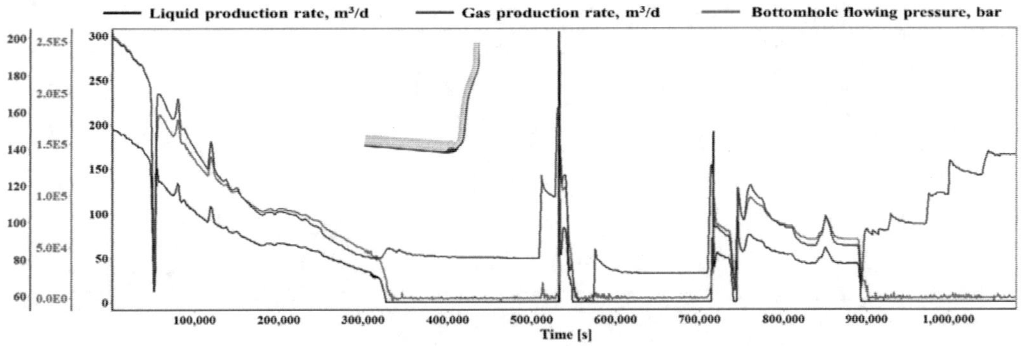

(b) Tubing runs into a deviated section

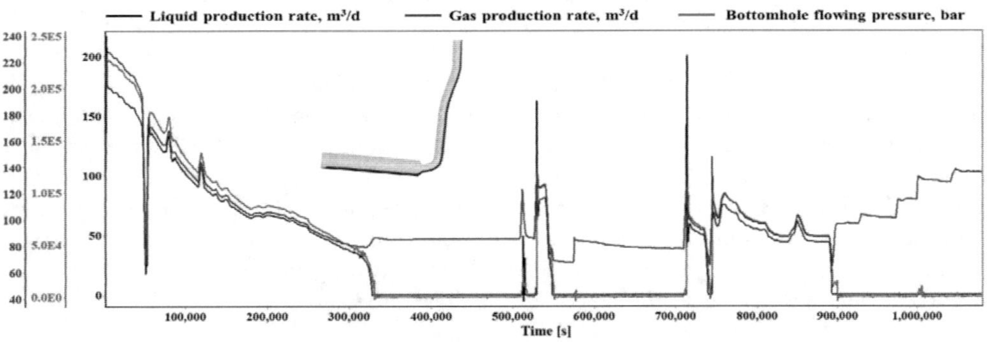

(c) Tubing runs into the heel end of the horizontal section

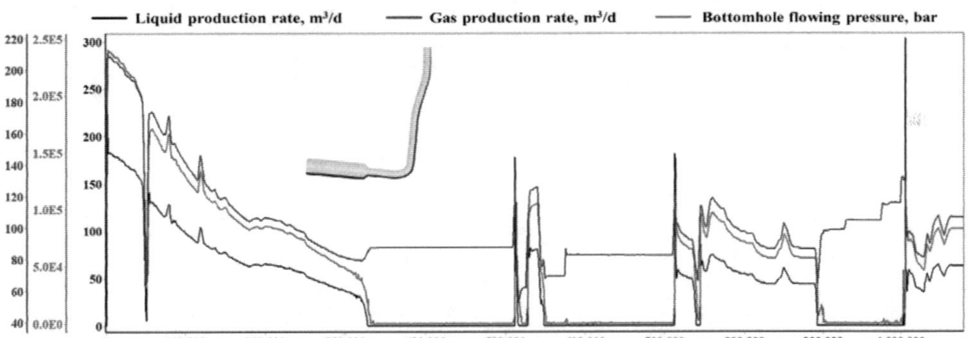

(d) Tubing runs into 1/3 of the horizontal section

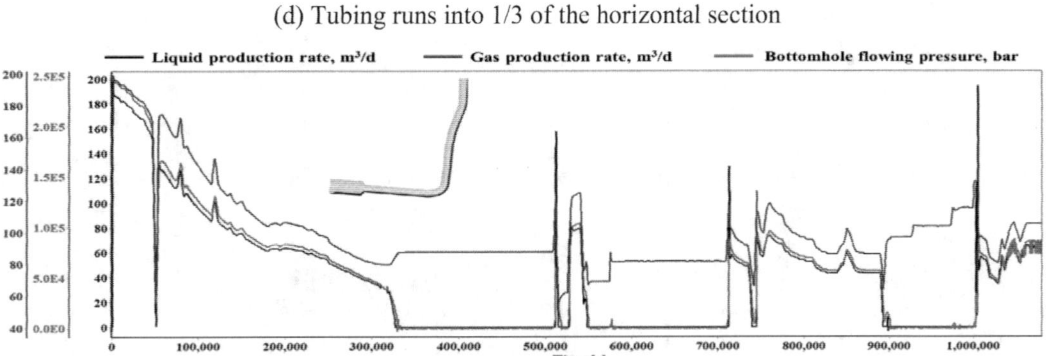

(e) Tubing runs into 2/3 of the horizontal section

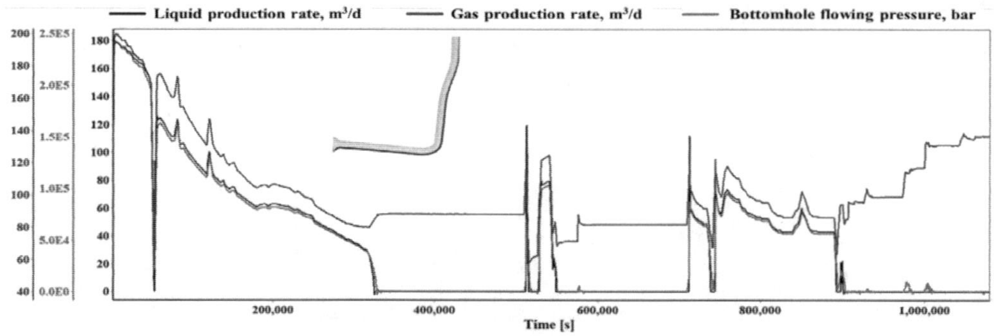

(f) Tubing runs into the toe end of the horizontal section

Figure 6. Production performance profiles corresponding to different running positions of tubing in the case of an up-warped horizontal section

The production performance profiles corresponding to 6 different running depths of tubing in the case of the up-warped horizontal section were placed in the same coordinate system for comparative analysis, with the results shown in Figure 7. As can be seen from Figure 7, there is a phase when the production rate is recovered when tubing is run into 1/3 and 2/3 of the horizontal section in the late stage of simulated production, while the gas production rate seldom fluctuates in the case that tubing is run into the other 4 positions. As can be seen from the comparative analysis of accumulative gas production in a single well, the accumulative gas production reaches the highest point when tubing is run into a region near 1/3 of the horizontal section, as shown in Figure 8.

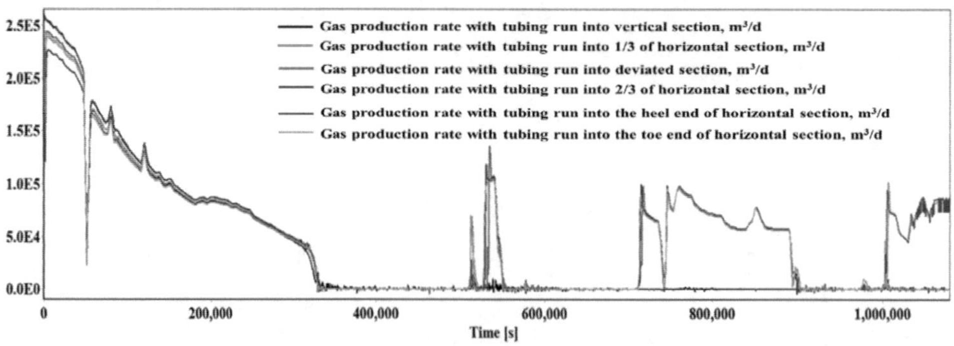

Figure 7. Comparison of gas production rate for well with an up-warped horizontal section in the case of different running depths of tubing

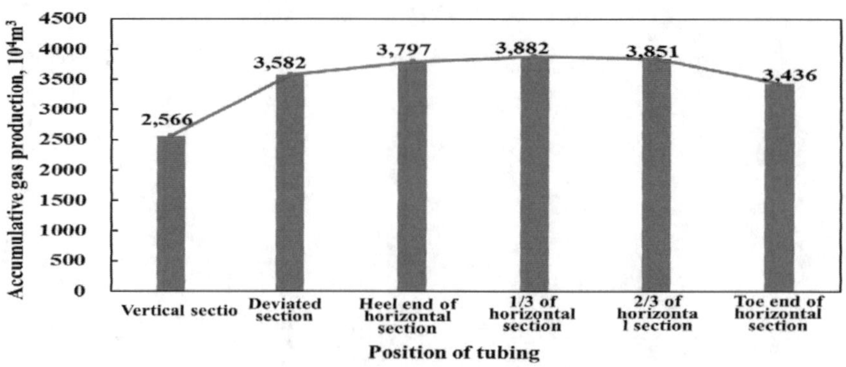

Figure 8. Comparison of accumulative gas production for well with an up-warped horizontal section in the case of different running depths of tubing

3.3. Numerical simulation analysis for well with a down-dipped horizontal section

To ensure that the two types of wells have the same liquid supply capacity, the productivity equation of Well Wei-204HX-1 was also used in the simulation of this type of well, but wellbore trajectory adopted the inclination data of Well Wei-204HX-4 to ensure that the horizontal section was in a down-dipped state. The production performance profiles obtained from the simulation are shown in Figure 9.

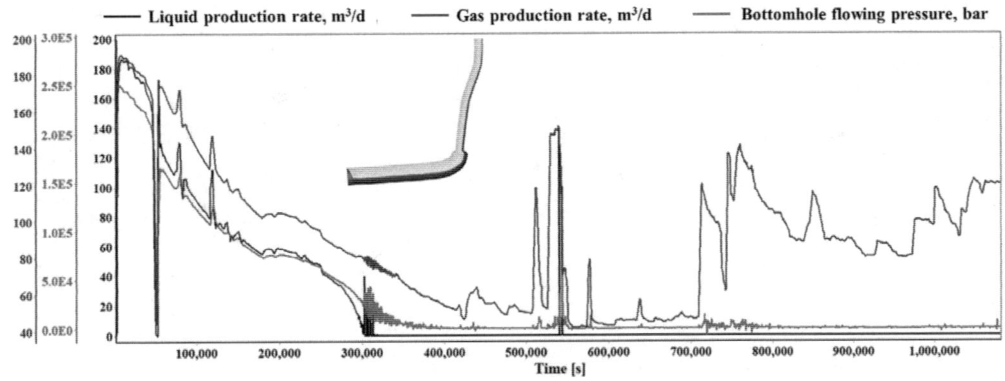

(a) Tubing runs into the vertical section

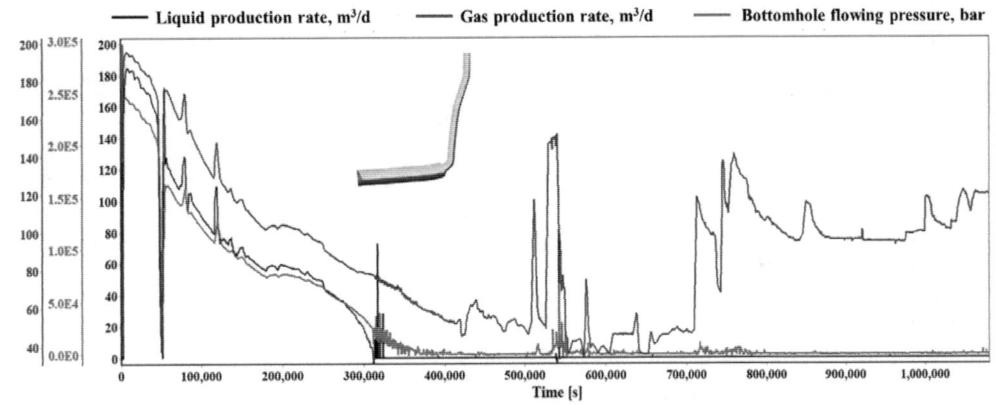

(b) Tubing runs into the deviated section

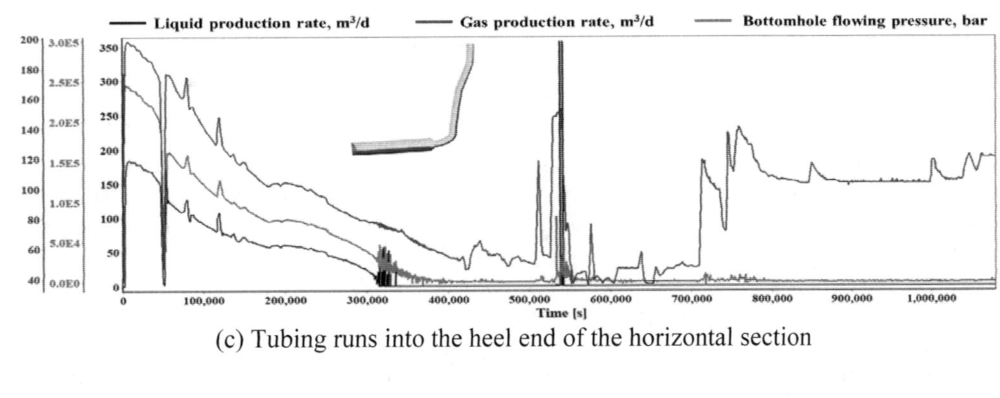

(c) Tubing runs into the heel end of the horizontal section

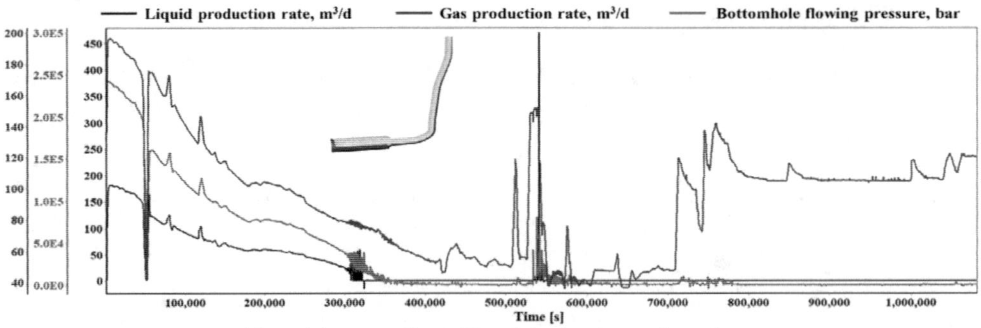

(d) Tubing runs into 1/3 of the horizontal section

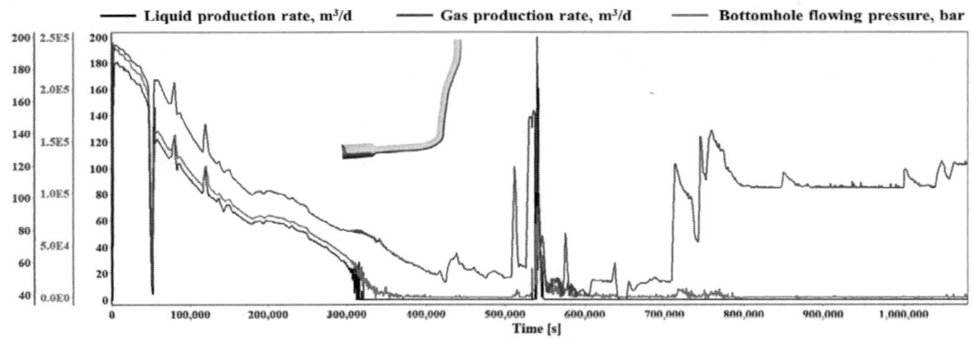

(e) Tubing runs into 2/3 of the horizontal section

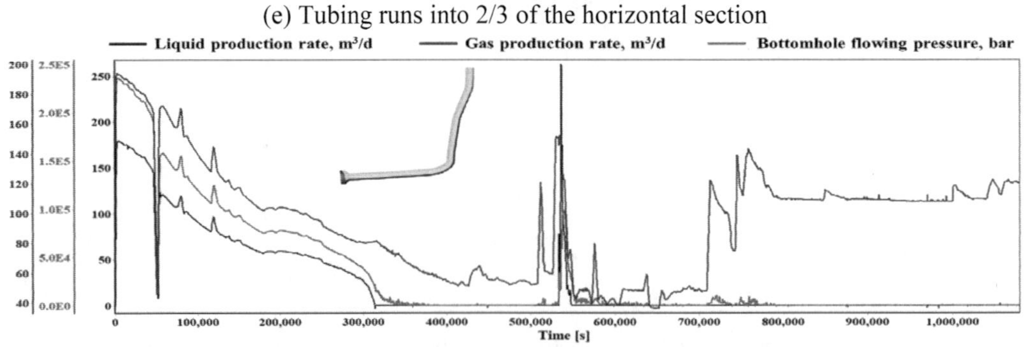

(f) Tubing runs into the toe end of the horizontal section

Figure 9. Production performance profiles corresponding to different running positions of tubing in the case of down-dipped horizontal section

The production performance profiles corresponding to 6 different running depths of tubing in the case of the down-dipped horizontal section were placed in the same coordinate system for comparative analysis, with the results shown in Figure 10. Different from the well with an up-warped horizontal section, there was no recovery of gas production rate when tubing was run into 1/3 and 2/3 of the horizontal section in the late stage of production. Production characteristics were similar for all the running depths of tubing, with the gas production rate fluctuating in the case of small gas volume. As can be seen from the comparative analysis of accumulative gas production in a single well, the accumulative gas production reached the highest point when tubing was run into the heel end of the horizontal section, as shown in Figure 11.

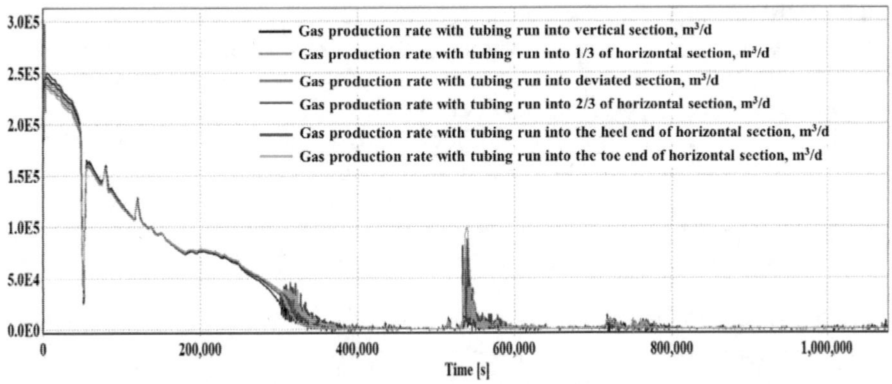

Figure 10. Production performance profiles corresponding to different running depths of tubing in the case of the down-dipped horizontal section

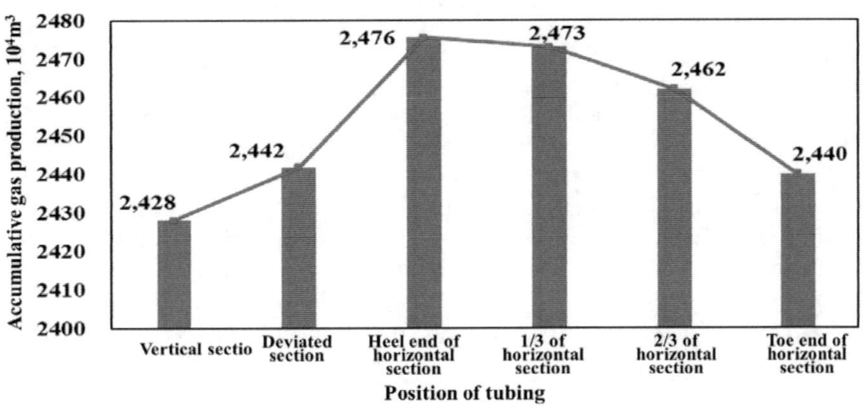

Figure 11. Accumulative gas production of well with the down-dipped horizontal section in the case of different running depths of tubing

4. Conclusions

The liquid-carrying gas flow is reduced by about 40% compared with a level horizontal section in the case of an up-dipped structure, and increased by about 5% compared with a level horizontal section in the case of a down-dipped structure.

The liquid-carrying gas flow is the minimum when tubing is run into deviated section in the case of a level or down-dipped horizontal section, and when tubing is run into 1/3 of a horizontal section in the case of an up-dipped horizontal section.

For different running depths of tubing, the accumulative gas production in the case of the up-dipped horizontal section is greater than that in the case of the down-dipped horizontal section. The accumulative production of a gas well is the maximum when tubing is run into 1/3 of the horizontal section in the case of the up-dipped horizontal section, and when tubing is run into the heel end of the horizontal section in the case of the down-dipped horizontal section.

References

[1] ZHANG H L, LIU H R. State of the art and effects of dewatering & recovery of shale gas in China [J]. Nature Gas Industry, 2012, 32(12): 49-51.

[2] WANG Q R, CHEN J X, XIANG J H, et al. Discussion on shale gas well effusion diagnosis and drainage gas production technique [J]. Oil & Gas Exploration and Development, 2020, 38(5): 83-87.

[3] SHANG S F, YAN H, WU J, et al. Production characteristics and development modes of shale-gas wells, Changning Gasfield, Sichuan Basin [J]. Nature Gas Exploration and Development, 2018, 41(4): 69-75.

[4] GUPTA SHANK, RAI CHANDRA, DEVEGOWDA DEEPAK, et al. Haynesville Shale: Predicting long-term production and residual analysis to identify well interference and frac hits[C]. SPE Oklahoma City Oil and Gas Symposium, Oklahoma, USA, 9-10 April 2019.

[5] KAUR BALPREET, HAQ BASHIRUL, DARYOUSH HABIBI, et al. A novel approach in gas well performance monitoring and forecasting using modified decline curve analysis[C]. International Petroleum Technology Conference held in Dhahran, Saudi Arabia, 13-15 January 2020.

[6] LI M. Coiled tubing of drainage gas recovery technology used in shale-gas downdip horizontal wells [J]. Oil Drilling & Production Technology, 2020, 42(3): 329-333.

[7] CAO M J, WU X D, AN Y S, et al. Optimal design of coiled tubing for production string of gas well in shale gas field[J]. Fault Block Oil & Gas Field, 2018, 25(6): 811-814.

[8] LUO X, WANG Y T, WANG X Q, et al. Shale gas wellbore liquid loading identification based on triangle plate[J]. Special Oil and Gas Reservoirs, 2020, 27(4): 168-174.

[9] DINATA R C, SARICA C, PEREYRA E. A methodology of end-of-tubing EOT location optimization for horizontal shale gas wells with and without DE liquification[C]. SPE North America Artificial Lift Conference & Exhibition held in Woodlands, Texas, USA, 25-27 October 2016.

[10] DINATA R, CARICA C, PEREYRA E. End of tubing EOT placement effects on the two-phase flow behavior in horizontal wells: Experimental study[C]. SPE Annual Technical Conference and Exhibition held in Dubai, UAE, 26-28 September 2016.

[11] CHENG Y. Study on pressure drop law of two-phase gas-liquid in horizontal well [D]. Jinzhou: Yangtze University, 2019.

[12] WANG Q. Experimental study on gas-liquid flowing in the wellbore of horizontal well[D]. Chengdu: Southwest Petroleum University, 2014.

Integrated Valve Product Fluid Simulation and Test Verification

Leilei Li*, Baorong Liu, Shengbo Gong, Jian Zhao, Qiang Tao, Xu Zhou, Yingli Xu, Menglei Guo

Beijing Spacecrafts, Beijing 100094, China

leilna@163.com

Abstract: The integrated valve product's main function is to electrolytic oxygen generation subsystem to produce hydrogen and oxygen to realize automatic discharge, to maintain the space station astronauts in orbit around in comfort and safety, high reliability, easy maintenance, and the device should be easy replacement of human-computer ergonomics requirements to ensure long-term on-orbit operation station. Based on the above requirements, an integrated valve product based on environmental control and a health protection system is designed. The integrated product includes hydrogen normal and backup emission branches, oxygen normal and backup emission branches. Through fluid simulation and experimental verification, it is verified that the flow rate and other key indicators of the integrated valve product meet the requirements, which has certain reference and guiding significance for the design of similar integrated products in the future.

1. Introduction

With the continuous breakthrough of new space technology, manned spacecraft are gradually developing to the moon base, Mars base, and deep space exploration [1-3]. The spacecraft is also transitioning from a non-regenerative system to a regenerative system and a controlled ecosystem to extend the service life of the spacecraft and reduce the supply cost of ground transportation [4-6]. As an important system in the environment control and life support subsystem of the space station, the electrolysis oxygen production subsystem is mainly used to electrolyze the clean water generated by the urine treatment subsystem to produce hydrogen and oxygen to realize the effective circulation of substances in the space station. Based on the service life of the space station of more than 15 years in our country's needs, taking into account the device with high reliability, easy maintenance, easy change of human-computer ergonomics requirements, gas emission control device for the space station electrolysis oxygen device to carry out the integrated design, and the device for flow calculation and experimental verification for subsequent space products provide certain reference of integrated design.

2. Integrated valve product design

As a key single machine in the electrolytic oxygen production subsystem, the integrated valve product needs two emission branches of hydrogen and oxygen to realize the automatic emission of hydrogen and oxygen in the system. The discharge branch needs to be equipped with a pressure sensing element, pressure regulating element, switch control element, overvoltage protection element, and connection structure. To make the substances in the cabin renewable and effectively utilized, CO_2 should be added to the hydrogen emission branch reduction port CO_2, if necessary, with H_2 reaction to form CH_4 and

H_2O, efficient design of environmental control, environmental protection, and renewable systems. The overall design of the product considers many factors such as mass, volume, flow rate, and leakage rate. The functional requirements of the product are shown in Table 1.

Table 1 Product demand-related indicators

Content	Indicators	Unit	Content	Indicators	Unit
Quality	≯ 7.0	kg	The oxygen circuit regulator element closes the pressure	≮ 90	kPa
Volume	≯ 230 * 200 * 180	mm	Oxygen road traffic	≮ 125	L/h
Opening pressure of hydrogen circuit regulating element	≯ 95	kPa	Overvoltage protection element opening pressure	150	kPa
Hydrogen circuit regulator element closing pressure	≮ 90	kPa	Overvoltage protection element closing pressure	≮ 140	kPa
Hydrogen road traffic	≮ 250	L/h	Leakage rate of the branch	≯ 1×10^{-5}	$Pa.m^3/s$
Oxygen circuit regulating element opening pressure	≯ 95	kPa	Stop valve leakage rate	≯ 1×10^{-5}	$Pa.m^3/s$

2.1. Overall design scheme

To meet the functional requirements of the electrolytic oxygen production subsystem for the automatic emission of hydrogen and oxygen, the workflow chart of the product is designed as shown in FIG. 1, in which the pressure sensing element monitors the exhaust pressure of the flow path, and the pressure regulating element maintains the oxygen and hydrogen emission pressure. The switch control element controls the automatic discharge of oxygen and hydrogen and is equipped with a manual valve as a backup to increase the reliability of flow control. The connection structure is used for the quick connection of the product to other devices in the electrolytic oxygen production subsystem.

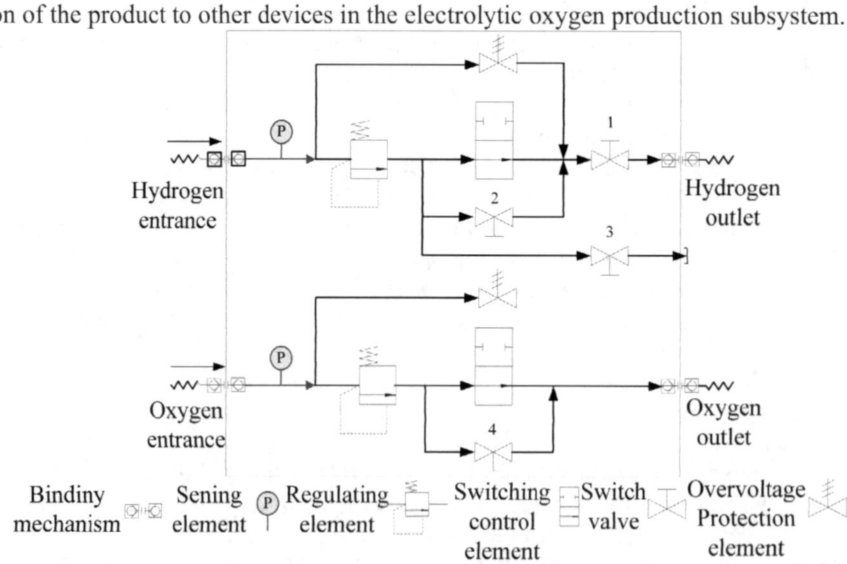

Fig 1. Flow chart of integrated valve product

When the product is working, the oxygen discharge branch flow: the oxygen produced by the oxygen production system enters the oxygen inlet. When the pipeline pressure reaches the set value, the pressure regulating element opens, and the oxygen is discharged from the oxygen discharge port through the switch control element. When the function of the control element fails, the cut-off Valve 4 is opened to ensure that oxygen is discharged from the oxygen discharge port through the manual cut-off valve. When the downstream of the oxygen supply line is blocked and the pressure in the line exceeds the design pressure, the pressure can be relieved automatically by the protection element.

The workflow of the hydrogen emission branch is the same as that of the oxygen supply line, where CO_2, the reduction interface was blocked with a plunger plug, and the on-orbit connection was made by astronauts when needed in orbit.

2.2. Detailed design of product structure

Based on the functional requirements of integrated valve products, the product is designed in three parts: fluid circuit structure; Manual cut-off valve, pressure sensor, pressure regulator, solenoid valve, safety valve, and other single products; Fast connection structure. The overall structure of the device is shown in Figure 2.

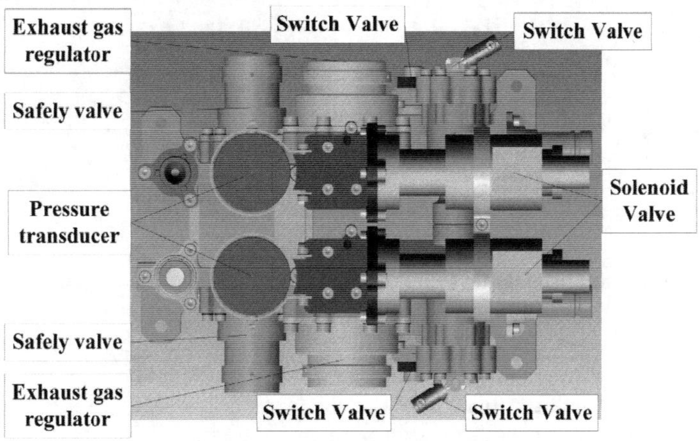

FIG. 2 Structure diagram of integrated valve product

2.3. Fluid circuit structure design

To achieve normal hydrogen and oxygen discharge and high reliability, normal hydrogen and oxygen discharge branches and backup branches are designed for the fluid circuit. To realize the lightweight, miniaturized, and integrated built-in design of the gas emission control device, four fluid loops are placed inside the support structure (hydrogen subgrade seat and oxygen subgrade seat assembly), and the normal hydrogen and oxygen emission is guaranteed through the effective communication of internal flow paths. At the same time, a safety valve, a solenoid valve, a pressure regulator, a pressure sensor, a cut-off valve, and an installation interface of a fast connection structure are arranged on the support structure. In the design of the flow path, based on the installation needs of multiple valves, the device mass volume limit, and the flow path size of the influence of convection resistance loss, the flow path needs to keep turning. Taking the normal hydrogen discharge branch as an example, the hydrogen flow sequence is as follows: quick connection structure (hydrogen inlet) → pressure regulating element (pressure regulator) → control element (solenoid valve) → manual cut-off valve → quick connection structure (hydrogen outlet), as shown in FIG. 3 (a). The hydrogen backup branch and the oxygen normal and backup discharge branches are shown in Figures 3 (b), 3 (c), and 3 (d), respectively.

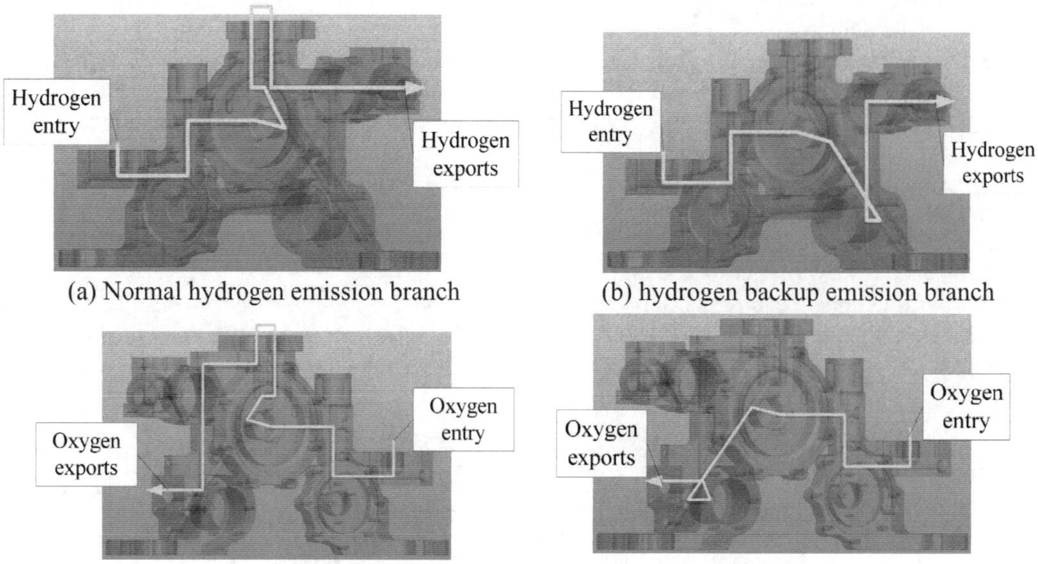

(a) Normal hydrogen emission branch (b) hydrogen backup emission branch

(c) Oxygen normal discharge branch (d) oxygen backup discharge branch

FIG. 3 Schematic diagram of a fluid circuit of integrated valve product

In addition, to improve the processing technology of the parts and the surface quality of the sealing surface, the support structure adopts a split design, that is, after all the internal fluid channels, internal installation interfaces, and sealing surfaces are processed. The hydrogen and oxygen subgrade seats are connected and fixed with four connecting plates, and then the installation surface and external interfaces of the support structure are processed together. The single-machine product involves a manual cut-off valve, safety valve, and pressure regulator. The sealed form of the manual stop valve adopts the end seal structure to meet the requirements of high reliability and long life; The safety valve and the pressure regulator both adopt a non-metallic diaphragm as the bearing element, through which the gas force can be amplified to improve the adjustment accuracy of the safety valve and the pressure regulator [9]. The diaphragm material is fluorine rubber F108, and the middle of the diaphragm is filled with silk material to increase the tear strength of the diaphragm, improve the elastic deformation ability, corrosion resistance, oil resistance, and fatigue resistance of the diaphragm. The import and export structure adopts the typical rapid connection design for aerospace products, which not only ensures the modular design but also realizes the rapid replacement and installation.

3. Simulation and test verification

Based on the above-integrated design, the flow index of the integrated valve product was simulated and tested, and the key parameter flow was analyzed in detail.

3.1. Flow path simulation

The FLUENT software in ANSYS is used to carry out simulation calculations on the designed fluid loop and verify the flow size of four branches under the pressure index of 105kPa. Taking the normal hydrogen emission branch as an example, the model is simplified and the k-ε model is adopted to make nitrogen equivalent to an ideal gas. The inlet pressure value of the hydrogen basin is set as 105kPa, the outlet pressure value is set as the default atmospheric pressure of 0Pa, and the rest of the wall is set as the wall. The corresponding boundary conditions and initial conditions are set as shown in Table 2.

Table 2 Simulation parameters setting the table of a fluid loop of the device

Working medium	Nitrogen
Import	105kPa (gauge pressure),
Exit	0kPa (gauge pressure),
The aperture	Phi is 4 mm
Model	k-ε model
The temperature	25 °C

The Watershed refers to the place where the fluid in the pipeline passes through, which is the core area for the numerical calculation of flow field characteristics. It needs to be divided to obtain an independent fluid area for calculation. The watershed segmentation principles of integrated valve products are as follows: (1) the valve is assumed to work in a fully open state; (2) the pipelines do not have a great impact on the overall watershed calculation and are difficult to mesh and reduce the calculation efficiency are simplified.

The watershed models of the hydrogen normal discharge branch, hydrogen backup discharge branch, oxygen normal discharge branch, and oxygen backup discharge branch are obtained as shown in Figs. 4 (a) to 4 (d), respectively.

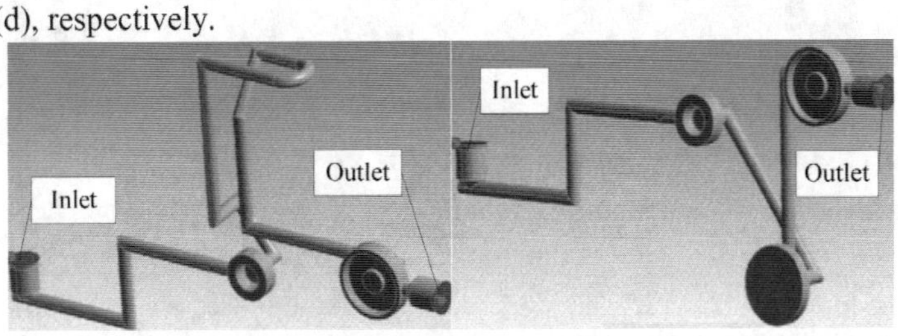

(a) Watershed model of normal hydrogen emission branch and (b) watershed model of hydrogen backup emission branch

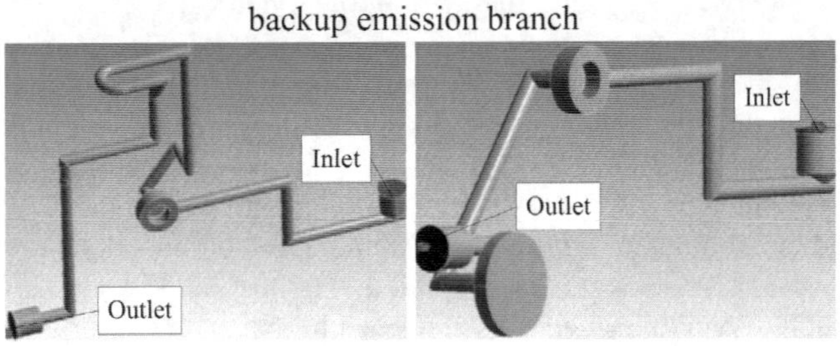

(c) Oxygen normal discharge branch watershed model (d) oxygen backup discharge branch watershed model

FIG. 4 Schematic diagram of the fluid circuit of integrated valve product

Considering that the local geometric structure of the watershed model is relatively complex, all the unstructured grids are used. The meshing results of the drainage basin with hydrogen normal discharge branch, hydrogen backup discharge branch, oxygen normal discharge branch, and oxygen backup discharge branch are shown in Figs. 5 (a) to 5 (d), respectively.

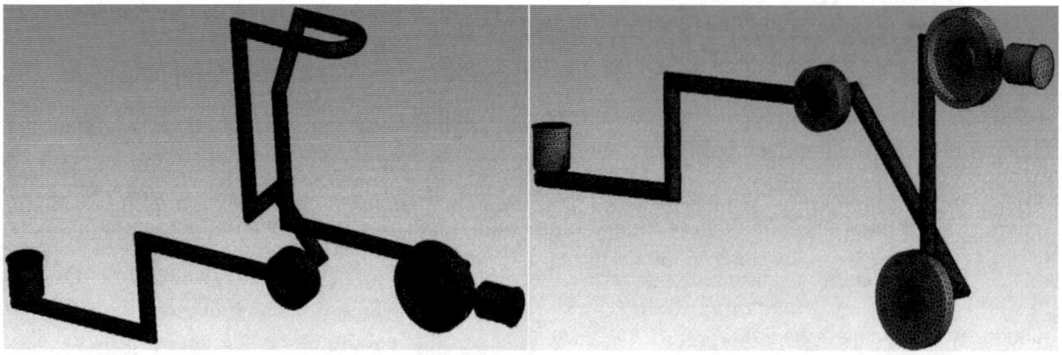

(a) Drainage grid of normal hydrogen emission branch (b) drainage grid of hydrogen backup emission branch

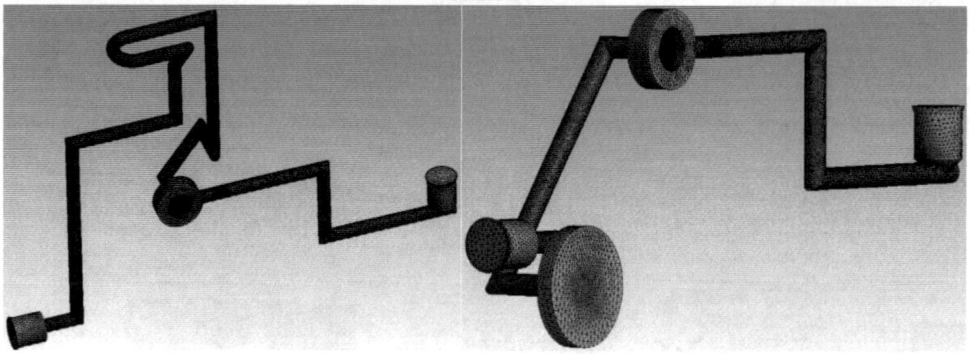

(c) Oxygen normal discharge branch drainage grid (d) oxygen backup discharge branch drainage grid
Fig 5. Drainage grid diagram of integrated valve product

Since each section of the watershed is like a circular pipeline, the Inflation method is used in grid division, which can generate a good boundary layer grid at the entrance and more accurately simulate the flow field characteristics of the fluid flow.

The SIMPLE algorithm was used to ensure the convergence of the calculation, and Hybrid Initialization was used to obtain a good initial flow field. Based on the above simulation methods and simulation conditions, the pressure cloud diagrams of the hydrogen normal emission branch, hydrogen backup emission branch, oxygen normal emission branch, and oxygen backup emission branch are shown in FIG. 6 (a) to FIG. 6 (d), and the velocity vector diagrams are shown in FIG. 7 (a) to FIG. 7 (d).

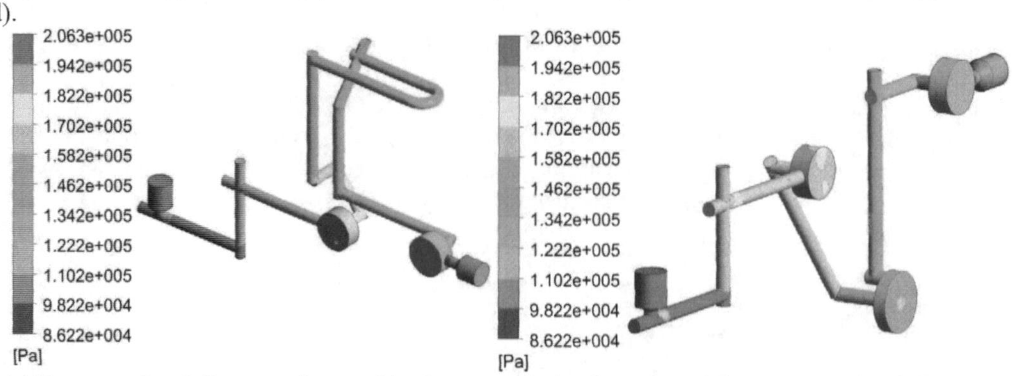

(a) Pressure cloud diagram of normal hydrogen emission branch and (b) pressure cloud diagram of hydrogen backup branch

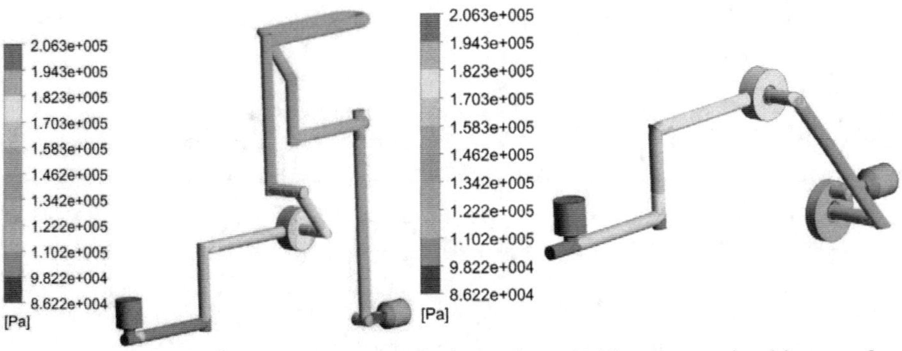

(c) Pressure cloud image of normal oxygen discharge branch (d) pressure cloud image of oxygen backup discharge branch

FIG. 6 Branch pressure cloud diagram of integrated valve product

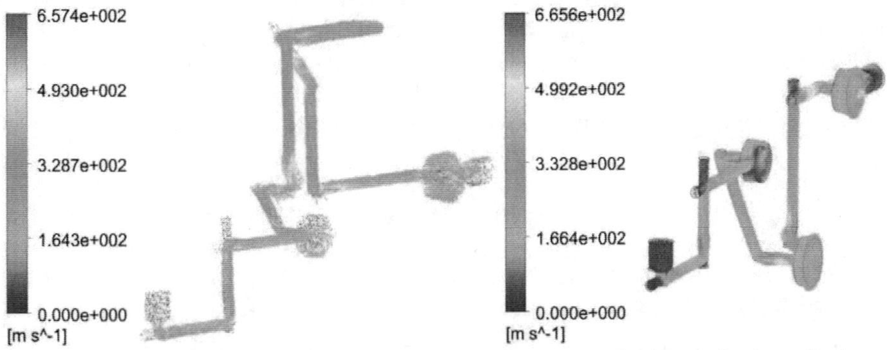

(a) Hydrogen normal discharge branch velocity vector diagram (b) hydrogen backup discharge branch velocity vector diagram

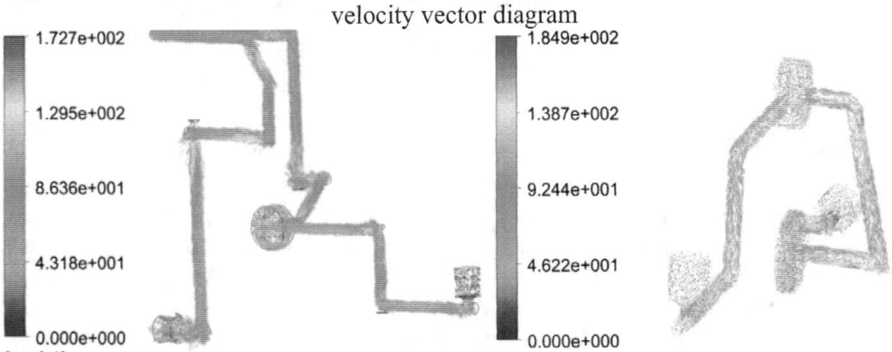

(c) Velocity vector diagram of normal oxygen emission branch (d) velocity vector diagram of oxygen backup emission branch

FIG. 7 Integrated valve product branch velocity vector diagram

It can be seen from FIG. 6 that the pressure drop is the largest when the flow path passes through the right-angle bend, which is caused by the sudden change of the flow path. Therefore, the design structure of the right-angle bend should be avoided as much as possible.

It can be seen from FIG. 7 that the gas flows through the valve and forms a regular vortex, which indicates that the gas flows smoothly at the valve and the flow loss is less. The designed flow path structure is reasonable to avoid the flow loss caused by the unreasonable structure.

Through the above fluid simulation, the flow at the inlet and outlet of the four branches under the pressure of 105kPa is shown in Table 3.

Table 3 Data Table of flow calculation results

The serial number	Branch	Volume flow rate (L/h)		Mass flow rate (x 10^{-4}Kg/s)	Index (L/h)
1	Hydrogen gas is normally vented off the branch	import	2240	4.09	≮ 250
		exit	4580	4.09	≮ 250
2	Hydrogen backup discharge branch	import	2530	4.16	≮ 250
		exit	5120	4.16	≮ 250
3	Oxygen is normally vented to the branch	import	2337	17.1	≮ 125
		exit	4686	17.1	≮ 125
4	Oxygen backup discharge branch	import	2653	19.5	≮ 125
		exit	5338	19.5	≮ 125

From Table 3, the mass flow rates at the inlet and outlet of the hydrogen and oxygen emission branch meet the conservation conditions, which verifies the correctness of the calculation. Because the simulation working medium nitrogen has compressibility, its density changes with the change in pressure, thus the import and export volume flow is different due to the different pressure. Hydrogen is normal and the backup discharge branch flow meets the requirement of not less than 250 L/h indexes, and oxygen and normal backup discharge branch flow meet the requirement of not less than 125 L/h index, which indicates that the fluid circuit structure and φ4mm flow path can meet the requirements of the index.

3.2. Flow performance test

According to the national standard "universal valve flow coefficient and flow resistance coefficient test method"[10] and the relevant test requirements for integrated valve products to carry out a flow test, the hydrogen branch is taken as an example to carry out the flow test, as shown in FIG. 8.

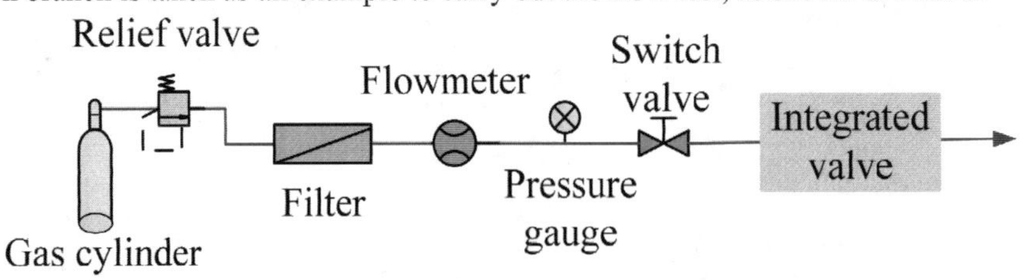

FIG. 8 Schematic diagram of flow test

The test system is connected according to Figure 8, and the hydrogen branch inlet to the pipeline is connected; the cut-off valve is closed, the pressure-reducing valve is opened, the pressure representation number is adjusted to 105kPa, and the flow meter is recorded. The other branches are tested by the same method, and the results are shown in Table 4.

Table 4 Data table of flow test results

The serial number	Branch	Test value/volume flow rate (L/h)		Simulation value/volume flow rate (L/h)	Percentage of deviation
1	Hydrogen gas is normally vented off the branch	import	2136	2240	4.64%
2	Hydrogen backup discharge branch	import	2448	2530	3.24%
3	Oxygen is normally vented to the branch	import	2250	2337	3.72%
4	Oxygen backup discharge branch	import	2567	2653	3.24%

Table 4 shows that the gas discharge control device's four branches under the pressure of 105 KPA and entrance flow can meet the requirements of indicators, including the normal discharge branch flow

Journal of Physics: Conference Series **2458** (2023) 012033 doi:10.1088/1742-6596/2458/1/012033

rate is less than the backup branch of hydrogen and oxygen normal discharge branch flow rate is less than the backup branch. This is because the backup discharge branch flow channel length is less than the normal branch flow channel length. The test and simulation values of the four branches flow are much higher than the index requirements, and the error between the test and simulation values is less than 5%, which reflects the rationality and superiority of the integrated embedded fluid loop design.

4. Conclusion

According to the functional requirements of the product, a modular integrated valve product is designed, and the relevant indicators are simulated and tested. The conclusions are as follows:

(1) Both normal and backup emission branches of hydrogen and oxygen are designed, which effectively improve the reliability and safety of the product;

(2) The modular design is adopted to realize the integrated and embedded design of the flow path, which improves the maintainability of the product;

(3) The simulation and experimental verification of the fluid circuit of the product show that the flow of four branches is much higher than the requirements of the index, and the error between the test value and the simulation value is less than 5%, which reflects the rationality and superiority of the integrated embedded fluid circuit structure design.

The design of integrated valve products has certain reference and guiding significance for the design of similar integrated products.

Reference

[1] Ye Pei-Jian, Peng Jing. Deep Space Exploration and Its Prospect in China[J]. Engineering Science, 2006, 10: 13-18.

[2] Deng Lian-yin, Guo Ji-feng, Cui Nai-gang. Progress and Prospects of Engineering for Lunar Bases[J]. MISSILEAND SPACE VECHILE, 2009, 2: 25-30.

[3] YU Deng-yun, GE Zhi-jiang, WANG Nai-dong, et al. Supposes for Structure Form of Lunar Base[J]. Journal of Astronautics, 2012, 12: 1840-1844.

[4] Ma Cong. The Design and Dynamics Analysis of the Orbital Replacement Unit Assembly[D]. Harbin Institute of Technology, 2014.

[5] HUANG Pan-feng, CHANG Hai -tao, LU Zhen-yu, et al. Key Techniques of On-Orbit Service-Oriented Reconfigurable Acellularized. Satellite and Its Prospects[J]. Journal of Astronautics, 2016, 37: 1-10.

[6] Li Yan, Dang Chang-ping. The Development of Orbital Servicing Technology in Space[J]. Ordnance Industry Automation, 2012, 31: 79-82.

[7] GB/T 30832-2014. Valves-Test method of flow coefficient and flow resistance coefficient[S].

Effects of structural elasticity on the flow-induced cavity tone in water

Wenwen Zhang

Institute of Vibration and Sound, Naval University of Engineering, Jiefang Boulevard 717, Wuhan 430033, China.

Corresponding author: zwwada@163.com

Abstract. Measurement of the flow-induced tones of cavities with three different structural elasticity was conducted in the water tunnel. The frequency and amplitude of the cavity tones varied with the water flow velocity are present and discussed. In addition, the modal jump phenomenon of flow-induced cavities for both elastic cavities and rigid cavities is compared and discussed. The influence law of structural elasticity on the characteristics of flow-induced tones in water cavities in the amplitude-frequency domain is analyzed.

1. Introduction

Flow-induced cavity tone has been an everlasting new problem in the area of aeroacoustics over the past 60 years since rich physical mechanisms were involved. Most previous studies of flow-induced cavity tones were focused on the engineering structures exposed to high-speed airflows, such as aircraft bomb bays [1], landing gear bays [2], and vehicle windows[3]. In recent years, with a focus on underwater noise, many studies have found that high-level tones exist in very low Mach cavity flow in water. Compared to the air cavity flow, the main difference in water cavity flow is that the influence of structural elastic vibration must be considered.

Flow-induced cavity tones are mainly excited from the fluctuating pressures generated by the instabilities of the shear layer in the cavity opening. The instability of the shear layer is initially produced by the separation of the boundary layer at the leading-edge of the cavity, and further enhanced by the interaction with the trailing-edge and the cavity feedback loop. The first well-known mechanism of flow-induced cavity tones is proposed by Rossiter[4], who used the acoustic feedback hypothesis successfully predict the frequencies of cavity tones in high Mach number flow. However, due to the very low Mach number in water cavity flow, many studies reported that acoustic feedback does not play an important role in the generation of cavity tones[5]-[7]. Rockwell and Naudascher[8] attributed the generation of cavity tones to three types of self-sustain oscillations: fluid-dynamic, fluid-resonant, and fluid-elastic. Of these, the fluid-elastic category was rarely discussed in cavity flow problems before, since the weak fluid-structure coupling in the air. Wang[9] conducted a wind tunnel experiment on the coupling of structural vibration and cavity tones and showed that the elastic vibration of cavity structures has little influence on cavity tones. However, since the compressibility of the structure is comparable to that of water, strong fluid-structure coupling exists in the water cavity flow and the effect of structural elastic vibration cannot be ignored. Burroughs[10] observed that the elastic vibration of structure may promote the generation of cavity resonance and significantly enhance the cavity tones.

However, the effect of structural elasticity on flow-induced cavity tone and its mechanism is still unknown. In this paper, the characteristics of water flow-induced cavity tones under different flow velocities and different structural elasticity are studied experimentally. The conclusions of this paper can provide beneficial technical support for the low-noise design of underwater cavity structures.

2. Experimental setup

The Shanghai Institute of Shipping and Transport Science's high-speed circulating water tunnel served as the test site for this experiment. The medium-sized circulating water tunnel with the best flow conditions in China is a scientific research facility made by the German company KEMPF & PEMMERS. Figure 1 displays the composition of the gadget. Special stainless steel is used to create the water cave. It has a pressure regulating box, a charging and degassing device, and an electronic flow velocity meter, all of which can precisely and steadily control the cave's water velocity, pressure, and gas content while maintaining the water's purity. The bottom of the water-circulating tunnel connects to a sizable water storage tank, and water from both the tank and the cave alternately circulates. Water in the cave is kept in the water storage tank when the water tunnel is not functioning. The water storage tank pumps water into the cave when the test is conducted in the water tunnel. The tunnel's working portion measures 2.6 meters in length and has a square cross-section with rounded corners. The water body in the tunnel can be circulated using a four-blade axial flow impeller, which can achieve a maximum flow rate of 12 m/s in the working section when powered by a 90-kW shunt motor. To maximize the stability of the flow state of the working section of the water tunnel and to ensure that the non-uniformity and instability of the working section's flow rate are less than 1%, the water tunnel's four corners are equipped with guide blades, and the front of the contraction section is equipped with a honeycomb flow stabilizer screen. To install the test model and witness the test phenomenon, two big plexiglass observation Windows of 0.35m × 1.0m, totaling 8 pieces, are installed on the upper, lower, left, and right sides of the working portion. The open cavity model is inserted in the observation window at the lower half of the working area for testing the fluid-induced oscillation properties of the open cavity in water.

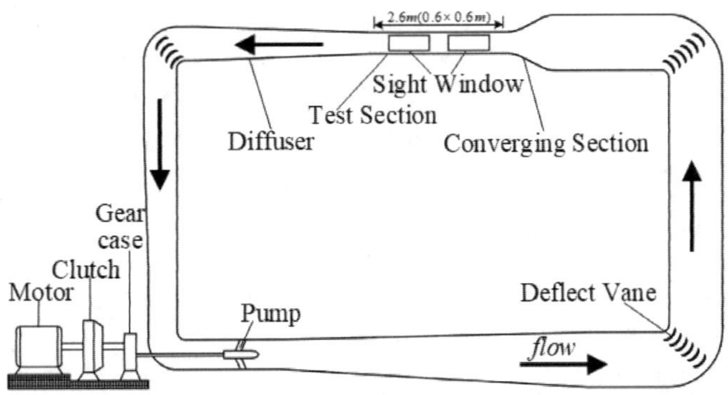

Figure 1. Configuration schematic of a water tunnel in the experiment

The experimental cavity model in this paper is a rectangular cavity with a small rectangular opening, as shown in Figure 2. The test cavity model is designed to be detachable, and the different components of the cavity are bolted together to facilitate adjustment of the structural elasticity of the cavity. The inside dimensions of the cavity model were fixed at L_x=320mm in length (streamwise), L_y=200mm in width (spanwise), and L_z=500mm in depth (cross-streamwise). The cavity opening sizes were a=90mm long and b=80mm wide, and the thickness of the open overhang plate is t=8mm. The four sidewalls and the overhanging plate of the cavity model were made of stainless steel and can be considered rigid structures. The adjustment of the structural elasticity of the cavity is accomplished by replacing the cavity bottom with different materials. The cavity model with stainless steel bottom, aluminum bottom, and plexiglass bottom were tested in the experiments of this paper, and the corresponding structural elasticity increases successively.

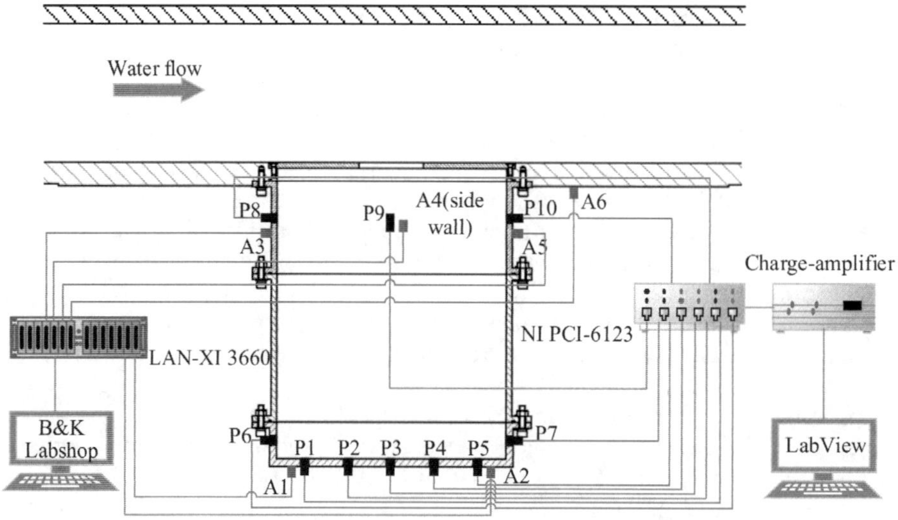

Figure 2. The installation of the cavity model in the test and its transducers arrangement

The flow-induced cavity tones in water can be used to measure the cavity's pure sound. The hydrophone, however, is unable to precisely detect the noise emitted by the cavity beneath the stimulation of the water flow because of the reverberation effect in the water tunnel. Instead, it is possible to determine the flow-induced cavity tones characteristics via pulsing pressure measurements. As shown in Figure 2, ten fluctuating pressure transducers, which were tagged by P, were used to measure the flow-induced cavity tones in this experiment. The 5mm by 15mm CYG505ALMF small dynamic diastolic sensor from Kunshan Shuangqiao is used in this test to measure the fluctuating pressure. The sensor's operating parameters are 0 to 500kPa, 0.5%FS or less sensitivity, 138kHz natural frequency, and a maximum adoptable frequency of 20kHz. The National Instruments PCI-6123 collects the pressure signal from the pulsing pressure sensor that is mounted on the cavity wall. After charge amplification, the computer terminal self-programs based on the LABVIEW platform to record and save the pulsing pressure time domain data. The position coordinates of the fluctuating pressure transducers are given in Table 1. The origin of the coordinates is the center of the opening sweeping surface, and the x-direction, y-direction, and z-direction represent the flow direction, spanwise direction, and the direction of opening towards the cavity bottom. Six accelerometers were also installed on the cavity model to measure the flow-induced structural vibrations in this experiment. The accelerometers are tagged by A in Figure 2. The signal sampling rate of fluctuating pressure transducers and accelerometer is set as 20.48kHz and 25.6kHz respectively in the experiments.

Table 1. Installation positions of fluctuating pressure transducers

Tags	P1	P2	P3	P4	P5	P6	P7	P8	P9	P10
x /mm	-120	-60	0	60	120	-160	160	-160	160	0
y /mm	0	0	0	0	0	0	0	0	0	100
z /mm	L_z	L_z	L_z	L_z	L_z	L_z-30	L_z-30	70	70	70

3. Measurement results

Water's fluid medium and structural materials share a similar characteristic impedance. Therefore, it is impossible to disregard the elastic vibration of the cavity structure caused by cavity pulsation pressure, including sound pressure. Cavity structural vibration affects the cavity's own Helmholtz resonance properties by creating volume fluctuation on the one hand, and the cavity shear layer through hydrodynamic waves on the other. As a result, the flow-induced cavity tones are influenced. The subsequent analyses in this section are based on the fluctuating pressure signal measured at the P10 measurement point on the rear wall of the cavity to eliminate the additional interference brought on by the vibration of the fluctuating pressure sensor at the bottom of the cavity.

Figure 3 shows the frequency distributions of the first two tones of three cavities with different structural elasticity under the water flow excitations. It can be seen that the structural elasticity has little effect on the tone frequencies of the cavity at a low flow velocity, and the tone frequencies of the cavity with three different structural elasticity increases linearly with the flow velocity at approximately the same slope. Since the main generating mechanism of cavity tone is self-sustaining oscillation at low flow velocity, the results indicate that the structural elasticity does not affect the frequency of self-sustaining oscillations in the shear layer. With the increase of flow velocity, when $U_\infty \geq 7.5$m /s, the tone frequency of the cavity model with plexiglass bottom gradually deviates downward from the constant *St* number of self-sustaining oscillation. That is, the shear tone gradually begins to couple the resonance mode of the cavity, while the tone frequency of the cavity at the bottom of the duralumin and stainless-steel cavity always maintains the self-sustaining oscillation state. This indicates that the influence of structural elasticity on the flow-induced cavity tone in water is mainly reflected in the resonance mode. In addition, it is noted that there is no obvious difference in tone frequency between the aluminum cavity bottom and the stainless-steel cavity bottom test models. This is due to the small scale of the cavity model in this experiment, and the natural frequency of the aluminum cavity bottom and stainless-steel cavity bottom is much larger than the acoustic modal frequency of the cavity, which is equivalent to the rigid structure compared with the experimental cavity model.

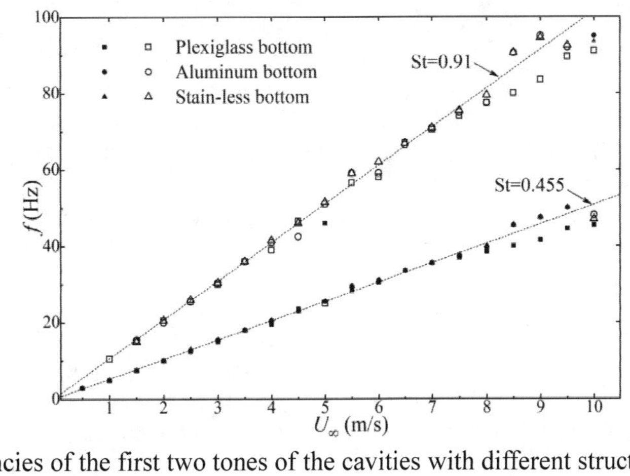

Figure 3. Frequencies of the first two tones of the cavities with different structural elasticity as a function of velocities. (Solid symbols: dominant tones; hollow symbols: secondary tones)

Figure 4 shows the amplitude of the dominant tone varied with the flow velocity of the cavity models with different structural elasticities. It can be seen that the tone amplitude is more sensitive to structural elasticity than the tone frequency. At U_∞=6m/s, the dominant tone of the cavity with the plexiglass cavity bottom exceeded that of the stainless-steel and aluminum cavity by about 5dB, while the tone frequency of the three cavity models with different elasticities did not change significantly. Compared to cavity flow studies in air medium that the elastic vibration of cavity structure does not enhance the amplitude of flow-induced tones, the experimental results in this paper further indicate that there is a strong coupling between the elastic and flow-induced tone of cavity structure in water. In addition, when U_∞= 7.5m /s, the amplitude of the dominant tone of the cavity with the plexiglass cavity bottom is about 10dB higher than that of other cavities, and the difference further expands with the increase of flow velocity. It is noted that when U_∞= 7.5m /s, the tone frequency of the cavity with the plexiglass bottom starts to deviate from the *St* number of self-sustaining oscillation. This indicates that the structural elastic effect makes the flow-induced resonance easier to occur in the water cavity and significantly increases the amplitudes of flow-induced tones.

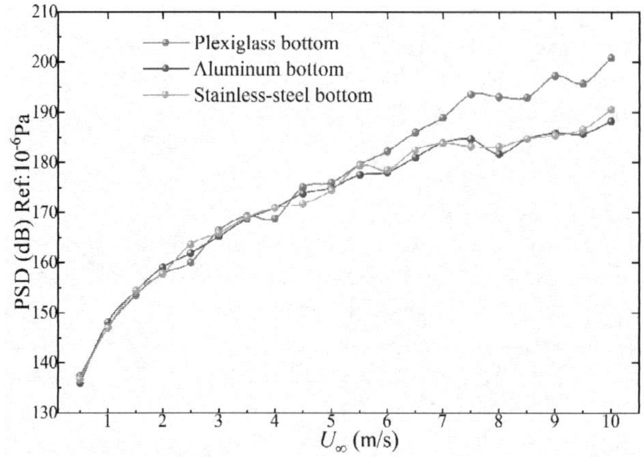

Figure 4. The amplitude of dominant flow-induced tones of aperture cavities with different structural elasticity as a function of velocities.

As can be seen from Figure 3, within the range of experimental flow velocity, the cavity models with different structural elasticity are dominated by the tone of order I, and the tone of order II is the secondary pure tone. However, the dominant mode transfer will occur under partial flow velocities, which is referred to as the "modal jump" of flow-induced cavity tone in this paper. A similar jump occurs when U_∞=10m/s for the cavities with aluminum bottom and stainless-steel bottom. The modal jump phenomenon is one of the basic characteristics of flow-induced cavity tones, which was previously discussed by Rossiter[4], Burroughs[10], and many other scholars in the study of water and air cavity flow respectively. Here, the pressure spectrum of the cavity near the flow velocity point of modal jump occurrence was further extracted to analyze the effect of structural elasticity on the modal jump of flow-induced cavity tones in water, as shown in Figure 5. Since the flow-induced tone of the cavity models with aluminum bottom and stainless-steel bottom are the same under this test condition, the rigid cavity wall in Figure 5 is represented by the cavity model with the stainless-steel bottom. The elastic cavity wall is represented by the cavity model with a plexiglass cavity bottom.

From Figure 5(a), it can be seen that the side lobe peak of 42.5Hz begins to appear on the right side of the second-order pure tone when U_∞=4m/s. The sidelobe peak is the non-resonance in the flow-induced oscillation shear mode of the open cavity. Then, when U_∞=5m/s, it further jumps to the dominant pure tone of 46Hz. After the modal step, the dominant pure tone and the secondary pure tone in the spectrum do not form a frequency-doubling relationship. However, there is also a side lobe peak with a similar frequency near the second-order pure tone in the rigid wall cavity, but its amplitude is always small. It can be seen from the subsequent analysis that the non-resonant peak is an oscillation mode mainly influenced by the feedback of the fluid in the cavity, while the resonant peak in the spectrum is an oscillation mode mainly influenced by the feedback of the pressure wave at the cavity mouth, which indicates that the step phenomenon in the open cavity with the elastic wall is caused by the amplification of the feedback of the fluid in the cavity. As can be seen from Figure 5(b), when U_∞=8.5m/s, a side lobe peak of 84Hz appears on the left side of the second-order pure tone in the rigid wall open-hole cavity. However, as the flow rate increases thereafter, the amplitude of its dominant pure tone (first-order) and its frequency doubling does not change significantly. When the flow velocity increases from U_∞=8.5m/s to U_∞=9.5m/s, the sidelobe peak of the original second-order pure tone increases from 172dB to 182dB and becomes a new secondary pure tone. When the flow velocity is U_∞=10m/s, the sidelobe peak further jumps to 93Hz dominant pure tone, and the amplitude increases to 190dB. It is noted that after the modal step occurs in the rigid wall cavity, the I-order pure tone reforms the half frequency of the dominant pure tone, which is lower than the frequency before the modal step. This phenomenon is similar to the dominant mode migration of the shear layer self-sustained oscillation in the air cavity flow, indicating that the modal step occurs in the rigid wall cavity at a higher flow rate due to the changes of the vortex structure in the shear layer.

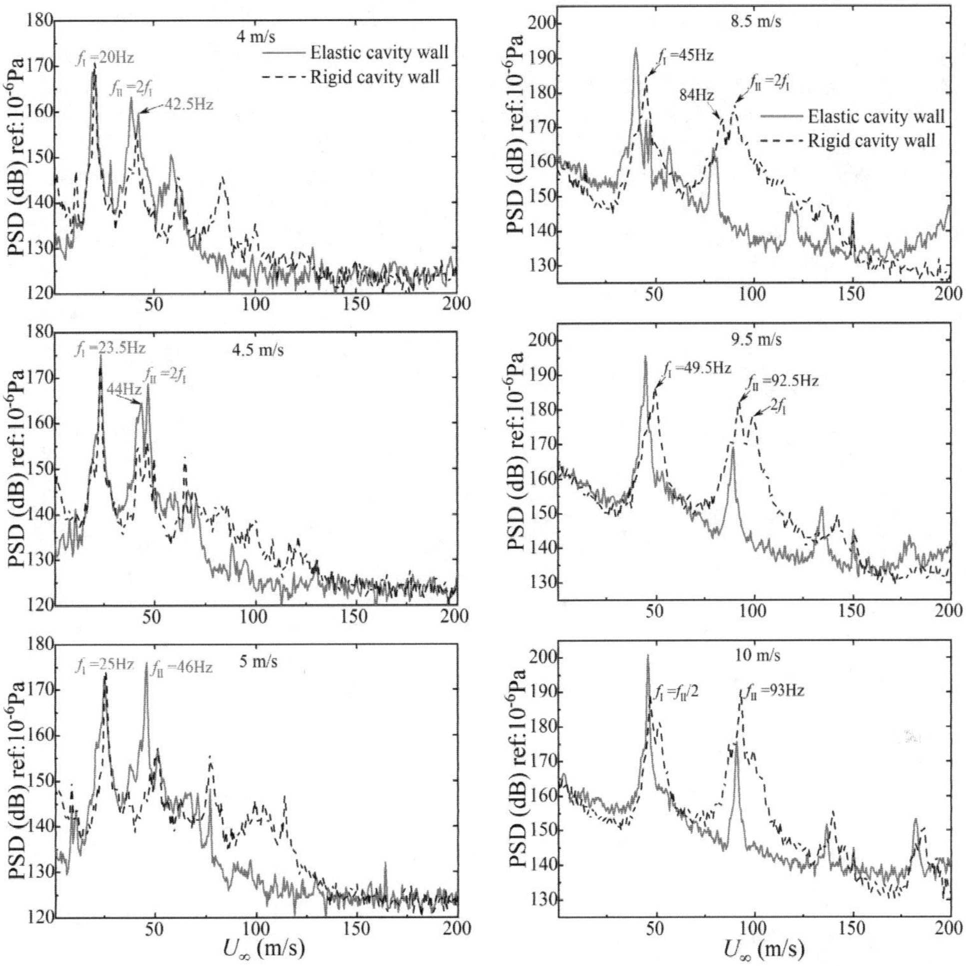

(a) Modal jump spectrum of the cavity with an elastic wall (b) Modal jump spectrum of the cavity with rigid wall

Figure 5. Spectral response of pressure fluctuations inside the cavity in the water near the modal jump point of flow-induced tones

4. Conclusion

In this paper, the flow excited pure tone characteristic test of the cavity in water is designed. The fluctuating pressure transducers were used to measure the fluctuating pressure of the cavity in water under different flow velocities and different structural elasticity. The influence law of structural elasticity on the characteristics of flow-induced tones in water cavities in the amplitude-frequency domain is analyzed. The main influence of structural elasticity on the flow-induced tones of the cavity in water is reflected in the reduction of the natural frequency of the cavity structure in water, to promote the flow-induced cavity resonance of the cavity in water. The structural elasticity does not affect the self-sustaining oscillation frequency of the shear layer but increases the amplitude of the self-sustaining oscillation correspondingly when approaching the resonance mode.

References

[1] Bacci D, Saddington A, and Bray D 2018 *Aerosp. Sci. Technol.* **77** 320-31
[2] Liang Y, Chen Y, Zhao K, Sun J, Lu X and Zhao Y 2019 *Acta Aeronaut. Astronaut. Sin.* **40** 22932 (in Chinese)
[3] Gu Z, Wang N, Wang Y, Zhang Y and Liu L 2014 *J. Vib. Eng.* **27**(3) 409-15 (in Chinese)
[4] Rossiter J E 1964 *Aeronautical Research Council* Technical Report No. R&M 3438
[5] Basley J, Pastur L R, Lusseyran F, Faure T M and Delprat N 2011 *Exp. Fluids* **50**(4) 905-918.
[6] Yamouni S, Sipp D, Jacquin L 2013 *J. Fluid Mech.* **717** 134-65
[7] Tuerke F, Sciamarella D, Pastur L R, Lusseyran F and Artana G 2015 *Phys. Rev. E* **91** 013005
[8] Rockwell D and Naudascher E 1978 *J. Fluids Eng.* **100** 152–65
[9] Wang X, Yang D, Liu J, Shi A, Zhou F and Lyu B 2017 *Acta Aeronaut. Astronaut. Sin.* **38**(7) 120873
[10] Burroughs C and Stinebring D 1994 *J. Acoust. Soc. Am.* **95**(3) 1256-63

A Combined Passive Automatic Detection Method of Constant False Alarms and Multiple Nodes

Di Huang*, Xian Zhu, Yi Gong, Fuhu Chen

Hangzhou Applied Acoustics Research Institute

798970918@qq.com

Abstract. Aiming at the problem of whether multi-array passive joint detection can improve the detection performance of sonar, this paper analyzes in detail the relationship between the detection probability, false alarm probability, the input signal-to-noise ratio, and integration time under the condition of single array detection, and then obtains the input signal-to-noise ratio and integration time required to achieve the same detection probability and false alarm probability under different joint detection rules. It is concluded that the multi-array joint detection can reduce the required input signal-to-noise ratio and integration time and provide detection performance, which is verified by simulation experiments.

1. Introduction

With the wide application of unmanned platforms such as UUVs, buoys, and submarines, underwater acoustic detection methods are gradually developing in the direction of automation, intelligence, and clustering. A single node is limited by the installation space, and its detection capability is limited, especially with the reduction of the radiation noise of the detection object and the deterioration of the underwater acoustic environment. The detection of weak targets is becoming more and more difficult[1]. In theory, multi-node (multi-array) joint detection can break through the limitation of single-node (single-array) array space, and become one of the important means to improve the comprehensive detection performance of underwater sound. At present, the means of multi-array joint detection are mainly divided into three categories: array-level joint detection, beam-level joint detection, and contact-level joint detection. Array-level fusion requires that the sampling clocks of each array are strictly synchronized, the phase and amplitude of the hydrophones are highly consistent, and the operating frequency and frequency resolution are exactly the same. Beam-level fusion includes coherent fusion and incoherent fusion. Coherent fusion and array-level fusion are equivalent and have the same requirements. Beam-level incoherent fusion requires that the pre-beam, main lobe width and matrix coordinate system of each array are basically the same. Otherwise, beam superposition may produce false targets, beam distortion, and other problems[2]-[4]. Contact-level fusion performs joint detection on the integrated data of each array. It has low requirements for multi-array sampling clock synchronization, hydrophone phase and amplitude consistency, array position, and formation. It also has the least amount of data interaction. It is most suitable for a multi-node joint probe. How to use multi-array contact-level joint detection to obtain better performance than single-array has always been a difficult research point in the field of underwater acoustics.

This paper analyzes the relationship between the input signal-to-noise ratio, integration time, detection probability, and false alarm probability in the case of single-array detection and multi-array detection. And for the joint detection of multiple nodes with the same detection performance, a joint passive constant false alarm multi-node is proposed. The automatic detection method takes the minimum product of input signal-to-noise ratio and integration time as the objective function and obtains the joint detection strategy and single-array detection threshold through an optimization solution to reduce the input signal-to-noise ratio and integration time required for single-array detection. The purpose is to improve the sonar action distance and response speed and provide a technical reference for the automatic detection of multi-node or multi-array unmanned platforms such as UUVs, submersibles, buoys, and manned platforms.

2. Theoretical analysis

2.1. The relationship between signal-to-noise ratio, integration time, detection probability, and false alarm probability in the case of single-array detection

In sonar passive target detection, both target radiation noise and background noise can be approximated by a zero-mean Gaussian random process. The variance of the target radiation noise is set. The variance of the σ_s^2 background noise is σ_n^2, the power spectral density is $\frac{N_0}{2}$, and the target radiation noise and the background noise are independent of each other. The best receivers for detecting Gaussian signals in Gaussian noise[5]-[8] are as follows.

$$\int_0^T x^2(t)dt \underset{H_2}{\overset{H_1}{\gtrless}} \frac{N_0}{2}\gamma \tag{1}$$

$x(t)$ is the received signal of the sonar (it can be the array element signal or the beam domain signal after beamforming). $s(t)$ is the target radiation noise, and $n(t)$ is the background noise. Its discrete form is given below.

$$\sum_{n=1}^{N} x^2(n) \underset{H_2}{\overset{H_1}{\gtrless}} \sigma_n^2\gamma \tag{2}$$

$$T(x) = \sum_{n=1}^{N} x^2(n)$$

$$\begin{cases} \dfrac{T(x)}{\sigma_s^2+\sigma_n^2} \sim \chi_N^2, \text{ in } \mathrm{H}_1 \ condition \\ \dfrac{T(x)}{\sigma_n^2} \sim \chi_N^2, \text{ in } \mathrm{H}_0 \ condition \end{cases} \tag{3}$$

The detection probability and false alarm probability are as follows[9]-[14].

$$\begin{cases} P_D = Q_{\chi_N^2}(\dfrac{\eta}{\sigma_s^2+\sigma_n^2}) \\ P_F = Q_{\chi_N^2}(\dfrac{\eta}{\sigma_n^2}) \end{cases} \tag{4}$$

in

$$Q_{\chi_N^2}(x)=\begin{cases} 2(1-\Phi(\sqrt{x})), N=1 \\ 2(1-\Phi(\sqrt{x}))+\dfrac{\exp(-\frac{1}{2}x)}{\sqrt{\pi}}\sum_{k=1}^{\frac{N-1}{2}}\dfrac{(k-1)!(2x)^{k-\frac{1}{2}}}{(2k-1)!}, N>1 \text{ \& N is odd} \\ \exp(-\frac{1}{2}x)\sum_{k=0}^{\frac{N}{2}-1}\dfrac{(\frac{x}{2})^k}{k!}, N \text{ is even} \end{cases} \quad (5)$$

$$\Phi(x)=\int_{-\infty}^{x}\frac{1}{\sqrt{2\pi}}e^{-\frac{t^2}{2}}dt$$

The functional relationship between detection probability P_D and false alarm probability P_F can be expressed as follows.

$$P_D=1-\Phi\left(\frac{\Phi^{-1}(1-P_F)-\sqrt{\frac{N}{2}\frac{\sigma_s^2}{\sigma_n^2}}}{1+\frac{\sigma_s^2}{\sigma_n^2}}\right) \quad (6)$$

$\frac{\sigma_s^2}{\sigma_n^2}$ is the input signal-to-noise ratio and N is the number of integration points. For passive target detection, usually $\frac{\sigma_s^2}{\sigma_n^2}\ll 1$, the detection probability at this time can be approximately expressed as follows[15]-[17].

$$P_D\approx 1-\Phi\left(\Phi^{-1}(1-P_F)-\sqrt{\frac{N}{2}\frac{\sigma_s^2}{\sigma_n^2}}\right) \quad (7)$$

Its graphical representation is the receiver operating characteristic curve (ROC curve).

Therefore, the detection probability and false alarm probability (or detection threshold) determine the signal-to-noise ratio and integration time required by the receiver, which in turn determine the passive detection distance and response time of the sonar to the target.

2.2. The relationship between signal-to-noise ratio, integration time, detection probability, and false alarm probability in the case of joint detection of multiple nodes with the same detection capability

There are sonars with the same K detection ability (same observation range and action distance) for joint passive detection[18]-[21]. The detection probability of each sonar is P_d, and the false alarm probability is P_f. The detection rule of multi-array joint passive detection is that if at least M part of the sonar detects a target, then there is a target. The detection probability of joint detection is P_D, and the false alarm probability is P_F, which can be expressed as follows.

$$P_D=\sum_{m=0}^{M}C_K^m P_d^m(1-P_d)^{K-m} \quad (8)$$

$$P_F = \sum_{m=0}^{M} C_K^m P_f^m (1-P_f)^{K-m} \tag{9}$$

The specified detection probability P_D and false alarm probability P_F are obtained. For single-array detection, the product factor of the required integration time and the input signal-to-noise ratio is as follows.

$$\chi_1 = \sqrt{\frac{N}{2}\frac{\sigma_s^2}{\sigma_n^2}} \approx \Phi^{-1}(1-P_F) - \Phi^{-1}(1-P_D) \tag{10}$$

The detection probability P_d and false alarm probability P_f of every single matrix can be obtained by solving Equations (8) and (9). In joint detection, the multiplication factor of the integration time required for every single array and the input signal-to-noise ratio is as follows.

$$\chi_2 = \sqrt{\frac{N}{2}\frac{\sigma_s^2}{\sigma_n^2}} \approx \Phi^{-1}(1-P_f) - \Phi^{-1}(1-P_d) \tag{11}$$

Therefore, in the process of joint detection, the appropriate decision rule and single-array detection threshold (or detection probability P_d, false alarm probability) P_f can be obtained by minimizing the objective function $\frac{\chi_2}{\chi_1}$ to reduce the required input signal-to-noise ratio, increase the operating distance, or reduce the required input signal-to-noise ratio. The required integration time increases the sonar response speed.

$$\frac{\chi_2}{\chi_1} = \frac{\Phi^{-1}(1-P_f) - \Phi^{-1}(1-P_d)}{\Phi^{-1}(1-P_F) - \Phi^{-1}(1-P_D)} \tag{12}$$

It can be achieved by optimization solutions[22]-[26].

3. Algorithm Description

The algorithm is as follows.

(1) The detection probability P_D and false alarm probability to be achieved P_F are outputted, and the number of K available arrays is inputted;

(2) The M values start from zero to 0 K, and the M Newton iteration method is used to calculate the detection probability P_d and false alarm probability P_f of the single-part sonar that satisfy Equations (8) and (9) respectively;

(3) The calculation is as follows.

$$\Psi_M = \Phi^{-1}(1-P_f) - \Phi^{-1}(1-P_d), M \in [0, K]$$

(4) The formula is taken as $S = \{s \mid \Psi_n = \min \Psi_M, M \in [0, K]\}$.

(5) P_f^S is the corresponding $M = S$ false alarm probability P_f, which is used to calculate the detection threshold $\frac{\eta}{\sigma_n^2} = Q_{\chi_N^2}^{-1}(P_f^S)$.

(6) The optimal detection method is obtained by calculating the multi-array $T(x)=\frac{\sum_{n=1}^{N} x^2(n)}{std(x^2(n))}$.
K at least S nodes in each $T(x) < Q^{-1}_{\chi^2_N}(P^S_f)$ node are judged to have no target. Otherwise, it is judged that there is a target.

4. Simulation results and analysis

In order to verify the effect of the algorithm, three simulation experiments are designed in this paper, namely the single-array detection experiment, the two-array joint detection experiment, and the triple-array joint detection experiment. In the single-array detection test, the ROC curve obtained by the Monte Carlo test is compared with the ROC curve obtained by the theoretical calculation to verify the correctness of the test. The two-array joint detection and the three-array joint detection test the effect of reducing the signal-to-noise ratio and shortening the integration time respectively.

It is assumed that the detection probability to be obtained is 0.9, the false alarm probability is 10-3, and the number of data points of the original integral is 512 points. It can be seen from Table 1 that the combined detection of two arrays can improve the detection capability by about 0.87 dB compared with the single-array detection. Or the integration time can be shortened to 2/3 of the single-array detection. It can be seen from Table 2 that the combined detection of three arrays is about 1.3% higher than that of single array detection. The integration time is shortened to about 1/2 of the single-array detection, and the detection capability of "at least two arrays in three arrays being determined to detect the target" are improved by about 1.8 dB compared with the single-array detection. And the integration time is shortened to the single-array detection of about 0.4.

Table 1 Effect analysis of two array joint detection

Joint detection rules	Detection probability	False alarm probability	Multiplier	Detection capability improvement /dB	Integration time ratio
Single Array Independent Detection	0.9	10^{-3}	4.7524	0	1
Two bursts at least once found	0.6838	5×10^{-4}	3.8848	0.8754	0.6682
Two bursts at least two bursts found	0.9487	0.0316	3.8864	0.874	0.6688

Table 2 Effect analysis of three array joint detection

Joint detection rules	Detection probability	False alarm probability	Multiplier	Detection capability improvement /dB	Integration time ratio
Single Array Independent Detection	0.9	10^{-3}	4.7524	0	1
Three bursts at least once found	0.5358	3.33×10^{-4}	3.5120	1.3130	0.5461
Three and at least two finds	0.8042	0.0184	3.1120	1.8388	0.4288
Three rounds at least three rounds found	0.9655	0.1 _	3.4976	1.3319	0.5416

4.1. Single-array detection test

The background noise is 0 mean Gaussian white noise. The variance is 1, and the target signal is also 0 Gaussian white noise. The signal-to-noise ratio SNR is -5.272 dB. The receiver is $\frac{1}{N}\sum_{n=1}^{N} x^2(n)$ 512 integration points. The detection threshold is 0 to 2, and the interval is a sequence of 0.005. The Monte Carlo test is used to count the times that the receiver is greater than different thresholds when there is a signal or no signal. And then the detection probability and false alarm probability corresponding to different thresholds are calculated. The relationship between the detection probability and the false alarm probability is drawn (ie ROC) into a curve. The ROC curve obtained by the experiment is compared with the ROC curve calculated by Equation (6). Here, the number of Monte Carlo trials is 800,000.

The figure shows that the ROC curve obtained by the experiment basically coincides with the ROC curve obtained by the theoretical calculation. The pair of false alarm probability and detection probability is (7.1842e-4, 0.8979). The corresponding detection threshold is 1.2.

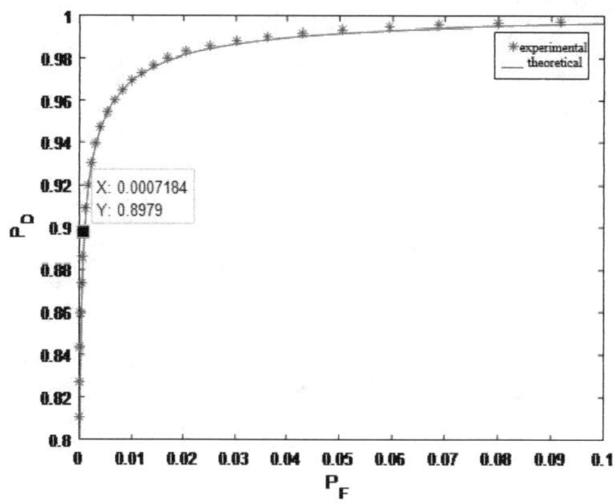

Figure 1 Comparison between experimental ROC curve and theoretical ROC curve

4.2. Two-array joint detection test

The background noise of the two receiving arrays is 0 mean Gaussian white noise. And the variance is 1. The target signal is also 0 mean white Gaussian noise. The joint detection rule is that if at least one of the two arrays detects a target, it is determined that the target is detected. The detection threshold is 0 to 2, and the interval is a sequence of 0.005. The Monte Carlo test is used to count the times that the receiver is greater than different thresholds when there is a signal or no signal. And then the detection probability and false alarm probability corresponding to different thresholds are calculated, and the relationship between the detection probability and the false alarm probability is drawn (ie ROC) into a curve. The ROC curve obtained by the experiment is compared with the single-array detection ROC curve calculated by Equation (6) with the signal-to-noise ratio of -5.272 dB. Here, the number of Monte Carlo trials is 800,000.

Figure 2 shows the comparison between the combined detection ROC curve and the single-array detection theoretical ROC curve with an input signal- SNR to-noise ratio of -6.1474 dB and 512 integration points.

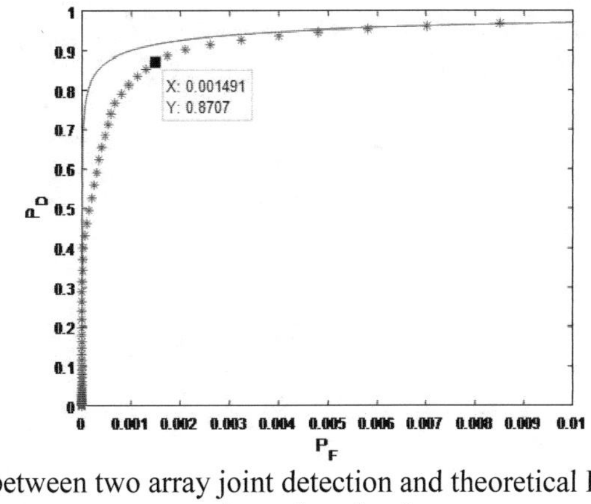

Figure 2 Comparison between two array joint detection and theoretical ROC curve after signal-to-noise ratio reduction

It can be seen from the figure that the ROC curves obtained by joint detection and single-array detection are very close, and the single-array input signal-to-noise ratio required for joint detection is 0.8754 dB lower than that of single-array detection. The pair of false alarm probability and detection probability is (0.001491, 0.8707), and the corresponding detection threshold is 1.205.

Figure 3 shows the comparison between the ROC curve of the joint detection with the input signal-to-noise ratio of -5.272 dB and the integration points of 343 points and the theoretical ROC curve of the single-array detection.

It can be seen from the figure that ROC curves obtained by the joint detection and the single-array detection are very close. And the integration time of the joint detection is 0.6682 of the integration time of the single-array detection. The pair of false alarm probability and detection probability is (0.001491, 0.8707), and the corresponding detection threshold is 1.245.

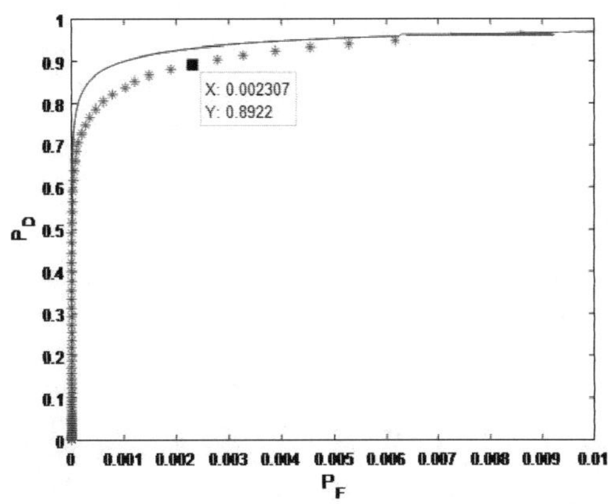

Figure 3 Comparison of two array joint detection and theoretical ROC curve after shortening integration time

Figure 4 shows that to achieve the same detection probability and false alarm probability, the detection capability of the two-array joint detection is improved compared to the single-array detection (ie, the input signal-to-noise ratio required for single-array detection minus the input signal-to-noise ratio required for joint detection). Among them, the input signal- *SNR* to-noise ratio is -10 dB, and the number of integration points is 512 points.

It can be seen from the figure that the performance of the two-array joint detection is improved by about 0.2 to 1.5 dB compared to the single-array detection. It is less than the gain brought by incoherent fusion (5lg2=1.5 dB). The reason is that the incoherent fusion requires each array. The orientation of the detection target and the processing frequency is completely coincidental. But the joint detection does not require this, resulting in performance loss.

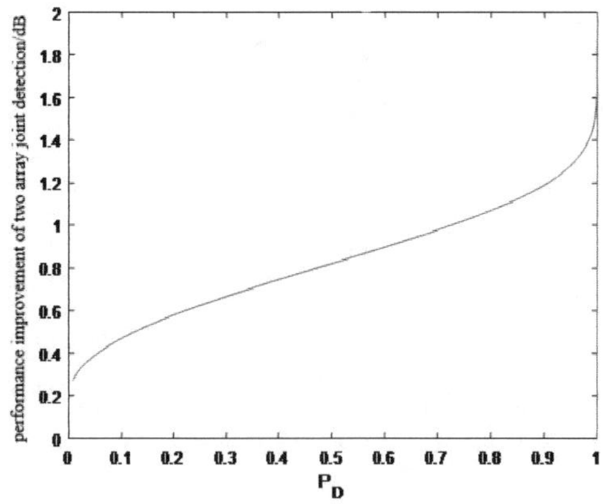

Figure 4. Relationship between performance improvement of two array joint detection and detection probability

4.3. Three-array joint detection test

The background noise of the three receiving arrays is all Gaussian white noise with 0 means, and the variance is 1. The target signal is also 0 mean white Gaussian noise. The joint detection rule is that if at least one of the three arrays detects a target. It is determined that the target is detected. The detection threshold is 0 to 2, and the interval is a sequence of 0.005. The Monte Carlo test is used to count the times that the receiver is greater than different thresholds when there is a signal or no signal. And then the detection probability and false alarm probability corresponding to different thresholds are calculated. And the relationship between the detection probability and the false alarm probability is drawn (ie ROC) into a curve. The ROC curve obtained by the experiment is compared with the single-array detection ROC curve calculated by Equation (6) with the signal-to-noise ratio of -5.272 dB. Here, the number of Monte Carlo trials is 800,000.

Figure 5 shows the comparison between the combined detection ROC curve and the single-array detection theoretical ROC curve with an input signal- *SNR* to-noise ratio of -6.585 dB and 512 integration points.

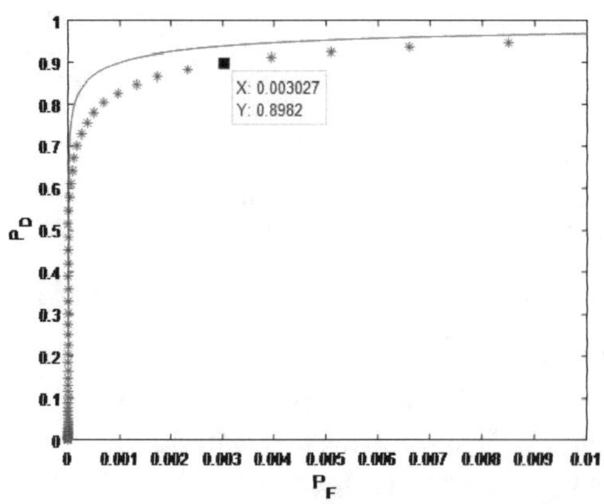

Figure 5 Comparison of three array joint detection and theoretical ROC curve after signal-to-noise ratio reduction

It can be seen from the figure that the ROC curve obtained by joint detection and single-array detection is very close. And the single-array input signal-to-noise ratio required for joint detection is 1.313 dB lower than that of single-array detection. The pair of false alarm probability and detection probability is (0.003027, 0.8982), and the corresponding detection threshold is 1.1950.

Figure 6 shows the comparison of the ROC curve of the joint detection with the input signal-to-noise ratio of -5.272 dB, the integration points of 280 points, and the theoretical ROC curve of the single-array detection. It can be seen from the figure that the ROC curve obtained by the joint detection and the single-array detection is very close. And the integration time of the joint detection is 0.5461 of the integration time of the single-array detection of 2650. The false alarm probability and detection probability pair are (0.003898, 0.8950). The corresponding detection threshold is 1.2650.

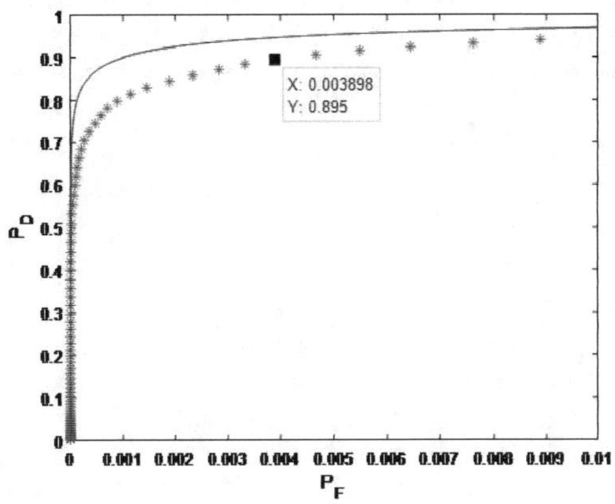

Figure 6 Comparison of three array joint detection and theoretical ROC curve after shortening integration time

Figure 7 shows that the three-array joint detection has improved its detection capability compared to the single-array detection (ie, the input signal-to-noise ratio required for single-array detection minus the input signal-to-noise ratio required for joint detection) to achieve the same detection probability and false alarm probability. Among them, the input signal-*SNR* to-noise ratio is -10 dB, and the number of integration points is 512 points.

It can be seen from the figure that compared with the single-array detection, the performance of the two-array joint detection is improved by about 0.2 to 2 dB, which is less than the gain brought by the incoherent fusion (5lg3 = 2.38 dB). The reason is that the non-coherent fusion requires the targets to be detected by each array. The orientation and processing frequency are completely coincidental. Joint detection does not require this, resulting in performance loss.

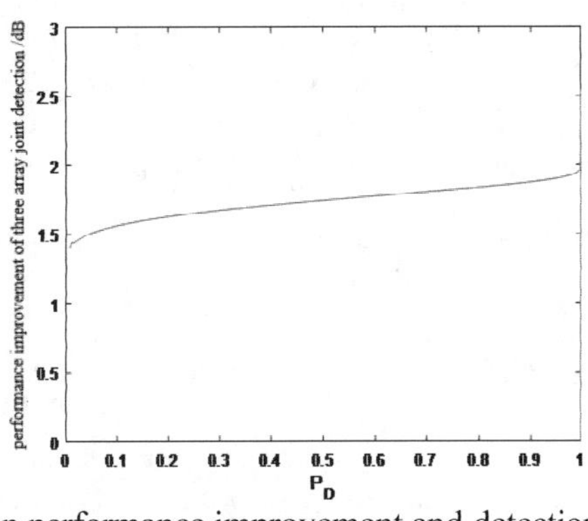

Figure 7 Relationship between performance improvement and detection probability of three array joint detection

From the above three simulation experiments under the condition that the integration time remains unchanged, the multi-array joint detection can reduce the input signal-to-noise ratio required by the single array, thereby improving the detection capability. But the improvement of the detection capability is smaller than that of the incoherent fusion band. The performance is improved by 5lgM (M is the number of arrays) under the condition that the input signal-to-noise ratio remains unchanged. The multi-array joint detection can shorten the integration time and improve the response speed.

5. Conclusion

The functional relationship between input signal-to-noise ratio, integration time, detection probability, and false alarm probability in single-array detection is analyzed in detail. For the joint detection of multiple nodes with the same detection performance, a combined passive automatic detection method of constant false alarm and the multi-node is proposed. The method takes the minimum product of the input signal-to-noise ratio and the integration time as the objective function and obtains the joint detection strategy and the single-array detection threshold through the optimization solution. Through the simulation experiment, it is found that the method can reduce the input signal-to-noise ratio required by the single array under the condition that the integration time remains unchanged, thereby improving the detection ability. But the improvement of the detection ability is less than the performance improvement brought by incoherent fusion by 5lgM (M is the number of arrays). Under the condition that the input signal-to-noise ratio remains unchanged, the integration time can be shortened, thereby increasing the reaction speed. This research provides a technical reference for multi-node or multi-array automatic detection of UUV, submersible, buoy, and other unmanned platforms and manned platforms.

References

[1] Yang Lujing, Geng Boying, et al. Handbook of Multi-sensor Data Fusion [M]. Electronic Industry Press, 2008.

[2] SUN Z, LI XL. YI W, et a1. Detection of weak maneuvering target based on keystone transform and matched filtering progress [J]. Signal Processing, 2017, 140(12): 127-138.

[3] HUANG P H, XIA X G, LIAO G S, et al. Long-time coherent integration algorithm for radar maneuvering weak target with acceleration rate [J]. IEEE Trans on Geoscience and Remote Sensing, 2019, 57(6): 3j28-3542.

[4] JIN K, LI G Q, LAI T, et a1. A novel long-time coherent integration algorithm In for Doppel-ambiguous radar maneuvering target detection[J]. IEEE Sensors Journal, 2020, 20(16): 9394-9407.

[5] Liu Hongliang, Chen Chao, Yue Kai. Multi-frame joint detection method based on tracking information [J]. Systems Engineering and Electronic Technology, 2021, 43(8).

[6] Lian Feng, Ma Dongdong, Han Chongzhao. Extending the error bound of joint detection and estimation of target [J]. Journal of Xi'an Jiaotong University, 2014, 48(4).

[7] Lian Feng, Ma Dongdong, Han Chongzhao. Joint detection and estimation error bound of multiple indistinguishable target groups [J]. Journal of Xi'an Jiaotong University, 2015, 49(11).

[8] Shi Shaoying, Du Pengfei, Zhang Jing. Multi-target joint detection, tracking, and classification method based on radar measurement [J]. Journal of Radio Wave Science, 2016, 31(1).

[9] Xu Yang, Xu Hui, Luo Shaohua. Joint detection and tracking algorithm of multi-sensor target based on random finite set theory [J]. Journal of National University of Defense Technology, 2013, 35(1).

[10] Xiao Aiming, Chen Zhaohui, Zhou Guohui. Target detection probability of multi-ship cooperative reconnaissance [J]. Electronic Information Countermeasure Technology, 2012, 27(1).

[11] Wu Guangxin, Liu Peng, Jiang Min, Shen Xueyong. Efficiency analysis of multi-radar cooperative guidance to detect stealth targets [J]. Modern Radar, 2016, 38(6).
[12] Qi Wei, Li Xia, Cai Wanyong. Collaborative detection capability and route planning of multiple early warning aircraft [J]. Systems Engineering and Electronic Technology, 2016, 38(12)
[13] THOMAS GH. Sensors cueing performance analysis [C] //IEEE National Aerospace & Electronics Conference. DaytonOH: IEEE Press, 1984: 262-265.
[14] Wang Guohong, Mao Shiyi. ESM to 2D radar guidance performance analysis [J]. Journal of Aeronautics and Astronautics, 2002, 23(4): 298-301.
[15] ESM guiding a 2D radar [J]. ActaAeronautica Et Astronautica Sinica, 2002, 23(4): 298-301.
[16] Wang Guohong, He You, Mao Shiyi. IRST to 3D radar guidance performance analysis [J]. Chinese Journal of Electronics, 2002, 30(12): 1737-1740.
[17] Wang Jianming, Liu Guochao. Research on the cooperative detection method of shipborne radar and ESM [J]. Ship Electronic Countermeasures, 2007, 30(6): 11-15.
[18] Peng Ruihui, Wang Shuzong, Lu Yongsheng, etc. Guidance analysis of ESM for 2D radar in off-site configuration [J]. Modern Radar, 2009, 31(1): 13-16.
[19] Lu Yongsheng, Wang Shuzong, Peng Ruihui, etc. Guidance analysis of IR ST for 3D radar in off-site configuration [J]. Infrared and Laser Engineering, 2008, 37(5): 911-915.
[20] Fan Hao, Huang Shucai, Li Yajun. 3D inter-radar guidance prompt probability analysis and simulation [J]. Modern Radar, 2012, 34(4): 9-15.
[21] BATON DK. Radar system analysis and modeling [M]. Norwood, MA: Artech House Inc., 2004.
[22] Dai Jun. Discussion on the feasibility of radar networking and anti-stealth technology [J]. Modern Radar, 1998, 20(3): 8-13
[23] Li Nengjing. Radar detectability of stealth aircraft [J].Modern Radar, 1992, 14(3): 17-23
[24] Zhang Kao, Zhang Yunfei, Ma Dongli, et al. Calculation and Analysis of Aircraft Stealth Performance to Radar [M]. Beijing: National Defense Industry Press, 1997
[25] Li Nengjing, Jin Zhaoxun. Introduction to Military Electronic Technology [M]. Beijing: National Defense Industry Press, 1997
[26] Liu Deshu. The Basic Theory and Technology of Radar Anti-Counterfeiting [M]. Beijing: Science Press, 1981

Three-Dimensional Computational Fluid Dynamics Simulation of the Turbine Blades of a Micro Turbojet Engine

Haotian Zhang*

Department of Aeronautics, Imperial College London, Exhibition Road, London SW7 2AZ, UK

*Corresponding author's e-mail address: haotian.zhang22@imperial.ac.uk

Abstract. Micro turbojet engine has been widely used in both civil and military applications and their turbine blades are the key components that determine their performance. The paper analyses the influence of adjusting rotational speed and attack angle on the pressure ratio and kinetic energy efficiency of a typical type of micro turbojet engine named KJ66 using computational fluid dynamics in three-dimensional. The objective is to analyse a basic research methodology using CFD for future micro turbojet engine research and workshops to maximize efficiency. Once the related statistic and the aerodynamic performance characteristic are obtained, the attack angle and rotational speed of the turbine blade can be set respectively according to the actual requirements to meet the needs of users. Based on the simulation results, it can be derived that the highest efficiency is 93.46% at the attack angle change of -10° and rotational speed of 110000 rpm. However, the numerical simulation of other conponents such as nozzle are not included in this paper. Therefore, the results would be more accurate and convictive if the object of research can be extended to the entire micro turbojet engine.

1. Introduction

The micro turbojet engine is widely used as a propulsion device and auxiliary power unit which has the advantages of a relatively simple structure, compact size and high thrust ratio. It is ideal for small aerial vehicles, cruise missiles and remote-piloted vehicles [1]. Among all components in a micro turbojet engine, the turbine blade is the key component to determine the overall performance including power output and fuel efficiency.

Since computer science and computer performance have developed rapidly over the years, the computational fluid dynamics technique is widely used in the numerical simulation of engineering problems related to fluid dynamics. The study aiming at the turbine blades of the micro turbojet engine is one of the viable ways to increase the overall efficiency because of the complexity and cost of the numerical simulation model of simulating the whole machine.

Micro turbojet engines have been extensively researched all over the world and have been highly valued by governments and related institutions of various countries. In terms of the micro turbojet engine field, many companies, universities and academic institutions from Europe and America are researching at the forefront of the industry, developing new products and technologies in a more advanced direction. Trebunskikh [2] used the software named FLoEFD to numerically analyse a certain type of micro turbojet engine and the conclusion demonstrates that the aerodynamic performance can be negatively affected by the uneven temperature distribution of the airflow directly from the combustor. Li et al [3] designed a micro turbojet engine independently with an expected thrust level of 510 N and an expected engine rotational speed of 76000 rpm and Zhang [4] designed and tested another micro turbojet engine with a diameter of 135 mm and the engine total length of 230 mm, which has the similar dimensions of the KJ66 model. Both the previous research proved that it is significant and feasible to improve the overall engine performance by adjusting the blade profile of the turbine blades. In terms of the internal components of the engine, Wang et al [5] carried out a three-dimensional CFD simulation of the engine nozzle and the turbine wheel of the KJ66 micro turbojet engine and verified that the calculated outcomes are like the experimental statistics obtained in the laboratory. According to the simulation results of Wang, the turbine can work normally in a large range of pressure ratios, making the CFD research on turbine blades in this paper reliable and viable.

However, since the above research hardly involves the effects of different angles of attack and rotational speeds on engine operation, many cases of such micro turbojet engines are used for model aircraft enthusiasts without any professional equipment. Therefore, it is necessary to simulate and analyze the influence of these two parameters on the entire engine efficiency. After obtaining the data and results of the engine operating under different conditions, users can easily adjust the operating parameters to suit different operating conditions, and researchers can also develop blade shapes that are more suitable for specific operating conditions based on these conclusions.

In this paper, aiming at the KJ66 micro-turbojet engine, the numerical simulation of the micro-engine is carried out based on the CFD simulation software ANSYS Fluent by simplifying its physical model. The paper is organized as follows: chapter 2 introduces the basic models, boundary conditions and parameter settings. Chapter 3 describes the simulation results and the corresponding analysis. The conclusion of the entire simulation is presented in chapter 4.

2. Numerical simulation method

2.1. Governing equations and turbulence model

All the derivatives to time from the universal control equations are zero since the numerical simulation used in this paper is steady state. It is also worth mentioning that the influence of viscous force requires consideration because of the compact internal structure and strong viscous effect in the engine. Therefore, the Reynolds Averaged Navier-Stokes (RANS) approach and Cartesian coordinate system are applied in the simulation process. Its equation is formulated as shown below.

$$\frac{\partial Q}{\partial t}+\frac{\partial(F-F_v)}{\partial x}+\frac{\partial(G-G_v)}{\partial y}+\frac{\partial(H-H_v)}{\partial z}=0 \tag{1}$$

where Q is a conserved variable, t denotes the time, F, G, H are inviscid fluxes in x, y, and z directions. F_v, G_v, H_v are the viscous fluxes in three directions. In terms of the turbulence model, the $k-\varepsilon$ model is relatively suitable for engineering turbulence models in these cases and is used in this CFD numerical simulation [6] because the various fluids can be mixed completely in the micro turbojet engine due to diffusion and dissipation phenomena.

2.2. Physical model

The physical model based on the real micro turbojet engine is the premise of obtaining accurate statistics and convictive conclusions. The KJ66 micro turbojet engine consists of a different number of rotors and stators from the turbocharger fixed in small road vehicles. It is equipped with a single-stage centrifugal compressor and the engine diameter is 66mm. The axial turbine has 23 blades and there are also 15 blades fixed on the stator [7]. The nickel-chromium iron alloy is the main material of the turbine stage, making it possible for the turbine to bear extremely high temperatures.

In the simulation process, the modelling of the turbine blade is carried out using Solidworks software based on the measured data of the real KJ66 engine. The turbine blade used in the numerical simulation is separated since all the blades have the same aerodynamic figure. Figure 1 shows the geometric model of the KJ66 engine and turbine blade. As shown in figure 1, the turbine blade used to simulate is extracted from the turbine wheel and the whole engine, which can save the calculation cost during the simulating process. Figure 2 introduces the definition of each face and the flow direction of the fluid domain. As shown in figure 2, the hot gas discharged from the combustor flows into the air inlet and drives the turbine blade to generate torque. The whole fluid domain is limited to the physical structure, which is fully based on the KJ66 micro turbojet engine.

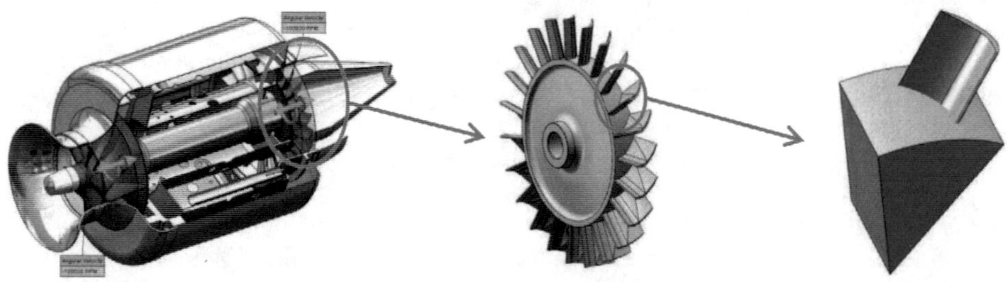

KJ66 micro turbojet engine Turbine wheel Blade model

Figure 1. The geometric model of the KJ66 engine and turbine blade.

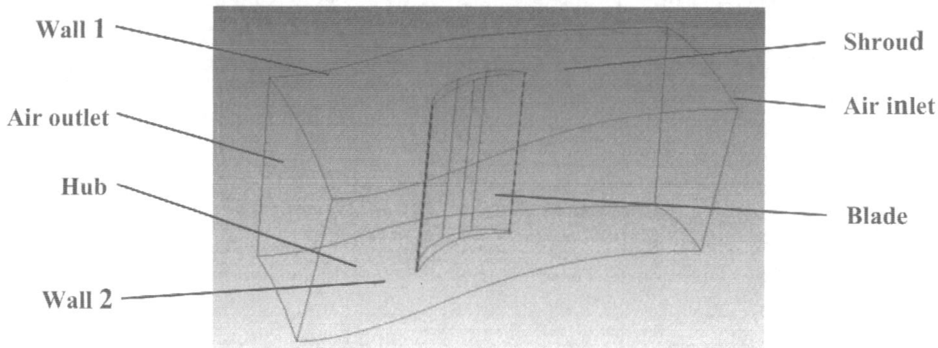

Figure 2. The definition of faces of the fluid domain

2.3. Periodic surface and inflation

The fluid domain of the experiment is a single flow channel, where hot air discharged from the combustor flows into the fluid domain at the air inlet and leaves at the air outlet. According to the high-rotational-speed and high-temperature working conditions of the turbine wheel, this CFD experiment is carried out by introducing the periodic surface to simulate the entire turbine in the actual working environment.

Since meshes automatically generated by ANSYS fluent software are not dense and precise enough to meet the desired requirement, five layers of boundary layers are introduced and applied to the turbine blade itself, air inlet and air outlet of the fluid domain. During the calculation process, the added boundary layers will significantly increase the accuracy of the calculated results and the final related statistics are relatively close to the experimental ones. It is found that the corresponding error is less than 5%.

2.4. Mesh independence verification

For most of the CFD physical models, once the mesh density of the model reaches a certain value, more accurate results will be much harder to achieve by adding more meshes into the physical model. Thus, mesh independence verification is necessary for the numerical simulation. To introduce a standard to measure the mesh number, the total pressure ratios at normal conditions are compared to find the appropriate mesh number. As an indicator of the overall performance, total pressure ratio is suitable to mesh independence verification. Normally, The mesh number increase by decreasing the mesh size. In this paper, 6 groups of mesh numbers are divided, varying from 300,000 to 1.3 million, as shown in figure 3.

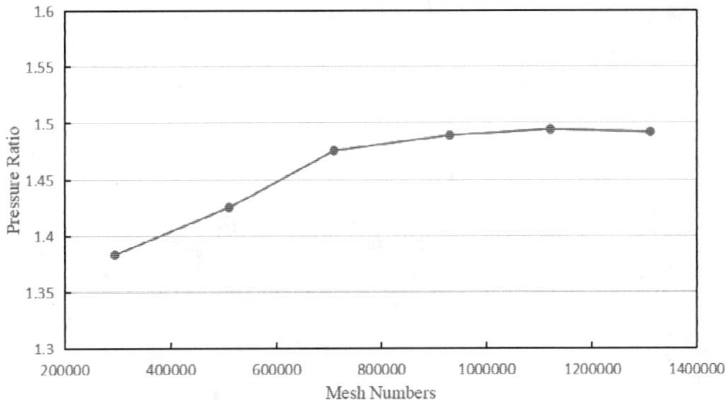

Figure 3. The mesh independence verification

According to figure 3, the calculated results with different mesh precision proved that the change in output total is negligible once the mesh number reaches 900,000. Therefore, the simulation in this condition satisfies the accuracy requirement as shown in the next chapter. In consideration of computational cost, the mesh number of 930,000 is set for the CFD calculation.

2.5. Boundary conditions and variables setting

Boundary conditions are the premise that the governing equations have a definite solution. For any numerical simulation operation, a given boundary condition is always required and its value directly affects the accuracy of the calculation results. The boundary conditions simulated in this paper are shown in Table 1, including the most critical inlet and outlet pressure and temperature.

Table 1. Boundary conditions of the turbine [8].

Boundary conditions	**Parameters**
Inlet total pressure	162120 Pa
Outlet total pressure	101300 Pa
Turbine rotational speed	100000 rpm
Inlet temperature	970 K
Wall surface	Adiabatic and Non-Slip

In this paper, there are two sets of related variables taken into consideration. The first variable is the attack angle of the turbine blade and another is the rotational speed. These two variables are independent of each other and are they adjusted within a certain range around a standard value to simulate the operation of the turbine blade under different working conditions. After that, this paper will make a detailed analysis of the simulation results and show the impact of these two variables on the turbine blade efficiency to a certain extent.

In terms of the attack angle, it can be adjusted by rotating the blade profile in Solidworks software and the attack angle changes (Hereinafter attack angle) are set as -10°, -5°, 0°, 5°, and 10°, where anticlockwise is considered as the positive direction. Based on the relevant literature written by Li T [8], the standard rotational speed is 100000 rpm and another five groups are set as 80000, 90000, 100000, 110000 and 120000 rpm.

3. Results and Discussion

3.1. Fundamental flow features

After the calculation, the pressure contour and the velocity streamline in figure 4 and figure 5 are obtained, which shows the fundamental flow features of the simulation. In this paper, 25 pressure contours and velocity streamlines are evidence to support the analysis of flow motion and various interactions.

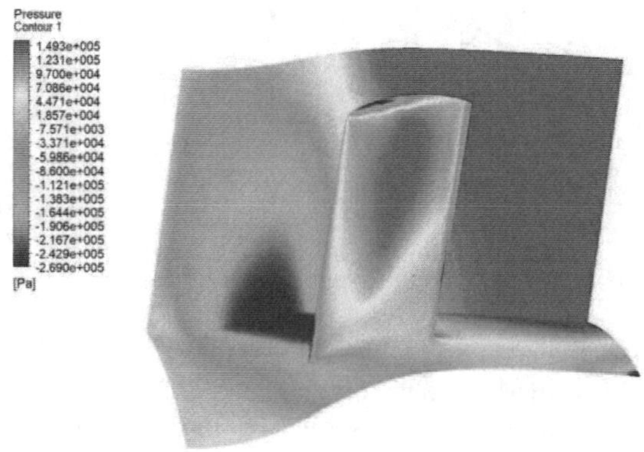

Figure 4. Pressure contours of the turbine blade.

Figure 5. Velocity streamline of the turbine blade.

According to the obtained pressure contours, the basic condition of the fluid domain with varying parameters can be obtained. It can be seen in the 25 groups of pressure contours that the increase in attack angle will lead to the increase of the high-pressure region on the lower side of the blade and significant resistance. It is worth noticing that the local high-pressure area at an attack angle of 0° is closer to the leading edge of the turbine blade than that of -10°, as shown in figure 6.

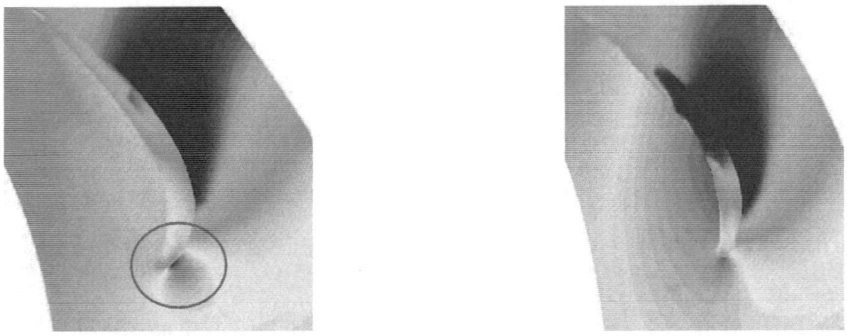

(a) Attack angle of -10° (b) Attack angle of 0°

Figure 6. Two types of pressure distributions on blade leading edges.

This undesirable condition may generate a reverse moment on the turbine wheel, decreasing the blade aerodynamic efficiency and the torque that is supposed to drive the compressor. When the attack angle is adjusted to -10°, it is obvious that there is a low-pressure region close to the upper side of the blade, which expands with the increase of rotational speed. Therefore, under the action of a larger pressure difference, the turbine blade generates a larger force, achieving higher aerodynamic efficiency.

In the case of 0°/10000 rpm, hot gas collides with the turbine blade, which blocks the motion of hot incoming flow to a certain extent. This may cause some of the original energy in the incoming flow to be lost in the form of heat dissipation, eventually leading to a decrease in efficiency.

In terms of the pressure distribution, it can change not only the magnitude but also the direction of the pressure and may cause a vibration of the turbine blade. The reason is that the hot air passes through the turbine blade creating a regular vortex behind it, which is known as vortex shedding. Once the frequency of the vortex shedding becomes closer to the blade's natural frequency, vibration will be generated which can enormously decrease the blade's efficiency.

3.2. Total pressure ratio

In this paper, the blades are explored under different rotational speeds and attack angle conditions. To calculate the related parameters such as the total pressure ratio and the kinetic energy efficiency, 25 sets of data from both inlet and outlet are obtained through the area integral using the fluent software.

The total pressure ratio is the ratio of the turbine at the turbine inlet to the total pressure at the turbine outlet [9]. It is a key parameter for micro turbojet engines to focus on to obtain a better output and efficiency and it is defined by the formula 2 as is shown below.

$$\pi = \frac{P_t^{out}}{P_t^{in}} \tag{2}$$

where P_t^{in} denotes the air inlet total pressure, P_t^{out} denotes the air outlet total pressure [10]. As shown in figure 7, the calculated results of different attack angles are collated into a line chart. According to formula 2, with the increase of the rotational speed and decrease of the attack angle, the total pressure ratio increases with a relatively stable trend. The total pressure ratio is calculated using the total pressure of air inlet and out, while the kinetic energy efficiencies in the next chapter are calculated using relative total pressure and static pressure.

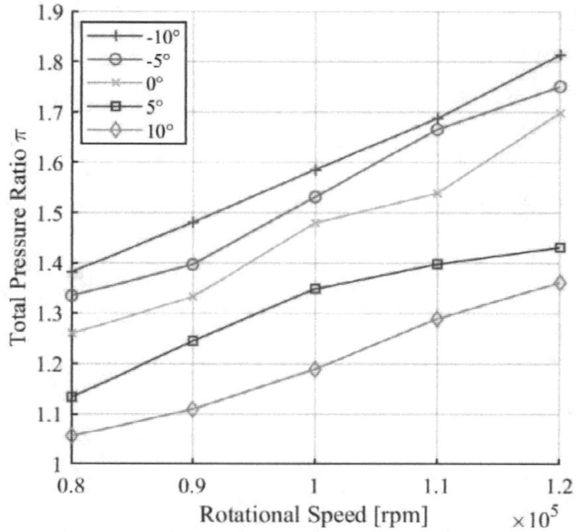

Figure 7. The total pressure ratio of the blade.

3.3. Kinetic energy efficiency

Aerodynamic efficiency is an important design index of the manufacture of the turbine blade. Therefore, the efficiency calculation formula of the turbine blade is also introduced in this section. According to the efficiencies in various situations, we can explore the best working conditions of blades under certain requirements and provide theoretical references for production and processing. Based on the publication of Yang Z et al [11], the equations of the ideal and actual exit velocities of the stator flow field of a turbine engine are shown as formula 3 and 4, where in this case C_1, C_2 can be recognized as the velocities of the air inlet of the model. Since the stators are fixed in turbine engines, where the stators do not generate work and the kinetic energy loss is converted to heat, so $T_0^* = T_1^*$ and the calculation formula of kinetic energy efficiency of the turbine blade can be expressed as formula 5.

$$C_1 = \sqrt{\frac{2k}{k-1} RT_0^*[1 - (\frac{P_{out}}{P^*_{Total,\ out}})^{\frac{k-1}{k}}]} \quad (3)$$

$$C_2 = \sqrt{\frac{2k}{k-1} RT_1^*[1 - (\frac{P_{out}}{P^*_{Total,\ in}})^{\frac{k-1}{k}}]} \quad (4)$$

$$\eta_r = \frac{C_1^2}{C_2^2} = \frac{1-(\frac{P_{out}}{P^*_{Total,\ out}})^{\frac{k-1}{k}}}{1-(\frac{P_{out}}{P^*_{Total,\ in}})^{\frac{k-1}{k}}} \quad (5)$$

where P_{out} denotes the outlet flow static pressure and $P^*_{Total,\ out}$, $P^*_{Total,\ in}$ denote the outlet and inlet relative total pressure [12]. The parameter k is the heat capacity ratio and R is the gas constant. C_1, C_2 are the ideal and actual velocities of the air inlet of the model respectively. According to literature written by Cengel et al [13], the k value is selected from 1.336 to 1.344 based on the specific turbine temperature. The results of kinetic energy efficiencies at different attack angles are arranged into figure 8.

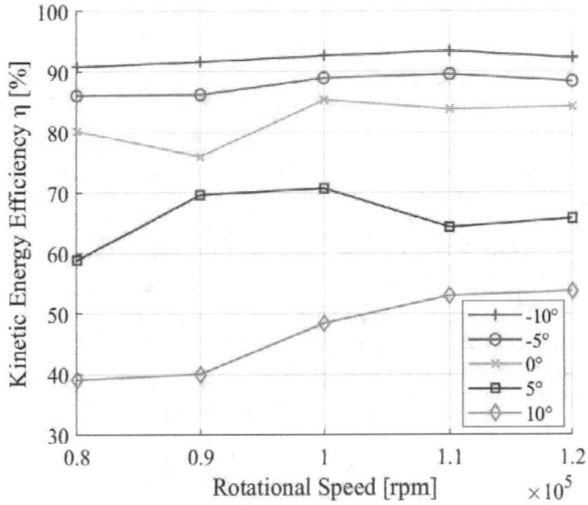

Figure 8. Efficiencies in different attack angles.

Based on the information shown in figure 8, the kinetic energy efficiencies change with varying degrees when rotational speed and attack angle are adjusted. According to this line chart, the variation of kinetic energy efficiency of the turbine blade with rotational speed when the attack angle is set to -5° and -10° is smaller. Therefore, it can be derived that the effect on efficiency caused by different rotational speeds is minor compared with the 0°, 5° and 10° groups. In comparison, the kinetic energy efficiency curve at the attack angle of 10° indicates a stable increasing trend with the rotational speed of the turbine, while that of attack angles of 0° and 5° show the tendency to fluctuate. The result indicates that the rotational speed affects the kinetic energy efficiency more significantly when the attack angle is set to 0°, 5° and 10°.

According to the analysis mentioned above, it is possible to maximize the kinetic energy efficiency at a given rotational speed by setting different attack angles. For the curves the attack angle is set to 0° and 5°, the efficiency curves have local minimum and maximum points. The 0° attack angle curve has a minimum point when the rotational speed is 90000 rpm. The sudden efficiency drops at this point proved that a certain level of energy loss occurs in this condition, showing that this combination of attack angle and rotational speed should be avoided in practical operation. By comparison, the 5° attack angle curve shows the opposite trend, which shows a high-efficiency level of about 70% when the rotational speed varies from 90000 to 100000 rpm. It can be found that there is a close relationship between figure 8 and pressure contours in the previous chapters, which indicates that the improper attack angle and rotational speed settings may cause huge efficiency loss or undesirable phenomena such as shedding vortex.

In terms of the variation tendency between different attack angle, the ability to compare data horizontally in figure 8 need to be improved. The two-dimensional line chart requires another axis to fully present the influences caused by varying attack angles and rotational speed. Therefore, the calculated data of kinetic energy efficiencies are imported into MATLAB software to produce three-dimensional images as shown in figure 9, where x, y, and z denote the rotational speed, attack angle and the kinetic energy efficiency of the turbine blade respectively.

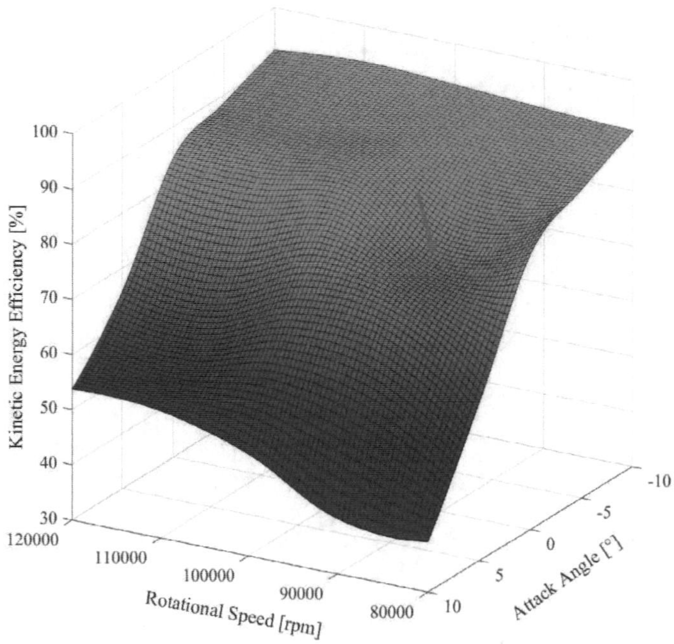

Figure 9. Efficiencies in Different Attack Angles (3D).

According to figure 9, there is an obvious low-efficiency region at an attack angle of 0° and rotational speed of 90000 rpm, which is consistent with the energy loss caused by the shedding vortex analyzed in previous sections. The highest point of kinetic energy efficiency can be found at an attack angle of -10° and rotational speed of 90000 rpm 110000 rpm so the highest turbine blade kinetic energy efficiency is 93.46% in this condition.

4. Conclusion

Based on the exact initial condition and numerical simulation, the pressure contour and the velocity streamline are obtained, showing the fundamental flow features around the turbine blade. All wall surfaces are set to adiabatic and non-slip conditions including the working conditions where the highest efficiency is achieved. After the calculation, all experimental groups' pressure ratios and kinetic energy efficiencies are obtained. Basic physical principles and some factors such as shedding vortex and regional high-pressure area that negatively affect the efficiency of the turbine blade are also introduced in detail. In terms of the total pressure ratio, it can be derived that the value of the pressure ratio increases when the rotational speed increases and the turbine blade attack angle decreases. It is also found that the magnitude of the increase also varies according to different variables.

The kinetic energy efficiency could be found using more parameters and corresponding formulas. The result is shown in detail which mainly discusses that the kinetic energy efficiency varies with the varying rotational speed and attack angles. At attack angle of -5° and -10°, the influence on efficiency caused by varying rotational speed reduce compared with other simulating groups and the 10° attack angle group indicate that the rotational speed affects efficiency more significantly. Apart from these, when the attack angle is adjusted to 0° and 5°, a sudden increase and drop can be found in kinetic energy efficiency. Moreover, combined with the pressure contours and analysis in the previous chapters, it can be derived that some improper combination of attack angle and rotational speed may cause huge efficiency loss or some undesirable phenomena, which deserve the attention of engineers and users.

Based on the three-dimensional kinetic energy efficiency graph, the highest efficiency can be found at an attack angle of -10° and rotational speed of 110000 rpm, at 93.46%. After the research on the turbine blade itself, it is possible to extend the simulation to other components even the entire engine to calculate more complete results.

References

[1] Turan O. Exergetic effects of some design parameters on the small turbojet engine for unmanned air vehicle applications. *Energy*, 2012, 46(1): 51-61.
[2] Trebunskikh T V, Ivanov A V, Dumnov G E. FloEFD simulation of micro-turbine engine. Proceedings of Applied Aerodynamics Conference on Modelling & Simulation in the Aerodynamic Design Process. New York: Curran Associates, Inc., 2012, 1: 51-64.
[3] Li M, et al, Design and Research of Micro Turbojet Combustion Chamber, *Journal of Aerospace Power*, 24 (1): 70-74, 2009.
[4] Zhang J. Preliminary Design and Experimental Analysis of Combustion Chamber for Missile micro turbojet Engine. Nanjing: School of Energy and Power, Nanjing University of Science and Technology, 2016. (In Chinese)
[5] Wang C, Zhang M, Li T, et al. THREE-DIMENSIONAL CFD SIMULATION OF A MICRO TURBOJET ENGINE, Proceedings of the Fourth Chinese International Turbomachinery Conference CITC, 2020.
[6] Xie L H, Zhao X Y, Zhang J M. ANSYS CFX fluid analysis and simulation. Publishing House of Electronics Industry: Beijing, China, 2013: 12-19.
[7] Jian X, Wu H, *Principles of Aero Engines*. Northwestern Polytechnical University Publishing, Xi'an, 2005
[8] Li T, Three-dimensional Numerical Simulation of An Entire Micro Turbojet Engine, Dalian University of Technology, 2019. (In Chinese)
[9] Cao Z, Wang H, Zhang S, et al. Effect of Drop Pressure Ratio on Turbine Blade Flow Characteristics of Aero-engine. *Aero-engine*, 2021, 47(S1):32-39
[10] Li C, Luo J, Xiao Z. A throughflow-based optimization method for the multi-stage axial compressor. *AIP Advances*, 2021, 11(11): 115207.
[11] Yang Z, Ni N, Yang Y. *Turbomachinery*. National Defense Industry Publishing, Beijing, 2007
[12] Peng M, Yang Z, Cao Y, Chu Z. Study on three-dimensional aerodynamic analysis of turbine blades. *Thermal turbine*, 2012,41(04):245-250.
[13] Cengel Y A, Boles M A, Kanoğlu M. *Thermodynamics: an engineering approach*. New York: McGraw-hill, 2011.

Study on Characteristics of Particle Dynamics with Coarse Particles in Vertical and Inclined Pipeline

Wanlong Ren[1, 2], Xuhui Zhang[1, 2], Yan Zhang[1, *] and Xiaobing Lu[1,2]

[1] Key Laboratory for Mechanics in Liquid Solid Coupling Systems, Institute of Mechanics, Chinese Academy of Sciences, Beijing 100190, China

[2] School of Engineering Science, University of Chinese Academy of Sciences, Beijing 100049, China

*zhangyan162@imech.ac.cn

Abstract: The pipeline hydraulic transport is an important component of the deep-sea mineral resources. The characteristics of particle dynamics with coarse particles is investigated by using the CFD-DEM method in the vertical and inclined pipeline. The normal pipeline and abnormal pipeline mentioned in this paper refer to vertical pipeline and inclined pipeline, respectively. The particles of the normal pipeline mainly move in the middle of the pipeline, while the particles mainly concentrate on the pipeline wall in the abnormal pipeline. The velocity difference of the abnormal pipeline between liquid and particle is much greater, which can be prone to cause particle aggregation. Finally, the pipeline to be blocked can be easy to be caused by the particle aggregation in the wall of abnormal pipeline. An appropriate increase in liquid velocity can improve the phenomenon.

1. Introduction

With the decreasing reserves of terrestrial mineral resources, marine-rich resources such as manganese nodules and hydrates are expected to become essential support for the future energy strategy of countries around the world. Therefore, it is crucial to master deep-sea mineral resources' mining and transportation technology. Efficient and environmentally friendly deep-sea mining projects have become an important part of the development of marine resources [1]. The normal pipeline and abnormal pipeline mentioned in this paper refer to vertical pipeline and inclined pipeline, respectively.

How to transport the rich mineral resources in the deep sea from the seabed of several kilometers to the land is one of the urgent problems faced by the current deep-sea mineral projects. It is recognized that the safest and most promising is the hydraulic conveying system. The core of the deep sea hydraulic conveying is to study the particle dynamics formed by the two phases flow in the pipeline [2]. Many researchers have conducted extensive experiments on the hydraulic conveying of normal pipelines. A device to simulate deep-sea mining systems was established, and obtained the sedimentation velocity of manganese nodules through experiments, and the relationship between the nodules concentration and the velocity difference between liquid and particle was found [3]. The velocity and particle volume fraction transitioning regimes were measured to study the stability of vertical tubes and provide recommendations for transport parameters [4]. Dai et al. [5] analyzed the particle dynamics with different liquid velocities and particle diameter es in the ascending process through a vertical hydraulic lifting pipeline. In addition, some researchers have studied the movement of particles in abnormal pipelines, but most studied fine particles [6].

Over the past decade, numerical simulation has become a meaningful way to obtain motion information of particles and fluids in normal pipelines [7]. Researchers began studying particle transport characteristics in normal pipelines using the CFD-DEM method. Motion characteristics of particles and flow field information in a normal pipeline are studied [8]. They found that particles tend to flow in the middle of the pipeline, which was not found in the experiment. The particle clustering on the suspension settlement of high-concentration particles was studied by Yao et al. [9]. The characteristics of the hydraulic conveying in normal pipelines was studied [10]. However, few researchers have studied the particle transport characteristics of abnormal pipelines. It is challenging to keep the pipeline vertical due to the influence of external forces in hydraulic conveying. Therefore, study the hydraulic transportation characteristics of is important to in abnormal pipelines.

The particles transported by deep-sea mining pipelines are coarse. Coarse particles transported by pipeline means $D/d < 10$ [11],where D and d mean the pipeline and particle diameter, respectively. Since the general CFD-DEM method requires a particle size smaller than the mesh size, it is difficult to solve the problem of coarse particle transport. Therefore, we proposed a new numerical method for solving coarse particles and successfully implement it on CFDEM platform [12].

This paper mainly studies particle dynamics and flow field information with coarse particles in normal and abnormal pipelines using the optimized CFD-DEM method in the context of deep-sea mining. In Sect. 2, the CFD-DEM method is introduced. The computational model is set up in Sect. 3. Sect. 4 is the analysis of results. Finally, the Sect.5 summarizes the particle dynamics and flow field information of normal and abnormal pipelines.

2. Numerical methods

2.1. Liquid phase description

Flow field is solved by the mass and momentum conservation equations [13]:

$$\nabla \cdot \mathbf{u}_l = 0 \tag{1}$$

$$\frac{\partial(\varepsilon_l \mathbf{u}_l)}{\partial t} + \nabla \cdot (\varepsilon_l \rho_l \mathbf{u}_l \mathbf{u}_l) = -\varepsilon_l \nabla P + \nabla \cdot (\varepsilon_l \boldsymbol{\tau}_l) + \varepsilon_l \rho_l \mathbf{g} + \mathbf{F}_{pl} , \tag{2}$$

where ε_l is the liquid volume fraction, $\mathbf{u}_l$ and ρ_l are the velocity and density of liquid, respectively, P means the liquid pressure, $\mathbf{F}_{pl}$ is the solid-liquid interaction force, and $\boldsymbol{\tau}_l$ is the liquid shear stress tensor.

2.2. Solid phase equations

Position and velocity information of a single particle is obtained by solving momentum and angular momentum equations [14]:

$$m_p \frac{d\mathbf{u}_p}{dt} = \mathbf{F}_{lp} + \mathbf{F}_{con} + m_p \mathbf{g} , \tag{3}$$

$$I_p \frac{d\omega_p}{dt} = \mathbf{M} , \tag{4}$$

where m_p and $\mathbf{u}_p$ mean the mass and the velocity of particles, respectively. $\mathbf{F}_{lp}$ and $\mathbf{F}_{con}$ mean the solid-liquid force and the particle-particle/wall collision force, respectively. I_p and ω_p mean the moment of inertia and the angular velocity, respectively. $\mathbf{M}$ is the particle-particle/wall torque. These physical quantities are commonly used in particle dynamics.

2.3. Liquid-solid interaction

There are many empirical force models between solid and liquid. In this paper, the main empirical force models are drag force model and pressure gradient force model [7]:

$$\mathbf{F}_{pl} = \mathbf{F}_d + \mathbf{F}_p , \tag{5}$$

where $\mathbf{F}_d$ and $\mathbf{F}_p$ mean the drag and the pressure gradient forces, respectively. More details can be found in [10]. An new numerical method can be found in Refs [12].

3. Numerical setup

A pipeline with a diameter of 100 mm inner diameter and 1.5 m long is formed in Figure.1. The essence of an abnormal pipeline is the angle between the pipeline and the direction of gravity. In this paper, the abnormal pipeline is formed by changing the direction of gravity by 45° based on the normal pipeline. The Table 1 is the parameters of the numerical simulation. The particles initially rest on the bottom, then rises with the liquid force.

Table 1. The numerical Parameters of hydraulic collection

Parameters	Values
Water properties	
Density ρ_l	1000 kg/m^3
Viscosity ν_l	1.0×10^{-6} m^2/s
Velocity u_0	2 - 3 m/s
Particle properties	
Density ρ_p	2450 kg/m^3
Diameter d_p	10 mm
Young's modulus E	1.0×10^9 Pa
Poisson's ratio ν	0.45

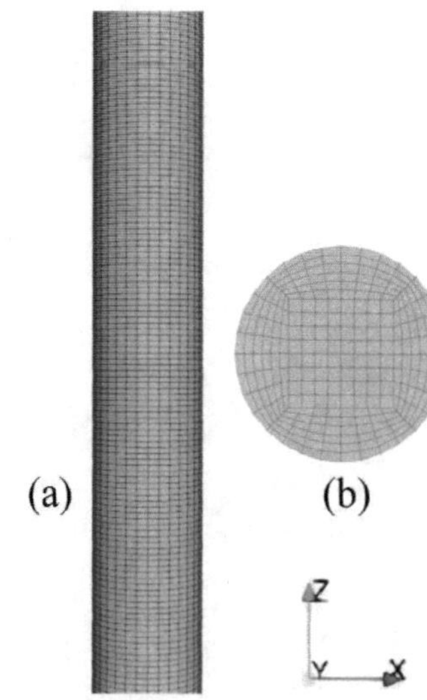

Figure 1. Geometry: (a) front meshes, and (b) inlet meshes.

4. Analysis of results

4.1. Method validation

The CFD-DEM method is verified based on the CFDEM platform in this part. The minimum fluidization velocity (MFV) is compared between the numerical and theoretical results with different viscosities [15]. Particle density is $\rho_p = 2200$ kg/m^3, and particle size is 10 mm. Liquid defaults to the normal temperature water and the viscosity of $\mu_l = 0.03, 0.04, 0.05, 0.06, 0.07, 0.08, 0.09$ kg/(m. s). Under the action of fixed fluid velocity at the inlet, particles at the bottom begin to fluidize. When the pressure change value in the pipeline is a constant, the fluid velocity is the MFV. The MFV between numerical and theoretical results is in Figure 2. The results of numerical simulation are close to the experimental results, so the new numerical method can be used to obtain particle dynamics with coarse particles.

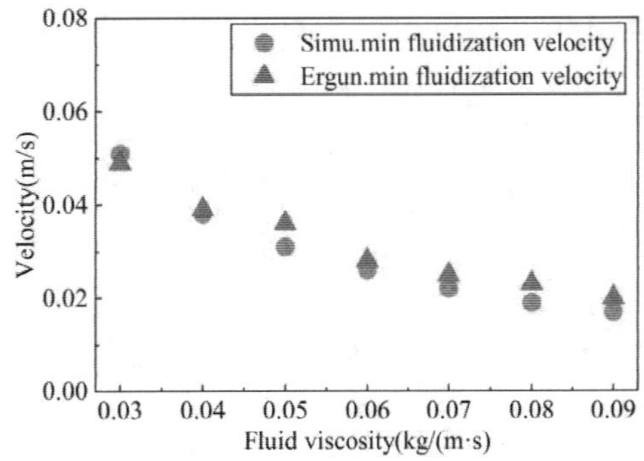

Figure 2. MFV between numerically and theoretically calculated.

4.2. Qualitative results

To measure the particle and liquid data at a steady state, we set up a monitoring area with a height of 0.1 m at 0.75 m long in pipeline. Parameter variations such as liquid, particle velocity, and local volume fraction in the region are obtained.

This part qualitatively analyzes the variation of particle local volume fraction at 0.75 m of normal and abnormal pipelines lengths. The contours of the local volume fraction distribution of particles at 0.75 m of the cross-section of the normal and abnormal pipelines in Figure 3 at an inlet liquid velocity of 2 m/s, respectively. Overall, the local concentration distribution of particles in the normal pipeline is uniform, and this transport state is what we hope in engineering in Figure 3a. The particles of the normal pipeline mainly move in the middle of the pipeline, so the local concentration distribution of particle in the wall is small in the normal pipeline. However, in the abnormal pipeline, the particles of the abnormal pipeline mainly concentrate on the pipeline wall in Figure 3b. The state in which particles concentrate on the pipeline wall is particularly easy to lead to pipeline damage, which is not an ideal conveying mode. Because of the complex marine environment, abnormal pipeline transportation is common. Therefore, how to ensure the uniform transport of particles in abnormal pipeline is a research topic to be solved at present.

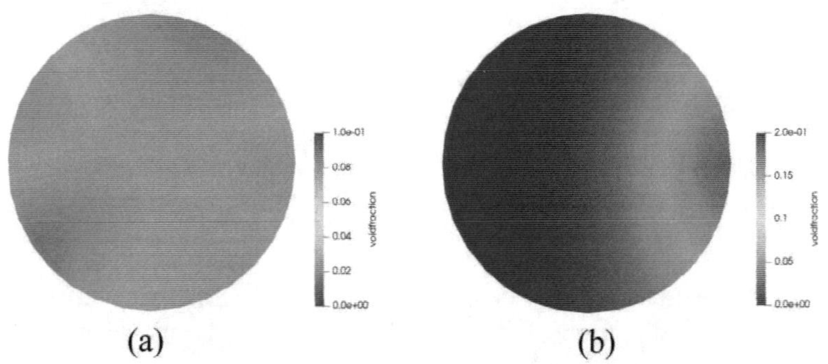

Figure 3. Particle volume fraction cloud map: (a) normal pipeline, and (b) abnormal pipeline.

4.3. Quantitative analysis of the particle

This section mainly analyzes the variation of particle local volume fraction and slip velocity at 0.75 m of normal and abnormal pipelines lengths. In this part, we investigate the variation in the velocity difference between liquid and particle by changing liquid velocities. We study the variation of the velocity difference at different liquid velocities in normal and abnormal pipelines, respectively. The velocity difference of the abnormal pipeline between liquid and particle is much greater, which can be easy to cause particle aggregation in Figure 4. Large slip velocity will lead to poor followability of particles, which can easily cause particle aggregation and cause pipeline blockage in the abnormal pipeline.

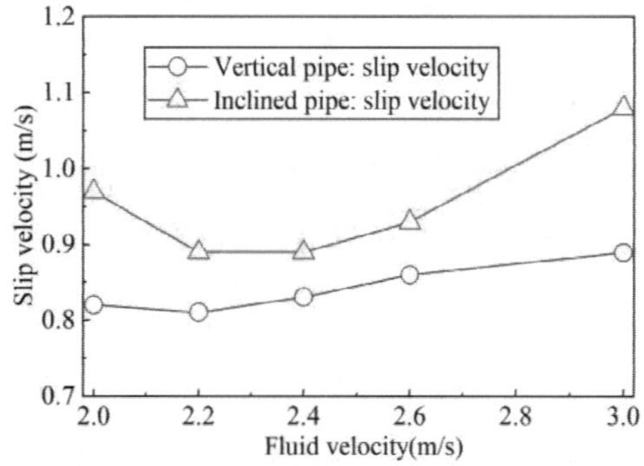

Figure 4. The variation of velocity at liquid velocity in the normal and abnormal pipelines.

The local concentration distribution of particles decreases with the increase of fluid velocity in Figure 5a. The local concentration distribution of particles is also more uniform in horizontal direction, and it is not easy to pipeline blockage. The particles gather at the wall is very high in Figure 5b. This is consistent with the phenomenon in Figure 3b. In addition, we find that the minimum inlet velocity required for abnormal pipelines is significantly greater than that of normal pipelines, and particle aggregation occurs easily in abnormal pipelines. Therefore, Once the pipeline is inclined, an appropriate increase in liquid velocity is required in practice.

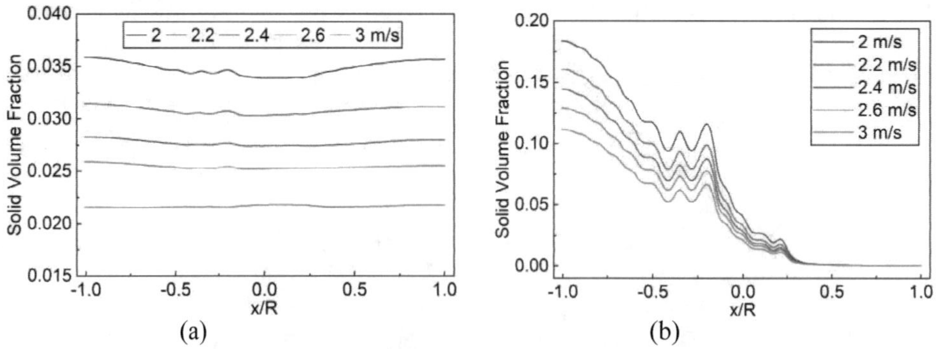

Figure 5. The changes of average local volume fraction of particles.

5. Conclusions

The characteristics of particle dynamics with coarse particles is investigated by using the CFD-DEM method in vertical and inclined pipelines. The normal pipeline and abnormal pipeline mentioned in this paper refer to normal pipeline and inclined pipeline, respectively.

- Firstly, the particles of the normal pipeline mainly move in the middle of the pipeline, while the particles of the abnormal pipeline mainly concentrate on the wall. Uniform distribution of particles in the pipeline is an ideal conveying condition.
- Secondly, the slip velocity of particles in abnormal pipelines is significantly greater than that in normal pipelines, which is easier to cause particle aggregation.
- Finally, the radial variation of the concentration distribution of particles in normal and abnormal pipelines is analyzed, respectively. Compared with normal pipelines, abnormal pipelines are prone to particle aggregation and cause pipeline blockage, which is a problem that we need to solve in engineering.

Acknowledgments

In the process of writing this article, I especially thank the National Natural Science Foundation of China (Grant No.12132018), the Strategic Priority Research Program of the Chinese Academy of Sciences (Grant No. XDA22000000), and the Youth Innovation Promotion Association of Chinese Academy of Sciences (No.2017027) for providing support.

References

[1] Glasby G. Lessons learned from deep-sea mining [J]. *Science*, 2000, **289**(5479): 551-3.
[2] Chung J S, Tsurusaki K.1994 Advance in deep-ocean mining systems research [C]. In: The Fourth International Offshore and Polar Engineering Conference. OnePetro.
[3] Xia J, Ni J R, Mendoza C. Hydraulic lifting of manganese nodules through a riser [J]. *J Offshore Mech Arct Eng*, 2004, **126**(1): 72-7.
[4] Van Wijk J, Talmon A, van Rhee C. Stability of vertical hydraulic transport processes for deep ocean mining: An experimental study [J]. *Ocean Engineering*, 2016, **125**: 203-13.
[5] Dai Y, Zhang Y, Li X. Numerical and experimental investigations on pipelineline internal solid-liquid mixed liquid for deep ocean mining [J]. *Ocean Engineering*, 2021, **220**: 108411.
[6] Archibong-Eso A, Aliyu A, Yan W, et al. Experimental study on sand transport characteristics in horizontal and inclined two-phase solid-liquid pipeline flow [J]. *Journal of Pipelineline Systems Engineering and Practice*, 2020, **11**(1): 04019050.
[7] Xie Z, Wang S, Shen Y. CFD-DEM study of segregation and mixing characteristics under a bi-disperse solid-liquid fluidised bed [J]. *Advanced Powder Technology*, 2021, **32**(11): 4078-4095..
[8] Zhou M, Wang S, Kuang S, et al. CFD-DEM modeling of hydraulic conveying of solid particles in a vertical pipeline [J]. *Powder Technology*, 2019, **354**: 893-905.
[9] Yao Y, Criddle C S, Fringer O B. The effects of particle clustering on hindered settling in high-concentration particle suspensions [J]. *Journal of Liquid Mechanics*, 2021, **920**.
[10] Sun D, Liu H, Guo R. A probability model for predicting the transport efficiency in vertical pipes considering the particle size distribution[J]. *Powder Technology*, 2023, **415**: 118104..
[11] Cúñez F D, Franklin E M. Mimicking layer inversion in solid-liquid liquidized beds in narrow tubes [J]. *Powder Technology*, 2020, **364**: 994-1008.
[12] Zhang Y, Lu X-B, Zhang X-H. An optimized Eulerian-Lagrangian method for two-phase flow with coarse particles: Implementation in open-source field operation and manipulation, verification, and validation [J]. *Physics of Liquids*, 2021, **33**(11): 113307.
[13] Zhao G, Xiao L, Yue Z, et al. Performance characteristics of nodule pick-up device based on spiral flow principle for deep-sea hydraulic collection [J]. *Ocean Engineering*, 2021, **226**: 108818.
[14] Guo Y, Curtis J S. Discrete element method simulations for complex granular flows [J]. *Annual Review of Liquid Mechanics*, 2015, **47**: 21-46.
[15] Ergun S, Orning A A. Liquid flow through randomly packed columns and liquidized beds [J]. *Industrial & Engineering Chemistry*, 1949, **41**(6): 1179-84.

Advances of asphalt pavement runoff models

WANG Zhenlong[1], WANG Xiao[2*], NI Dong[2], SHAO She-gang[2]

[1]Guangdong Huiqing Expressway Co., Ltd., Guangzhou 510900, China

[2]Research Institute of Highway Ministry of Transport, Beijing 100088, China

*wang.xiao@rioh.cn

Abstract: It is of great significance to study the formation mechanism of pavement runoff for pavement drainage optimization and traffic safety. Based on the research achievements of pavement runoff at home and abroad, this paper summarizes the existing runoff models into three types: empirical model, theoretical model and hydrodynamic model, and the theories and methods of the above models are introduced and analyzed in detail. The hydrodynamic model allows for the pavement attributes and achieves a dynamic simulation of runoff hydraulic parameters. It effectively overcomes the disadvantages of the dependence of basic data by empirical model and the complicated solution by theoretical model and provides a more accurate method for studying the mechanism of runoff generation and flow concentration on asphalt pavement.

1. Introduction

The extreme precipitation events caused by global climate change increase year by year. Rainfall flows along the synthetic slope of the highway pavement under the action of gravity, which will form water film and even cause serious flooding phenomenon. When the rainwater cannot drain through tire grooves in time, large hydrodynamic pressure will be generated, which will induce the phenomenon of hydroplaning and bring great safety risk to the highway traffic. In addition, long-term and repeated water will aggravate highway water damage. This problem has already attracted the attention of many researchers at home and abroad, and a series of studies have been performed around the pavement water film thickness, and the research results obtained have been widely used in practice.

According to the theory and method of model building, the existing models about the rainfall runoff on asphalt pavements can be divided into three types: (1) empirical model; (2) theoretical model; (3) hydrodynamic model. This paper analyzes each model in detail and points out each advantages and problems.

2. The empirical models

The empirical models are mainly built through regression analysis on the experiment data obtained by a series of artificial rainfall experiments to predict the water film thickness. This paper lists several frequently-used models for predicting pavement water film thickness.

Ross and Russam[1]proposed the formula for calculating water film thickness based on the measured data, which allows for rainfall intensity, drainage length and slope. The expression is as follows:

$$d = 0.017\left(L \cdot I\right)^{0.47} S^{-0.2} \tag{1}$$

where, d is the thickness of water film, mm; L denotes the drainage length, m; I is rainfall intensity, cm/h; S is the pavement slope.

Anderson[2] also studied and obtained the relationship between pavement water film thickness and rainfall intensity, drainage length and slope through experiments, which is expressed as:

$$d = 0.15\left(L \cdot I\right)^{0.5}\left(\frac{1}{S}\right)^{0.5} \quad (2)$$

In addition to the above influencing factors (rainfall intensity, drainage length and slope), the pavement roughness has also been identified and quantified[3-6]. The prediction model of water film thickness presented by Gallaway[3] is:

$$d = 0.00338\left(T^{0.11}L^{0.43}I^{0.59}S^{-0.42}\right) - T \quad (3)$$

where, T is the average pavement texture depth, in.

The prediction formula of water film thickness obtained by Wambold[4] is

$$d = j_1 T^{j_2} I^{j_3} S^{j_4} - T \quad (4)$$

where, j_1、j_2、j_3 and j_4 are empirical coefficients. The recommended values are 0.005979, 0.11, 0.59 and -0.42 respectively, when the flow path length is 11m.

The domestic scholar Ji developed the prediction formula of water film thickness as[5]:

$$d = 0.1258L^{0.6715}I^{0.7786}S^{-0.3147}T^{0.7261} \quad (5)$$

The domestic scholar Luo provided the prediction formula of water film thickness as[6]:

$$d = 0.068\frac{L^{0.32}I^{0.11}T^{1.17}}{S^{0.31}} \quad (6)$$

By comparison among the above models, the formulas proposed by Ross and Anderson are similar in form, but the difference lies in the index of each parameter; the formulas proposed by Gallway and Wambold are similar in form, but the constant coefficients in the model are different. In addition, the empirical coefficients in the calculation formula proposed by Wambold are only applicable when the drainage length is 11m, which is useful but has obvious limitations. The formulas developed by Ji and Luo are similar in the structural form, however, the parameters and their exponents differ greatly. In order to compare the above models, the rainfall intensity is uniformly taken as 50 mm/h, the pavement slope is taken as 2%, and the pavement structure depth is 0.91 mm. Figure 1 shows the variation of water film thickness of each model under the drainage length of 5~50 m. It shows that under the same rainfall intensity and slope, the pavement water film thickness calculated by each model increases with the increase of drainage length. The water film thickness calculated by the model proposed by Anderson is the largest and that calculated by the model proposed by Ross is the smallest. The calculation results of the models proposed by Gallaway, Ji and Luo are relatively close.

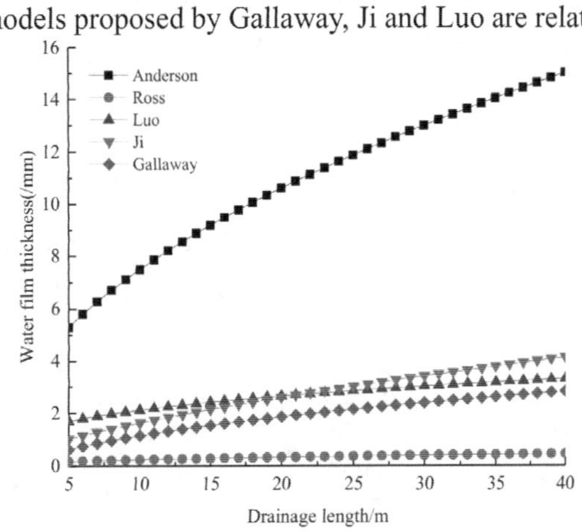

Fig.1 Comparison of water film thickness calculated by different empirical models

The differences among the above models are mainly caused by different test conditions and methods. In addition, the regression formulas are also different with different physical variables considered. Therefore, these empirical models are simple and easy to understand, but poor universality, because the model parameters are empirical and statistical, lack of clear physical significance.

3. The theoretical models

The influence of cross slope and longitudinal slope on pavement water film thickness was distinguished by Zhang[7], thus the water film thickness calculation formula was derived under different rainfall intensity, slope length, cross slope and longitudinal slope conditions by using Xie Cai formula and Manning formula:

$$d = 0.13589\left[\frac{InL_x\left(S_x^2+S_y^2\right)^{1/4}}{S_x}\right]^{3/5} \tag{7}$$

where, n is the roughness coefficient; L_x is the length of cross slope; S_x and S_y are the cross slope and the longitudinal slope respectively.

Chen[8] derived the basic differential equation of water film thickness based on the continuity equation and momentum equation of overland flow:

$$\left(h^4+a_1hL^2\right)\frac{dh}{dL}+ih^3+a_2h^3-a_3Lh^2-a_4L=0 \tag{8}$$

where, $a_1=\frac{I^2\cos^2\alpha}{g}$, $a_2=\frac{I\cos\alpha\sin\left(\alpha+\beta\right)}{g}$, $a_3=\frac{2I^2\cos^2\alpha}{g}$, $a_4=\frac{3\upsilon I\cos\alpha}{g}$, α is the angle between the pavement slope and the horizontal plane; β is the Angle between the raindrop and the vertical direction.

Equation (8) is not a simple first-order differential equation and only approximate numerical solutions can be solved due to complex variable structure. The calculated results are compared with those of Anderson's empirical formula and the measured data and the comparisons show that although the trend is consistent, there are still some differences, especially significant differences with the measured data. In addition, the formula mainly considers the action of raindrops, and does not take into account the pavement width, roughness and other properties of the pavement itself.

4. The hydrodynamic models

According to the simulation dimension, hydrodynamic models can be divided into one-dimensional, two-dimensional and three-dimensional. The difficulty and workload of three-dimensional hydrodynamic model implementation are relatively large due to its complex algorithm, thus it is rarely used in the simulation of rainfall runoff on asphalt pavement.

The one-dimensional hydrodynamic model takes the Saint-Venant equation as the governing equation. According to the simplification degree of the momentum equation, the models of kinematic wave (ignoring the inertia term and pressure term), diffusion wave (ignoring the inertia term) and dynamic wave are obtained, as shown in Equation (9-10) below. Most overland flow problems are simulated by one-dimensional simplified model due to the simulation instability caused by small water depth. Cristina[9] simulated 11 rainfall runoff events based on the kinematic wave model of Lax-Wendroff finite-difference scheme, and studied the influence of urban pavement and traffic on storm runoff. Staufer[10] applied the complete dynamic wave model to simulate the water level of a runway with transverse grooves at an airport. Naian[11] proposed a hydrologic-hydrodynamic model to simulate urban dual drainage system, in which the rainfall runoff module was solved by the kinematic wave approximation, and the road network module was solved by the complete Saint-Venant equation. The one-dimensional hydrodynamic model has the advantages of both high computational efficiency and less data requirements, however, it cannot describe the two-dimensional characteristics of runoff on the pavement surface.

The continuity equation:

$$\frac{\partial Q}{\partial x}+B\frac{\partial h}{\partial t}=0 \tag{9}$$

The momentum equation:

$$\frac{\partial Q}{\partial t}+\frac{\partial}{\partial x}\left(\frac{Q^2}{A}\right)+gA\left(I_E-I_S\right)=0 \tag{10}$$

where, Q is the discharge; B is the section width; h is water depth; t is time; g is the gravity acceleration; A is the area of water; I_E is hydraulic slope; I_S is friction gradient.

The two-dimensional hydrodynamic model is developed based on the two-dimensional shallow water equation, as shown in Equation (11) and usually simulated by the following steps: (1) geometric modeling; (2) grid generation; (3) equation discrete processing; (4) set the boundary conditions, initial conditions, calculation time and time step, and apply the power source; (5) the post-processing of the result; (6) model verification.

$$\frac{\partial U}{\partial t}+\frac{\partial F}{\partial x}+\frac{\partial G}{\partial y}=S \tag{11}$$

where, $U=\begin{pmatrix} h \\ hu \\ hv \end{pmatrix}, F=\begin{pmatrix} hu \\ hu^2+gh^2/2 \\ huv \end{pmatrix}, G=\begin{pmatrix} hv \\ huv \\ hv^2+gh^2/2 \end{pmatrix}, S=S_o+S_f=\begin{pmatrix} 0 \\ ghS_{ox} \\ ghS_{oy} \end{pmatrix}+\begin{pmatrix} 0 \\ -ghS_{fx} \\ -ghS_{fy} \end{pmatrix}$, u and v denote velocities in x and y directions; S_{ox} and S_{oy} are source terms of bottom slope in x and y directions; S_{fx} and S_{fy} are friction source terms in x and y directions, the specific expression is:

$$S_{fx}=\frac{n^2u\sqrt{u^2+v^2}}{h^{4/3}}, S_{fy}=\frac{n^2v\sqrt{u^2+v^2}}{h^{4/3}}$$

This model has been widely used in the urban flooding simulation, but is still rarely used in rainfall runoff on highway pavement. It takes into account the geometric parameters of the pavement, which can visually describe the two-dimensional geometric characteristics of the pavement and exactly reflect the movement characteristics of the pavement runoff under the influence of geometric alignment, pavement width, slope, roadside drainage and other factors. It reduces the dependence of empirical model on basic data and makes up for the lack of accuracy caused by the complicated solving of the theoretical model. Therefore, it has also attracted the attention of more and more researchers. Chen[12] introduced the pore source term into the momentum equation of two-dimensional shallow water equation to describe the influence of pavement macro structure on rainfall runoff. Ma et al.[13] developed a two-dimensional hydrodynamic model of asphalt pavement, which had been applied to study the influence of pavement width and slope on the water depth distribution and concentrated drainage. Wang et al.[14] adopted the DPM model and EWF model of computational fluid dynamics software Fluent to simulate the water depth distribution on the pavement surface under rainfall conditions.

Inspired by this, a hydrodynamic model is developed for rainfall–runoff on asphalt pavement using two-dimensional shallow water equations by our research group[15], which has been applied to analysis the temporal and spatial characteristics of pavement runoff. Fig.2 shows the spatial distribution of water depth on both straightline and circular curve segments. It shows that the pavement runoff presents a significant two-dimensional characteristics in spatial. The water depth is increasing along the longitudinal and cross slope of the pavement and the obvious ponding zone is formed in both straightline and circular curve segments.

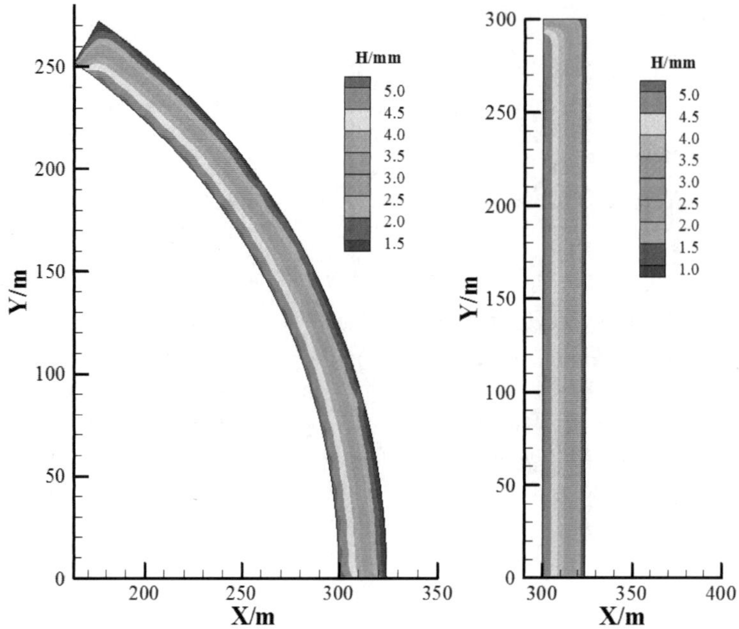

Fig.2 Spatial distribution of water depth on both straightline and circular curve segments

5. Conclusions

(1) The empirical models are simple, but lack of adaptability due to the dependence on a large number of measured data. Therefore, when choosing empirical model to predict water film thickness on the pavement, we should first verify whether the model can meet the prediction accuracy requirements. The theoretical model is difficult to solve and can only approximate numerical solutions can be obtained. How to improve the accuracy of the theoretical model is the direction of future research.

(2) The hydrodynamic models allow for the attributes of the pavement itself, which are mainly used to study the dynamic variations of the hydraulic parameters of the pavement rainfall runoff. The modeling is relatively complex, but it can describe the time-varying process of rainfall runoff on pavement more precisely, which provides a more accurate method for studying the mechanism of rainfall runoff on asphalt pavement.

(3) Pavement flooding is the result of the comprehensive action of multiple factors such as pavement alignment, width, slope, drainage facilities and rainfall process. Most of the existing studies are only focused on the variation of pavement runoff in straight sections or under a single line type. In the future, attention should be paid to the analysis of rainfall runoff on complex pavement alignment, which will provide technical support for the optimization of pavement drainage structure and the utilization of rainfall resources in the region of highway.

References

[1] Ross N.F. & Russam K. (1968) The depth of rain water on road surfaces. England: Ministry of Transport, RRL (Road Research Laboratory) report LR 236.

[2] Anderson, J.A. (1995) Depth of rain water on road surfaces. Highway & Transportation, 42(5): 45-49.

[3] Gallaway B M, Schiller R E, Rose J G. (1971) The effects of rain fall intensity, pavement cross slope, surface texture, and drainage length on pavement water depths. College station: Texas Transportation Institute, 1971.

[4] Wambold J C, Henry J J, Hegmon R R. (1983) Evaluation of pavement surface texture significance and measurement techniques. Wear, 83(2):351-368.

[5] Ji Tianjian. (2004) Influence of rainfall on tyre and road adhesion coefficient. Southeast University.
[6] Luo J, Liu J B, Wang Y Q. (2015) Validation test on pavement water film depth prediction model. China J. Highw. Transp. 28(12): 57-63.
[7] Zhang L, Zhang Z. (2013) Impact of road slope on water film thickness. Journal of Chongqing Jiaotong University(Natural Science), 32(3): 404-406, 423.
[8] Chen F P. (2014) Study on the theoretical calculation model of rain water depth on road surface. Modern Transportation Technology, 11(3): 12-15.
[9] Cristina, C., Sansalone, J. (2003) Kinenatic wave model of urban pavement rainfall-runoff subject to traffic. J. Environ. Eng. 129(7): 629-636.
[10] Staufer, P., Siekmann, M., Loos, S., et al. (2012) Numerical modeling of water levels on pavements under extreme rainfall. J. Transp. Eng. 138(6): 732-740.
[11] Nania, L.S., Leon, A.S., Garcia, M.H. (2015) Hydrologic-hydraulic model for simulating dual drainage and flooding in urban areas: application to a catchment in the metropolitan area of Chicago. J. Hydrol. Eng. 20(5):04014071.
[12] Chen X H, Geng Y F, Jiang Q L, et al. (2017) Innovative approach for pavement runoff characterization. J. Perform. Constr. Fac. 31(5): 04017047.
[13] Ma Y L, Chen X H, Geng Y F, et al. (2019) Influence factor and response identification for asphalt pavement central drainage. China J. Highw. Transp. 32(04): 122-129.
[14] Wang H L, Guo Z H, Li Y G, et al. (2018) Analysis of the influence factors on the characteristics of thin layer flow on the road surface. Transportation Science and Technology, 2: 1-4.
[15] Wang X, Chen H, Ni D, et al. (2021) Study on hydrodynamic characteristics and influence factors of asphalt pavement runoff. Water Sci. Technol. 84(12): 3928-3940.

Broadband measurement method for acoustic performance parameters of large sample materials

Shang Shi, Shijin Lyu

Nation Key Laboratory on Ship Vibration and Noise, China Ship Science Research Center, Wuxi, 21400, China

*Corresponding author: shawnshangs@163.com

Abstract: In order to measure the acoustic performance parameters of large sample acoustic materials under water and eliminate the influence of wall reflection in limited space, the principle of inverse filtering and transfer function method is used to establish a wide frequency transfer function method for underwater materials under the vertical incidence of sound waves. The acoustic performance test of the 10mm uniform steel plate model shows that the broadband transfer function method is used to improve the measurement, and in the frequency band above 500Hz, the reflection coefficient test deviation is less than 0.1, and the sound insulation test deviation is less than 2dB. Effectively reduced the test limit of the broadband pulse method for underwater measurements.

1.INTRODUCTION

The design of mechanical vibration isolation equipment for submarines and the application of surface acoustic coverage are the main design directions of submarine underwater acoustic stealth. For the performance test of acoustic overlay materials applied to the surface of submarines, it is possible to provide a reference for material design and evaluate the application effect of the material.

Currently, underwater acoustic material measurements typically cut the material into small samples for measurement in the water-filled tube. Traditional measurement approaches and approaches for acoustic performance parameters widely adopted today, including pluse-separation method[1] and transfer function method[2,3]are carried out in the water-filled tube. In order to design the low target strength and the vibration isolation structure of the acoustic material, the overall performance cannot be measured by using a small sample size, so it is necessary to measure and analyze the large sample. In addition, to reduce the influence of sample edges on the measurement, the sample size is at least five times the wavelength in water[4]. The measurement of large samples of underwater acoustic materials is generally carried out in anechoic tank or pressure tank. However, the sound-absorbing material laid in the anechoic tank does not absorb 100 % of the sound energy[5], and its reflection has a great influence on the free sound field in the anechoic tank, which affects the accuracy of acoustic performance measurement. Problems of transmitting signal distortion caused by the transient process of the transducer in-field measurement of a large sample by traditional pulse method.

Wideband pulse method is a wideband test method that can obtain measurement information from all frequency points in the frequency band at once and solve the interference problem from the wall by sending wide frequency narrow pulse measurement. Successfully generating a controllable pulsed acoustic measurement signal is a key issue in the wideband pulse method. Piquette[6] proposed a transient suppression method to improve the transient process of the pulse signal transmitted by the transducer so that the pulse source can emit short-period CW pulses, but it is necessary to know the

Journal of Physics: Conference Series **2458** (2023) 012039 doi:10.1088/1742-6596/2458/1/012039

internal electro-acoustic performance parameters of the transducer in advance. In recent years, the broadband pulse method has been studied a lot[7,8,9]. These studies are mostly carried out in the acoustic tube, using the principle of inverse filtering, a stable broadband pulse is generated in the acoustic tube, and robust results can still be obtained in the low frequency band of 500-2000Hz measured in practice. At present, there are fewer wideband measurement studies on large samples, and considering the complexity brought about by the size of large samples, the effective frequency of testing is still high

In this paper, the broadband pulse method is used to measure the acoustic material of underwater large samples. The broadband short pulse combined with inverse filtering technology is used to generate a steady-state broadband pulse. The broadband transfer function method is proposed to measure the performance of acoustic materials., the method is theoretically deduced and simulated, and the experimental verification is carried out in the acoustic pressure barrel.

2. MEASUREMENT PRINCIPLE

When using the traditional pulse method and the transfer function method for acoustic performance parameter measurements, the lower limit of the test frequency is higher due to the limited space. As shown in Figure 1, the sound source emits pulse waves, and the hydrophone will receive direct waves, specimen reflected waves, and wall reflected waves at the same time, which has a greater impact on the accuracy of the test. Furthermore, broadband measurements are time-consuming. Using a broadband pulse as the driving signal, due to its short-term characteristics of the time domain and the rich frequency domain information, it can effectively overcome the interference of limited space, and can measure all the frequency data in the frequency band at one time to improve the detection efficiency.

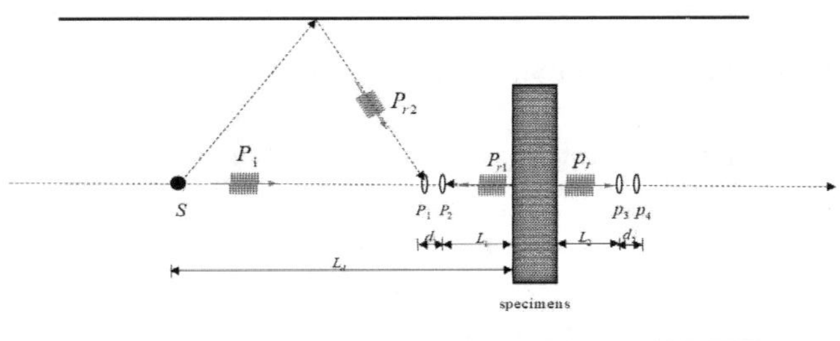

Figure 1. Measurement model of the broadband transfer function method

The first problem to be solved in the measurement of the broadband pulse method is the generation of the pulse. The impulse response of the Butterworth signal is used as the excitation signal to drive the transducer. First, the signal is amplified by the power amplifier and then input into the pulsating sphere source to generate the signal. The Butterworth frequency domain signal has the characteristics of flat in the frequency band and low-pass characteristic that the amplitude is cut off with the increase of frequency. The Butterworth pulse can be obtained by inverse Fourier transform of the Butterworth low-pass frequency domain signal.

The system response can be calculated by the ratio of the received signal spectrum $H_y(\omega)$ to the computer-transmitted signal spectrum $H_x(\omega)$

$$H(\omega) = \frac{H_y(\omega)}{H_x(\omega)} \tag{1}$$

To avoid instabilities at the notches of $H(\omega)$, a positive constant, τ , must be added to the denominator (regularization). Thus,

$$G(\omega)=\frac{H^{*}(\omega)}{\left|H(\omega)^{2}\right|+\tau} \tag{2}$$

$$H_{x}(\omega)=H_{y}(\omega)G(\omega) \tag{3}$$

Where $H^{*}(\omega)$ stands for the conjugate of $H(\omega)$. In general, the small values of τ (typically less than 5% of the transfer-function peak) are usually enough to stabilize the method.

By applying an inverse Fourier transform to Eq. (3), the driving signal of the transducer-pressure vessel system is obtained, after which it is fed back into the transducer and a single pulse wave is obtained. In addition, it should be noted that by using this process, any deficiencies in the frequency response of the sound generation system will be compensated for, which should improve the signal-noise ratio for measurements at lower frequencies.

Based on the broadband pulse separation method, this paper proposes the broadband transfer function method. When the sound source is close to the 1m of the specimen, and two hydrophones are placed at 5cm and 10cm on the surface of the specimen, the hydrophone is closer to the test sample, and the received pulse waveform can be superimposed in the form of a plane wave, and the incident sound field is (ignoring the time factor):

$$p=Ae^{-ikx}+Be^{ikx} \tag{4}$$

where A is the incident wave amplitude and B is the reflected wave amplitude.The transmitted sound field is:

$$p_{t}=A_{t}e^{-ikx}+B_{t}e^{ikx} \tag{5}$$

where A_t is the transmitted wave amplitude and B_t is the transmitted wave by the interface reflected wave amplitude.

Then the 4 position hydrophone receives the sound pressure as:

$$p_{1}=Ae^{ikL_{1}}+Be^{-ikL_{1}} \tag{6}$$

$$p_{2}=Ae^{ik(L_{1}+d_{1})}+Be^{-ik(L_{1}+d_{1})} \tag{7}$$

$$p_{3}=A_{t}e^{-ikL_{2}}+B_{t}e^{ikL_{2}} \tag{8}$$

$$p_{4}=A_{t}e^{-ik(L_{2}+d_{2})}+B_{t}e^{ik(L_{2}+d_{2})} \tag{9}$$

Perform self-spectral and cross-spectrum analysis of the test pulse signals of the test positions p_1, p_2, p_3 and p_4 sensors to define the transfer function:

$$H_{1}=G_{p_{1}p_{2}}/G_{p_{1}p_{1}},H_{2}=G_{p_{1}p_{3}}/G_{p_{1}p_{1}},H_{3}=G_{p_{1}p_{4}}/G_{p_{1}p_{1}} \tag{10}$$

Among them, $G_{p_1p_2}$ is the mutual spectrum of the pulse signal received by p_1 and p_2 sensors.

The material reflectance coefficient is $R=B/A$, and the transmission coefficient is $T=A_t/A$. The available material reflection coefficient R, transmission coefficient T and sound absorption coefficient α and sound insulation amount TL are:

$$R=\frac{e^{ikd_{1}}-H_{1}}{H_{1}-e^{-ikd_{1}}}e^{i2kL_{1}} \tag{11}$$

$$T = \frac{H_2 e^{ikd_2} - H_3}{H_1 - e^{-ikd_1}} \frac{\sin(kd_1)}{\sin(kd_2)} e^{ik(L_1+L_2)} \tag{12}$$

$$TL = -20\log_{10}|T| \tag{13}$$

3. EXPERIMENTAL RESULTS

According to the scheme of Figure 1, the experimental facilities are arranged, and the 10mm uniform steel plate is tested by driving the transmission transducer with a 5th order cutoff frequency Butterworth pulse signal. The hydrophone can receive the direct wave emitted by the transducer, the reflected wave of the material specimen, the wall reflected wave, and the received waveform is shown in Figure 2, the system response of the inverse filter, as well as the filtered signal spectrum. The incident wave is separated from the time domain and compared with the excitation signal, as shown in Figure 2(a), due to the transient response of the transducer, it takes a certain amount of time to form a steady-state waveform, and the distortion is more serious in the early stage of the waveform. The signal-to-noise ratio is very low, and the coincidence with the excitation signal is higher after the transducer reaches a steady state. There is a certain amount of interference between the incident wave of the transducer and the reflected wave of the specimen, which is caused by the inherent interference of the experimental facility and the edge reflection of the material itself. Hence, the error is large when directly separating the pulse wave to calculate the acoustic properties of the material, so it is necessary to calculate the deconvolution calculation of the reverse filter to form a stable Butterworth pulse wave in the anechoic pressure barrel.

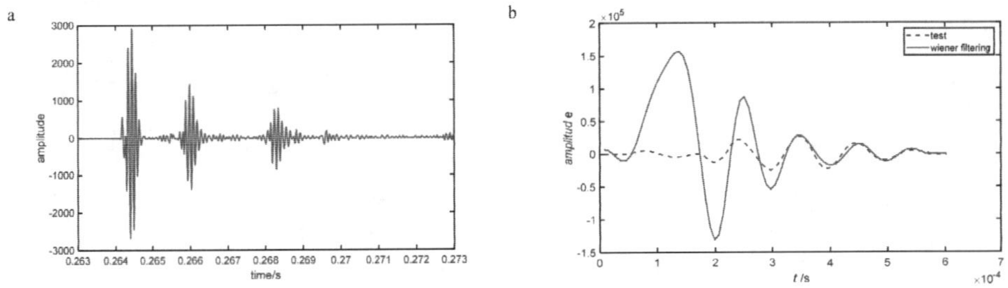

Figure 2. Hydrophones using the broadband pulse method receive signals(a)Direct waves and reflected waves received by hydrophones.(b)The drive signal is compared to the actual received signal.

The direct and reflected waves of the test are separated using a time window. The inverse filter of the deconvolution is calculated using the direct wave signal and the excitation signal, and the transient response of the transmitting transducer and the edge reflection of the material is compensated by inverse filtering. By selecting the regularization parameter (usually a small constant) and giving a deconvolution filter, the frequency domain signal calculated by deconvolution is inversely transformed into Fourier to obtain the time domain signal.

The receiver signal is used to make measurements of the broadband transfer function method. First of all, the driving signal of the broadband pulse method is used as the excitation, and the signals of hydrophones p_1 and p_2 are used as a transfer function to calculate the corresponding acoustic parameters. The experiment uses the compensation signal of the broadband pulse method to continue to drive the transducer, because the distance between the adjacent two hydrophones is only 5cm. The transmission loss of sound waves can be approximately ignored, and due to the characteristics of the transfer function calculation, it has a better measurement effect in the low-frequency band. Figure 3 shows the signals received by hydrophones p_1 and p_2 .The measurement results of the broadband transfer function method as shown in Figure 4, in the low-frequency band of 500Hz, the traditional single-frequency pulse transfer function method has been unable to accurately measure, and the broadband transfer function method can still obtain results consistent with the theoretical value. The average error between the reflection coefficient and the theoretical value measured by the broadband transfer function method is less than 0.1, and the average error of the sound insulation test deviation is less than 2dB.

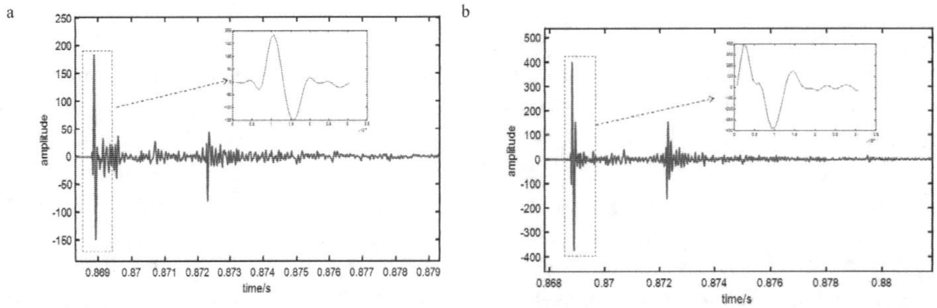

Figure 3. The drive signal using the broadband pulse method, the broadband transfer function method is measured, and the hydrophone on the surface of the specimen receives the signal.(a)A 5 cm hydrophone on the surface of the specimen receives a signal.(b)A 10 cm hydrophone on the surface of the specimen receives a signal.

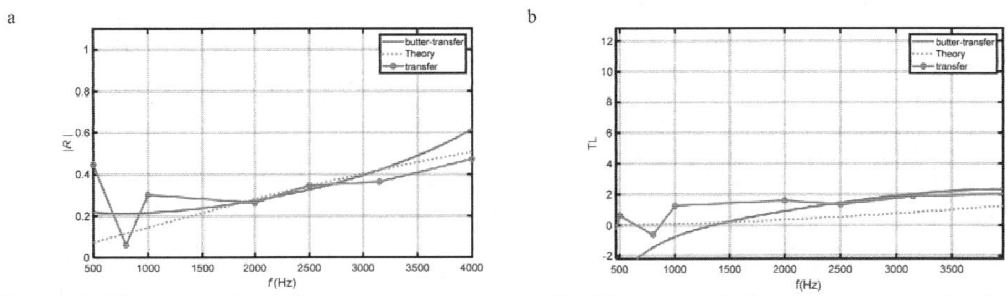

Figure 4. The acoustic performance parameters of a 10 mm steel plate are measured using the broadband transfer function method.(a)Specimen reflection coefficient measured using the broadband transfer function method.(b)The amount of sound insulation of the specimen measured using the broadband transfer function method.

4. CONCLUSIONS

After theoretical and experimental proof, the broadband pulse method can be applied to the field measurement of large samples of acoustic materials, and the measurement data of all frequency points in the frequency band can be obtained at one time, which greatly improves the measurement efficiency.

REFERENCES

[1]SEYBERT A F, ROSS D F. Experimental determination of acoustic properties using a two-microphone radon-excitation technique[J]. The Journal of the Acoustical Society of America, 1977, 61(5): 1362-1370.

[2]CHUNG J Y, BLASER D A. Transfer function method of measuring in-duct acoustic properties[J]. The Journal of the Acoustical Society of America, 1980, 68(3): 907-921.

[3]CHUNG J Y, BLASER D A. Transfer function method of measuring acoustic intensity in a duct system with flow[J]. The Journal of the Acoustical Society of America, 1980, 68(6): 1570-1577.

[4]Bobber R J. Underwater electroacoustic measurements[M]. Naval research laboratory, 1970.

[5]Corsaro R D, Klunder J D, Jarzynski J. Filled rubber materials system: application to echo absorption in waterfilled tanks[J]. The Journal of the Acoustical Society of America, 1980, 68(2): 655-664.
[6]Piquette J C. Method for transducer transient suppression. I: Theory[J]. The Journal of the Acoustical Society of America, 1992, 92(3): 1203-1213.
[7]Cobo P. Application of shaping deconvolution to the generation of arbitrary acoustic pulses with conventional sonar transducers[J]. Journal of sound and vibration, 1995, 188(1): 131-144.
[8]Sun L, Hou H. Measurement of sound absorption by underwater acoustic material using pulse-separation method[J]. Applied acoustics, 2014, 85: 106-110.
[9]Wei Z, Hou H, Gao N, et al. Sound absorption coefficient measurement by extracting the first reflected wave in a short tube[J]. Applied Acoustics, 2020, 159: 107087.

A Control Algorithm of Active Wave Compensation System Based on the Stewart Platform

Huiyuan Wu[1], Wenlin Yang[1]*, Muk Chen Ong[2], Tianhui Fan[3], Guang Yin[2], Weixiang Zeng[1], Weilun Situa[1], Yunting Wang[1]

1. Guangdong Institute of Intelligent Unmanned System, Guangzhou, China;
2. Department of Mechanical and Structural Engineering and Materials Science, University of Stavanger, Stavanger, Norway
3. School of Civil Engineering and Transportation, South China University of Technology, Guangzhou, China.

*Correspondence: yangwenlin@gis.sia.cn

Abstract. Aim at the actual engineering requirements of wind power operation and maintenance under complex sea conditions, a control method of the active wave compensation system for maintenance ships based on the Stewart platform is presented. The kinematics of the platform is analyzed, and the coordinate transformation, pose, and inverse solutions are analyzed and calculated. The multi-body dynamics simulation model is established by using MATLAB. For the problem of the load nonlinearity and strong coupling of the nonlinear Stewart platform, an active wave compensation active disturbance rejection control (ADRC) is built to attain the high-precision control of the six-degree-of-freedom (6-DOF) Stewart platform. The numerical simulation shows that the proposed control scheme has good tracking accuracy and strong anti-interference ability.

1. Introduction

In recent years, under the requirements of carbon peaking and carbon neutrality, the offshore wind power industry has received increasing attention. Wind power operation and maintenance are an important part of the industrial chain. It is also favored by the market[1-3]. However, offshore weather changes rapidly, the natural conditions are complex. Therefore, effective solutions need to be developed to address the difficult challenges of offshore wind operation and maintenance[4].

When the operation and maintenance ship sends professionals to the operation site, the ship will have six degrees of freedom movement under the influence of the Marine environment. If not handled properly, it is easy to have safety accidents[5]. The boarding trestle has a relatively large movement under high sea conditions, which will cause serious safety hazards to the boarding operators. To ensure the safety of personnel, the ship needs to be equipped with a device that can actively compensate the six degrees of freedom motion. Compared with the ordinary gangway, the active wave compensation trestle can remain relatively stable under complex ocean conditions, meeting the safety requirements of personnel boarding, leaving the wind tower, and effectively extending the operation and maintenance window period[6].

However, the control system of the active wave compensation trestle has problems such as large inertia, large time delay, and multi-parameter coupling[7-9]. In response to these problems, engineers and technicians have carried out a lot of research work. In some studies[10], a solution using wave compensation value was proposed, and a compensation device based on a micro inertial navigation sensor was designed. Aim at decreasing the influence of sea wind and waves on ship operation and maintenance, a wave compensation platform was proposed[11]. A boarding trestle with an active compensation function was explored to meet the commuting needs of operation and maintenance under high sea conditions[12]. An intelligent pseudo-differential control algorithm was studied to overcome the influence of wind and waves[13]. The author improved the particle swarm method to enhance the controller parameters of the wave compensation control system[14]. In this paper, a parallel 6-DOF Stewart platform is selected as the compensation device. The platform is a nonlinear system with complicated coupling relationship between inputs and outputs[15-17]. And the traditional PID algorithm is difficult to meet the control requirements in the face of nonlinear systems. For the problem of parameter uncertainty and unknown interference in the Stewart platform, a sliding mode control method with a boundary layer was presented[18]. ADRC has great advantages in dealing with nonlinear systems and has been favored by researchers. The authors designed a permanent magnet synchronous motor servo control system by combining ADRC with PID control[19]. A strategy of combining a new nonlinear function and an extended state observer to form an ADRC system was designed to overcome the problem of low control accuracy for coal mine robots working in complex coal mine environments[20].

However, there is still a lack of research on applying ADRC to the active wave compensation control of the 6-DOF Stewart platform with nonlinear characteristics. The Stewart platform is adopted in this paper to meet the engineering requirements of the ship in the course of operation. The active wave compensation mechanism of the boarding trestle adopts the ADRC method to realize the tracking control of the 6-DOF platform. The result of simulation show the tracking effectiveness and robustness of the strategy.

The remainder of this article is organized as follows. In section 2, the kinematic analysis of the Stewart platform is described. Section 3 presents the single-channel model of hydraulic servo channels. In section 4, the controller is designed. In section 5, a simulation is conducted to verify the effectiveness of the proposed control, and a summary is provided in section 6.

2. KINEMATIC ANALYSIS OF THE STEWART PLATFORM

2.1. Position and attitude description of the platform

The model of the 6-DOF Stewart platform is displayed in Figure 1. The platform includes an upper platform, a lower platform, an upper and lower Hooke hinge, a servo actuator, and a base. The servo actuator receives the control signal of the servo control system and drives the electric cylinder or hydraulic cylinder to move. The upper platform can realize the movement in the roll, pitch, yaw, X, Y, and Z directions, thus providing six degrees of freedom motion compensation for the active wave compensation boarding trestle installed on the upper platform.

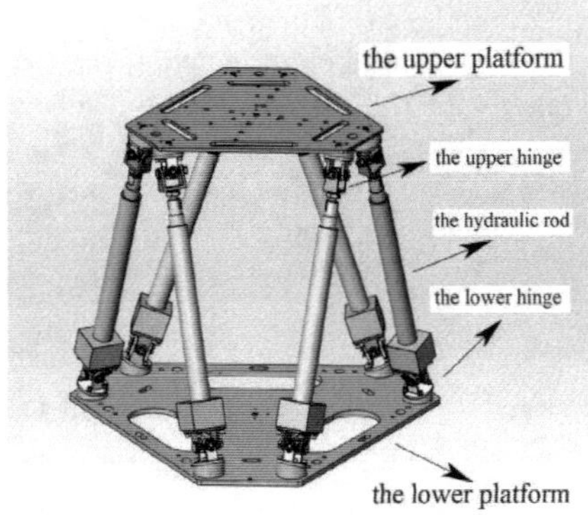

Figure 1 Stewart platform schematic

The platform coordinate system is established to show the mobile platform's position and altitude, as shown in Figure 2. A static coordinate system {B} $(O_b - X_b Y_b Z_b)$ is settled on the base and a dynamic coordinate system {P}$(O_p - X_p Y_p Z_p)$ on the upper platform. The position description of the Stewart platform is represented by the relative position of the moving coordinate system of the upper platform relative to the static coordinate system of the base. It is expressed as $t = [X_1, Y_1, Z_1]^T$. Where, X_1 , $Y_1$and Z_1 represent the longitudinal, lateral, and vertical motions of the mobile platform relative to the static coordinate system respectively. The altitude of the platform is described by the method of rotating the coordinate system. The Euler angle $[\phi, \theta, \psi]^T$ are obtained by three consecutive rotations of the upper platform, among which $\phi,\ \theta,\ and\ \ \psi$ respectively define the rolling, pitching, and yaw rotational motions of the platform around the three axes of the static system. The rotation matrix can be obtained according to the following equations.

$$R = \begin{bmatrix} c\psi c\theta & c\psi s\theta s\phi - s\psi c\phi & s\psi s\phi + c\psi s\theta c\phi \\ s\psi c\theta & c\psi c\phi + s\psi s\theta s\phi & s\psi s\phi c\phi s\phi \\ -s\theta & c\theta s\phi & c\theta c\phi \end{bmatrix} \tag{1}$$

Where, $s = sin$, and $c = cos$.

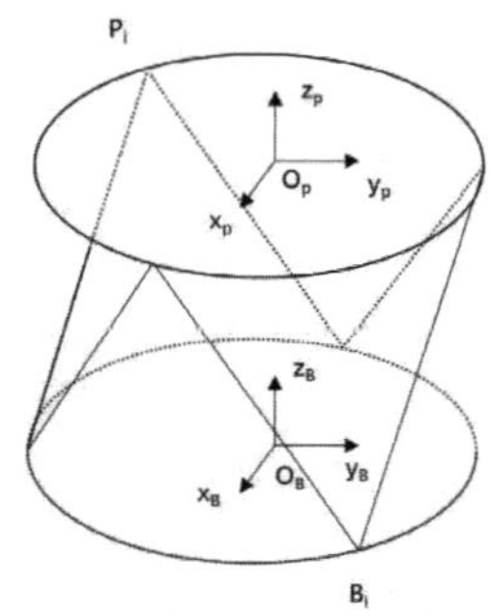

Figure 2 Schematic diagram of platform coordinates

2.2. Position inverse analysis

In the case of obtaining the parameters of the platform structure and the position and altitude of the upper platform, the process of solving the telescopic lengths of all its driving mechanisms belongs to the inverse solution process of solving the kinematics of the 6-DOF platform. According to the geometric significance, the vector of the driving mechanism can be expressed to analyze the *i* platform drive mechanism shown in Figure 3.

$$\vec{l_\iota} = R\vec{d_\iota} + \vec{t} - \vec{b_\iota} (i = 1, \dots, 6) \tag{2}$$

Among them, *d* represents the coordinates of the hinge points of each driving mechanism in the moving coordinate system, and *b* represents the coordinates of the lower hinge points of each driving mechanism in the static coordinate system. The length of the driving mechanism can be from the Euclidean norm.

$$\|\vec{l_\iota}\| = \left(\vec{l_\iota}^T \vec{l_\iota}\right)^{\frac{1}{2}} \tag{3}$$

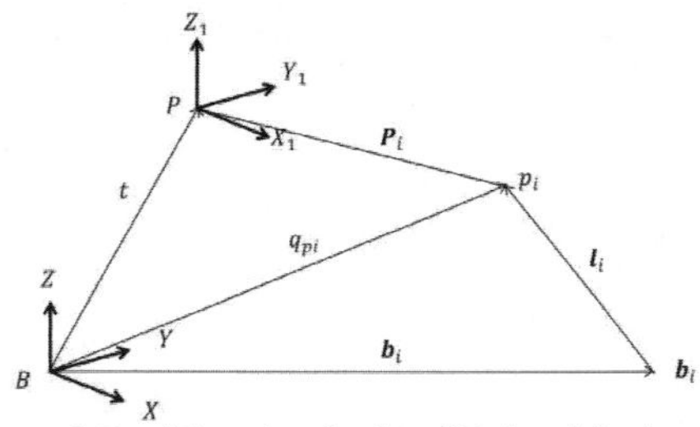

Figure 3 Position Analysis of Drive Mechanism

3. ACTIVE DISTURBANCE REJECTION CONTROLLER

For high-frequency signals, the ADRC has designed a transition process method, which is similar to low-pass filtering a given command to obtain an easier-to-implement command, thereby greatly reducing overshoot while sacrificing part of the rapidity. In the PID controller, one of the main functions of the integral is to eliminate disturbances, but the integral works relatively slowly and can also cause an overshoot. To improve this shortcoming, the ADRC uses an extended state observer to observe target. The estimated value is used as compensation to cancel out the interference. The traditional linear feedback method has shortcomings in the convergence speed and anti-interference ability. Active disturbance rejection control replaces the traditional gain with a nonlinear function and replaces the linear feedback with nonlinear feedback. There are six driving mechanisms on the 6-DOF platform, and the control principles of these six driving mechanisms are the same. The control algorithm of the ADRC controller is as follows.

(1) The tracking differentiator (TD) for the transition process is as follows.

$$v_1(k+1) = v_1(k) + hv_2(k) \tag{4}$$

$$v_2(k+1) = v_2(k) + hfst(v_1(k) - v_0(k), v_2(k), r, h_0) \tag{5}$$

The output v_1 of the TD tracks the input v. And v_2 tracks the differential of v. The parameters that need to be tuned include the speed factor r, the integration step size h, and the filter factor h_0. And the function $fst(v_1(k) - v_0(k), v_2(k), r, h_0)$ is expressed as follows.

$$fst(x_1, x_2, r, h) = -\begin{cases} rsgn(a), |a| > b \\ r\dfrac{a}{d} \quad , |a| < d \end{cases} \tag{6}$$

Where, sgn is a symbolic function, and a and d can be designed.

$$a = \begin{cases} x_2 + \dfrac{a_0 - d}{2} sgn(y), |y| > d_0 \\ x_2 + \dfrac{y}{h}, |y| < d_0 \end{cases} \tag{7}$$

$$\begin{cases} d = rh \\ d_0 = hd \\ y = x_1 + hx_2 \\ a_0 = (d^2 + 8r|y|)^{\frac{1}{2}} \end{cases} \tag{8}$$

(2) The extended state observer is used.

$$\begin{cases} \varepsilon = z_1(k) - y(k) \\ z_1(k+1) = z_1(k) + h[z_2(k) - \rho_1\varepsilon] \\ z_2(k+1) = z_2(k) + h[z_3(k) + \rho_2 fal(\varepsilon, a_1, \delta) + b_0 u(k)] \\ z_3(k+1) = z_3(k) + h\rho_3 fal(\varepsilon, a_1, \delta) \end{cases} \tag{9}$$

Among them, ρ_1, ρ_2, and ρ_3 are called system error correction gains, which are the observer coefficients of extended state observer. Selecting appropriate parameters can well estimate the disturbance. The expression of the function *fal*() is established below.

$$fal(\varepsilon, a_1, \delta)\begin{cases} |\varepsilon|^{a_1} + sgn(\varepsilon), |\varepsilon| > \delta \\ \dfrac{\varepsilon}{\delta^{1-a_1}}, |\varepsilon| < \delta \end{cases} \tag{10}$$

(3) The nonlinear state error feedback is given.

$$\begin{cases} e_1 = v_1(k) - z_1(k) \\ e_2 = v_2(k) - z_2(k) \\ u_0 = \rho_1 fal(e_1, a_1, \delta) + \rho_2 fal(e_2, a_2, \delta) \\ u(k) = u_0(k) - \dfrac{z_3(k)}{b_0} \end{cases} \tag{11}$$

Where, u_0 is the control quantity, $u(k)$ is the total output, and z_3 is the disturbance estimator in the control system.

4. Simulation analysis

The structure sketch map of the simulation system is displayed in Figure 4. The desired pose of the moving platform is obtained according to the desired trajectory, and the desired displacement of each driving mechanism paired with the desired trajectory of the upper platform is obtained by the inverse kinematics solution. Then the expected displacement, actual displacement, and speed of each driving mechanism are used as the input of the controller. And the output of the controller is the force acting on each driving mechanism.

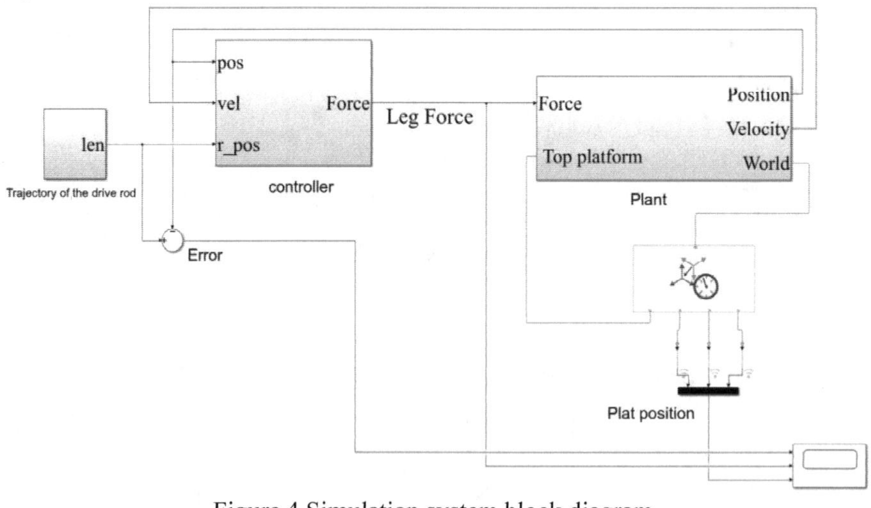

Figure 4 Simulation system block diagram

The driving mechanism structure is described below. The upper and lower ends are linked to the upper and lower platforms by a Hook hinge. There is a hydraulic cylinder in the middle of the rod that connects the upper and lower parts of the rod. As the upper platform moves, causing the length of the rod to change and the sensor to measure the displacement and speed of the rod.

Figure 5 shows the tracking trajectories of the six driving mechanisms using the ADRC method, where, the abscissa is time, the ordinate is the drive mechanism displacement, the blue lines are the desired trajectory, and the red lines are the actual trajectory. Figure 6 demonstrates the tracking error of the driving mechanism by using the ADRC algorithm, and the tracking error is close to zero. It can be seen that the ADRC method can obtain high-precision tracking performance.

Journal of Physics: Conference Series **2458** (2023) 012040 doi:10.1088/1742-6596/2458/1/012040

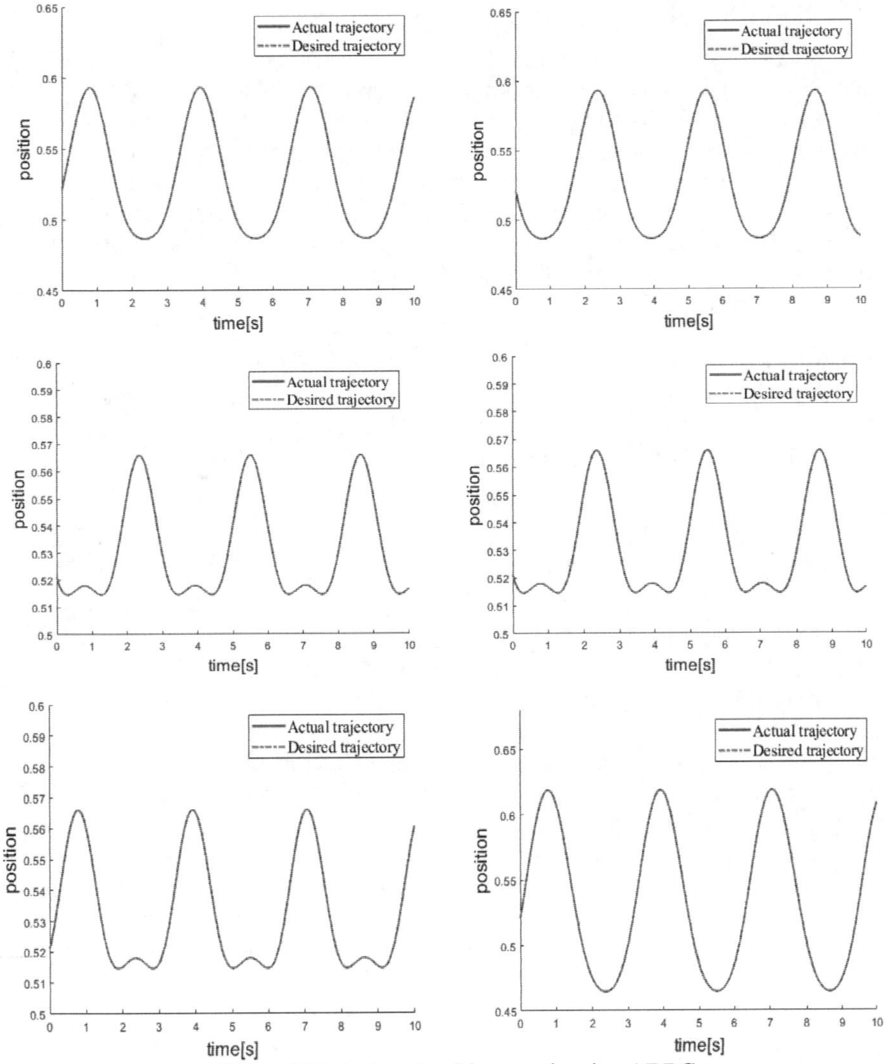

Figure 5 Trajectory tracking graph using ADRC

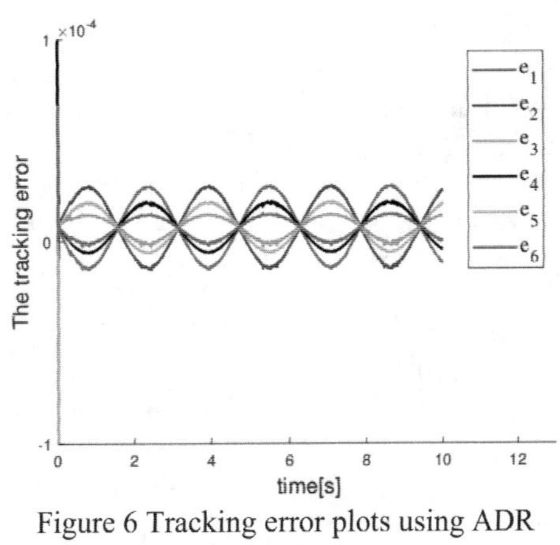

Figure 6 Tracking error plots using ADR

5. CONCLUSION

In this paper, a 6-DOF active wave compensation control algorithm of marine operation and maintenance ship boarding trestle based on the Stewart platform is developed and studied. Firstly, the kinematics analysis and calculation of the Stewart platform are carried out through MATLAB/ Simulink. Secondly, based on the nonlinearity and strong coupling problems of the Stewart platform and combined with the active wave compensation automatic disturbance rejection control algorithm, a six-degree-of-freedom active wave compensation simulation experiment was carried out. The simulation experiments show that the active wave compensation active disturbance rejection control strategy has excellent tracking property and well anti-interference ability, which offers an effective solution for the high-precision active wave compensation control of wind power operation and maintenance ship boarding trestle bridges.

Acknowledgments

This work was supported in part by the International Science and Technology Cooperation Project of Guangdong Province under Grant 2022A0505050027, and the Provincial Basic and Applied Basic Research Fund Project of Guangdong Province under Grant 2021A1515012552.

References

[1] X. Yibo. Research on the watertight performance of offshore wind turbines. *Nanfang Agricultural Machinery* 2022.
[2] L. Yang, G. Li, Z. Zhang, X. Ma, and Y. Zhao. Operations and maintenance optimization of wind turbines integrating wind and aging information. *IEEE Transactions on Sustainable Energy*; 2021; pp. 211-221.
[3] A. Kisvari, Z. Lin, and X. Liu. Wind power forecasting a data-driven method along with the gated recurrent neural network. *Renewable Energy*; 2021; pp. 1895-1909.
[4] K. Y. Yan Xu and G. Zhao. The influencing factors and hierarchical relationships of offshore wind power industry in china. *ENVIRONMENTAL SCIENCE AND POLLUTION RESEARCH*; 2021; pp. 329-344.
[5] M. Li, X. Jiang, and R. R. Negenborn. Opportunistic maintenance for offshore wind farms with multiple-component age-based preventive dispatch. *Ocean Engineering*; 2021; pp. 109-115.
[6] L. Tang and Y. Chen. Dynamic model of four degrees of freedom rope-driven rigid-flexible hybrid wave compensation device. *Journal of Mechanical Science and Technology*; 2020; pp. 1937-1948.
[7] S. F. Lu Dongqing. Simulation Research on Active Wave Compensation Control Technology. *Ship*; 2012.
[8] C. Y. Chen. An Wanping, Research on hydraulic control principle of active wave compensation based on boarding trestle. *Equipment Manufacturing Technology*; 2019.
[9] J. Leng, C. Xu, M. Li, H. Du, and S. Liu. Control strategy for performing predictions for a semi-active compensation system. *Ocean Engineering*; 2021; pp. 137-148.
[10] G. W. Lu Daohua. Design and Simulation of Active Wave Compensation Platform and Its Test System. *Ship Engineering*; 2015.
[11] H. C. Tang Gang. Modeling and Simulation of Three-DoF Parallel Wave Compensation Platform. *journal of Shanghai Maritime University*; 2020.
[12] Z. W. Su Changqing. An active wave compensation method for boarding trestle bridges. *Ship and Ocean Engineering*; 2017.
[13] M. X. Yan Hualin. Application of Active Wave Compensation in Position Control System. *Manufacturing Automation*; 2017.
[14] T. G. P. Control system research in wave compensation based on particle swarm optimization. *Scientific Reports*; 2021.
[15] C. H. Zhou Ting. Electro-hydraulic servo six-degree-of-freedom composite loading device control system. *Modern Manufacturing Engineering*; 2021.

[16] Y. Cai, S. Zheng, W. Liu, Z. Qu, J. Zhu, and J. Han. Sliding-mode control of ship-mounted Stewart platforms for wave compensation using velocity feedforward. *Ocean Engineering*; 2021.
[17] Y. Cai, S. Zheng, W. Liu, Z. Qu, J. Zhu, and J. Han. The adaptive robust dual-loop control scheme of ship-mounted Stewart platforms for wave compensation. *Mechanism and Machine Theory*; 2021.
[18] T. L. Hu Yongpan. Robust control method for parallel wave compensation system. *Journal of National University of Defense Technology*; 2014.
[19] H. L. Wang Zhida. An active wave compensation method for boarding trestle bridges. *Computer and Digital Engineering*; 2022.
[20] J. X. Xu Weifeng. Research on Active Disturbance Rejection Control of Coal Mine Robot Trajectory Tracking. *Coal Mining Machinery*; 2021.

Motion Response of Floating Platform Hoisting Impeller Based on Fully Coupled Model

Xinghua Shi, Wei Zhang*, Qingqing Ben, Jing Zhang

College of Ship and Marine Engineering, Jiangsu University of Science and Technology, Zhenjiang, 212003

Email: sh28710512@163.com

Abstract: Given the construction efficiency and operation safety of offshore wind turbine hoisting, it is necessary to study the motion response characteristics of a floating platform-hoist coupling system. Using SESAM hydrodynamic calculation software, a fully coupled model of a floating crane ship and fan impeller is established. The accuracy of the calculation model is verified by comparing the numerical simulation results with the pool experimental results. Based on this research, the sensitivity factors of floating crane ship motion are analyzed. The results show that the floating crane should choose the bow-top wave operation to minimize the rolling motion of the hull. The wind speed and the elevation angle of the boom have a great influence on the motion of the floating crane. The larger the wind speed, the greater the motion response of the hull, the smaller the elevation angle of the boom, and the larger the roll value of the hull. The higher the impeller height, the greater the motion displacement of the floating crane, but the influence is different on the hull motion under different working conditions. The results of the simulation analysis can provide a reference and basis for offshore wind turbine hoisting.

1. Introduction

The coupling analysis model of the floating platform is composed of a hull, lifting system, mooring system, and other parts. The large-scale motion of each component of the system is coupled with the vibration of the system itself, which makes the motion of the system show nonlinear characteristics. At present, the research on the motion response of crane ship-lifting in waves has not reached a very mature stage. Xu XiuJun et al. [1] combined the mathematical model analysis method and software analysis method to establish the analytical model of crane ship motion under various environmental excitation loads, which improved the convergence and accuracy of the simulation. Based on the theory of multi-body dynamics, Zhu Ming et al. [2] simulated the motion response of the fan during the release process and analyzed the influence of factors such as sling release speed on the dynamic characteristics of the fan. Based on the potential flow theory, Sun Lei et al. [3] verified the reliability by comparing the simulated data with the measured data. Luo HanBing et al. [4] used SIMO software to establish a coupled analysis model to study the characteristics of low-frequency second-order motion and wave-frequency motion of the 'Blue Whale' crane ship in three degrees of freedom of sway, heave, and roll in regular waves. Chen Jin et al. [5] established a time-domain coupling analysis model of the crane ship-lifting system and discussed the influencing factors that cause the large-scale resonance motion of the lifting object. Zhang DaPeng et al. [6] established the dynamic model of the rotary crane ship and analyzed the support force of the deck of the crane ship when the lifting object is lifted, the impact force between the lifting object and the deck of the receiving ship, and the dynamic response of the lifting cable. Han DuanFeng and Zhou Tong et al. [7] used MATLAB/Simulink software to simulate the ship motion under multi-system coupling and analyzed the influence law of ship motion parameters, which provided a scientific basis for accurately predicting the motion response of a floating crane ship. Huang et al. [8] established a physical experimental model of a 1600t crane ship, tested the motion response of the crane ship model with a scale ratio of 1:50 in both lifting and non-lifting conditions, and analyzed the working wave conditions of the crane ship. Wang Juanjuan and Huang Yanshun et al. [9] took the crane ship as the model and analyzed the influence of different lifting weights, lifting distances, and lengths of lifting rope on the swing of the crane ship and lifting an object. Li Liang [10] calculated the hydrodynamic load and analyzed the overall structural strength response of the wind power installation vessel by SESAM software.

In the offshore hoisting project, the large floating crane is difficult to operate at sea due to the environment, and the environmental requirements are also very strict. Wind turbine hoisting is the most difficult operation in the offshore hoisting project. The shaking of the hull and the lifting object not only leads to the low efficiency of the wind turbine hoisting operation but also endangers the safety of the working ship. Therefore, it is of great engineering significance for offshore hoisting operations to study the influence of the coupling effect of lifting objects on hull motion under different working conditions by establishing a simulation model. Based on the background of the floating crane hoisting fan, the full coupling model of the floating platform and lifting object is established by using SESAM hydrodynamic software. Numerical simulation is carried out for different working conditions to explore the sensitivity factors of floating crane movement.

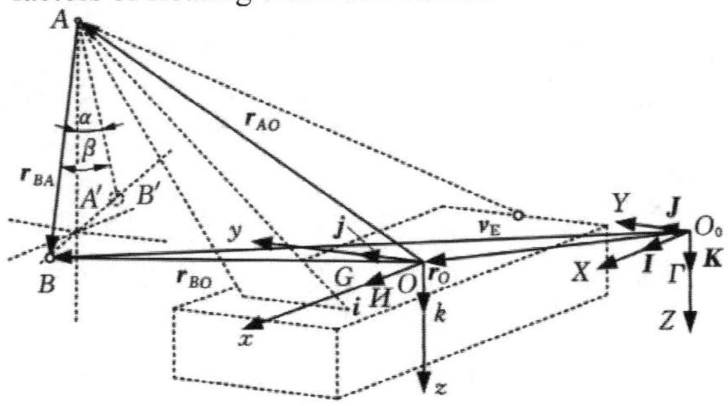

Figure 1. Schematic diagram of lifting object model

2. Computational theory

A coordinate system is established as shown in Figure 1, $OoXYZ$ is a fixed coordinate system, and Oo is the origin. The $OoXYZ$ coordinate system is also an inertial coordinate system, denoted by Γ, I、J、K respectively, representing the unit vectors of the three axes of the fixed coordinate system. The dotted line in Figure 1 shows the structure of the floating crane and the crane boom. The A point is the top point of the crane boom, and the B point represents the lifting point. G point is the center of mass of the hull. A' is the projection of the A point in the XY plane, that is, AA' parallel to the Z axis. The OAA' plane composed of three points is defined as the boom plane. B' is the projection of B. In the plane of the boom, α and β are the in-plane angle and the out-of-plane angle. r_0 is the position vector between the origin O point of the hull coordinate system and the fixed coordinate system O_0 point. r_{A0} is the position vector between the endpoint A of the boom and the origin O of the hull coordinate system. r_{B0} is the position vector between the lifting point B and the origin O of the hull coordinate system. r_{BA} represents the position vector between the lifting point B and the endpoint A of the boom. r_{GO} is the position vector from the center of gravity G to the origin O of the hull coordinate system. The equation of translational motion of the hull is derived according to the momentum theorem and vector derivation rule:

$$\begin{cases} m[\dot{v}_O + \Omega \times v_o + \dot{\Omega} \times r_{GO} + \Omega \times (\Omega \times r_{GO})] = F_{c/s} + F_{en} \\ \dot{v}_o = [\ddot{x} \quad \ddot{y} \quad \ddot{z}]^T \\ \dot{\Omega} = [\ddot{\phi} \quad \ddot{\theta} \quad \ddot{\psi}]^T \end{cases} \tag{1}$$

$$v_0 = \frac{{}^{\Gamma}dr_0}{dt} = [\dot{x} \quad \dot{y} \quad \dot{z}]^T \tag{2}$$

In the formula, $\dot{x}, \dot{y}, \dot{z}$ denotes the projection of v_0 on the x, y, and z axes of the hull coordinate system. Ω is the rotational speed of the hull $\Omega = [\dot{\phi} \quad \dot{\theta} \quad \dot{\psi}]^T$; m is the hull mass; $F_{c/s}$ is the coupling force of the lifting system to the hull; F_{en} is the environmental load force, including wind, wave, current, and other environmental loads on the hull force, hydrodynamic load, and mooring system force.

The total moment of momentum L of the hull is

$$L = L_G + L_H = J_G\Omega + r_{GO} \times (m\Omega \times r_{GO}) + r_{GO} \times mv_o \tag{3}$$

In the formula, J_G is the inertia matrix of the coordinate system of the hull to the origin in the center of gravity; J is the total moment of inertia of the hull to the origin; L_H is the moment of momentum H to the origin. According to the theorem of momentum and the rule of vector derivation, we can deduce:

$$J\dot{\Omega} + \Omega \times J\Omega + r_{GO} \times m\dot{v}_o + r_{GO} \times m(\Omega \times v_o) = M_{c/s} + M_{en} \tag{4}$$

In the formula $J\Omega = J_G\Omega + r_{GO} \times (m\Omega \times r_{GO})$, $M_{c/s}$ is the moment of action of the lifting object on the hull; M_{en} is the action torque of environmental load on the hull.

According to the above derivation, the motion of the hull can be determined by Equations (1) and (3). The coupling effect between the sling and the hull is reflected by the force and moment of the sling system on the hull.

3. Floating crane system calculation model establishment

3.1. Validation of numerical calculation method

To verify the reliability of the numerical calculation method, the results are compared with the experimental results of the pool model. The simulation model of double-ship mooring is established, which is the same as the test model. The JONSWAP wave spectrum is selected. The significant wave height is 2.5 m, the spectral peak period is 10 s, the wind speed is 13 m/s, and the flow velocity is 0.81 m/s. The wind, wave, and current are in the same direction and the incident angle is 180° as the same environmental parameters. The motion amplitude response operator RAOs is calculated respectively, and the numerical results are compared with the pool model test data. The test model and simulation model are shown in Figure 2. The results are shown in Figure 3.

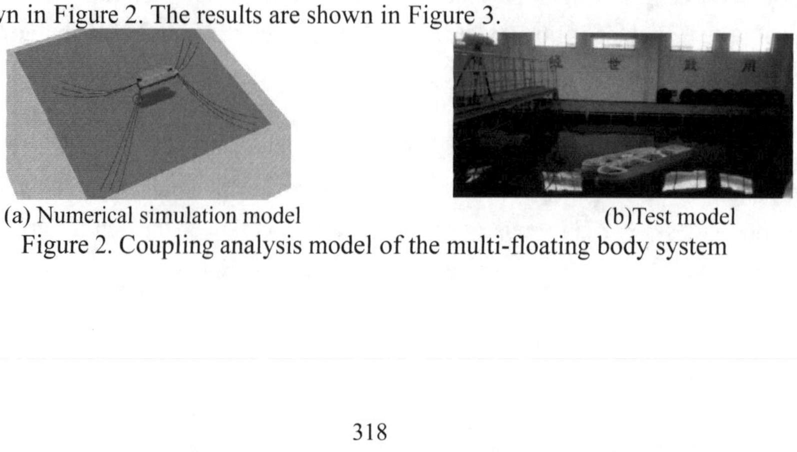

(a) Numerical simulation model (b)Test model

Figure 2. Coupling analysis model of the multi-floating body system

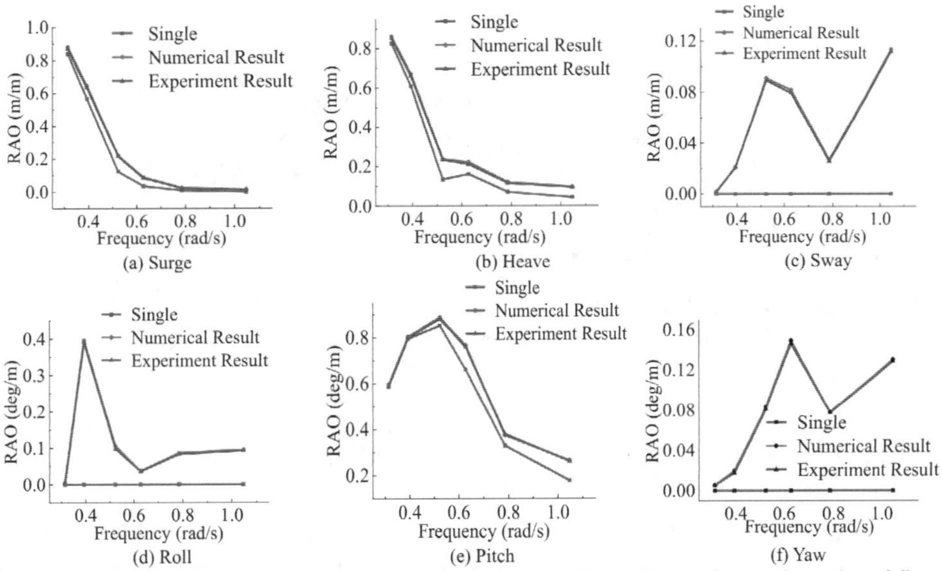

Figure 3. Comparison between numerical calculation results and experimental results of floating crane's RAO

By comparing the RAO obtained by the pool model experiment and the numerical simulation calculation under the same environmental conditions, it can be found that the maximum error between the numerical simulation and the experiment is 3.05 %. The error between the two is very small and within the allowable range. The experimental results and the numerical calculation have high consistency, which verifies the reliability of the numerical calculation model.

3.2. Full coupling model of the floating platform and lifting objects

In the lifting operation of offshore wind turbines, the impeller and blade of the wind turbine will shake under the action of wind load. The force generated by the shaking of the impeller is transmitted to the lifting vertex through the sling and finally transmitted to the hull of the floating crane through the lifting arm. The transfer process and simplification of the coupling force between the crane and the object are shown in Figure 4. The floating crane ship is affected by the coupling force of the wind, wave, and current. The force of the impeller sloshing on the hull makes the motion of the hull more complicated. Based on the above situation, the time-domain analysis model of the full coupling of the floating platform and the lifting object is established in the SIMO module. The fan impeller [11] is used as the lifting object, and the corresponding mass and inertia moment is applied to the center of gravity of the fan impeller. It is equivalent to replacing the actual fan impeller and simulating and analyzing different working conditions. The water depth is set at 50 m, and floating crane ship-specific parameters are shown in Table 1. Figure 5-7 shows the floating crane system, mooring mode, and fan impeller model.

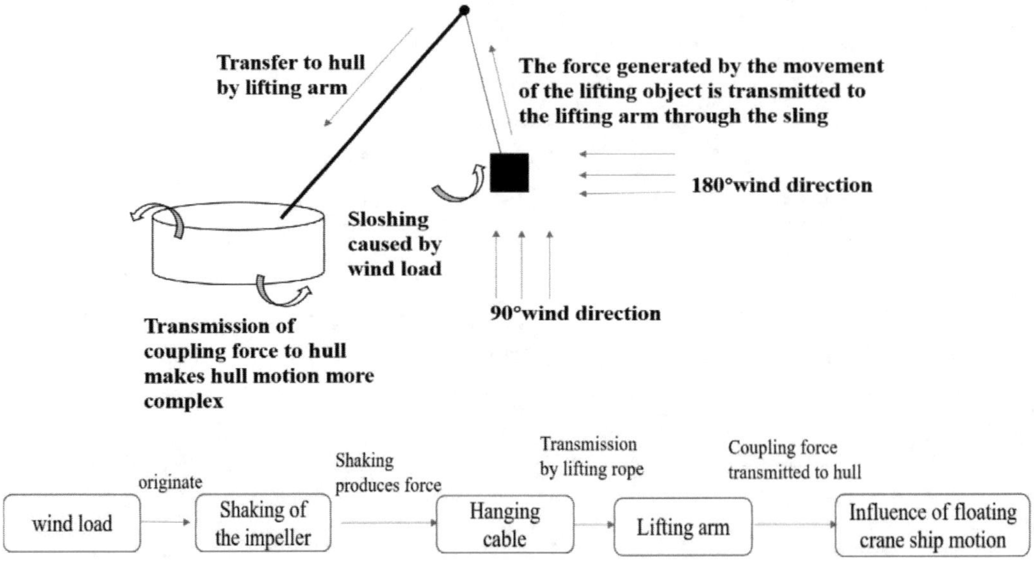

Figure 4. Transfer process and simplified schematic diagram of coupling force between crane and object

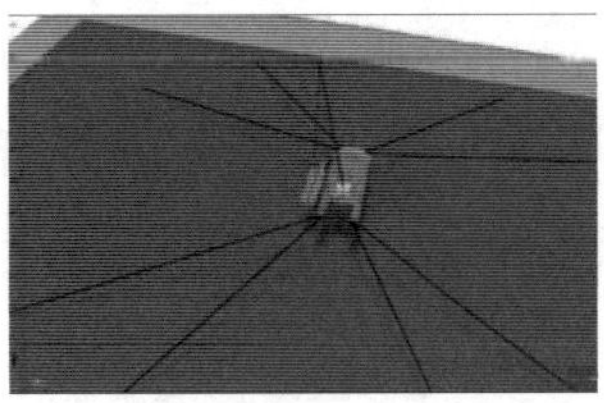

Figure 5. Schematic diagram of full coupling of the impeller

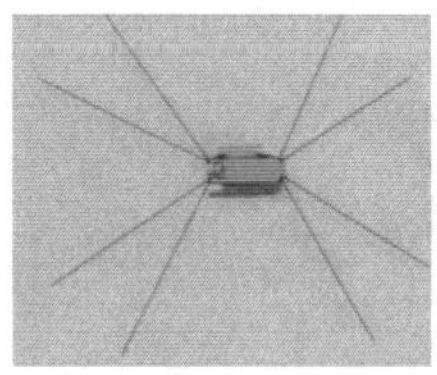

Figure 6. Mooring system layout floating platform and lifting material

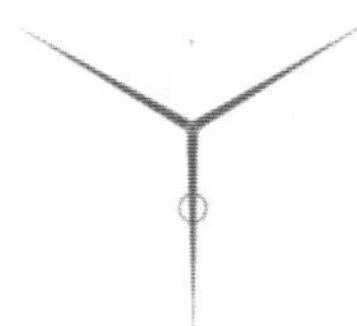

Figure 7. Model of fan

Table 1 Basic parameters of floating crane

Project		Unit	data	Project		Unit	data
Total length	L_{oa}	m	207	Eating water	d	m	8.7
Type width	B	m	36	The vertical position of the center of gravity	KG	m	11.8
Type deep	D	m	16	The longitudinal position of the center of gravity	LCG	m	1.8

4. Analysis of simulation results

4.1. Effect of wave direction angle on motion of the floating crane

The floating crane will encounter waves in all directions during the operation, and the wave loads in different wave directions will have different effects on the floating crane system. Five wave directions of 0°, 45°, 90°, 135°, and 180° are selected to simulate the motion of the floating crane in six degrees of freedom under five wave directions, and the influence of different wave directions on the motion of the floating crane is analyzed.

The lifting height is set at 70 m, the boom elevation angle of 75°, the crane rotation speed of 0.2 rad/ 4s, the wave height of 2.5 m, the wind speed of 6 m/s, the flow rate of 0.5 m/s, the sling stiffness of 1.62E+6 kN/m, and the irregular waves. JONSWAP spectrum is used for irregular waves, and the total time of numerical simulation is 3600 s. The motion time history data of the 1000-2000 s floating crane is selected for comparative analysis. Table 2 shows the maximum six degrees of freedom motion of the floating crane in different waves.

It can be seen in Table 2 that under the set mooring state, five kinds of waves are downward, 90° wave angle of the floating crane sway, and surge. The roll degree of freedom on the motion displacement is the largest, the pitch is smaller. 0° and 180° wave are down, the surge and pitch motion displacement is larger, and other degrees of freedom is smaller.

Table 2 Motion values of a floating crane with 6 degrees of freedom under different wave angles

Parameter	Maximum displacement										
	0°	45°	90°	135°	180°		0°	45°	90°	135°	180°
Sway/m	0.49	0.37	0.49	0.57	0.34	Roll/°	0.46	0.42	0.41	0.63	0.19
Surge/m	0.42	0.60	0.36	0.49	0.48	Pitch/°	0.26	0.26	0.37	0.20	0.39
Heave/m	0.25	0.30	0.57	0.512	0.39	Yaw/°	0.132	0.21	0.30	0.39	0.19

4.2. Influence of wind speed on the motion of the floating crane

Because the rolling motion of the hull is the largest under the wave, the irregular wave 90° and 180° wave directions are selected to compare and analyze the dynamic motion response of the floating crane ship. The wind speed is set to 6 m/s, 9 m/s, and 12 m/s, the fan impeller height is 70 m, the crane rotation speed is 0.2rad/s, the lifting arm elevation angle is 75°, the flow rate is 0.5 m/s, the wind wave flow is in the same direction, and the crane is in the side lifting state (crane rotation 90°). The time history curves of the floating crane under different wind speeds are shown in Figure 8 and Figure 9. The specific motion values are shown in Table 3.

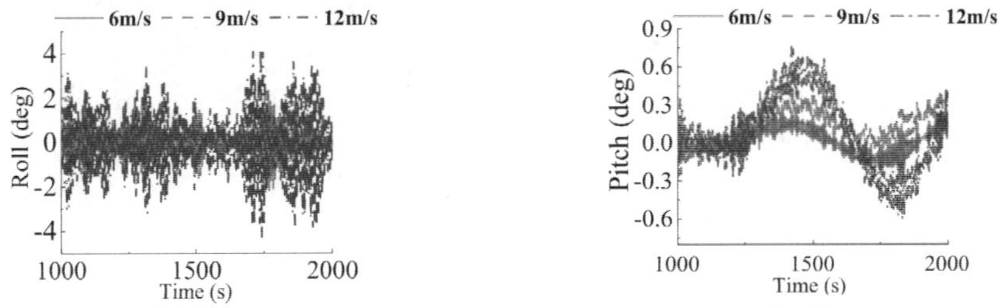

(a) Roll (b) Pitch

Figure 8. The time-history curve of the floating crane moving with 90° wave direction

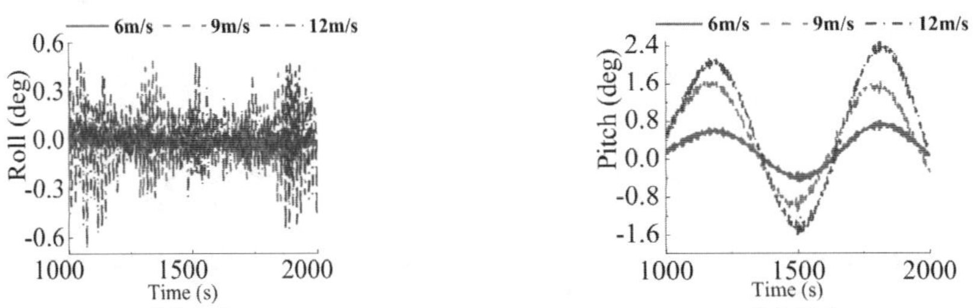

(a) Roll (b) Pitch

Figure 9. The time-history curve of the floating crane moving with 180° wave direction

Table 3 Movement value of floating crane under different wind speeds

90 ° Wave Direction						
Parameters	Maximum displacement			Minimum displacement		
Wind Speed(m/s)	6	9	12	6	9	12
	1.646	2.521	4.218	-1.646	-2.296	-4.352
	0.248	0.457	0.751	-0.229	-0.258	-0.605

180 ° wave direction						
	Maximum displacement			Minimum displacement		
Wind Speed(m/s)	6	9	12	6	9	12
Roll	0.126	0.465	0.487	-0.146	-0.387	-0.671
Pitch	0.807	1.729	2.494	-0.484	-1.098	-1.638

It can be seen in Figure 8, Figure 9, and Table 3 that the increase in wind speed leads to an increase in impeller sloshing. Due to the influence of impeller weight, under the action of multi-body coupling, the wind speed of 90° wind direction mainly affects the roll of the hull, and the wind speed of 180° wind direction mainly affects the pitch motion of the hull. The motion response of the floating crane increases with the increase of wind speed in both wind directions.

4.3. Effect of impeller height on the motion of the floating crane

During the operation of the floating crane, there are two common working conditions: side crane (crane rotation 90°) and stern crane (crane rotation 180°). The motion response of the floating crane will change accordingly under different working conditions. Based on the two working conditions, the 90° and 180° wave directions are selected, and the motion response of the floating crane is compared and analyzed when the lifting height is 50 m, 70 m, and 90 m. The boom elevation angle is set as 75°, sling stiffness is 1.62E+6 KN/ m, and the crane rotation speed is 0.2 deg/s. The peak period is selected T_P = 4s, 6s, 8s... the 20s. Figures 9 - 12 are the motion response of the crane ship under different lifting heights.

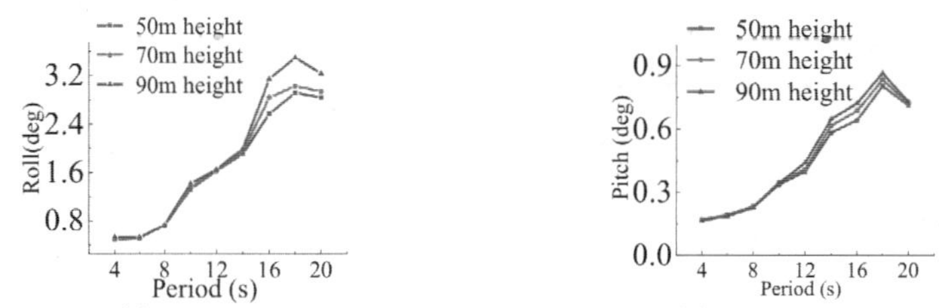

(a) Roll (b) Pitch

Figure 10. Motion response of floating crane under the condition of 90° wave to side lifting

(a) Roll (b) Pitch

Figure 11. Motion response of floating crane under the condition of 90° wave to bow lifting

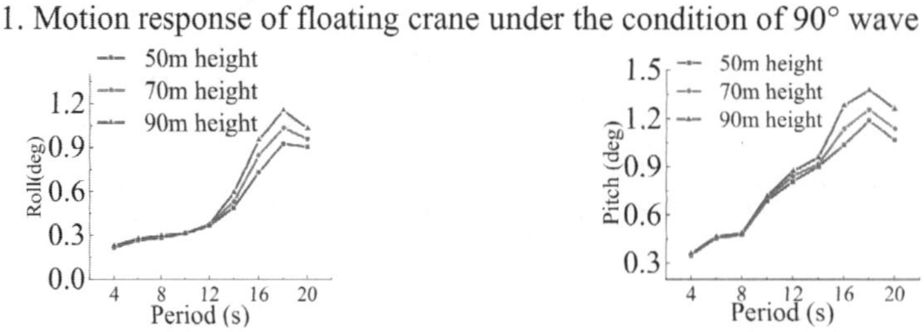

(a) Roll (b) Pitch

Figure 12. Motion response of the floating crane under the condition of 180° wave to the side

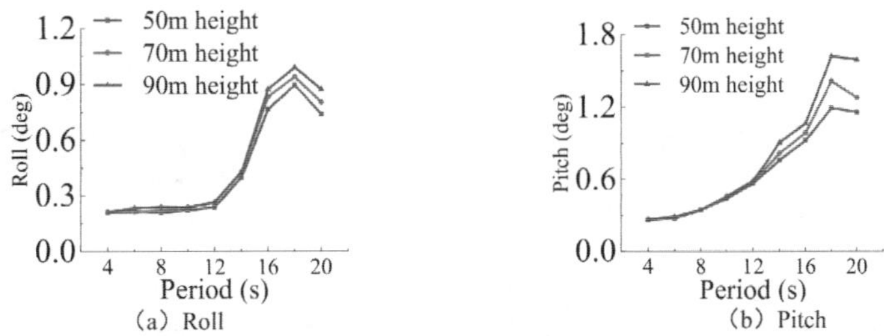

Figure 13. Motion response of the floating crane under the condition of 180° wave to the bow

It can be seen in Figures 10-13 that under the two working conditions of side lifting and bow lifting, with the increase of the spectral peak period, the extreme value change trend of the floating crane ship in roll and pitch is the same. The dynamic response values of roll and pitch increase with the increase of impeller height, but the numerical growth changes differently under different wave directions. Under the side lifting condition: 90° wave downward, because the boom and the lifting object are parallel to the horizontal section and wave direction of the floating crane, with the increase of the height of the impeller, the movement of the impeller in the air reacts to the hull, resulting in the increase of the rolling motion of the hull with little effect on the pitching motion. Under the bow crane condition and the same wave direction, the change of impeller height has little effect on the roll motion of the floating crane ship, mainly on the pitch motion. Under the 180° wave direction, because the impeller and boom are parallel to the longitudinal section of the hull and the wind wave, and the crane is close to the bow and the mutual coupling between the impeller and the hull, the pitch motion of the hull increases with the increase of the impeller height.

4.4. Influence of boom elevation on the motion of the floating crane

Three wave directions of 45° (oblique wave), 90° (transverse wave), and 180° (top wave) are selected, and the elevation angles of the boom are 80°, 70°, and 60° for numerical simulation analysis. The specific working conditions are shown in Table 4. Based on the time domain simulation, the influence of the change in the elevation angle of the boom on the motion of the floating crane is compared and analyzed. Figures 14 - 16 show the influence of the elevation angle of the boom on the motion of the floating crane in the three degrees of freedom of surge, sway and heave.

Table 4 Specific operating parameters

Project		Numerical value	
Wave spectrum peak period	4s, 6s, 8s, 10s.....20s	Wave direction	45°, 90°, 180°
Wave height	1.25m	Arm elevation angle	80°, 70°, 60°
Wind speed	6m/s	Sling stiffness	1.62E+6kN/m
Fan impeller height	70m	Crane rotation speed	0.2deg/s

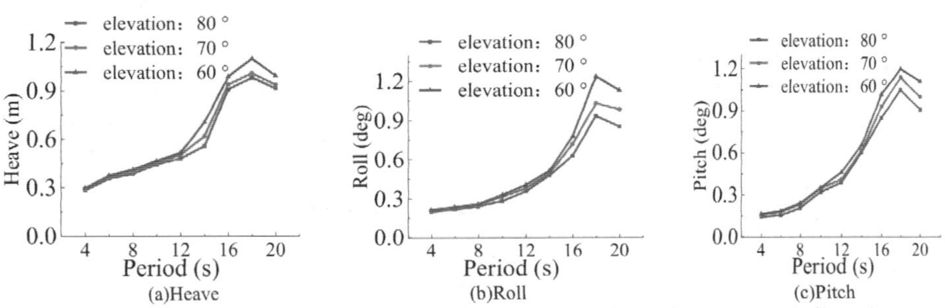

Figure 14. Motion response analysis of a crane ship under 45° wave direction

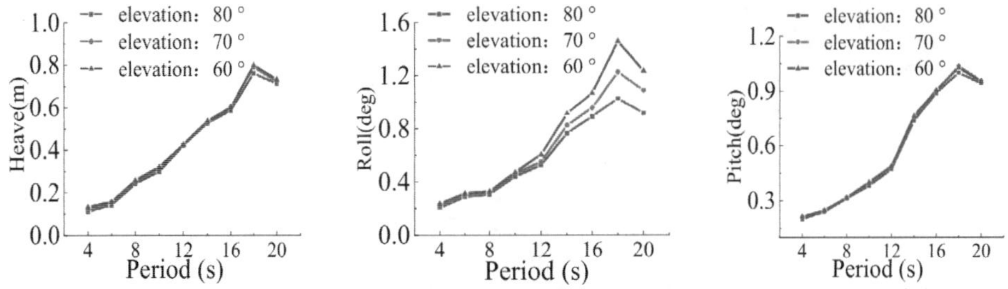

(a) Heave (b) Roll (c)Pitch

Figure 15. Motion response analysis of a crane ship under 90° wave direction

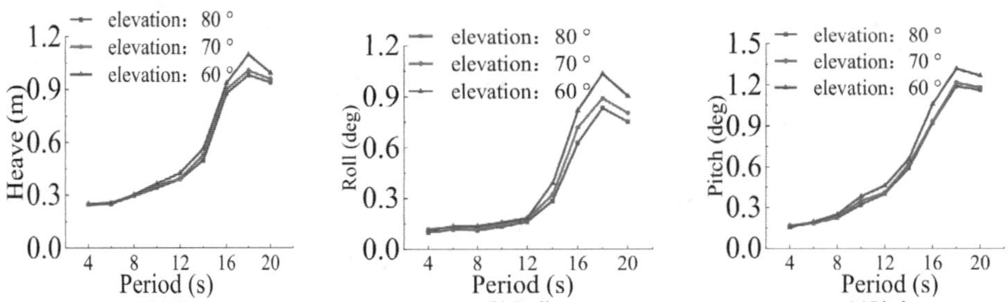

(a) Heave (b) Roll (c)Pitch

Figure 16. Motion response analysis of the crane ship under 180° wave direction

It can be seen from Figures 14 to 16 that the motion trends of the floating crane under different wave directions are the same in the three degrees of freedom of surge, sway, and heave. At the same spectral peak period, with the decrease of the elevation angle of the boom, the motion values of the three degrees of freedom show an increasing trend. With the increase of the spectral peak period and the decrease of the elevation angle of the boom, the value of the roll of the floating crane shows an increasing trend.

5. Conclusion

Through the above analysis, this paper draws the following conclusions:

(1) When the floating crane is lifting, the bow top wave operation is helpful to improve the stability and accuracy of the offshore lifting operation, and improve the lifting efficiency and safety of the floating crane. The wind speed and the elevation angle of the boom have a great influence on the hull motion. As the wind speed increases, the motion displacement of the floating crane increases. The 90 ° wind direction mainly affects the rolling motion of the hull, and the 180 ° wind direction mainly affects the pitching motion of the hull. The decrease in the elevation angle of the boom will lead to an increase in the rolling value of the hull, and the rolling motion of the hull changes the most under a 90° wave.

(2) The change in impeller height has different effects on the hull motion under different working conditions. Under the same wave direction, the dynamic response value of the hull in the roll and pitch degrees of freedom increases with the increase of the impeller height. The role value of the hull in the side lifting condition is larger than that in the bow lifting condition, and the pitch value in the bow lifting condition is larger.

The floating crane ship can adjust the stability of the hull by changing the wave direction, adjusting the lifting height of the impeller, and the elevation angle of the boom.

References

[1] Xu Xiujun, Li Zhen, Wang L, et al. Modeling and simulation analysis of offshore lifting operation

process[J]. Computer simulation, 2019, 36(06): 236-241+325.
[2] Zhu Ming, Zhang Peng, Zhu Changming. Dynamic model simulation analysis of crane ship-fan hoisting system[J]. Mechanical design and manufacturing, 2019(10): 50-53.
[3] Sun Lei, Zhao Juan. The influence of the lifting process on the coupling motion of the lifting system of the lifting vessel[J]. Port Science and Technology, 2020(8): 31-37.
[4] Luo Hanbing, Liu Fusheng, Luo Xiao, et al. A study on the motion response of the upper components of the 'Blue Whale ' lifting vessel in beam seas[J]. Ship Mechanics, 2021, 25(1): 43-51.
[5] Chen Jin, Yang Shu, Shuai Lihua, et al. Numerical simulation of coupled motion response of crane ship-hanging system under wave action[J]. Ship Engineering, 2021, 43(5): 60-65,78.
[6] Zhang Dapeng, Zhu Keqiang. Dynamic analysis of rotary lifting process of large rotary crane ship under different waves[J]. Journal of Ocean University of China (Natural Science Edition), 2017, 47 (5): 128-134.1.
[7] Han Duanfeng, Zhou Tong, Zan Yingfei, et al. Simulation analysis of multi-system coupling motion response of full-revolving crane ship[J]. Ship Engineering, 2018, 40(1): 48-53,87.
[8] Huang Xuanjun, Lin Chaoming, Sun Yunjia, et al. Experimental study on the operation capacity of 1600t crane ship under wave action[J]. China Harbour Construction, 2018, 38(10): 41-44.
[9] Wang Juanjuan, Huang Yanshun, Li Huailiang, et al. Motion response of lifting vessels in waves[J]. China Ship Research, 2013, 8(03): 50-57.
[10] Li Liang. Structural strength analysis of wind power installation platform based on SESAM software[J]. China Shipping (Second Half Month), 2021, 21(07): 9-11.
[11] Jonkman J, Butterfield S, Musiai W, et a1. Definition of a 5-MW reference wind turbine for offshore system development[R]. Colorado, US: National Renewable Energy Laboratory, 2009: 2-16.

Numerical Simulation for Nonlinear Interaction between an Attached Air Bubble and Nearby Oscillating Bubble

Mingzuo Jiang[1] and Jiaxia Wang[2*]

[1] College of Automation, Jiangsu University of Science and Technology, Zhenjiang, Jiangsu 212003, China.

[2] School of Naval Architecture and Ocean Engineering, Jiangsu University of Science and Technology, Zhenjiang 212003, Jiangsu, China.

[*]E-mail: jxwang66@yeah.net

Abstract. The numerical simulations of a non-spherical evolution of pulsating bubble interacting with a stationary air bubble attached to a fixed structure were performed by using the three-dimensional boundary integral method which is based on the mirror image method to simplify the processing of the numerical model. Validation of the code is presented by comparison of numerical predictions with laboratory experimental data obtained by our previous research. Complex phenomena are observed such as three kinds of bubble jet forms: upward bubble jet, bubble splitting phenomenon, and downward bubble jet. The present simulations provide insight into the interesting bubble jetting phenomena in detail, including bubble splitting, jet away from the plate, and bubble shedding.

1. Introduction

Oscillating bubbles are highly relevant to several industrial applications[1] ranging from micrometer-sized bubbles in bio-medical related applications, millimeter-sized bubbles in cavitation erosion on a ship propeller[2], structure cleaning[3], to meter-sized underwater explosion bubbles[4]. In both these areas, the final phase of bubble collapse is associated with a high-speed liquid jet directed away from[5] the nearby boundary[6].

Earlier studies have concentrated on the problem of studying the bubble dynamics near a solid wall through various degrees of sophistication both experimental and numerical works. Reuter[7] conducted experimental studies on the non-spherical bubble collapse, the near-wall disturbance caused by the annular vortex, and the damage to the aluminum sheet induced by bubble collapse. Dular[8] experimentally studied the damage induced by a cavitation bubble close to the aluminum foil material and pointed that a high-speed micro-jet cause the damage. Cui[9] pointed out that the shock wave created by the bubble jet may casuse the ice cracking.

In recent years, an increasing interest has been shown that the behavior of an oscillating bubble in the vicinity of a non-oscillating bubble is attached to a rigid/ flexible structure as its prospect in reducing cavitation erosion with a non-oscillating bubble. Goh[10] investigated the interaction between a cavitation bubble and an attached air bubble through an experimental numerical study and summarized the law of the bubble jet direction. Wang[11] studied the motion characteristics of a bubble close to a rigid/ flexible structure with an attached air bubble. Complex bubble jetting trends such as bubble splitting, jets away from the plate, variable directional jets, and multidirectional jets are caught, and five types of two bubbles behaviors are summarized in detail.

The dynamical characteristics of pulsating bubbles near an air bubble attached to a fixed structure are numerically studied based on the 3D BIM with a mirror image theory to simplify the numerical treatment of the gas-solid interface.

2. Numerical model

A three-dimensional boundary integral method based on potential flow theory was used to study the evolution of the shape of the bubble by giving other valuable information such as the pressure counter. It is not obtained from experimental data. In the multiple bubble interaction study, the flow field is considered inviscid, irrotational, and incompressible[12], and the governing equation according to mass conservation satisfies the Laplace equation $\nabla^2\phi=0$. The problems in the three-dimensional computational domain can be transformed into two-dimensional surface integral problems by using the free surface Green's function G.

$$c(\mathbf{P})\phi(\mathbf{P})=\int_S\left[\frac{\partial\phi(\mathbf{Q})}{\partial n}G(\mathbf{P},\mathbf{Q})-\phi(\mathbf{Q})\frac{\partial G(\mathbf{P},\mathbf{Q})}{\partial n}\right]dS(\mathbf{Q}) \tag{1}$$

Here, the free space green function is imported.

$$G(\mathbf{P},\mathbf{Q})=\frac{1}{|\mathbf{Q}-\mathbf{P}|}=\frac{1}{r_{PQ}} \tag{2}$$

Where, $P=(x, y, z)$ and $Q=(x',y',z')$ are the field point and the source point, respectively. S is the boundary surface which includes three parts, the oscillating bubble surface S_b, the attached bubble surface S_a, and the plate surface S_t. $c(P)=\int_S \partial G/\partial n dS$ is the solid angle. $\partial/\partial n=n\bullet\nabla$ is the normal derivation, and n is the unit normal vector.

When it is processed from a physical problem to a numerical model, the mirror image method is commonly used in the study of bubble dynamics. Then, a negative image of two bubbles is placed behind the plate with the similar length. Essentially, a three bubbles system is conducted in the numerical simulation. The related coordinate system used here is shown in Figure 1.

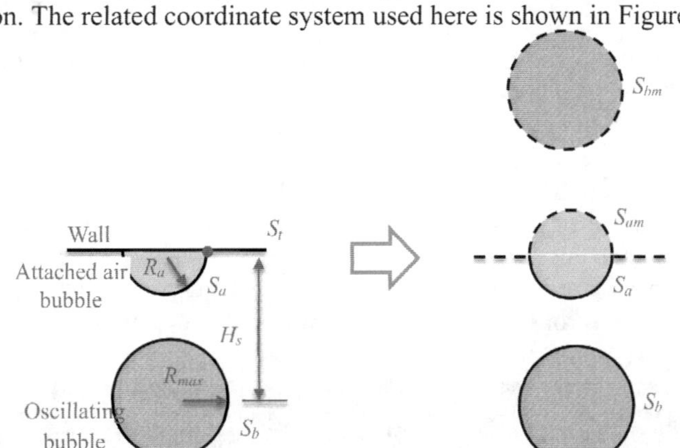

Figure 1. The coordinate system and three bubbles system information

The planer triangular elements is imported to discrete the boundary S of the computation zone. The linear interpolations are used for the position function **r**, potential ϕ, and normal velocity $\partial\phi/\partial \boldsymbol{n}$. Therefore, the integral form of Equation (1) leads to the following equation.

$$c(P)\phi_j(P)+\sum_{m=1}^{M}\sum_{i=1}^{3}\hat{H}_{m,i}\phi_{m,i}(Q)=\sum_{m=1}^{M}\sum_{i=1}^{3}G_{m,i}\frac{\partial}{\partial n}\phi_{m,i}(Q), j=1,2,\cdots N \tag{3}$$

Here, the definition is as follows.

$$\hat{H}_{m,i} = \int_0^1 \int_0^{1-\eta} \frac{\partial}{\partial n}\left(\frac{1}{r_{PQ}}\right) N_i J_m d\xi d\eta \tag{4}$$

$$G_{m,i} = \int_0^1 \int_0^{1-\eta} \frac{1}{r_{PQ}} J_m N_i d\xi d\eta \tag{5}$$

Where, J is the Jacobian matrix, ξ and η are the normalized coordinate systems, and M is the number of all discreted elements on the surface S. The numerical calcualaion of the singularity in BIM and the bubble topological changement to transform a singly connected bubble to annular bubble can be found in our previous studies[13].

All variables are non-dimensionalized in the numerical process. It can be seen that the length can be scaled as Rm (the maximum bubble radius), the pressure as Δp=p∞–pc, and the time as $R_m\sqrt{\rho/\Delta p}$. The dimensionless variables are yielded as follows.

For oscillating bubbles, the specifics are given below.

$$\gamma_s = \frac{H_s}{R_{\max}} \quad \varepsilon = \frac{P_{g0}}{\Delta p} \tag{6}$$

Where, ε is the strength parameter, γ_s is the non-dimensional distance parameter, P_{g0} is the initial pressure of non-condensing gas for the bubble, and H_s is the distance of the bubble center from the solid surface at generation.

For the attached air bubble, the specifics are as follows.

$$\varepsilon_a = \frac{P_\infty}{\Delta p} \tag{7}$$

ε_a is the air bubble strength parameter, and r_0 and r_{a0} are the initial radii of the oscillating bubble and air bubble. For more detail about the two bubbles' information, please refer to the past study[10].

3. Validation

To help ascertain the validity of the present numerical model, a comparison with the result in our previous experimental data[11] is done for a cavitation bubble in the proximity of an attached air bubble. Figure 2 shows a comparison of spark-generated experimental bubble and numerical results at selected times. The upper part is the experimental result, and the lower part is the numerical calculation. It can be seen that in the first pulsation period of the bubble, the predicted profiles of the bubble, and the attached air bubble are in good agreement with those of the experimental results except for some small differences in the timings when the bubble shrinks to the minimum volume. This little discrepancy may be caused by the neglect of viscosity, surface tension, and some other factors. The changing feature of two bubble volumes, of which the attached air bubble expands with the contraction of the bubble and has been reproduced in our numerical simulation.

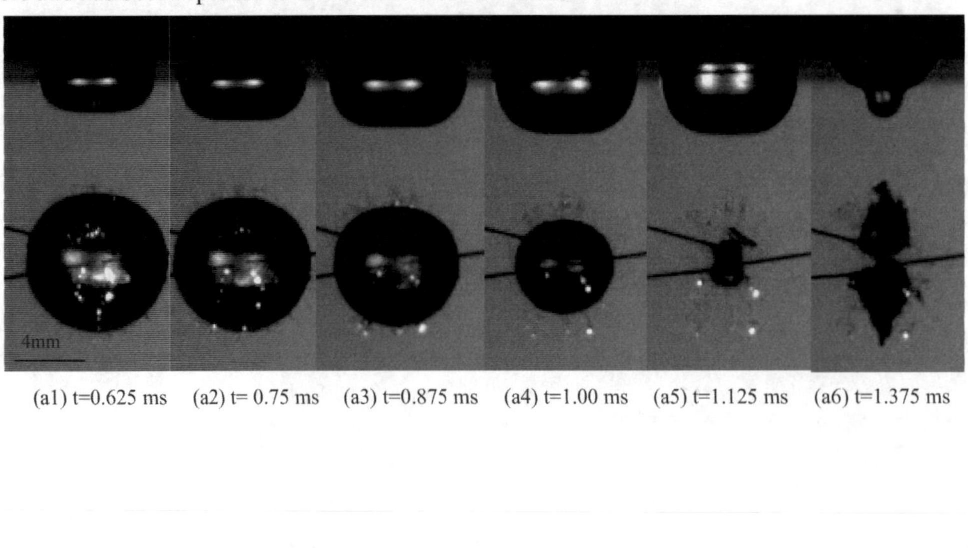

(a1) t=0.625 ms (a2) t= 0.75 ms (a3) t=0.875 ms (a4) t=1.00 ms (a5) t=1.125 ms (a6) t=1.375 ms

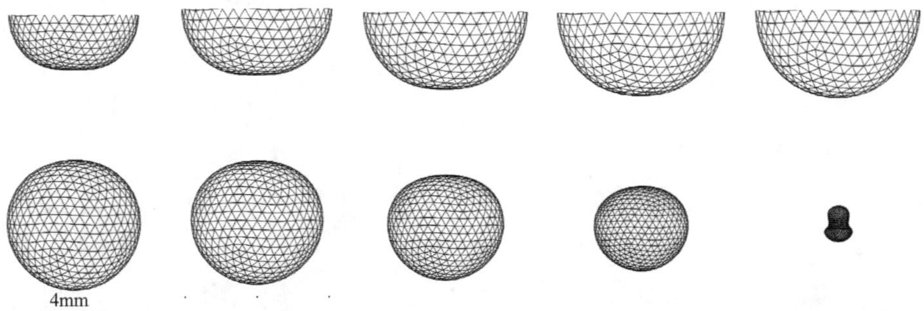

(b1) t=0.603 ms (b2) t=0.797 ms (b3) t=1.016 ms (b4) t=1.162 ms (b5) t=1.349 ms
Figure 2 Comparison of the experimental results with the numerical results obtained by the 3D BIM

4. Numerical results

As opposed to two oscillating bubbles interaction, passive pulsation of a surface-attached bubble caused by a nearby oscillating bubble can exhibit much more complicated behavior. In the following calculation, the related parameters are listed in Table 1. And three typical bubble jetting directions, which are respectively upward jetting, splitting phenomenon, and downward jet, are given in sections 4.1-4.3.

Table 1 The calculation parameters for three typical bubble jetting directions

	γ_s	Bubble r_0	Bubble ε_b	Attached bubble r_{a0}	Attached bubble ε_a
Case 1	2	0.1485	100	0.4	2.798
Case 2	3	0.1485	100	0.632	2.798
Case 3	1.5	0.1485	100	0.632	2.798

4.1. The upward bubble jetting

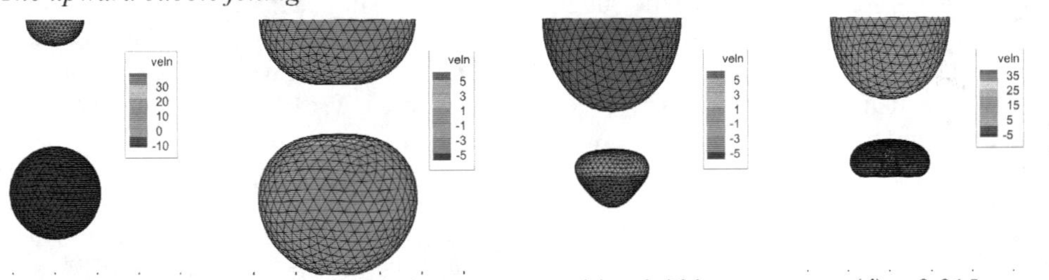

(a) t=0.146 (b) t=0.958 (c) t=2.089 (d) t=2.215
Figure 3 The upward bubble jetting when the attached cavitation bubbles interact with pulsating bubbles in case 1

Figure 3 shows the upward bubble jetting between the attached bubble and pulsating bubble when γ_s=2, ε_b=100,ε_a=2.798,r_{a0}=0.4, and r_{b0}=0.1485. The bubble shapes with velocity counter are given in sequence a. What needs special attention is that the wall surface is not directly reflected in the result as in the imported mirror image method. At the beginning of the bubble expansion phase (t<0.146), the volume of the air bubble decreases, and the internal pressure of the air bubble increase with the bubble expansion. At t=0.146, the internal pressure of the air bubble reaches its maximum. In the subsequent bubble expansion process, the oscillating bubble reaches the maximum radius induced by the larger internal pressure and the inertia effect when t=0.958, as shown in Figure 3b. At this moment, the top of the oscillating bubble and the bottom of the air bubble tend to be flat. In the early stage of oscillating bubble contraction, the air bubble will continue to expand until t=1.6225. It is mainly caused by the attraction of the shrinking bubble. Subsequently, the two bubbles contract synchronously. For an oscillating bubble, the contraction speed at the bottom obviously increases, then a "strawberry-shaped" bubble appears. For an air bubble, the vertical length increases, and the width decreases under the attraction of the bubble (as shown in Figure 3c). An upward and narrow bubble jetting forms in Figure 3d, and the vertical length and width of the air bubbles decrease compared with Figure 3c.

4.2. Bubble splitting phenomenon

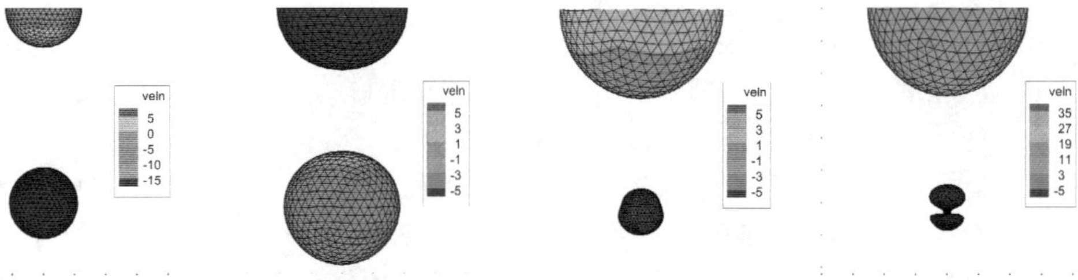

(a) t=0.151 (b) t=0.877 (c) t=1.896 (d) t=2.014

Figure 4 The bubble-splitting phenomenon when the attached cavitation bubbles interact with pulsating bubbles in case 2

The coupling interaction between the attached bubble and the pulsating bubble when γ_s=3, ε_b=100, ε_a=2.798, r_{a0}=0.632, and r_{b0}=0.1485 are presented in Figure 4, including two bubble shapes. Compared with case 1, the distance parameter and the initial cavitation radius of case 2 are increased to some extent. The two bubble's shapes during the initial expansion phase are similar to that of Figure 3. The air bubble is compressed to the minimum volume at t=0.151, and the pressure inside is large. The maximum radius of the oscillating bubble reaches when t=0.889, and the other information will not be repeated here. The present focus is on the interactions of two bubbles in the contraction stage. The upper part of the oscillating bubble contracts faster, making it forms a "gourd" shaped at t=1.896 (as shown in Figure 4c). A "dumbbell-shaped" bubble with a slender middle part can be clearly observed at t=2.014, and there should be a high-pressure area in the middle depression part of the bubble. Under this high pressure, the central junction part of the bubble continues to shrink, indicating an obvious bubble-splitting tendency. The particularity of this case lies in the appearance of "dumbbell-shaped" bubbles and the splitting trend.

4.3. A downward bubble jetting

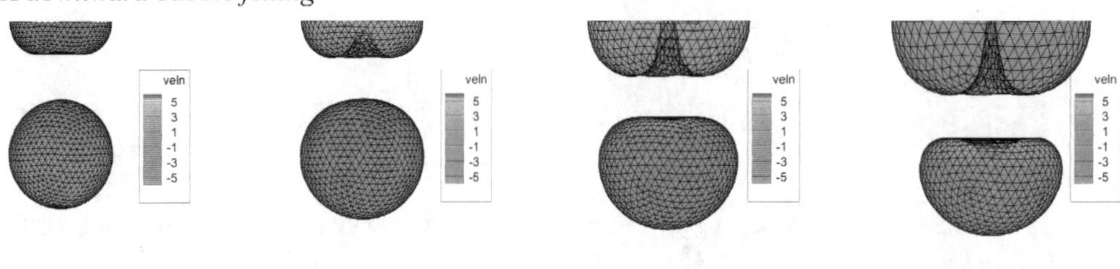

(a) t=0.194 (b) t=0.327 (c) t=0.521 (d) t=0.749

Figure 5. The downward bubble jetting when the attached cavitation bubbles interact with pulsating bubbles in case 3

Figure 5 shows the coupling evolution phenomenon between two bubbles when γ_s=1.5, ε_b=100, ε_a=2.798, r_{a0}=0.632, and r_{b0}=0.1485. The variation trend of the bubble radius in Figure 5 is easier than that of cases 1 and 2. For an air bubble, it is compressed to its minimum radius of 0.539 at t=0.194, and the lower surface is flat. Then it began to rebound. Its volume continued to increase in the next moment. The most significant is a vertical upward water jet emerges (shown in Figure 5b). Then, an annular air bubble is forming, and the penetration of the air bubble was also observed in experiments, such as in Figure 8(b4) in the literature[11]. With the expansion of the annular air bubble, the top of the bubble is compressed into a flat shape (shown in Figures 5c and 5d). The special feature of this case is the two special phenomena, an annular air bubble and a downward jetting.

5. Conclusion

In this paper, the pulsation characteristics of the oscillating bubble close to the attached air bubble are numerically studied. The three-dimensional BIM method with a mirror image method is imported to calculate multiple bubble dynamics. The validity of the numerical model is verified by comparing it with the experimental results obtained by our previous research. The special shapes of two bubbles and various bubble jet directions are numerically observed. Three types of bubble jet directions, such as upward bubble jet, bubble splitting phenomenon, and downward bubble jet, are captured numerically.

Acknowledgments

This research is financially supported by the National Natural Science Foundation of China (Grant No. 52171311; 51809122), the Natural Science Foundation of Jiangsu Province (Grant No. BK20200998), and the Natural Science Foundation of the Jiangsu Higher Education Institutions of China (Grant No. 20KJB510046).

References

[1] Blake J R, Leppinen D M and Wang Q X. 2015 *Interface focus* 5(5): 20150017.
[2] Hsiao C T and Chahine G L. 2015 *J Fluid Struct* 54: 760-783.
[3] Reuter F, Lauterborn S, Mettin R and Lauterborn W. 2017 *Ultrason Sonochem* 37: 542-560.
[4] Duplat J. 2019 *J Fluid Mech* 859: 677-690.
[5] Liu L T, Wang J X, Zhang Y F and Ma T. 2021 *Ocean Eng* 237: 109577.
[6] Goncalves Da Silva E and Parnaudeau P. 2021 *Phys fluids* 33(11): 113318.
[7] Reuter F, Cairós C and Mettin R. 2016 *Ultrason Sonochem* 33: 170-181.
[8] Dular M, Požar T and Zevnik J. 2019 *Wear* 418: 13-23.
[9] Cui P, Zhang A-M, Wang S and Khoo B C. 2018 *J Fluid Mech* 841: 287-309.
[10] Goh B, Ohl S, Klaseboer E and Khoo B C. 2014 *Phys fluids* 26(4): 042103.
[11] Wang J X, Liu K, Yuan S J, Jiang M Z and Wang Z L. 2020 *Phys fluids* 32(6): 067101.
[12] Brujan E-A, Takahira H and Ogasawara T. 2019 *Exp Therm Fluid Sci* 101: 48-61.
[13] Wang J X, Zong Z, Sun L, Li Z R and Jiang M Z.2016 *J Hydrodyn* 28(1): 52-65.

Numerical Simulation of Particle Filtration by Rectangular Fibres within a Parallel and Staggered Array

Shixian Yue

School of Aeronautics, Northwestern Polytechnical University, Xi'an 710072, PR China
yuesxian@126.com

Abstract: In this research, two-dimensional numerical simulation is used to examine the filtration performance of rectangular fiber filters. Ansys Fluent is employed to obtain the stable flow field distribution and the DPM model is used to calculate the particle trajectories in the flow field. The effect of pertinent factors such as fiber arrangement, particle-to-fiber diameter ratio, and Stokes number on the performance of rectangular fiber filters are analyzed. The simulation results indicate that both the parallel and staggered arrays of rectangular fibers have a significant influence on the capture efficiency and pressure drop. At the same flow velocity, the pressure drop of the fibers with a parallel array is lower than that of the staggered array fiber model. The capture efficiency of the staggered array model and the parallel array model increases with the increase of the ratio of particle to fiber diameter and the Stokes number. When the interception and inertial impaction mechanisms domiant, the filter with a staggered fiber array structure performs batter than the parallel array in terms of capture efficiency. The research results can provide a theoretical basis for the structure optimization and design of the air filter with multi-layer fiber.

1. Introduction

In recent years, the aggravation of global air pollution has seriously endangered human health and brought an antagonistic effect on our life such as leading to respiratory, lung, thoracic, and bronchial diseases. Exhaust gas and particles produced by human activities and industrial production need to be removed from their sources. Thus, treating this particulate matter makes contemporary society need air filters with high efficiency and low resistance characteristics. At present, fiber filter devices are widely used in fine particle filtration. According to the single fiber filtration theory, the capture efficiency and pressure drop that can be derived when evaluating filtration performance.

Based on the theory of flow around a single cylinder proposed by Kuwabara [1] and Happel [2], the research on the characteristics of flow around a single fiber and its surface particle capture has gradually developed and improved. The basic dynamic equation of size distribution function and its use in emission sources and plumes are explored by Friedlander et al [3]. They studied the deposition morphology of particles on fibers at high Reynolds numbers. Davies [4] obtained the empirical correlation of fiber filtration experiment at a low Reynolds number by correlating the three filtering effects of deposition diffusion, direct interception, and inertial collision. Magnetic resonance imaging (MRI) technology was used by Lehmann et al. [5] to non-invasively identify the local fiber structure of a single filter. They discovered that MRI is an excellent tool for gathering the necessary structural data for filtration modeling due to the good agreement between the simulated pressure drop and the MRI data. Hou et al. [6] revealed the quantitative influence of coulomb force on aerosol particle filtration efficiency. Karadimos et al. [7] provided a simple model reflecting the flow field analysis within a single fiber with particle loading. The relationship between capture efficiency of a single fiber and its surface area as well as other morphological factors was investigated by Wang et al. [8].

Computational fluid dynamics (CFD) has gradually been used to study the filtration process as a result of fast development in computer technology. This has allowed researchers to better understand dynamic information such as trajectory and instantaneous force that are difficult to measure experimentally in the dynamic movement of particles. Based on the CFD approach, Yu et al. [9] numerically studied the gas-solid two-phase flow behavior of fine particles in the fiber filtration system. During the simulation, factors such as particle group composition, particle size distribution, rebound in particle-particle and particle-fiber, and particle agglomeration were fully considered. Hosseini [10] investigated the filtration performance of fiber media without the need of empirical correction factors, and simulated the particle-loaded flow in a single fiber using Fluent CFD code, then calculated the capture efficiency and pressure drop. Maddineni et al. [11] found that the filtration efficiency was reduced due to particle collision and rebound each other, before being captured. Flow velocity, particle diameter, and fiber porosity are the decisive factors for filtration efficiency. The key influencing factors of the dendritic fiber structure. Wang et al. [12-13] simulated the particle capture process of the two-phase flow via lattice Boltzmann model, and they found that the key influencing factors of the dendritic fiber structure are the shielding effect and the "river diversion" effect in the parallel and staggered model, respectively. The capture contribution of the front fiber, particularly the first row in the parallel array model and the initial two rows in the staggered array model, is greater than the rear one when the inertial impaction mechanism is dominant. At the particle scale, the distribution of particles in the filter is used to obtain internal details.

In this paper, particle filtration in a filter with parallel and staggered arrangements of rectangular fibers are simulated using the fluent-DPM model. Firstly, Section 2 introduces theoretical models. Following that, Section 3 discusses the impact of different fiber array configuration on pressure drop and capture efficiency. Section 4 presents the conclusion and the prospects for future study.

2. Computational model of rectangular filter

2.1. Geometric model

Two classical arrangement types including rectangular parallel array and staggered array are designed to represent fibers. To simplify the calculation, a two-dimensional model is used for analysis. In the rectangular parallel array model, 54 fibers are arranged regularly in 9 rows and 6 columns. As shown in Figure 1, the staggered array model is the same as the rectangular parallel array model, with Columns 2, 4, and 6 moved vertically based on the rectangular parallel array model. The flow field boundary dimensions of the rectangular parallel fiber model and staggered fiber model are 3 mm long and 2 mm high.

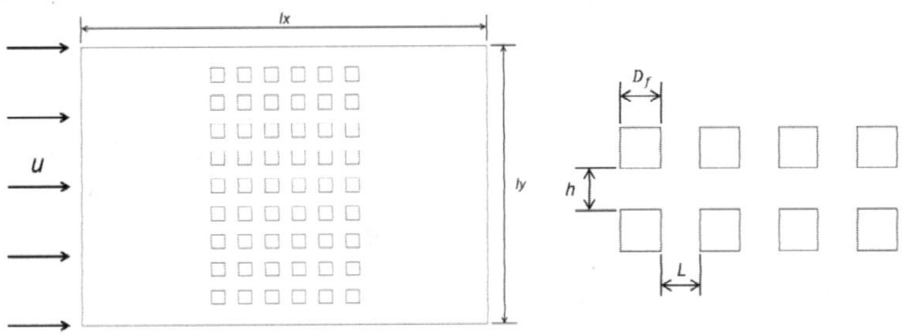

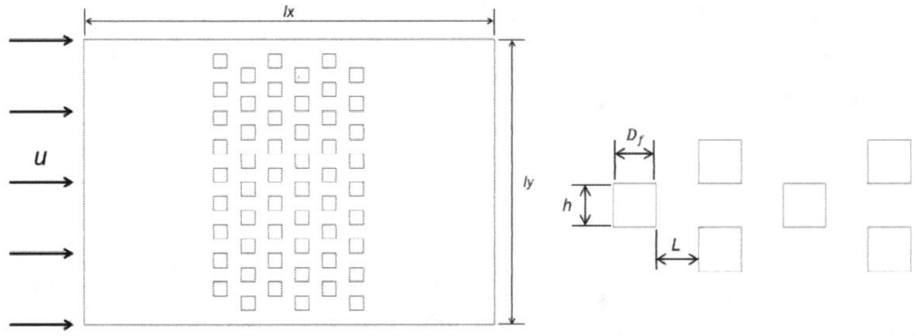

Figure.1 Two-dimensional filtration model for parallel and staggered array fibers.

2.2. Governing equations for gas flow and particle motion

In the fiber filter, the flow of air through the fiber media is viewed as a steady-state laminar flow, and the flow field is computed using the conservation of momentum equation and the two-dimensional continuity equation (Equations. 1 and 2).

$$\frac{\partial \rho}{\partial t} + \rho\left(\frac{\partial u}{\partial x} + \frac{\partial v}{\partial y}\right) = S_m \tag{1}$$

$$\frac{\partial V}{\partial t} + (\vec{U} \cdot \nabla)V = f - \frac{1}{\rho}\nabla p + \frac{\mu}{\rho}\nabla^2 V \tag{2}$$

ρ — Fluid density, $\mathrm{kg \cdot m^{-3}}$.
S_m — Mass source term.
f — External force on unit volume fluid.
$\vec{U}$ — Velocity vector.
μ — Dynamic viscosity, Pa·s.

2.3. Equation of particle motion

Particle injection is considered as a discrete phase when the volume fraction of the injected particle is less than 10%. In the process of solving the Fluent-DPM model, the Lagrange algorithm is used to calculate the force applied on a particle and motion trajectory of each particle in combination with the obtained flow field. In this model, only air resistance is considered as the main force exerted by particles. The equations of motion of spherical particles in the flow field can be described as follows (Equation. 3).

$$\frac{d\overrightarrow{U_p}}{dt} = \frac{18\mu}{\rho_p d_p^2 C_c}(\vec{U} - \overrightarrow{U_p}) \tag{3}$$

d_p — The particle diameter.
$\vec{U}$ — The fluid velocity.
$\overrightarrow{U_p}$ — The particle velocity.
ρ_p — The particle density.

2.4. Boundary Conditions

The fluent-DPM model was employed to simulate a two-phase flow. The fluid phase is set to air. The inlet is subject to the velocity boundary condition, while the outlet is set to the pressure boundary condition. The boundary of the top and bottom sides of the flow field is set to a periodic boundary condition (periodic boundary). The reference variables particle diameter to fiber diameter ratio ***R*** and Stokes number ***St*** are defined, respectively, which are expressed as follows (Equations. 4 and 5).

$$R = \frac{d_p}{D_f} \tag{4}$$

$$St = \frac{\rho_p d_p^2 U}{18\mu D_f} \tag{5}$$

R — The ratio of the particle diameter to the fiber size.
d_p — The diameter of the injected particles, m.

Journal of Physics: Conference Series **2458** (2023) 012043 doi:10.1088/1742-6596/2458/1/012043

D_f — The fiber diameter, m.
St — Stokes number.
ρ_p — Density of injected particles, $kg \cdot m^{-3}$.
U — Fluid velocity, $m \cdot s^{-1}$.
μ — Dynamic viscosity, Pa·s.

The DPM model was used to inject particles with a generation speed of 12000 cells per second. Once the flow field has attained a steady state, the particles are stater to inject into the computational domain with the same velocity as the fluid. The parameters of particle and fiber materials are depicted in Table 1.

Table 1. Model materials and parameters.

Parameter	Value
Length of the flow field Lx(mm)	2
Height of the flow field Ly(mm)	3
The fiber diameter(m)	1×10^{-4}
Particle density ($kg \cdot m^{-3}$)	2500
The diameter of the injected particles(m)	1×10^{-6}~1×10^{-5}
Flow field velocity u ($m \cdot s^{-1}$)	0.02~0.4
Dynamic viscosity (Pa· s)	1.87×10^{-5}
Air temperature(K)	300
Atoms (Pa)	1.013×10^{5}

3. Results and discussions

3.1. Pressure drops

In the DPM model, the particle phase has no influence on the flow field distribution and the pressure drops in the flow field keep unchanged once the flow arrives at a steady state. The pressure differences at the inlet and outlet of the parallel array fiber model and staggered array model are calculated respectively for different inlet velocities and injected particle diameters.

Figure. 2 shows the pressure distribution of the filter for inlet velocity v=0.2 m/s and 0.4m/s. It is found that after the flow field reaches steady state, the pressure at the inlet is significantly higher than the pressure at the outlet during the process of the fluid passing through the filter, and the pressure distribution is in a stepped shape. The distribution characteristics do not change with the progress of the filtering process.

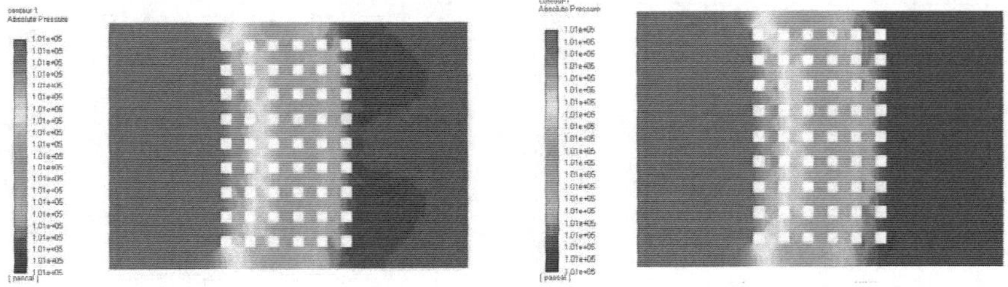

(a) (b)

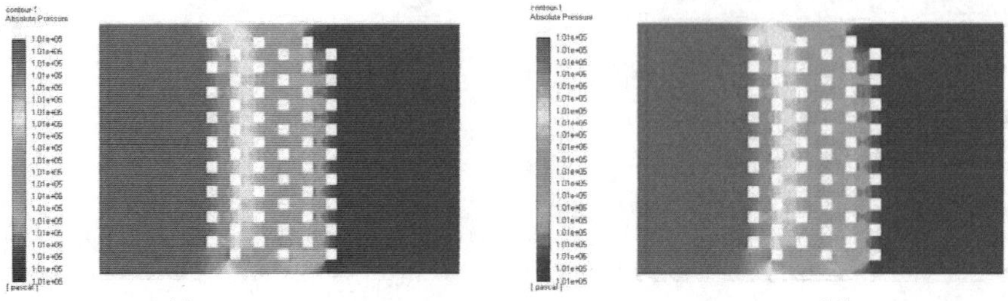

(c) (d)

Figure. 2 Pressure distribution of filter with parallel array (a) V=0.2m/s (b) V=0.4m/s and staggered array fiber model (c) staggered array V=0.2m/s (d) V=0.4m/s

Under different velocities, the pressure gradient distribution of the same fiber array model is similar, while at higher flow velocity, the pressure difference between the inlet and outlet is higher. Under the same flow velocity, the inlet and outlet pressure difference of the parallel array fiber model is lower than that of the staggering array fiber model. It is found that the staggered fiber model has a more significant effect on fluid resistance.

By adjusting Stokes number St, the pressure difference ΔP between the inlet and outlet of the parallel array fiber model and staggered array fiber model is recorded respectively as shown in Figure. 3. It is discovered that as the Stokes number rises, ΔP for both fibre configurations increases. Under the same Stokes number, the pressure drop of the parallel array shows to be lower than that of the staggered array.

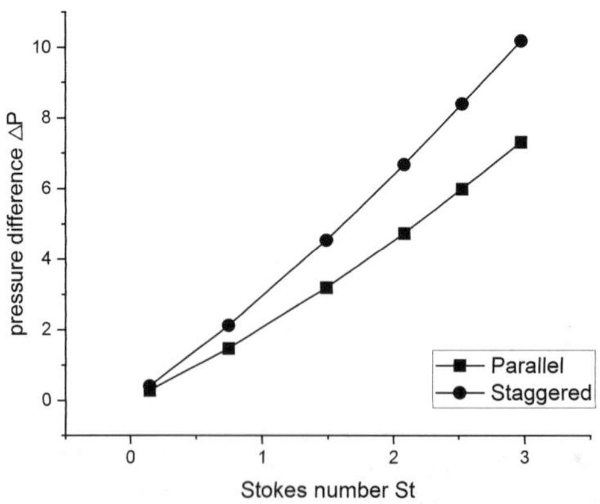

Figure. 3 Pressure difference ΔP of filter with parallel and staggered Fiber Array Modes versus Stokes Number St

For a fixed Stokes number, the relationship between the pressure differences at the inlet and outlet of the filter is obtained. Since the Fluent-DPM model is a single-phase coupling of the particle-flow field, i.e., particle injection does not affect the characteristics of the flow field, or the pressure difference between the two models remains unchanged with the ratio ***R***.

3.2. Capture efficiency

The results of the parallel array fiber model and the staggering array fiber model are calculated respectively. The effect of the filter and the track of particles is shown in Figure. 4. The flow field velocity V=0.2m/s. In the initial stage, particles flow freely and evenly in the normal phase direction of the inlet surface. After that, the flow field is altered by the fiber array and interference. Most particles are captured by the collision with the fiber array when they pass through the fiber area, and accumulate on the windward side of the fiber. A small number of particles pass through the gap of the fiber array along the flow direction. The majority of particles are caught by the fibre array as the filtration process progresses. Particles are mainly captured and stacked with the first three columns of the fiber array, and a small amount of particles are captured by the windward side of the fiber receiving the flow field. By about 0.05s, the distribution cloud of particles has stabilized.

According to the parallel array fiber model, as shown in Figure. 4, when the flow field tends to be stable, most of the particles are intercepted by the first column of fibers. The primary intercept function in the filtration process is performed by the first fibre column. After the first column of fibers, because of the flow field by the fiber array, the flow line is deflected, a small number of particles are captured by subsequent fiber arrays, and some particles are not captured through the gap of the fibers. According to the interlaced array fiber model, the flow field around t=0.05s tends to be stable. Most of the particles are captured by the first two fibers, which have an obvious interception effect on the particles. The flow field is deflected by staggered fibers, and only a few particles are collided with subsequent columns 3 and 4 of fibers. The particle distribution cloud chart shows that the staggering array fiber model captures more particles. In terms of intercepting and capturing particles, the staggered array fibre model outperforms the parallel array fibre model.

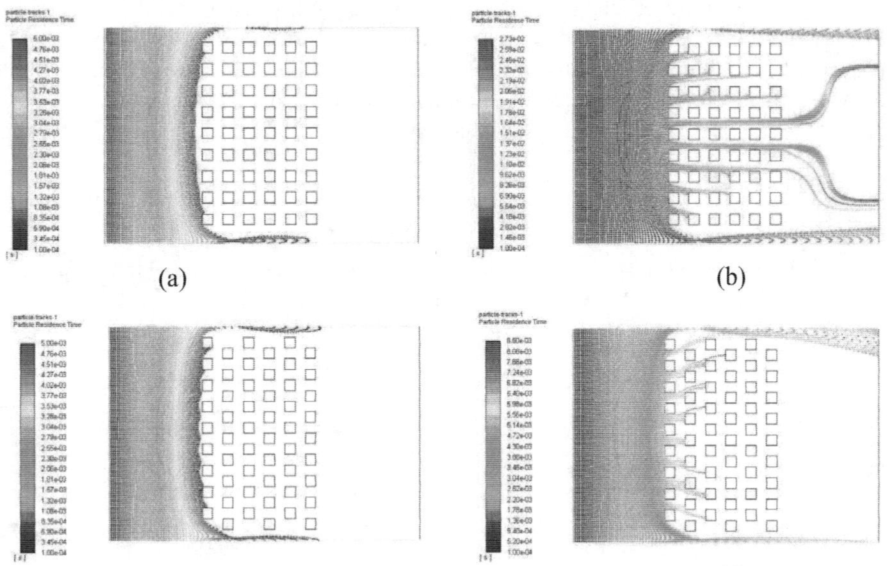

(a) (b) (c) (d)

Figure. 4 Particle motion track for a parallel model at (a) t=0.005s (b) t=0.05s, and staggered model at (c) t=0.005s (d) t=0.05s

Based on the Fluent-DPM model, tracing samples of particles are in the flow field. The capture of particles is simulated under different Stokes numbers *St* and different particle diameter to fiber diameter ratio *R*. We obtained the total amount of particles that were caught by the fibre throughout the entire process, as well as the quantity of particles that escaped through the outlet. The capture efficiency in different states was obtained according to the capture efficiency definition [Equation. 6].

$$E = \frac{n_{fiber}}{n_{fiber}+n_{outlet}} \quad (6)$$

E — Capture efficiency.

n_{fiber} — The total quantity of particles captured by the fiber.

n_{outlet} — The total quantity of particles that escaped from the outlet.

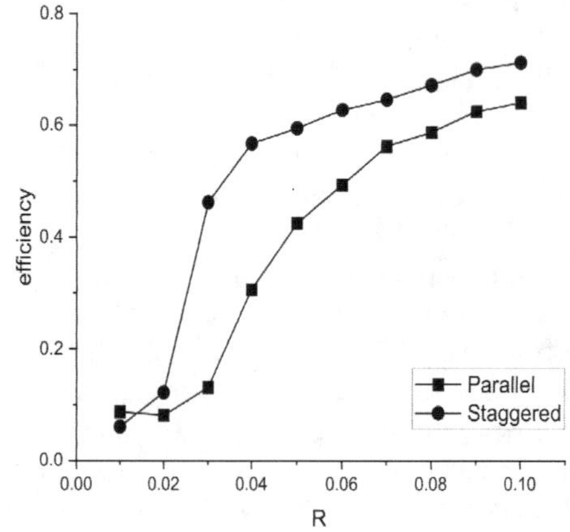

Figure. 5 Capture efficiency as a function of ***R*** (V=0.4m/s) for parallel and staggered models

Figure. 5 shows the capture efficiency *E* of the parallel array fiber model and the staggering array fiber model as a function of *R*. *E* increases with the increase of the ratio *R*, which is most significant in the range of R=0.02~0.04. Finally, the capture efficiency of the staggered array fiber model tends to be around 70%, while that of the parallel array fiber model tends to be 63%. The capture efficiency of the staggered array fibre model is consistently greater compared to the parallel array fibre model as *R* increases.

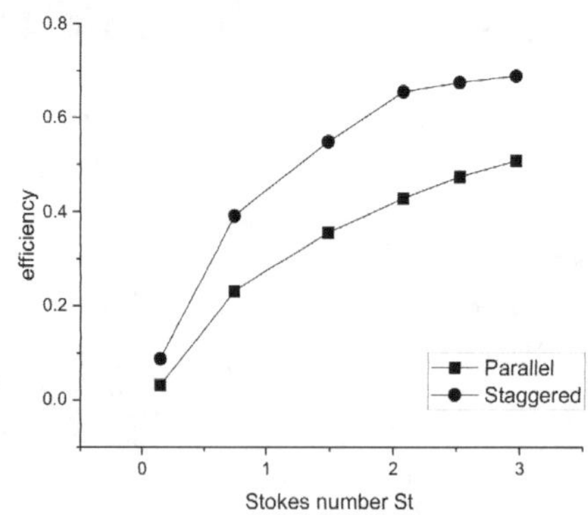

Figure. 6 Graph of capture efficiency varying with Stokes number St (R=0.1)

Figure. 6 presents the capture efficiency of the filter versus Stokes number St., which has been discovered with the increase of Stokes number *St*. The capture efficiency of both fiber models also increases. The capture efficiency *E* eventually tends to a stable value as *St* reaches a large number. The staggered array fibre model has a considerably better capture efficiency than the parallel array fibre model for the same St value, indicating that it has a higher capture performance.

4. Conclusions

In this paper, a rectangular fiber array filtration model is established, and the Fluent DPM coupling method is used to conduct a numerical simulation of the fiber array filtration process with different arrangements under the same fiber volume fraction. The following conclusions are generally attained from the analysis of drop in pressure and particle capture throughout the filtration process.

① Comparing the pressure drop between the inlet and outlet of the filter with the parallel array fiber model and the staggered array fiber model, the pressure drop of the staggered array fiber model is greater than that of the parallel array fiber model, indicating that the staggering array fiber model has a stronger effect on the flow field.

② The capture efficiency of the filter is controlled by the R and Stokes number, which increases with the increase of the R and St, and finally tends to be stable. The staggered array fibre model has a substantially greater capture efficiency stability value than the parallel array fibre model. Furthermore, the captured particles on the surface of the fibers in the staggered array are more favorable than on fibers in a parallel arrangement. In the initial stage, the simulations of the particle filtration process are investigated without taking into account how the particle phase will affect the fluid. The presented model is helpful to predict the capture efficiency and pressure distribution at the particle scale. By taking into account the back influence of particles on the fluid phase, future research will concentrate on the study of the particle-loaded flow.

Reference

[1] KUWABARA S. The forces experienced by randomly distributed parallel circular cylinders or spheres in a viscous flow at small Reynolds numbers *J. Journal of the Physical Society of Japan*, 1959, 14(4): 527-532.
[2] HAPPEL J. Viscous flow is relative to arrays of cylinders *J. AIChE Journal*, 1959, 5(2): 174-177.
[3] FRIEDLANDER S K, MARLOW W H. Smoke, dust, and haze: fundamentals of aerosol behavior *J. Physics Today*, 1977, 30(9): 58-59.
[4] DAVIES C N. Air filtration *M. London: Academic Press*, 1973:45-48.
[5] M.J. Lehmann, E.H. Hardy, J. Meyer, G. Kasper, Fibrous filters: non-invasive determination of local 3D fiber structure by MRI, *Filtration 5* (2005) 62–67.
[6] HOU L, ZHOU A, HE X, et al. CFD simulation of the filtration performance of fibrous filter considering fiber electric potential field *J. Transactions of Tianjin University*, 2019, 25(5): 437-450.
[7] A. Karadimos, R. Ocone. The effect of the flow field recalculation on fibrous filter loading: a numerical simulation, *Powder Technol*. 137 (2003) 109–119.
[8] WANG K, ZHAO H B. The influence of fiber geometry and orientation angle on filtration performance *J. Aerosol Science and Technology*, 2015, 49(2): 75-85.
[9] X.K. Yu, F.P. QIAN, J.L. LU, CFD-DEM Simulation to Study the Filtration Characteristic of the Fine Particle in the Filter Media *J. Journal of Civil and Environmental Engineering*, 2012,34(S2).
[10] S.A. Hosseini, H.V. Tafreshi. 3-D simulation of particle filtration in electro-spun nanofibrous filters, *Powder Technol*. 201 (2010) 153–160.
[11] MADDINENI A K, DAS D, DAMODARAN R M. Air-borne particle capture by fibrous filter media under collision effect: a CFD-based approach *J. Separation and Purification Technology*, 2018, 193: 1-10.
[12] H.M. Wang, H.B. Zhao, Z.L. Guo, C.G. Zheng, Numerical simulation of particle capture process of fibrous filters using Lattice Boltzmann two-phase flow model, *Powder Technol*. 227 (2012) 111–122.
[13] H.M. Wang, H.B. Zhao, K. Wang, Y.X. He, C.G. Zheng, Simulation of filtration process for the multi-fiber filter using the Lattice-Boltzmann two-phase flow model, *J. Aerosol Sci*. 66 (2013) 164–178.

Schwinger-Dyson Equations-of-Motion Method for Quantum Spin-1/2

Chuanru Dai[1], Wenxin Ding[1*]

[1]School of Materials Science and Engineering, School of Physics and optoelectronics engineering, Anhui University, Hefei 230601, China

[*]wenxinding@gmail.com

Abstract: The essential difficulty of a large portion of the strongly interacting systems can be traced down to the "noncanonicality" of the underlying operators which are used to formulate the theory, like t-J model, Hubbard-type models in the large-U limit, quantum spin, etc. We proposed a new Schwinger-Dyson equations-of-motion method to compute the dynamical correlation functions of quantum spins at zero temperature. This method can also compute all the static observables from the dynamical correlation functions. We propose that this method can formulate dynamical perturbation calculations for quantum spin-1/2 models.

1. Introduction

Quantum spin systems can display rich and vastly different behaviors depending on the dimensions, lattices, and frustration. In condensed matter physics, they are the foundation for the microscopic understanding of the magnetic properties of insulators with localized magnetic moments. Theoretically, they provide various more or less solvable models as paradigms, such as the universality class of criticalities, quantum criticalities, and topological phases.

In most cases, people must resort to canonical auxiliary operators, either bosonic or fermionic, which inevitably enlarges the Hilbert space, except for a few exactly solvable cases. Despite the tremendous success of auxiliary operator theories, like the Dyson-Maleev bosons [1], Holstein-Primakoff bosons [2], Abrikosov pseudo-fermions [3], etc., such constructions each has its limitations, typically more convenient for certain types of problems and not so convenient for others. For example, the convenient choice of the auxiliary operators strongly relies on the nature of the underlying ground state. For example, bosonic constructions are used more often; for frustrated quantum magnets, there is no census of which is more fundamental. Although in different problems, the appearance of the failure of canonical perturbation theories is different, the essential difficulty can summarize as noncanonicality, i.e., either the operators that are used to formulate the problem, like the quantum spin operators, or the operators that we need to use as the basis of perturbative calculations, like the Hubbard operators for Hubbard models in the strongly interacting limit, are noncanonical [4]. The term "noncanonicality" means that the commutation/anticommutation relations/algebra of these operators are NOT simple Kronecker- nor Dirac-δ-functions. Such noncanonical operator algebras immediately invalidate Wick's theorem, which is the very foundation of the Feynman diagram technique and hence the whole framework [5].

In this work, we formulated a new noncanonical Schwinger-Dyson equations-of-motion (SDEOM) theory for the quantum spin-1/2 system and demonstrated that for a simple two-spin problem, it could give the correct and exact results for the Ising limit. As a proof-of-principle, we treated the Ising limit

Journal of Physics: Conference Series **2458** (2023) 012044 doi:10.1088/1742-6596/2458/1/012044

as the "free" theory, and with the study of this work, we proposed this noncanonical Schwinger-Dyson equations-of-motion theory.

2. FROM HEISENBERG EOM TO SCHWINGER-DYSON EOM

While quantum spins have both nontrivial commutation relations and anticommutation relations locally, they are traditionally treated as bosons since spin operators on different sites commute. Therefore, only bosonic type Green's Functions (GFs) are considered for spin systems. It can be proved that physically we also require that either Green's Function, if solved correctly, should yield the same observables, i.e., (retarded) spectral functions or Dynamical Structure Factors (DSF). The spectral functions of both definitions of GFs satisfy a sum rule.

However, as we will show later, there is a practical benefit to considering both simultaneously. Let us start with the quantum spin SU (2) Lie-algebra on a lattice:

$$\left[S_i^\alpha, S_j^\beta\right] = i\epsilon^{\alpha\beta\gamma}\delta_{ij}S_i^\gamma , \tag{1}$$

$$\left[S_i^+, S_j^-\right] = 2\delta_{ij}S^z , \tag{2}$$

$$\left[S_i^z, S_j^\pm\right] = \pm\delta_{ij}S^\pm , \tag{3}$$

and also the fermionic part of the algebra

$$\left\{S_i^+, S_j^-\right\} = \delta_{ij} + \left(1-\delta_{ij}\right)\hat{D}_{ij} , \tag{4}$$

$$\left\{S_i^z, S_j^\pm\right\} = 0 , \tag{5}$$

where

$$S^+ = S^x + iS^y , S^- = S^x - iS^y , \tag{6}$$

where $\epsilon^{\alpha\beta\gamma}$ is the Levi-Civita symbol of rank 3, and

$$\hat{D}_{ij} = 2\left(S_i^x S_j^x + S_i^y S_j^y + i\boldsymbol{S}_i \times \boldsymbol{S}_j \cdot \hat{e}_z\right) . \tag{7}$$

where $\widehat{D}_{ij}$ is the sum of an in-plane XY dimer-type term plus a $\hat{z}$-direction chirality term.

In many situations, the first and the most important correlation function, especially for what we shall begin with, the Ising-type magnetic states, is the correlator of the spin ladder operators $S^\pm$. As noted in the previous section, we consider both bosonic and fermionic causal (time-ordered) GFs, which are defined as the following:

$$\left\langle\left\langle \hat{O}_i[t_i]\hat{O}'_f[t_f]\right\rangle\right\rangle_\eta = \left\langle \mathcal{T}_\eta\left[\hat{O}_i[t_i]\hat{O}'_f[t_f]\right]\right\rangle - A_\eta\left\langle\hat{O}_i\right\rangle\left\langle\hat{O}'_f\right\rangle, \tag{8}$$

where T_η denotes time-ordering with the sign $\eta = \pm$, $A_+ = 1$ and $A_- = 0$. Specifically, we start with consideration of

$$iG_B^{\alpha\alpha'}[i,f] = \left\langle\left\langle S_{x_i}^\alpha[t_i] S_{x_f}^{\alpha'}[t_f]\right\rangle\right\rangle_+ , \tag{9}$$

$$iG_F^{\alpha\alpha'}[i,f] = \left\langle\left\langle S_{x_i}^\alpha[t_i] S_{x_f}^{\alpha'}[t_f]\right\rangle\right\rangle_- , \tag{10}$$

where $\alpha = \pm$ and the expectation value $\langle\rangle$ is computed for arbitrary states $|\psi\rangle$. Here we adopt a short notation that abbreviates the initial-final space-time coordinates as $[x_i, x_f; t_i, t_f] \rightarrow [i, f]$. When encountering intermediate spatial coordinates other than the initial one, we abbreviate the notation as $[x_i, x_f; t_i, t_f] \rightarrow [i, f]$, as no intermediate time coordinate shall appear in this work. In general, it would be our convention when defining all dynamical correlation functions.

In many cases, the retarded or advanced GFs are also needed, which are defined similarly as

$$iG_\eta^{R\alpha\alpha'}[i,f] = \left\langle \theta\left(t_i - t_f\right)\left[S_{x_i}^\alpha[t_i], S_{x_f}^{\alpha'}[t_f]\right]_\eta\right\rangle , \tag{11}$$

$$iG_{\eta}^{A\alpha\alpha'}\left[i,f\right]=-\left\langle \theta\left(t_f-t_i\right)\left[S_{x_i}^{\alpha}\left[t_i\right],S_{x_f}^{\alpha'}\left[t_f\right]\right]_{\eta}\right\rangle. \quad (12)$$

The SDEOM of the correlator $G_{B(F)}^{\alpha\alpha'}[i, f]$ can be formally written as

$$-i\partial_{t_i}G_B^{\alpha\alpha'}\left[i,f\right]=\left\langle A_B^{\alpha\alpha'}\left[i,f\right]\right\rangle\delta\left(t_i,t_f\right)+\left\langle\left\langle\left[H,S_i^{\alpha}\right]S_f^{\alpha'}\right\rangle\right\rangle_+, \quad (13)$$

$$-i\partial_{t_i}G_F^{\alpha\alpha'}\left[i,f\right]=\left\langle A_F^{\alpha\alpha'}\left[i,f\right]\right\rangle\delta\left(t_i,t_f\right)+\left\langle\left\langle\left[H,S_i^{\alpha}\right]S_f^{\alpha'}\right\rangle\right\rangle_-, \quad (14)$$

where $A_{\eta}^{\alpha\alpha'}\left[i,f\right]=\left[S_i^{\alpha},S_f^{\alpha'}\right]_{\eta}$ and the HEOM

$$-i\partial_t S_i^+=\left[H,S_i^+\right],-i\partial_t S_i^-=\left[H,S_i^-\right] \quad (15)$$

have been applied. The last terms are commonly known as the vertex functions

$$i\Gamma_{B(F)}^{C_{\alpha};\alpha'}\left[i,f\right]=\left\langle\left\langle\left[H,S_i^{\alpha}\right],S_f^{\alpha'}\right\rangle\right\rangle_{+(-)}, \quad (16)$$

where C_{α} indicates any sequence of operators generated by $[H, S_i^{\alpha}]$.

3. THE FORMALISM

In the previous section, we introduced two types of GFs. Normally, for a particular type of particle, only one of them needs to be considered: GB for bosons and GF for fermions. For quantum spins, and in general, these two GFs can be proven to be not independent [6]. It can be proved that their spectral functions sum up to a unique function. Therefore, conventionally, only GB is considered for quantum spins. However, practically, there is tremendous convenience when both are considered. Below, we discuss the generic system. For a generic system, we generally need to study a full set (at the single particle level) of correlation functions. One convenient choice would be

$$\mathbf{G}_{B(F)}\left[i,f\right]=\left\langle\left\langle\begin{pmatrix} S_i^+S_f^- & S_i^+S_f^+ & S_i^+S_f^z \\ S_i^-S_f^- & S_i^-S_f^+ & S_i^-S_f^z \\ S_i^zS_f^- & S_i^zS_f^+ & S_i^zS_f^z \end{pmatrix}\right\rangle\right\rangle_{+(-)} =\left[\mathbf{G}_{B(F)}^{\alpha\alpha'}\right], \quad (17)$$

where α = +, −, z.

Without losing generality, we first discuss $G_{B(F)}^{+-(-+)}[i, f]$. $G_{B(F)}^{+-(-+)}$ is commonly considered for systems with magnetic order along $\hat{z}$-axis:

$$-i\partial_{t_i}G_B^{+-}\left[i,f\right]=\left\langle 2S_i^z\right\rangle\delta\left[i,f\right]+i\Gamma_{B(F)}^{C_+;-}\left[i,f\right], \quad (18)$$

$$-i\partial_{t_i}G_B^{-+}\left[i,f\right]=\left\langle 2S_i^z\right\rangle\delta\left[i,f\right]+i\Gamma_{B(F)}^{C_-;+}\left[i,f\right]. \quad (19)$$

There are two apparent difficulties within a framework which only considers $G_{B(F)}^{+-(-+)}$. First, the most important observable, $\langle S_i^z\rangle$, cannot be straightforwardly computed. In previous works, self-consistency is used to achieve the goal. Secondly, when the system becomes paramagnetic, i.e. $\langle S_i^z\rangle \to 0$, the SDEOM is out of control as all poles now stem from the vertex functions. The method by Kondo and Yamaji (KY)[7] overcomes this problem by considering the SDEOM of the vertex functions. It turns out that the SDEOM of the vertex functions can be controlled by static correlators $\langle S_i^zS_j^z\rangle$. But we would use a simple example to illustrate that this approach crucially relies on the existence of the XY term in the Hamiltonian and is incapable of describing singlet or triplet correlations in general.

In a self-sustained theory, we argue that, for a given state or wavefunction, observables like $\langle S_i^z\rangle$ must be fully determined and computable from the GFs. We note that the following relations conveniently solve the aforementioned difficulties:

$$G_B^{+-}\left[\boldsymbol{x}_i,\boldsymbol{x}_j;t_0,t_0\right]=\delta_{ij}+\left(1-\delta_{ij}\right)\left\langle\hat{D}_{ij}\right\rangle \tag{20}$$

$$G_F^{+-}\left[\boldsymbol{x}_i,\boldsymbol{x}_j;t_0,t_0\right]=\delta_{ij}\left\langle 2S_i^z\right\rangle. \tag{21}$$

Similar relations also exist for $G_{B(F)}^{-+}$. Taking the above equations into account, we can unambiguously compute the observables from the two types of GFs in a dual manner without resorting to any additional self-consistency conditions.

In addition to the determination of observables, considering both GFs at the same time also helps with describing the very different behaviors of dynamics in the quantum spin systems on equal footing. $G_{B(F)}^{+-(-+)}$ is naturally suited for describing magnetism as $\langle S^z\rangle$ generally dictates its poles. However, when $\langle S^z\rangle \rightarrow 0$, and SDEOM of G_B is out of control and all dynamical poles must enter through the vertex function. In contrast, G_F becomes naturally suited for describing the $\langle S^z\rangle \rightarrow 0$ states. The SDEOM for G_F is controlled by the static, nonlocal dimer correlations aside from the local δ-function. So we have a unified theory for describing both bosonic and fermionic type correlation within a single framework, without resorting to any potentially biased pure bosonic or fermionic representations for the quantum spins.

This description of the paramagnetic states also provides a sharp distinction between different types of states. $\langle\hat{D}_{ij}\rangle$ can be considered an order parameter of the quantum state. $\langle S_i^{x(y)}S_j^{x(y)}\rangle \neq 0$ reflects the existence of resonating valence bonds (RVBs). It is easy to verify that for a triplet/singlet state, $\langle S_i^x S_j^x + S_i^y S_j^y\rangle = \langle 2(S_i^+S_j^- + S_i^-S_j^+)\rangle = \pm 1$, which is otherwise 0 for unentangled states like $|\uparrow\downarrow\rangle$. If it is short ranged, we have a valence bond solid (VBS). If the distribution is a power law, we may have a quantum spin liquid state. The last piece $\chi_{ij} = i\langle \mathbf{S}_i \times \mathbf{S}_j \cdot \hat{e}_z\rangle$ is associated with the breaking of time-reversal-symmetry/parity-symmetry, hence the chiral spin liquid state.

Let us first consider sufficiently simple cases where $\Gamma_{B(F)}[i, f] \propto G_{B(F)}[i, f]$ so that the SDEOM can be easily inverted to obtain a closed-form solution. Then the low energy dynamics, i.e., the poles of the GFs, so as the DSF, are dictated by the static order parameters: for G_B it is the local magnetization, and for G_F it is the two-point dimer-type correlators.

4. A PROBLEM OF TWO SPINS

4.1. Free theory: Ising limit.

In this section, we demonstrate the validity and the advantages of this new approach by studying a two-spin problem. We first start with an easy-component-only Hamiltonian

$$H_0 = JS_a^z S_b^z - h\left(S_a^z + S_b^z\right), \tag{22}$$

which contains two competing terms: i) an antiferromagnetic Ising coupling ($J > 0$) and ii) a longitudinal field term. The ground-state solution of this model is the following:

$$\left(E,|\psi\rangle\right)=\begin{cases}\left(0,\alpha|\uparrow\downarrow\rangle+\beta|\downarrow\uparrow\rangle\right), & \text{if } J>|h|,\\ \left(h,|\downarrow\downarrow\rangle\right), & \text{if } h<-J,\\ \left(-h,|\uparrow\uparrow\rangle\right), & \text{if } h>J.\end{cases} \tag{23}$$

The interesting point to note is that for $J > |h|$, the ground state can be an arbitrary linear combination of the two states $|\uparrow\downarrow\rangle$ and $|\downarrow\uparrow\rangle$, reflecting a residual U (1) symmetry. Each state is characterized by

$$M = \langle \hat{M} \rangle = \langle S_a^z + S_b^z \rangle, \tag{24}$$

$$m = \langle \hat{m} \rangle = \langle S_a^z - S_b^z \rangle, \tag{25}$$

the local total and staggered magnetization. In general, we need $(M, m, \Delta m)$ to fully characterize an arbitrary wavefunction of the form

$$|\psi(M, m, \Delta m)\rangle = a_1|\uparrow\uparrow\rangle + a_2|\downarrow\downarrow\rangle + a_3|\uparrow\downarrow\rangle + a_4|\downarrow\uparrow\rangle. \tag{26}$$

Then we have

$$M = a_1^2 - a_2^2, m = a_3^2 - a_4^2,$$
$$\Delta M^2 = \langle \hat{M}^2 \rangle = a_1^2 + a_2^2, \Delta m^2 = \langle \hat{m}^2 \rangle = a_3^2 + a_4^2. \tag{27}$$

Since $\Delta M^2 + \Delta m^2 \equiv 1$, according to the normalization condition of the wavefunction, there are only three independent parameters for an arbitrary state.

The corresponding HEOM read

$$-i\partial_t S_\sigma^\alpha = \alpha\left(J S_{\bar{\sigma}}^z S_\sigma^\alpha - h S_\sigma^\alpha\right), \tag{28}$$

where $\sigma = a, b$, and $\alpha = \pm$.

Using the HEOM, we can write down the following SDEOM of the spin GFs:

$$-i\partial_{t_i} G_{B,\sigma\sigma'}^{\alpha\bar{\alpha}'}[i,f] = \alpha\left(2\langle S_\sigma^z\rangle \delta_{\alpha\alpha'}\delta_{\sigma\sigma'}\delta[i,f] - hG_{B,\sigma\sigma'}^{\alpha\bar{\alpha}'}[i,f] + J\Gamma_{B,\bar{\sigma}\sigma;\sigma'}^{z\alpha\bar{\alpha}'}[i,f]\right), \tag{29}$$

$$-i\partial_{t_i} G_{F,\sigma\sigma'}^{\alpha\bar{\alpha}'}[i,f] = \delta_{\sigma\sigma'}\delta_{\sigma\sigma'}\delta[i,f] + \alpha\left(-hG_{F,\sigma\sigma'}^{\alpha\bar{\alpha}'}[i,f] + J\Gamma_{F,\bar{\sigma}\sigma;\sigma'}^{z\alpha\bar{\alpha}'}[i,f]\right), \tag{30}$$

where $\Gamma_{B(F)\sigma\sigma';\sigma''}^{\alpha\alpha';\alpha''}$[i, f] denotes the vertex functions defined as

$$i\Gamma_{B(F),\sigma\sigma';\sigma''}^{\alpha\alpha';\alpha''}[i,f] = \langle\langle S_\sigma^\alpha[t_i] S_{\sigma'}^{\alpha'}[t_i] S_{\sigma''}^{\alpha''}[t_f]\rangle\rangle^{\pm}. \tag{31}$$

Here we use $\sigma = a, b$ subscript indices to label the sublattices in the presence of antiferromagnetic interactions, which in this simple case are also the only spatial coordinates. This is for our future convenience when discussing antiferromagnetic systems on a lattice.

4.1.1 Vertex functions and the solution of $G_{B(F)}^{\alpha\alpha'}$ for product states.

The most important information is apparently contained by the vertex functions $\Gamma_{B(F)\sigma\sigma';\sigma''}^{\alpha\alpha';\alpha''}$[i, f]. For this simple problem, we can verify by direct computation with the wavefunctions to determine the vertex function. We find that only for the fully polarized states (along $\hat{z}$ direction) can the vertex function be decoupled as

$$\Gamma_{B(F),\bar{\sigma}\sigma;\sigma'}^{z\alpha\alpha'}[i,f] \to \langle S_{\bar{\sigma}}^\alpha\rangle G_{B(F),\sigma\sigma'}^{\alpha\alpha'}[i,f]. \tag{32}$$

which is termed "random phase approximation" (RPA) in earlier literature [8]. Therefore, we first examine these SDEOMs for the fully polarized states for which the vertex functions can be trivially decoupled. For notational simplicity, we denote $G_{B(F)\sigma\sigma'}^{+-(-+)}$ as $G_{B(F)}^{\alpha\alpha'}$ with $\alpha = \pm$, and the physical spin index is ignored and $G_{B(F)}^{\alpha\alpha'}$ is interpreted as a 2×2 matrix. Then the above SDEOM can be reduced to

$$-i\partial_{t_i} G_B^{\alpha\bar{\alpha}}[i,f] = \alpha\left(I_+\delta[i,f]\delta_{\alpha\alpha'}\sigma_0 - hG_B^{\alpha\bar{\alpha}'}[i,f] + \frac{J}{2}I_- G_B^{\alpha\bar{\alpha}'}[i,f]\right), \tag{33}$$

$$-i\partial_{t_i} G_F^{\alpha\bar{\alpha}'}[i,f] = \delta[i,f]\delta_{\alpha\alpha'}\sigma_0 + \alpha\left(-hG_S^{\alpha\bar{\alpha}'}[i,f] + \frac{J}{2}I_- G_S^{\alpha\bar{\alpha}'}[i,f]\right), \tag{34}$$

where

$$I_+ = M\sigma_0 + m\sigma_z, I_- = M\sigma_0 - m\sigma_z, \tag{35}$$

$\sigma_{0,x,y,z}$ denote the 2 × 2 identity matrix and Pauli matrices. In the current solution, we can only take M = ±1, m = 0 or M = 0, m = ±1 as we pointed out above that Eq. (32) is only correct for fully polarized states. (We consider M = 0 and m = ±1 also fully polarized in a staggering way.) Now we perform Fourier transform and obtain the expression for *G*'s in the frequency space:

$$G_B^{\alpha\bar{\alpha}'}[\omega] = \frac{\alpha\delta_{\alpha\alpha'}(M\sigma_0 + m\sigma_z)}{\omega - \alpha(-h + J(M\sigma_0 - m\sigma_z)/2)}, \tag{36}$$

$$G_F^{\alpha\bar{\alpha}'}[\omega] = \frac{\delta_{\alpha\alpha'}\sigma_0}{\omega - \alpha(-h + J(M\sigma_0 - m\sigma_z)/2)}. \tag{37}$$

So the retarded and advanced GFs are

$$G^{R\alpha\bar{\alpha}'}[\omega] = \frac{\delta_{\alpha\alpha'}((1+\alpha M)\sigma_0 + \alpha m\sigma_z)/2}{\omega - \alpha(J(M\sigma_0 - m\sigma_z)/2 - h) + i\eta}, \tag{38}$$

$$G^{A\alpha\bar{\alpha}'}[\omega] = \frac{\delta_{\alpha\alpha'}((\alpha M - 1)\sigma_0 + \alpha m\sigma_z)/2}{\omega - \alpha(J(M\sigma_0 - m\sigma_z)/2 - h) - i\eta}, \tag{39}$$

where the limit $\eta \to 0^+$ is always assumed.

We can verify that we already have a nontrivial property for the polarized states, G^{R+-} is asymmetric/independent of G^{R-+}. For example, if we take m = 1 so |ψ(ab)⟩ = | ↑↓⟩, we have

$$G_{aa}^{R+-}[\omega] = (\omega + J/2 + i\eta)^{-1}, G_{aa}^{R-+}[\omega] = 0, \tag{40}$$

$$G_{aa}^{A+-}[\omega] = 0, G_{aa}^{A-+}[\omega] = (\omega - J/2 - i\eta)^{-1}. \tag{41}$$

4.1.2 Solutions of arbitrary states and vertex functions beyond KY-RPA.

In general, for any other states |ψ (M, m, Δm) as given in Eq. (26), we find that the arbitrary $G_{B(F),para}$ can construct as

$$\begin{aligned} G_{B(F),para} &= \\ a_1^2 G_{B(F),M=0,m=1} &+ a_2^2 G_{B(F),M=0,m=-1} \\ +a_3^2 G_{B(F),M=1,m=0} &+ a_4^2 G_{B(F),M=-1,m=0}, \end{aligned} \tag{42}$$

where *para = (M, m, Δm)* is the parameters that completely specify the underlying state. With Eq. (42), we can plug the $G_{B(F)}$ back into the SDEOM to obtain solutions for the vertex functions. We will examine the most interesting case for which both *M = 0* and *m = 0* to illuminate the complex nature of the vertex function problem.

For *M = 0* and *m = 0*, we use $a_1 = 0$, $a_2 = 0$, $a_3 = 1/\sqrt{2}$, and $a_4 = -1/\sqrt{2}$. From the polarized states |*m* = 1⟩ = |↑↓⟩ and |*m* = −1⟩ = |↓↑⟩, we can obtain

$$G_{m=\pm1}^{R+-}[\omega] = \frac{(\sigma_0 + m\sigma_z)/2}{\omega + m\sigma_z J/2}, \tag{43}$$

$$G_{m=\pm1}^{A+-}[\omega] = \frac{(-\sigma_0 + m\sigma_z)/2}{\omega + m\sigma_z J/2}. \tag{44}$$

Therefore, we have

$$G_B^{\alpha\bar{\alpha}'}\Big|_{M=0,m=0} = \frac{-\sigma_0 J/2}{\omega^2 - J^2/4}, \tag{45}$$

$$G_F^{\alpha\bar{\alpha}'}\Big|_{M=0,m=0} = \frac{\omega\sigma_0}{\omega^2 - J^2/4}. \tag{46}$$

Now we go back to the SDEOM with $M = 0, m = 0$:

$$-i\partial_{t_i} G_{B,\sigma\sigma'}^{\alpha\bar{\alpha}'}[i,f] = \alpha J \Gamma_{B,\sigma;\sigma;\sigma'}^{z\alpha;\bar{\alpha}'}[i,f], \tag{47}$$

$$-i\partial_{t_i} G_{F,\sigma\sigma'}^{\alpha\bar{\alpha}'}[i,f] = \delta[i,f]\delta_{\alpha\alpha'}\delta_{\sigma\sigma'} + \alpha J \Gamma_{F,\sigma;\sigma;\sigma'}^{z\alpha;\bar{\alpha}'}[i,f]. \tag{48}$$

Using Eq. (45) and Eq. (46), we obtain the vertex function from the SDEOM:

$$\Gamma_{B,\sigma\sigma;\sigma'}^{z+;-}[\omega] = \frac{-\omega\sigma_0}{\omega^2 - J^2/4}, \tag{49}$$

$$\Gamma_{F,\sigma\sigma;\sigma'}^{z+;-}[\omega] = \frac{2}{J}\left(\frac{\omega^2\sigma_0 J/2}{\omega^2 - J^2/4} - \sigma_0\right). \tag{50}$$

To prepare for the perturbation calculation for the next section, we write down the explicit expressions for arbitrary states with physical parametrization (use $(M, m, \Delta m)$ instead of a_is). First, the solution for real a_is is not unique. For later purposes, here we pick a solution that gives us a positive and uniform $\langle S^x \rangle$:

$$a_1 = \sqrt{\frac{1+M-\Delta m^2}{2}}, a_2 = \sqrt{\frac{1-M-\Delta m^2}{2}}, \tag{51}$$

$$a_3 = \sqrt{\frac{m+\Delta m^2}{2}}, a_4 = \sqrt{\frac{\Delta m^2 - m}{2}}, \tag{52}$$

which gives

$$\begin{aligned} \langle S_a^x \rangle &= \langle S_b^x \rangle \\ &= 1/2\Big(\sqrt{\Delta m^2 + m}\sqrt{1-\Delta m^2 - M} . \\ &+ \sqrt{\Delta m^2 - m}\sqrt{1-\Delta m^2 + M}\Big) \end{aligned} \tag{53}$$

The $G_{\mathrm{B(F)}}^{\alpha\alpha'}$ for arbitrary state with parameters $(M, m, \Delta m)$ reads

$$G_B^{\alpha\bar{\alpha}}[\omega] = \frac{\alpha\left((1-\Delta m^2)\sigma_0 + m\sigma_z\right)(\omega+\alpha h) + (M+\alpha\Delta m^2)J\sigma_0/2}{(\omega+\alpha h)^2 - J^2/4}, \tag{54}$$

$$G_F^{\alpha\bar{\alpha}}[\omega] = \frac{(\omega+\alpha h)\sigma_0 + \alpha(M\sigma_0 - m\sigma_z)J/2}{(\omega+\alpha h)^2 - J^2/4}. \tag{55}$$

4.1.3 Calculation of Observables.

With the solution of the GFs, we discuss the calculation of static observables like $\langle S_a^z \rangle$, $E_0 = \langle H_0 \rangle$ via the GFs. We first consider $\langle S_a^z \rangle$. Since $[S_i^+, S_j^-] = 2\delta_{ij}S_i^z$, naively we have

$$\langle S_a^z \rangle = \frac{i}{2} G_{F,aa}^{+-}[t_i, t_i] = -\frac{i}{2} G_{F,aa}^{-+}[t_i, t_i]. \tag{56}$$

However, when computed via the GFs in frequency space, the equal-time limit is reached by

$$G_{F,aa}^{\alpha\bar{\alpha}'}[t_i, t_i] \equiv \frac{1}{2}\left(G_{F,aa}^{\alpha\bar{\alpha}'}[t_i, t_i+\eta] + G_{F,aa}^{\alpha\bar{\alpha}'}[t_i, t_i-\eta]\right), \tag{57}$$

where the sign of η determines whether to take poles above or below the real axis when computing the frequency integral. (* In the canonical case, this is not necessary since the retarded and advanced GFs are not independent. In the current case for noncanonical theory, such construction is generically necessary.) Throughout this manuscript, we always consider the retarded case, i.e., take poles above the real axis. Therefore, we compute $\langle S_a^z \rangle$ as

$$\begin{aligned}\langle S_a^z \rangle &= \frac{i}{4}\left(G_{F,aa}^{+-}[t_i, t_i+\eta] - G_{F,aa}^{-+}[t_i, t_i+\eta]\right) \\ &= \frac{i}{4}\int \frac{d\omega}{2\pi}\left(G_{F,aa}^{+-}[\omega] - G_{F,aa}^{-+}[\omega]\right),\end{aligned} \tag{58}$$

and the specification of η (for t) shall be omitted. Similar consideration also applies to the computation of other observables, such as $\langle S_a^z \rangle$ etc., as equal-time correlators. We can verify that Eq. (58) does give the correct values of $\langle S_a^z \rangle$ for all cases.

5. Conclusion

In work, we formulated a new Schwinger-Dyson equation of motion for quantum spin-1/2 systems. Then, we proved that for a simple two-spin problem, the Schwinger-Dyson equation of motion could give an exact solution about the Ising limit for arbitrary states. These results can serve as foundations for dynamical perturbative computations for interacting perturbations such as a transverse field term or XY-term. More importantly, our method can also be applied to lattice models, which would provide a novel way to study quantum spin-1/2 models.

References

[1] Dyson, F. J. (1956). General theory of spin-wave interactions. Physical review, 102(5), 1217.
[2] Holstein, T., & Primakoff, H. (1940). Field dependence of the intrinsic domain magnetization of a ferromagnet. Physical Review, 58(12), 1098.
[3] Abrikosov, A. A. (1965). Electron scattering on magnetic impurities in metals and anomalous resistivity effects. Physics Physique Fizika, 2(1), 5.
[4] Ovchinnikov, S. G., & Val'kov, V. V. (2004). Hubbard operators in the theory of strongly correlated electrons. World Scientific.
[5] Spencer, H. J. (1968). A Wick theorem for spin-one operators. Physical Review, 167(2), 430.
[6] Fröbrich, P., & Kuntz, P. J. (2006). Many-body Green's function theory of Heisenberg films. Physics Reports, 432(5-6), 223-304.
[7] Kondo, J., & Yamaji, K. (1972). Green's-function formalism of the one-dimensional Heisenberg spin system. Progress of Theoretical Physics, 47(3), 807-818.
[8] Tiablikov, S. V. (2013). Methods in the quantum theory of magnetism. Springer.

Semi-supervised Learning Method for Source Range Estimation in Shallow Water

Lebo Li[1,*], Fusheng Gong[1] and Shijing Liu[1]

[1] Hangzhou Applied Acoustics Research Institute, Hangzhou, China

* Correspondence: yuebo715@163.com

Abstract: Data-driven machine learning methods in recent years have shown great vitality in underwater acoustic, especially in the localization of underwater sound sources. Meanwhile, there exists a great difficulty in the acquisition and processing of underwater acoustic data, which caused the lack of label data. This paper proposes a two-step semi-supervised learning-based range estimation method for shallow ocean sound sources, which utilizes semi-supervised learning in deep learning to alleviate the problem of limited access to underwater acoustic label data and relatively abundant unlabelled data in practical scenarios. The performance of the method has been validated on SWellEx-96 experiment data.

1. Introduction

The accurate localization of acoustic sources in a shallow waveguide remains a burning problem in the field of underwater acoustic. Over the past two decades, data-driven machine-learning methods have been employed in underwater acoustic passive localization in various ways [1-4]. Niu et al. [5] proposed a class of feasible methods based on machine learning algorithms for underwater source range estimation, after they systematically investigated the performance of three different machine learning models, named Feedforward Network (FFN), Support Vector Machines (SVM), and Random Forests (RF), for source range estimation, and verified the localization performance of the machine learning algorithms firstly using experiment data from sea trials. In the same year, Huang et al. [6] attempted to localize sound sources via a multilayer deep convolutional neural network with simulated sound fields as training data. Meanwhile, there are many challenges, including the limited access to label datasets [7] and the relatively rich unlabelled data generated by passing ships and other underwater sound sources (seismic waves, volcanic eruptions, etc.) phenomena remaining to be exploited.

This paper proposes a two-step semi-supervised learning (SSL) framework [8-9] to transform sound source range estimation into a classification problem. Given that the available dataset consists of unlabelled and label data, at the first step, a convolutional self-encoder with an attention mechanism is applied to extract features from the label dataset and unlabel dataset. At the second step, a multilayer perceptron (MLP) and label data are used to learn the features extracted previously to achieve classification, and sound source range estimation ultimately.

2. Data pre-processing and SWellEX-96 data analysis

2.1. Receiver array data pre-processing

In array signal processing methods for waveguides, such as Matched-Field Processing (MFP) , and Sample Covariance Matrices (SCM) of the receiver array are the most widely used features, which contain the source location parameters and the waveguide parameters. The sample covariance matrices which calculated from the data of a VLA array are used as the feature input of the network.

The lth measurement of the VLA [10] can be modeled as follows:

$$\mathbf{y}_l(f) = P(f) \bullet \mathbf{g}(f, \mathbf{r}_s, \mathbf{r}_m) + \mathbf{n} \tag{1}$$

where $\mathbf{y}_l(f) = [y_1(f), y_2(f), ..., y_Q(f)]^T$ consists of Fourier coefficients of Q elenment of the VLA at a single frequency f , which is obtained via an FFT. $P(f)$ is the complex source amplitudes at a single frequency f and $g(f, \mathbf{r}_s, \mathbf{r}_m)$ is the Green's function(calculated by sound propagation model), $\mathbf{r}_s$ is the position of source and $\mathbf{r}_m$ is the position of VLA. $\mathbf{n}$ is the ambient noise.

Data normalization:

$$\tilde{\mathbf{y}}_l(f) = \frac{\mathbf{y}_l(f)}{\|\mathbf{y}_l(f)\|_2}, \tag{2}$$

Then the sampling covariance matrix calculated by L snapshots

$$SCM(f) = \frac{1}{L} \sum_{l=1}^{L} \tilde{\mathbf{y}}_{l,r_s}(f) \tilde{\mathbf{y}}_{l,r_s}{}^H(f). \tag{3}$$

We extract the real and imaginary parts of the SCM and normalize them to the interval (0,1] as the input of the $Q \times Q \times 2$-dimensional network. When using sound pressure data of multiple-frequency points for processing, the multiple frequencies are processed to obtain the input superimposed with the input dimension of $Q \times Q \times 2K$, K being the number of frequency points.

2.2. Performance indicators

The data of the VLA yielded by a moving source is discretized by equal-width bins according to the range, and the ranges are equally divided into F classes:

$$\Delta r = \frac{r_{\max} - r_{\min}}{F}. \tag{4}$$

Assuming that r_i is the actual range of the ith sample, then the label of the ith sample is built as:

$$\text{label}_i = \left\lceil \frac{r_i - r_{\min}}{\Delta r} \right\rceil. \tag{5}$$

2.3. Label acquisition

The performance metrics use Mean Absolute Error (MAE) and the localization statistic P_{CL} :

$$MAE = \frac{1}{N} \sum_{i=1}^{N} |r_i - f(x_i)|. \tag{6}$$

$f(x_i)$ is the predicted range.

$P_{CL-5\%}$ calculated as follows:

$$P_{CL-5\%} = \frac{\sum_{i=1}^{N} \eta(i)}{N}, \quad \eta(i) = \begin{cases} 1, \dfrac{|r_i - f(x_i)|}{r_i} \times 100\% \le 5\% \\ 0, otherwise \end{cases}. \tag{7}$$

2.4. SWellEX-96 data analysis

The VLA data of the SWellEX-96 experiment's event S5 [11] is used in our semi-supervised learning method, and the waveguide parameters of the experimental sea region are shown in Figure 1. The source ship (R/VSproul) sailed from south to north, and the towed source transmitted CW signals at multiple-frequency points. We used narrowband data {163Hz} and broadband data {109, 127, 145, 163, 198,

232, 280Hz}. The range estimation result of the matched-field process (MFP) [12-14] based on the bartlett processor is given in Figure 2.

The original sampled data is processed through FFT to obtain the frequency domain sound pressure with 1500 single FFT points, 50% overlap, and a snapshot number of 2. Then 4500 SCM samples are obtained by the pre-processing step of 2.1, where the range of the tag data is calculated from the shipboard GPS data.

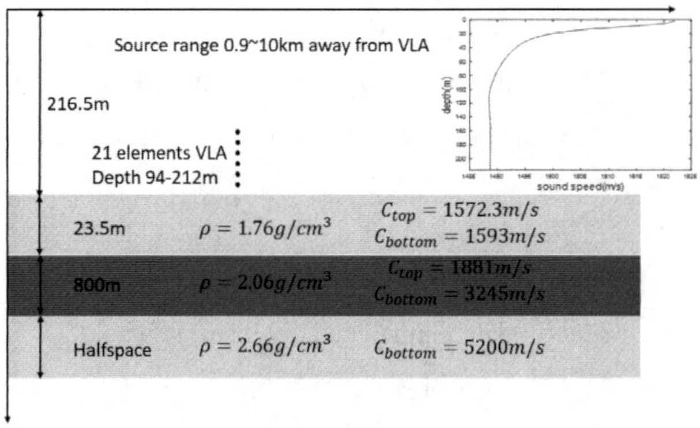

Figure. 1 SWellEX-96 experimental environment parameters

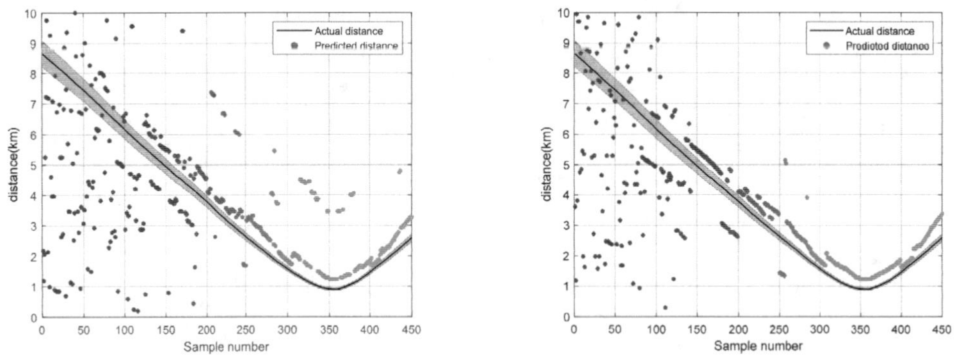

(a) (b)

Figure. 2 MFP results: (a) Narrowband {163} Hz localization results.
(b)Broadband {109, 127, 145, 163, 198, 232} localization results

3. Framework for underwater source range estimation models

3.1 Step 1

Step 1: Unsupervised learning in self-attentive mechanism-convolutional self-encoder (SA-CAE) models.

The SA-CAE model consists of a self-attentive mechanism [15], an encoder, and a decoder [16], as shown in the blue and purple parts of Figure 3. Its unsupervised learning model enables coverage of the entire dataset (4500), and more efficient and accurate feature extraction can be achieved by introducing the self-attentive mechanism into the convolutional autoencoder. The loss function is the mean square error (MSE).

3.2 Step 2

Step 2: Supervised learning in combination with a multi-layer perceptron model (MLP) for source range estimation.

The label data is used in this step. After the SA-CAE model has been trained, the hyperparameters in the blue part of Figure 3 are frozen as a feature extraction network and connected to a 4-layer MLP classification network, the yellow part of the figure, to form the SA-CAE-SSL model. Classification learning is performed using a label dataset for the source range estimation task and the loss function is the Cross-Entropy Loss Function.

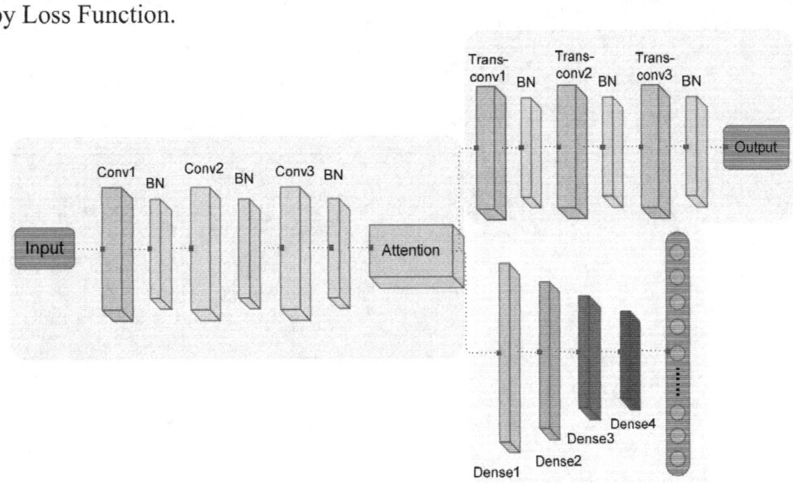

Figure 3. Semi-supervised learning framework

4. Experiment setup and results

4.1. Style and spacing
Two hundred samples were selected as the test sample, and we divided the remaining samples into label and unlabelled samples according to 20%, 50%, and 80% of the label data respectively.

4.2. Control group setting
Control Group I (CAE-SSL): Just get rid of the self-attentive mechanism module from our semi-supervised learning framework, preserve the rest structure of our framework.

Control Group II (SA-SL): A network with the same structure as the semi-supervised range estimation model (SA-CAE-SSL) proposed in this paper is trained using a supervised learning approach.

Control Group III (SL): Just get rid of the self-attentive mechanism module from Control Group II (SA-SL), preserve the rest structure of SA-SL.

4.3. Experiment results and discussion
The experiment data in this section uses the dataset partitioning proposed in Section 4.1. To verify the localization performance of our framework especially when the label data is not sufficient, We selected 80%, 50%, and 20% of the entire data to be label data. The experiment results are shown below. The following figure shows the prediction results for label data accounting for 20% of the entire dataset.

Narrowband: {163} Hz; Broadband: {109,127,145,163,198,232} Hz

Table 1. Experiment results

Percentage of label data		80%		50%		20%	
Frequency band	Localization framework	MAE (km)	$P_{CL-5\%}$ (%)	MAE (km)	$P_{CL-5\%}$ (%)	MAE (km)	$P_{CL-5\%}$ (%)
Narrowband	SA-CAE-SSL	0.1893	90	0.2305	90	0.4210	76
	CAE-SSL	0.2072	89	0.2580	89	0.5016	71

	SA-SL	0.2297	87.5	0.3196	84.5	0.5001	77
	SL	0.2475	87	0.3397	85.5	0.5413	74.5
Broadband	SA-CAE-SSL	0.0136	99.5	0.0170	98.5	0.0089	99
	CAE-SSL	0.0070	99	0.0221	97	0.0999	91
	SA-SL	0.0186	99	0.0852	93.5	0.2045	86.5
	SL	0.0387	98	0.0550	93	0.3870	80.5

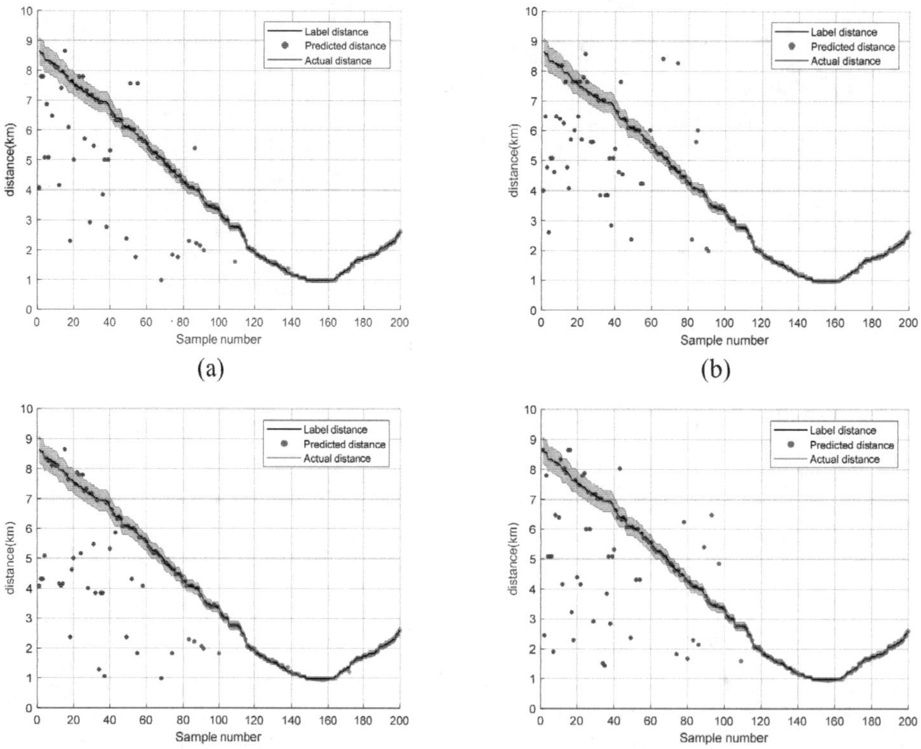

(a) (b)

(c) (d)

Figure 4. Range estimation results of narrowband dataset: (a) RA-CAE-SSL model; (b) CAE-SSL model; (c) RA- SL model; (d) SL model

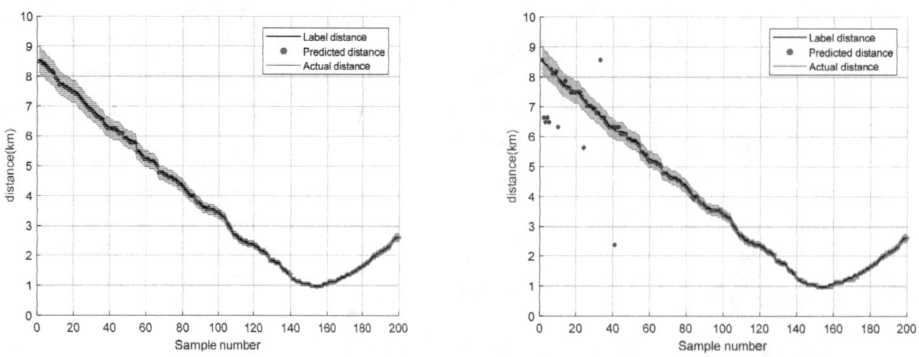

(a) (b)

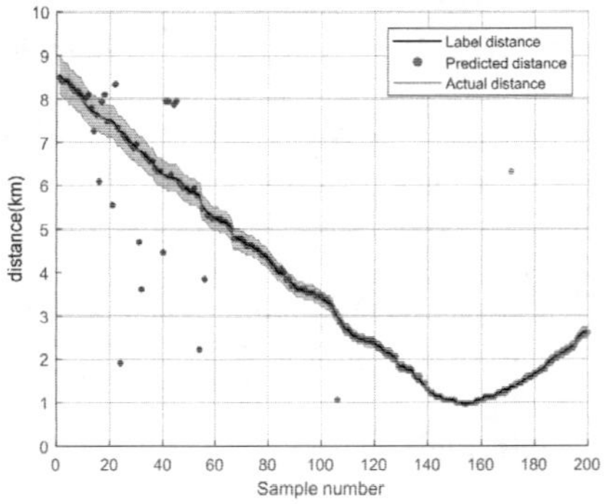

(c)

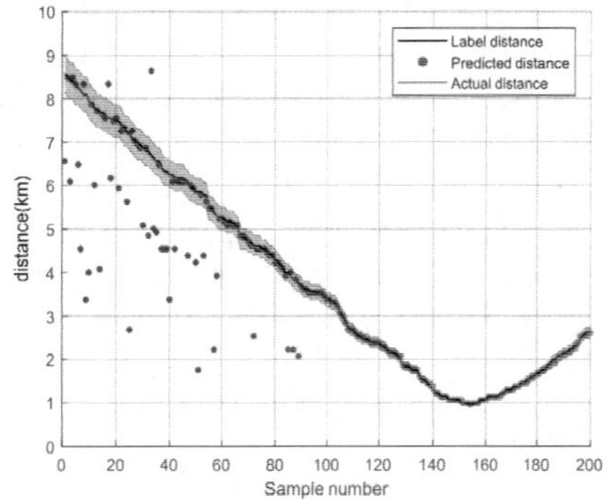

(d)

Figure 5. Range estimation results of broadband dataset: (a) RA-CAE-SSL model; (b) CAE-SSL model; (c) RA- SL model; (d) SL model

Though the contrast of the four models predicted result of narrowband is not apparent due to the loss of the gain of broadband, from the results of the broadband dataset in Table 1 and Figure 5, we can see clearly that the performance advantage of our framework over the control group becomes more and more apparent as the number of training sets decreases. As expected, the contrast is relatively less pronounced when the proportion of the dataset increases. This indicates that the unsupervised learning module mines the structural knowledge of the data through convolutional self-encoders, effectively using unlabelled data to alleviate the traditional supervised learning process which requires a large amount of label data. Comparing the localization results of SA-SL model and SL model, we can see the SA mechanism also contributes to the performance advantage, which indicates that the self-attentive mechanism can focus on the features that are more relevant to the source location. In addition, the trajectory prediction plots show that this paper outperforms the model-driven MFP approach because of the mismatch between the actual waveguide and the sound field model.

5. Conclusion

In this paper, a semi-supervised learning framework was proposed; it consists of a convolutional autoencoder structure with a residual self-attention mechanism module and an MLP. In the case of insufficient label data, we divide the source range estimation into two steps. We verified our framework is more adaptable to the underwater acoustic problem than the supervised learning framework in situations where label data acquisition is difficult, especially with the help of SA mechanism.

It is important to note that experiments are limited by the size and richness of datasets, and it is more convincing to use datasets from non-target sound sources. In addition, this paper does not consider that the test samples and training samples are collected in two waveguides with specific differences, that is, they obey different data distributions, which can be the direction of further research.

References

[1] Yang, H; Lee, K; Choo Youngmin, et al. Underwater Acoustic Research Trends with Machine Learning: Passive SONAR Applications. IEEE J. Ocean. Eng. 2021, 34(3),227-236

[2] Wenbo W.; Haiyan N.; Lin S.; Tao H. Deep transfer learning for source ranging: Deep-sea experiment results. J. Acoustic. Soc. Am.2019. 146(4), EL317–EL322.

[3] Niu H, Gong Z and Ozanich E et al. Deep-learning source range estimation using multi-frequency magnitude-only data. J Acoust Soc Am. 2019 Jul;146(1):211.

[4] Niu H, ZaiXiao G and Emma Ozanich et al. Source Ranging Using Ensemble Convolutional Networks in the Direct Zone of Deep Wate. J Acoust Soc Am.2019

[5] Niu H, Reeves E, Gerstoft P. Source range estimation in an ocean waveguide using supervised machine learning. J Acoust Soc Am. 2017 Sep;142(3):1176.

[6] Huang Z, Xu J, Gong Z, et al. Source localization using deep neural networks in a shallow water environment[J]. J Acoust Soc Am. 2018,143(5): 2922-2932.
[7] Yang, H; Lee, K.; Choo, Y. Underwater Acoustic Research Trends with Machine Learning: General Background. IEEE J. Ocean. Eng.2020,34(2),147-154.
[8] Bianco, M. J.; Gannot, S.; Gerstoft, P. Semi-supervised source localization in reverberant environments with deep generative modeling. In IEEE Access, 2021,09, 84956-84970.
[9] Zhou, M.; Tang, Y.; Tian, Z. Robust Neighborhood Graphing for Semi-Supervised Indoor Localization with Light-Loaded Location Fingerprinting. IEEE Internet Things J. 2018, 05,3378-3387.
[10] F. B. Jensen, W. A. Kuperman, M. B. Porter, and H. Schmidt, Computational Ocean Acoustics. New York: Springer-Verlag, in press, 1993.
[11] Murray, J; Ensberg, D. The Swellex-96 Experiment 1996. Available online: http://http://swellex96.ucsd.edu/index.htm
[12] Bucker, H. P. Use of calculated sound fields and matched field detection to locate sound sources in shallow water. J. Acoust. Soc. Am. 1976, 59 (2), 368—373.
[13] Baggeroer, A. B.; Kuperman, W. A.; Mikhalevsky, P. N. An overview of matched field methods in ocean acoustics. IEEE J. Ocean. Eng.1993, 18 (4) , 401—424.
[14] Westwood, E. K. Broadband matched-field source localization. J. Acoust Soc. Am. 1992,91:2777-2789.
[15] S. Noh; D. J. Ji; D. -H. Cho. A Self-Attention-Based I/Q Imbalance Estimator for Beyond 5G Communication Systems. IEEE Commun. Lett. 2021,25,3262-3266.
[16] Min, C.; Xiaobo, S.; Yin, Z. Deep Feature Learning for Medical Image Analysis with Convolutional Autoencoder Neural Network. IEEE Trans. Big Data, 2021,07,750-758.

 2458 (2023) 012046 doi:10.1088/1742-6596/2458/1/012046

Structure design of a radar display console muffler

Ji Kang*, Liang Juan Zhang, Sheng Sheng Liu and Zhao Hui Lv

Nanjing Institute of Electronic Technology, Nanjing, China

953725243@qq.com

Abstract. Radar display console is an equipment which is closely associated with people as a medium of human-computer interaction. However, for the sake of heat dissipation, draught fans in display console make continuous noise when radar is on operation, the noise dose harm to human health. In order to reduce the noise, an impedance compound muffler is designed in this paper, and then the acoustic characteristics of muffler is studied by simulation method. In the end, the simulation results are verified by means of noise spectrum analyser, as a result, the muffler obtains good silencing effect and the maximum sound level of the radar display console is reduced by 3dB.

1. Introduction

As the main medium of man-machine interaction, the display console is a radar control equipment with the closest relationship with people. It mainly undertakes the functions of monitoring radar state, monitoring targets and extracting target parameters such as distance, azimuth, elevation and so on. The main heat dissipation of radar display console is forced convection cooling technology using draught fans to generate high-speed air which flows through the heating equipment to dissipate heat. At the same time, the fans become the main noise source of the display console due to the high speed when the radar is working. Working in the noisy environment for a long time will have a serious impact on physical and mental health of operators [1, 2].

In order to reduce the noise of the display console, an impedance composite muffler for radar display console is designed in this paper, and then, the pressure and velocity field in the muffler are analysed by FLUENT and COMSOL finite element analysis software, the distribution of sound pressure and sound pressure level in the muffler and transmission loss at different frequencies are obtained. Finally, the noise spectrum analyzer is used to measure the noise of display console, the test results are in accord with the simulation.

2. Structure design of muffler

On the basis of the principle of noise reduction, mufflers can be divided into resistive mufflers, reactive mufflers and resonant mufflers [3]. The resistive muffler mainly arranges sound-absorbing materials in the gas channel to increase the friction between the fluid and the muffler, and converts sound energy into heat energy which contributes to reduce noise. However, the resistive muffler has poor absorption effect on low-frequency noise, it is suitable for medium and high-frequency noise elimination. The reactive muffler reflects sound waves by changing the air duct, so as to achieve the purpose of noise elimination. The reactive muffler requires a large size of the air duct section, and its absorption effect on high-frequency noise is poor, so it is suitable for medium and low frequency noise elimination. Resonant muffler is a resonant cavity formed by porous metal plate and muffler. Sound waves move back and forth in the metal hole under the action of pressure, a part of sound energy is consumed under the action of friction. Nevertheless, resonant muffler has obvious effect on low-frequency noise.

Noise curve of the display console was obtained by the noise spectrum analyzer in Figure 1. It can be seen that the noise of the display console is mainly concentrated at the intermediate frequency of 1500Hz.

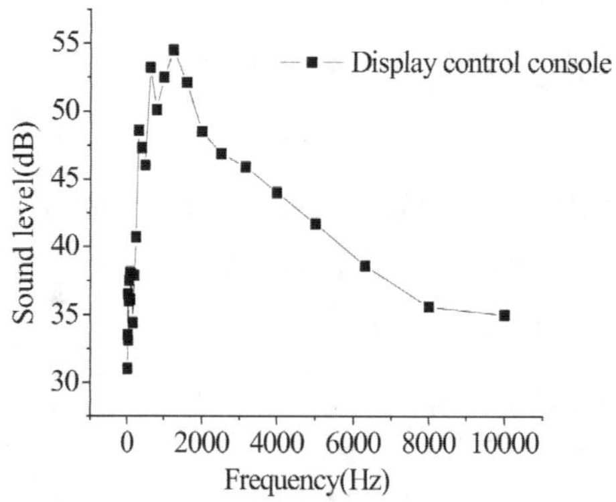

Figure 1. Noise spectrum of console.

In this paper, a composite muffler with characteristics of both resistive muffler and resonant muffler is designed. The muffler is installed at the air outlet of the display console to reduce the noise generated by draught fans. The installation schematic diagram is shown in Figure 2.

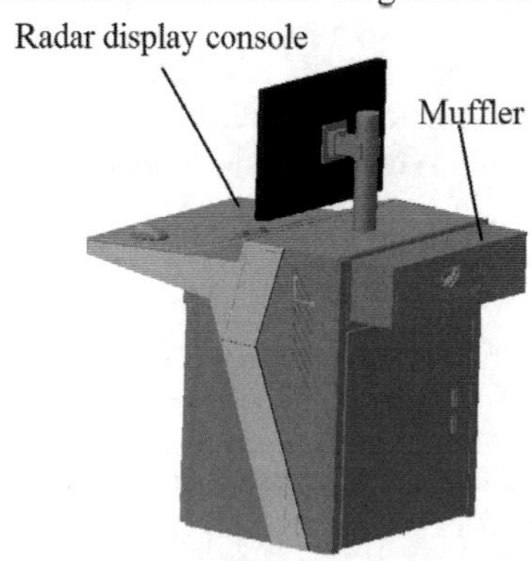

Figure 2. Muffler installation diagram.

Composite muffler is composed of porous metal plate, deflector and muffler shell. There is a gap between the porous metal plate and muffler shell forming a resonant cavity to absorb part of acoustic energy. A layer of sound-absorbing cotton is attached to the deflector, as shown in Figure 3.

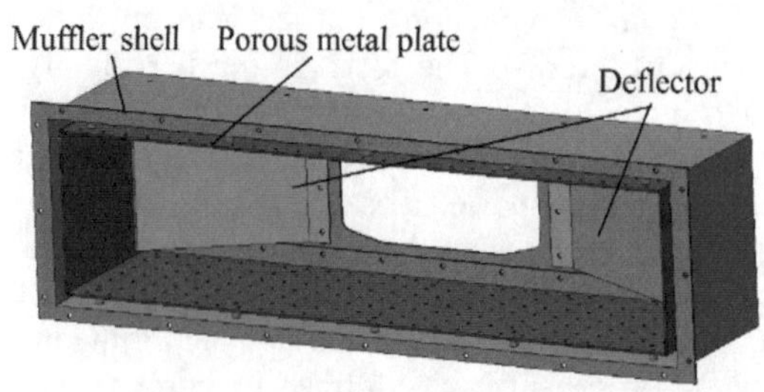

Figure 3. Muffler schematic diagram.

3. Simulation analysis of flow field

FLIENT software is used to simulate and analyze the gas velocity and pressure loss in the muffler with porous metal plate.

3.1. Simulation model

The standard K-Epsilon turbulence model is adopted in this paper, which is widely used with accurate calculation in the field of fluid analysis [4,5]. In simulation mode, the internal fluid material is set to air, the air inlet is set to the velocity inlet, and the air outlet is set to the pressure outlet boundary condition. The muffler shell and porous metal plate are set to aluminium alloy.

3.2. Simulation result analysis

Figure. 4 shows the distribution of gas velocity field in the muffler equipped with porous metal plate by means of simulative calculating. It can be seen in Figure. 4 that a series of vortexes appears in the area near the porous metal plate, the flow rate in the rest of the muffler is uniform with no drastic change. The reason is that the reflection of the airflow is beneficial to weaken the noise energy and achieves the silencing effect after hitting the porous metal plate.

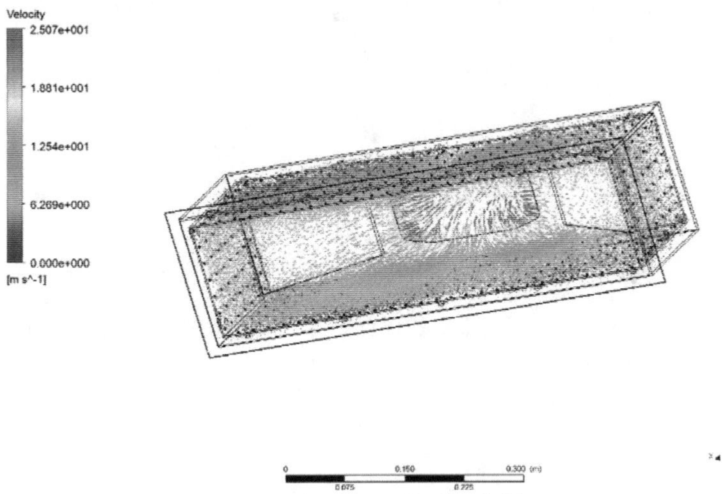

Figure 4. Muffler velocity field.

Pressure distribution diagram of the inner section in the muffler as shown in Figure. 5. Just as described in Figure. 5, the pressure distribution in the muffler is relatively uniform, and the pressure at the outlet of the muffler will be reduced by about 30Pa, so that the muffler has little effect on heat dissipation.

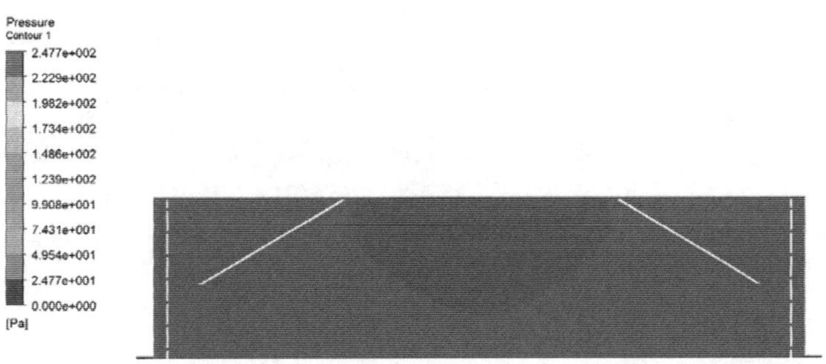

Figure 5. Pressure distribution of internal airflow in muffler.

4. Acoustic characteristic analysis

By using COMSOL acoustic simulation software and based on modified Helmholtz equation [6, 7], the acoustic characteristics of muffler are simulated and analyzed by pressure setting. Hard sound field boundary conditions are used at all solid boundaries of the muffler. It is assumed that under this boundary condition, the normal velocity of the object is zero, so boundary condition for the mixing of incident and radiated plane waves is adopted at the inlet boundary of the muffler [8], the outlet of the muffler is set to plane wave radiating outward.

The isosurface distribution of the total sound pressure field in the muffler at the frequency of 1500Hz is shown in Figure. 6. It is significant that the propagation mode of sound is the superposition of longitudinal standing wave and transverse mode at the frequency of 1500Hz.

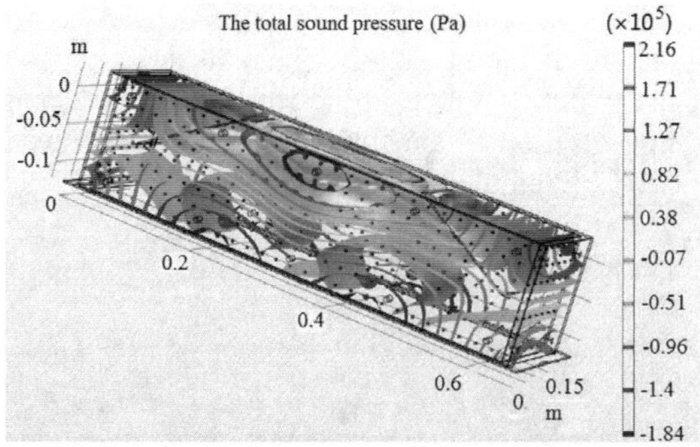

Figure 6. Total sound pressure of the muffler at 1500Hz.

The average energy flow density of sound waves is the sound intensity, which is used to measure the property of sound absorption. The distribution of sound intensity varies with frequency. Figure. 7 shows the distribution of sound intensity streamline in muffler at 1500Hz. Figure 7 shows that the energy flow of sound waves decreases due to the action of porous metal plates at the frequency of 1500Hz, the sound intensity increases due to the action of the baffle at the outlet.

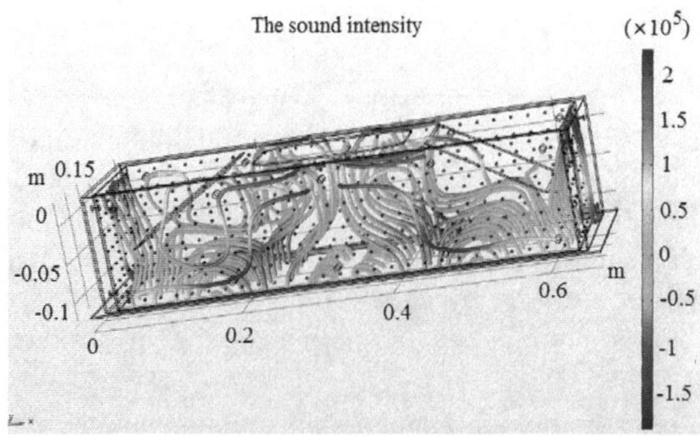

Figure 7. Streamlines of muffler sound intensity at 1500Hz.

Sound pressure level is used to indicate the sound pressure and measure the noise. Figure. 8 shows the distribution of sound pressure level in the muffler at 1500Hz. The maximum sound pressure level in the muffler is 62.5dB. The sound pressure level behind the resonant cavity and the baffle is higher with more sound loss which contributes to absorb noise.

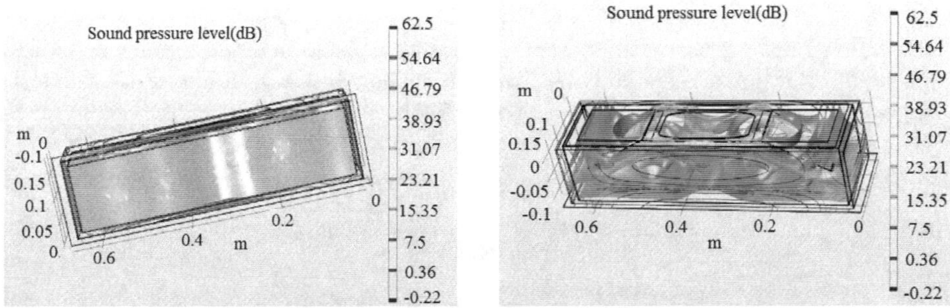

Figure 8. Muffler sound pressure level distribution at 1500Hz.

The transmission loss of muffler is an important physical quantity to measure the performance of muffler [9]. The greater the transmission loss, the better the silencing effect of muffler. Transmission loss (dB) is defined as the ratio of the incident power of sound waves to the outlet power, as shown in the following expression [10, 11, 12]:

$$d_w = 10\log(\frac{w_{in}}{w_{out}}) \tag{1}$$

Where w_{in} and w_{out} are the inlet incident power and outlet output power of sound wave respectively:

$$w_{in} = \int_{\partial\Omega} \frac{P_0^2}{2\rho c} dA \tag{2}$$

$$w_{out} = \int_{\partial\Omega} \frac{|p|^2}{2\rho c} dA \tag{3}$$

Where P_0 is the amplitude of sound pressure at the entrance, P is the amplitude of sound pressure at the exit; ρ is the gas density, which is 1.293kg/m^3.

Figure 9 shows the transmission loss of muffler at different frequencies. It can be seen from the figure that the transmission loss is about 3dB at the frequency lower than 1500Hz. When the frequency is greater than 1500 Hz, the transmission loss of the muffler is complicated with greater attenuation fluctuates on account of the resonance frequency in different parts of the muffler, which causes a deep valley of attenuation in the transmission loss curve. In addition to the longitudinal resonance transmission mode, there is also a cross section transmission mode in the muffler. So, it is difficult to predict the loss properties because of complex transmission patterns at frequencies higher than 1500 Hz.

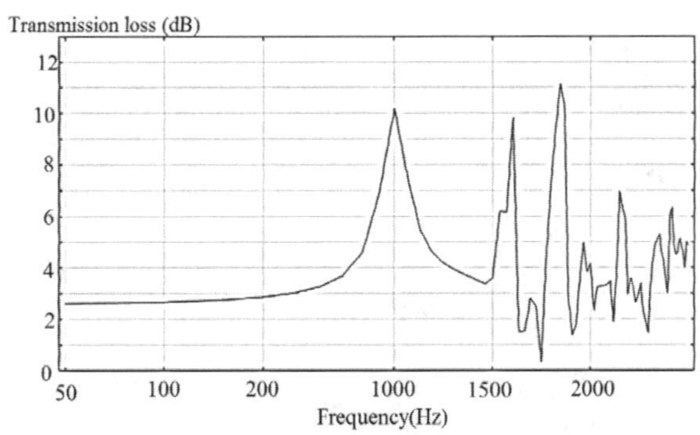

Figure 9. The transmission loss of muffler.

5. Noise testing

The noise spectrum analyzer is used to test the display console equipped with the muffler. The comparison result on two situations, with and without muffler, is shown in Figure 10. The test results show that the maximum sound level of the radar display console can be reduced by about 3dB installed with the muffler.

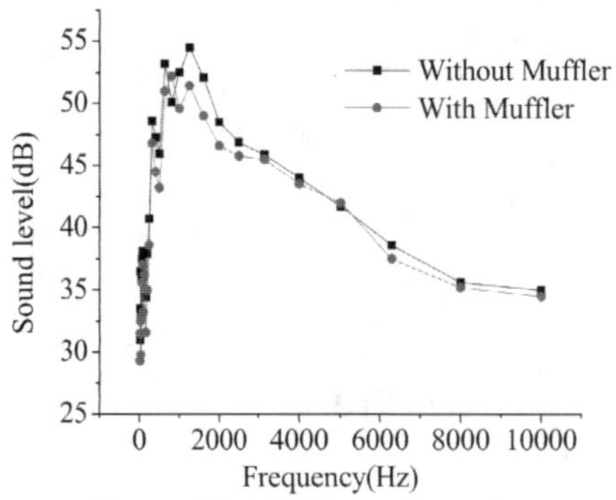

Figure 10. Muffler test results.

6. Conclusion

In this paper, an impedance compound muffler for radar display console is designed, and then the acoustic characteristics of muffler is analysed by using CAD software FLUENT and COMSOL. Simulation results show that the muffler has little influence on heat dissipation of display console. Finally, the simulation results are verified by means of noise spectrum analyser, as a result, the muffler obtains good silencing effect and the maximum sound level of the radar display console is reduced by 3dB.

References

[1] Jin X X and Hu Z G 1997 *Noise Control of Construction Machinery* (Tongji University Press) 24-28

[2] Cai J. 2011 *Noise Pollution Control Engineering* (China Environmental Science Press) 1-10

[3] Xu J and Gao H W 2018 *Noise control technology* vol 2 (Wuhan University of Technology Press) 163

[4] Yakhot V and Orszag S A Renormalization group analysis of turbulent solar convection. 2002 *J. International Journal for Numerical Methods in Fluids*, 39(9) 855-864

[5] Wang F J 2004 *Computational Fluid Dynamics Analysis: Principle and Application of CFD Software* (Tsinghua university press)

[6] Helmholtz H 1885 *On the sensations of tone as a physiological basis for the theory of music* Second English Edition, translated by Alexander J. Ellis. London: Longmans, Green, and Co.: 44-45

[7] Zhang X H 2001 Application of Gibbs-Helmholtz equation *Journal of Liaoning Teachers College: Natural Science Edition* 3(1) 4.

[8] Dan G and Neta B 2003 High-order non-reflecting boundary scheme for time-dependent waves. *Journal of Computational Physics* 186(1) 24-46.

[9] Munjal M L 1987 *Acoustics of ducts and mufflers* (Aoboken: John Wiley & Sons)

[10] Prasad A and Thiagarajan R C 2019 Acoustic performance design of automotive muffler. Technologies Atoa

[11] Kore S, Aman A and Direbsa E 2011 Performance Evaluation Of A Reactive Muffler Using Cfd Journal of EEA Vol 28

[12] Vasile O and Gillich G 2012 Finite Element Analysis Of Acoustic Pressure Levels And Transmission Loss Of A Muffler 6th Wseas European Computing Conference (ECC '12) Prague

Underwater source localization based on a matrix distance measurement

Yuyan Guo[1,2*] and Xianyi Gong[1,2,3]

[1]Hangzhou Applied Acoustics Research Institute, Hangzhou, China

[2]Science and Technology on Sonar Laboratory, Hangzhou, China

[3]Zhejiang University, Hangzhou, China

[*]Correspondence: guoyuyan09230@163.com

Abstract: In this paper, a matrix distance measurement is introduced to underwater source localization. Matched field processing (MFP) and beamforming(BF) are widely used methods that both make the correlation between data and copy vectors. We proposed statistical matched field processing (SMFP) and statistical beamforming (SBF) in this paper that measure the distance between data and copy cross-spectral density matrices (CSDMs) in place of the correlation of the two vectors. As CSDM is Hermitian and positive semidefinite, Riemann distance is suitable. Matrix-matrix comparison increases the dimension of the system compared with vector-vector correlation, so as to include environmental uncertainty. SMFP and SBF are compared with conventional matched field processing (CMFP) and conventional beamforming (CBF) respectively in simulations. The results show that SMFP and SBF achieve satisfactory localization performances.

1. Introduction

The underwater ocean waveguide is a complex thermodynamic system with various dynamic processes, such as internal waves [1], eddies and circulation [2]. These processes and the coupling between them, together with their interaction with the seabed, result in uncertainty in underwater environmental parameters. Internal waves are one of the main sources of uncertainty in the underwater environment. The ultimate effect of internal waves is to introduce fluctuations into acoustic signals that travel long distances in the ocean. In addition to this, waves and currents may lead to the random positions of the array hydrophones, thus affecting the localization performance.

MFP [3] and BF [4] are common methods for sources localization. MFP is a generalization of BF. What they have in common is that both are the correlation between data and copy. However, in the CMFP (Bartlett processor), the copy vector is deterministic and cannot contain the environmental uncertainty, which leads to the mismatch [5]. For complex dynamic processes such as the ocean, the randomness cannot be described by traditional Newtonian mechanics, so the concepts of probability and entropy should be added from the viewpoint of statistical mechanics. In this paper, SMFP and SBF is proposed under the category of statistical mechanics. Riemann distance [6] measurement is used to characterize the correlation between the random copy CSDM and the random data CSDM. The most common measure of the distance between matrices is the Euclidean distance. Since the CSDM is Hermitian matrix and is positive semidefinite to form a Riemannian manifold in the signal space, so the Riemannian distance is suitable. The smaller the Riemannian distance means the greater the correlation between the matrices. In the field of statistical mechanics, the Riemann geometry method is very effective. Statistical models are subsets of all possible probability distributions and often form finite-dimensional manifolds embedded in the set of all possible probability distributions. However, the real distribution may not be in the statistical model, but very close to the statistical model, so it is necessary to know the proportion and shape of the statistical model in all possible distribution sets and form the distance or difference measure between the two probability distributions, which is the geometric method in statistical mechanic [7]. Riemann distance measurement method has been widely applied in various fields. Li et al. use Riemann distance to classify signal [8]; Aubry et al. use Riemann distance to estimate covariance matrix in radar signal processing [9]; Finette et al. use Riemann distance to realize passive localization of sound sources [10].

In this paper, we propose SMFP and SBF based on a matrix distance measurement. The minimum Riemann distance between the data and the copy matrix is found to locate the underwater sources. Two scenarios with environmental uncertainty are considered respectively to demonstrate the validity of SMFP and SBF, one for the presence of internal solitary waves in which data and copy are mismatched with respect to internal solitary waves positions, and the other for the array geometry mismatch. Simulation results show that SMFP and SBF can effectively narrow the main lobe and lower the side lobes compared with CMFP and CBF.

2. Underwater source localization using SMFP and SBF

2.1. CMFP

The following is an introduction to CMFP. Consider the acoustic pressure field is initiated with a far field narrowband source with frequency ω_f at position $\mathbf{r}_s$. It is measured by a vertical linear array composed of M hydrophones at position $r_m\left(m=1,2,...,M\right)$.The pressure field can be expressed as

$$y_l\left(\omega_f,\mathbf{r}_s,\mathbf{r}_m\right)=s_l\left(\omega_f\right)g\left(\omega_f,\mathbf{r}_s,\mathbf{r}_m\right)+n_l\left(\omega_f,\mathbf{r}_m\right),l=1,2,...,L \tag{1}$$

where $y_l\left(\omega_f,\mathbf{r}_s,\mathbf{r}_m\right)$ is the acoustic pressure field data of the l-th snapshot received by the vertical linear array; $g\left(\omega_f,\mathbf{r}_s,\mathbf{r}_m\right)$ is Green function, calculated by simulation software according to the environment file. $n_l\left(\omega_f,\mathbf{r}_m\right)$ is the additive noise of the l-th snapshot.

Then, CMFP can be written as

$$P\left(\omega_f,\mathbf{r}_s,\mathbf{r}_m\right)=\frac{1}{L}\sum_{l=1}^{L}\left|\mathbf{a}\left(\omega_f,\tilde{\mathbf{r}}_s,\mathbf{r}_m\right)^{\mathrm{H}}\mathbf{y}\left(\omega_f,\mathbf{r}_s,\mathbf{r}_m\right)\right|^2. \tag{2}$$

where $P\left(\omega_f,\mathbf{r}_s,\mathbf{r}_m\right)$ is the ambiguity surface, $\mathbf{a}\left(\omega_f,\tilde{\mathbf{r}}_s,\mathbf{r}_m\right)=\mathbf{g}\left(\omega_f,\tilde{\mathbf{r}}_s,\mathbf{r}_m\right)/\left\|\mathbf{g}\left(\omega_f,\tilde{\mathbf{r}}_s,\mathbf{r}_m\right)\right\|_2$, $\|\cdot\|$ represents the ℓ_2 norm and $\tilde{\mathbf{r}}_s$ is the possible position of the source.

2.2. SMFP

As mentioned previously, in order to obtain the estimate of the source localization, it is to find a copy matrix that is closest to the data matrix in SMFP. Matrix-matrix comparison can be understood as a measure of the distance between matrices. The most common measure of distance is Euclidean distance, which is defined as follows:

The $M \times M$-dimensional matrix **A** and **B** can be regarded as points in the M^2 complex signal space, so the Euclidean distance between **A** and **B** is

$$D_{\mathrm{E}}\left(\mathbf{A},\mathbf{B}\right)=\sqrt{\mathrm{tr}\left[\left(\mathbf{A}-\mathbf{B}\right)\left(\mathbf{A}-\mathbf{B}\right)^{\mathrm{T}}\right]} \tag{3}$$

where T is the transpose operator, $\mathrm{tr}(\cdot)$ is the trace operator.

The estimate of the data CSDM is obtained by averaging the received data vectors:

$$\hat{\mathbf{K}}_{\text{data}} = \frac{1}{L}\sum_{l=1}^{L} \mathbf{y}_l\left(\omega_f, \mathbf{r}_s, \mathbf{r}_m\right)\mathbf{y}_l^{\mathrm{H}}\left(\omega_f, \mathbf{r}_s, \mathbf{r}_m\right), \tag{4}$$

where H is the conjugate transpose operator. Like equation (4), the estimation of the copy CSDM is written as

$$\hat{\mathbf{K}}_{\text{copy}} = \frac{1}{L}\sum_{l=1}^{L} \mathbf{a}\left(\omega_f, \tilde{\mathbf{r}}_s, \mathbf{r}_m\right)\mathbf{a}^{\mathrm{H}}\left(\omega_f, \tilde{\mathbf{r}}_s, \mathbf{r}_m\right), \tag{5}$$

Assume that the sampling is sufficient, so the estimation sign is removed. $\mathbf{K}_{\text{data}}(\omega)$ and $\mathbf{K}_{\text{copy}}$ are normalized to obtain

$$\mathbf{K}_{\text{data}} = \hat{\mathbf{K}}_{\text{data}} \Bigg/ \left(\sum_{m=1}^{N}\sum_{n=1}^{N}\left|\left(\hat{\mathbf{K}}_{\text{data}}\right)_{m,n}\right|^2\right)^{1/2} \tag{6.1}$$

$$\mathbf{K}_{\text{copy}} = \hat{\mathbf{K}}_{\text{copy}} \Bigg/ \left(\sum_{m=1}^{N}\sum_{n=1}^{N}\left|\left(\hat{\mathbf{K}}_{\text{copy}}\right)_{m,n}\right|^2\right)^{1/2} \tag{6.2}$$

CSDM is Hermitian matrix and positive semidefinite, so the data and copy CSDM of different frequencies are points on the Riemann manifold to form the Riemann manifold curve. The distance between two points on the manifold is

$$\ell(\mathbf{K}) = \int_{\theta_{\text{data}}}^{\theta_{\text{copy}}} g_{\mathbf{K}}^{1/2}\left(\dot{\mathbf{K}}, \dot{\mathbf{K}}\right)\mathrm{d}\theta, \tag{7}$$

where $\dot{\mathbf{K}} = \mathrm{d}\mathbf{K}/\mathrm{d}\theta$, $g_{\mathbf{K}}\left(\dot{\mathbf{K}}, \dot{\mathbf{K}}\right)$ is an inner product metric defined on the manifold, called a Riemann metric. The line with the shortest distance in the curve connecting $\mathbf{K}_{\text{data}}$ and $\mathbf{K}_{copy}$ is called geodesic, and the length of geodesic is the Riemann distance between the two matrices,

$$D_R\left(\mathbf{K}_{\text{data}}, \mathbf{K}_{\text{copy}}\right) \triangleq \min\left\{\ell\left(\mathbf{K}(\theta)\right)\right\}. \tag{8}$$

Different solving methods of the above equation can produce different Riemann distance estimators. Therefore, we give three kinds of statistical matched field processors [11]

$$\mathrm{SMFP}_1 = \left[\mathrm{tr}\left(\mathbf{K}_{\text{data}}\right) + \mathrm{tr}\left(\mathbf{K}_{\text{copy}}\right) - 2\mathrm{tr}\left(\left(\mathbf{K}_{\text{data}}\mathbf{K}_{\text{copy}}\right)^{1/2}\right)\right]^{1/2}. \tag{9.1}$$

$$\mathrm{SMFP}_2 = \left[\mathrm{tr}\left(\mathbf{K}_{\text{data}}\right) + \mathrm{tr}\left(\mathbf{K}_{\text{copy}}\right) - 2\mathrm{tr}\left(\mathbf{K}_{\text{data}}^{1/2}\mathbf{K}_{\text{copy}}^{1/2}\right)\right]^{1/2}. \tag{9.2}$$

$$\mathrm{SMFP}_3 = \left[tr\left(\left(\log \mathbf{K}_{\text{data}}^{-1/2}\mathbf{K}_{\text{copy}}\mathbf{K}_{\text{data}}^{-1/2}\right)^2\right)\right]^{1/2} = \left[\sum_{i=1}^{N}\lg^2\left(\lambda_i\right)\right]^{1/2}. \tag{9.3}$$

where λ_i is the eigenvalue of $\mathbf{K}_{\text{data}}^{-1}\mathbf{K}_{\text{copy}}$.

2.3. SBF

BF is a versatile approach to spatial filtering. Consider a far field narrowband source $s(t)$ with frequency f which is measured by a horizontal linear array with M hydrophones. Then the signal received by the m-th element is:

$$x_m(t) = s(t - \tau_{m0}) + n_m(t) \tag{10}$$

where $n_m(t)$ is the additive noise on the m-th hydrophone, and τ_{m0} represents the time delay between the m-th hydrophone and the reference hydrophone. For a uniform linear array with array spacing of d,

$$\tau_{m0} = md \sin\theta / c, \tag{11}$$

where θ is the source incidence angle. A uniform linear array is shown in figure 1.

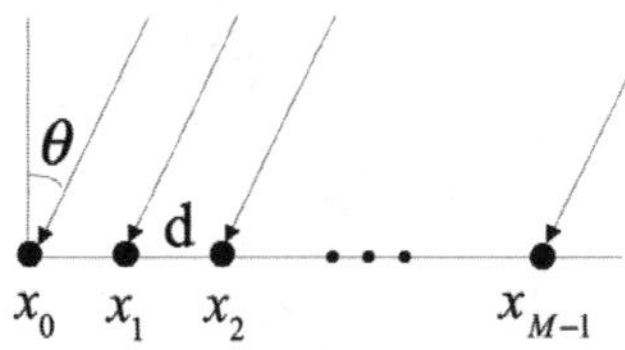

Figure 1. Uniform line array.

With the x_0 as the reference element, Fourier transform is performed on the received data, which can be expressed as

$$\mathbf{x}(\omega) = \left[x_1(\omega), x_2(\omega), \cdots, x_M(\omega)\right]^T = s(\omega)\left[1, e^{j2\pi d\sin\theta/\lambda}, \cdots, e^{j2\pi(M-1)d\sin\theta/\lambda}\right]^T = s(\omega)\mathbf{f}(\theta). \tag{12}$$

where $s(\omega)$ represents the Fourier transform of the source signal waveform and $\mathbf{f}(\theta)$ represents the phase compensation vector of each array element.

Beamforming is a weighted summation of each array element. Assuming the beam direction is θ_e, the conventional beamforming is:

$$y(\theta, \theta_e, \omega) = s(\omega)\sum_{m=0}^{M-1} \mathrm{e}^{\mathrm{j}2\pi md\sin\theta_e/\lambda} \mathrm{e}^{-\mathrm{j}2\pi md\sin\theta/\lambda} = \sum_{n=0}^{M-1} \mathrm{e}^{-\mathrm{j}2\pi md\sin\theta_e/\lambda} \mathbf{x}(\omega) = \mathbf{a}(\theta_e)^{\mathrm{H}} \mathbf{x}(t). \tag{13}$$

where $\mathbf{a}(\theta_e)$ is the copy vector in a certain beam direction θ_e.

The CBF correlates the data vector $\mathbf{x}(t)$ with the copy vector $\mathbf{a}(\theta_e)$ in equation (13). Similar to SMFP, the matrix-matrix correlation can contain more degree of freedom compared with the vector-vector correlation. Therefore, SBF correlates the data matrix with the copy matrix.

According to the expression in Equation (4), (5), (6.1), (6.2), the data and copy CSDM is normalized to obtain

$$\mathbf{X} = \hat{\mathbf{X}}(\omega) \Big/ \left(\sum_{m=1}^{N}\sum_{n=1}^{N}\left|\left(\hat{\mathbf{X}}(\omega)\right)_{m,n}\right|^2\right)^{1/2}. \tag{14.1}$$

$$\mathbf{A} = \hat{\mathbf{A}}(\theta_e) \Big/ \left(\sum_{m=1}^{N}\sum_{n=1}^{N}\left|\left(\hat{\mathbf{A}}(\theta_e)\right)_{m,n}\right|^2\right)^{1/2}. \tag{14.2}$$

Using equation (9.1), (9.2), (9.3), SBF is written as

$$SBF_1 = \left[\mathrm{tr}(\mathbf{X}) + \mathrm{tr}(\mathbf{A}) - 2\mathrm{tr}\left((\mathbf{XA})^{1/2}\right)\right]^{1/2}. \tag{15.1}$$

$$SBF_2 = \left[\mathrm{tr}(\mathbf{X}) + \mathrm{tr}(\mathbf{A}) - 2\mathrm{tr}\left(\mathbf{X}^{1/2}\mathbf{A}^{1/2}\right)\right]^{1/2}. \tag{15.2}$$

$$SBF_3 = \left[tr\left(\left(\log \mathbf{X}^{-1/2}\mathbf{A}\mathbf{X}^{-1/2}\right)^2\right)\right]^{1/2} = \left[\sum_{i=1}^{N} \lg^2(\lambda_i)\right]^{1/2}. \tag{15.3}$$

where λ_i is the eigenvalue of $\mathbf{X}^{-1}\mathbf{A}$. Traversing the beam direction θ_e, the smaller the Riemann distance is, the greater the correlation between the data and the copy. Therefore, the minimum value of the above processors is the estimated value of the azimuth angle of the source.

3. Simulations

3.1. Environmental uncertainty and internal waves

An important factor causing environmental uncertainty is the random perturbation of sound velocity field. In CMFP, there is a deterministic sound velocity profile which simplifies the sound velocity field into a constant or positive/negative gradient profile. However, the actual sound velocity field is random, so random perturbation should be added to the traditional deterministic sound velocity profile to describe the environmental uncertainty. The sound speed is written as [12]

$$c(r,z;t) = \bar{c}(z) + \delta c(r,z;t) \tag{16}$$

where z is the depth, r is the range. $\bar{c}(z)$ is the mean sound-speed profile

$$\bar{c}(z) = \begin{cases} c_s - \Delta c\left[1 + \tanh\left((z - z_{th})/\Delta_{th}\right)\right]/2, 0 \le z \le D \\ c_b, z > D \end{cases} \tag{17}$$

where the water depth D=100 m, the surface velocity c_s=1520 m/s, the thermocline velocity disturbance Δc=40 m/s, the thermocline depth z_{th}=40 m, the thermocline width Δ_{th}=15 m, and the seafloor velocity c_b=1650 m/s. The deterministic background sound velocity field is shown in figure 2(a). $\delta c(r,z;t)$ is a random parts of the sound velocity field. In the shallow water environment, we consider this perturbation to be caused by a internal solitary wave,giving

$$\delta c(r,z) = -\frac{1}{c_0}\frac{d\bar{c}}{dz}\sin(\pi z/D)\sum_{p=1}^{N_s}\zeta_0(p)F(r,\Delta_p,r_0(p)), z \quad D, \tag{18}$$

where the depth structure of the internal solitary wave is assumed to be sinusoidal, $\zeta_0(p)$ is the vertical displacement of each internal solitary wave, Δp is the width of the internal solitary wave. $F(r,\Delta_p,r_0(p))$ is approximated by

$$F(r,\Delta_p,r_0(p)) = e^{-(r-r_0(p))^2/\Delta_p^2}. \tag{19}$$

The position of internal solitary wave $r_0 = 300m$, the vertical displacement of internal solitary waves $\zeta_0 = 10m$, and the width of internal solitary wave $\Delta p = 100m$. The sound velocity profile is shown in figure 2(b).

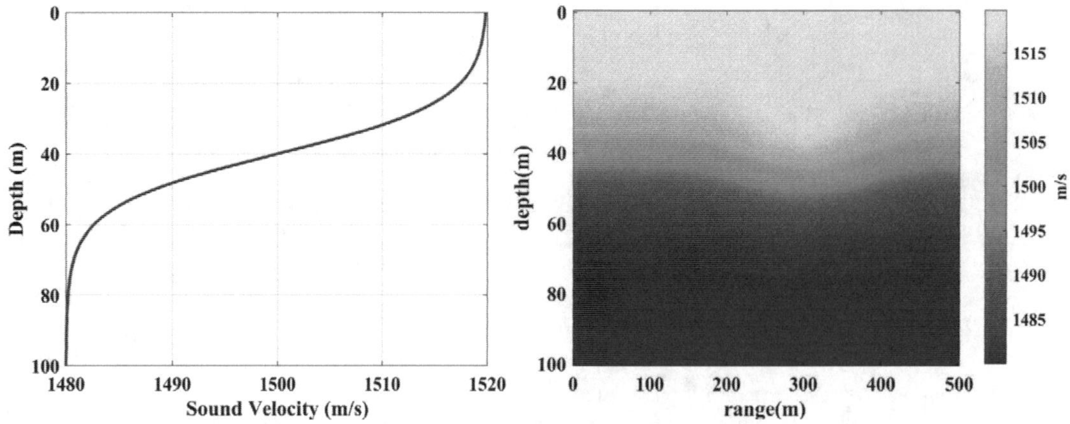

(a) Deterministic background sound speed profile; (b)sound speed profile when internal wave at 300m.
Figure 2. (a)Deterministic background sound speed profile
and (b) random sound speed profile when internal wave at 300 m

3.2. Main results of SMFP

Consider that the position of the internal solitary wave is randomly between 50 and 350m, so there is a random mismatch between the data field and the copy field. The remaining parameters of the internal wave are the same as in figure 2. The source transmits a 100Hz narrow-band signal which depth is 20m and the range is 0m. The vertical receiving array is located at 350m, the array depth is 0~100m, one array is spaced every 1 m, and the signal-to-noise ratio is 10dB. Range-dependent Acoustic Model (RAM)[13] was used for simulation. Figure 3 and figure 4 show the localization results of CMFP and SMFP. For simplicity, only the results of $SMFP_1$ are shown.

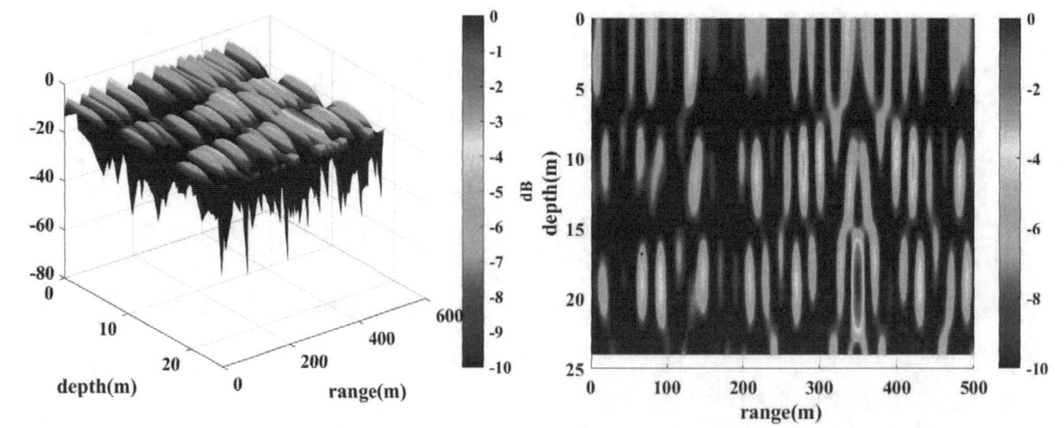

(a) three-dimensional diagram (b) range-depth ambiguity surfaces
Figure 3. Source localization using CMFP processors.

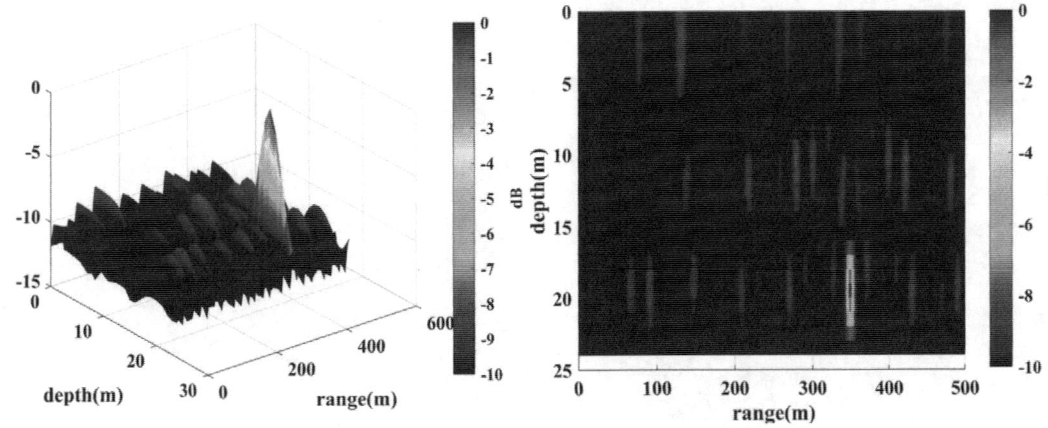

(a) three-dimensional diagram (b) range-depth ambiguity surfaces

Figure 4. Source localization using SMFP processors.

It can be seen that SMFP has improved source localization resolution and effectively reduced side lobes in the case of random perturbation in sound velocity field and random mismatch between data field and copy field.

3.3. Main results of SBF

As for beamforming, the randomness of array hydrophone positions caused by possible waves, tides and currents is considered. It is possible for a linear array to bend into an approximate sinusoidal array due to the action of waves and currents. The horizontal sinusoidal array of 32 elements with aperture of 1m and amplitude of 0.5m is shown in figure 5.

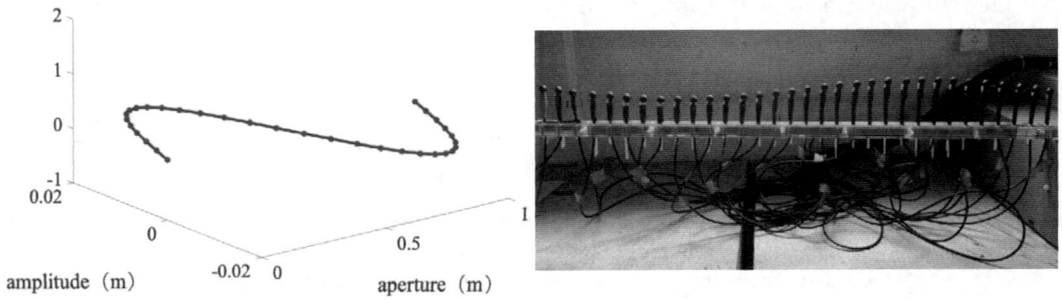

(a) Simulation diagram (b) Experiment diagram

Figure 5. Diagram of horizontal sinusoidal array

The transmitted signal is PCW, amplitude $A = 1\text{Vpp}$, signal center frequency $f_0 = 18\text{kHz}$, sampling frequency $f_s = 100\text{kHz}$, pulse width $\tau = 50ms$, and period $T = 100\text{ms}$. The target azimuth angle is 0°, and the SNR is 10dB. In the simulation experiment, the data is generated by sinusoidal array, and the copy is generated by linear array, so there is a mismatch between the two. The simulation results of CBF and SBF are shown in Figure 6. It can be seen that the resolution of SBF is higher than that of CBF. The -3dB main lobe width of SBF is 1.8°, the -3dB main lobe width of CBF is 4.5°, the maximum side lobes level of SBF is -16.01dB, and the maximum side lobes level of CBF is -10.05dB. It is seen that compared with CBF, SBF can reduce the width of the main lobe and lower the side lobes, but the average energy is increased with the lowest output concentrated around -18dB.

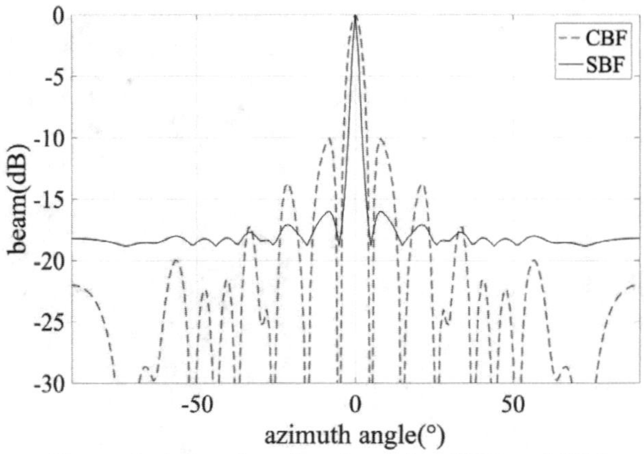

Figure 6. Azimuth estimation using CBF and SBF

4. Conclusion

In this paper, we have proposed an approach for underwater source localization based on data and copy matrix distance measurements. Since the CSDM is Hermitian and positive semidefinite, the estimation of the source localization is obtained by solving the minimum Riemann distance between the data cross-spectral density matrix and the copy cross-spectral density matrix. Since matrix-matrix comparison increases the degree of freedom of the system compared with vector-vector correlation and environmental uncertainty and random mismatch are included, thus improving the localization performance. Simulations show that SMFP and SBF achieve better localization performances when environment mismatch and array mismatch exist respectively.

We focus on the environmental uncertainty caused by internal waves. In fact, environmental uncertainty can be caused by many other factors, such as surface waves, depth mismatch, and so on. In the future, we will explore the localization performance of SMFP in the presence of uncertainty in different underwater environments.

References

[1] Yang C.T. "Acoustic mode coupling induced by nonlinear internal waves: Evaluation of the mode coupling matrices and applications." *Journal of the Acoustical Society of America* 135.2(2014):610-25.

[2] Holloway G. "Eddies, Waves, Circulation, and Mixing: Statistical Geofluid Mechanics." *Annual Review of Fluid Mechanics* 18.1(1986):91-147.

[3] Baggeroer, B. and Kuperman W. A. "An overview of matched field methods in ocean acoustics." *IEEE Journal of Oceanic Engineering* 18.4(1993):401-424.

[4] Veen, Van B, and Buckley M. "Beamforming: a versatile approach to spatial filtering." *IEEE Assp Magazine* 5.2(2002):4-24.

[5] Shang E C. and Wang Y Y. "Environmental mismatching effects on source localization processing in mode space" *JASA* 1991, 89:2285-2290

[6] Jost J. *Riemannian geometry and geometric analysis* (Springer) 2008

[7] Amari S. *Differential-geometrical methods in statistics, lecture notes in statistics thesis.* (Springer-Verlag) 1985

[8] Wong, K. M., Li, Y. and Wong, K.M. "Riemannian Distances for Signal Classification by Power Spectral Density," IEEE Journal of Selected Topics in Signal Processing, vol.7, no.4, pp.655,669, Aug. 2013." *IEEE Journal of Selected Topics in Signal Processing* 7.4(2013).

[9] Aubry, et al. "Covariance matrix estimation via geometric barycenters and its application to radar training data selection." *IET Radar, Sonar & Navigation* 7.6(2013):600-614.

[10] Finette, et al. "Stochastic matched-field localization of an acoustic source based on principles of Riemannian geometry." *The Journal of the Acoustical Society of America* 143.6 Pt.1(2018):3628-3638.

[11] Li, Y., K. M. Wong, and H. D. Bruin. "Electroencephalogram signals classification for sleepstate decision - A riemannian geometry approach." *IET Signal Processing* 6.4(2012):288-299.

[12] Morozov, A. K. , and J. A. Colosi. "Entropy and scintillation analysis of acoustical beam propagation through ocean internal waves. " *Journal of the Acoustical Society of America* (2005).

[13] Ocean Acoustics Library. http://oalib.hlsresearch.com.

Study on the Self-Pressurization Laws of Cryogenic Liquid Krypton Tank for the Electric Propulsion System

Ji Dong Chen*, Jian Bing Zhu, Ji Peng Zhao, Sen Dong Gu, Bin Yu, Yu Qi Yuan

Lanzhou Institute of Physics

chenjidongyx@163.com

Abstract: To meet the application requirements of zero boil-off storage of cryogenic liquid krypton propellant in the process of completing deep space exploration missions by electric propulsion system, it is important to study the thermodynamic changes in cryogenic liquid krypton tanks under different environments. The gas phase and liquid phase material model of krypton working fluid, the gas-liquid phase change model at the interface layers, and the heat transfer process between the wall and the internal fluid are established by ANSYS-Fluent. The energy source term, gas phase, liquid phase mass source term, and phase change saturation temperature in the liquid krypton cryogenic tank are defined. This paper's three influencing factors of gravity, initial liquid filling rate, and wall heat leakage are simulated and analysed. The results show that: (1) The self-pressurization rate significantly rises as the wall heat flux rises, and the gas-liquid interface drops less quickly. The pressure and temperature in the tank are also increasing at the same liquid level. (2) In the normal gravity environment, the bigger the filling rate, the lower the corresponding self-pressurization rate. (3) In the process of variable gravity, the gas pressurization rate in the cryogenic liquid krypton tank falls with the decrease of gravity.

1. Introduction

The concept of electric propulsion was first proposed by Robert H. Goddard of the United States in the early 20th century [1]. An electric propulsion system uses electromagnetic energy to heat or ionize propellant to produce plasma and then accelerate its high-speed jet to generate thrust. After years of research and technological development, it has been widely used due to its long life, high specific impulse, and good controllability. The electric propulsion system has been used in various space missions many times, such as NASA JPL's explorer "Deep Space 1", ESA's lunar explorer "SMART-1", NASA's asteroid explorer "Dawn" and Japan's asteroid explorer "Hayabusa". As traditional electric propulsion gas propellants, Xenon and krypton have good thrust performance. Jacobson [2] and Welle [3] compared and analysed the properties of xenon and krypton. The results show that xenon has lower ionization energy and higher atomic mass, and the ionization and focusing process in the discharge channel is better controlled. It is slightly better than krypton in propellant utilization and thrust, but the specific impulse of krypton is higher than that of xenon at the same voltage and power. In-depth comparisons of xenon and krypton's performance characteristics, including propellant dose, specific impulse, thrust, and efficiency for the 12 kW electric thruster used in exploration satellites, were made by Duchemin et al. [4]. According to the results, krypton outperforms xenon in propellant dosage and specific impulse. The demand for gas propellants is greatly increased with the application of electric propulsion systems in deep space exploration missions. Considering the performance and market price of both xenon and krypton gas propellants, krypton is more suitable for such tasks than xenon.

As a cryogenic propellant, krypton is gaseous at room temperature, and the boiling point temperature is 120 K. There are three storage methods for gas krypton [5-7]: high-pressure gaseous storage, supercritical storage, and cryogenic liquefaction storage. Compared with high-pressure gaseous and supercritical storage, cryogenic liquefaction storage has higher storage efficiency and can realize the miniaturization and lightweight design of a large-capacity krypton propellant tank. However, the low storage temperature of liquid krypton brings a series of difficulties to the thermal management of cryogenic liquid krypton tanks. The self-pressurization phenomenon is a process in which the heat leakage from the cryogenic tank wall and the support cause the liquid phase evaporation and the gas phase expansion to increase the pressure in the tank. The long-term storage of the cryogenic liquid krypton propellant will go through different stages, such as the ground parking stage, the launching stage, and the space application stage. It is necessary to monitor the pressure change in the tank at all times. Otherwise, it may seriously affect the safety management of the cryogenic propellant.

In recent years, domestic and foreign scholars have carried out a lot of simulation analysis works on the heat and mass transfer process of self-pressurization in chemical cryogenic propellant tanks such as liquid hydrogen, liquid oxygen, and liquid methane based on CFD (Computational Fluid Dynamic) method. To study the phase change process at the gas-liquid interface and compare the pressurization process in the tank under various gravity conditions, Fu et al. [8,9] built a two-dimensional axisymmetric model for cylindrical tanks. Sun et al. [10,11] discussed Fluent software's phase transition model. Using numerical simulation, Choi et al. [12] investigated the tank's thermal stratification and natural convection of cryogenic propellant. They compared the effects of various working fluids, boundary conditions, and liquid filling rates on the tank's self-pressurization and thermal stratification. Ovidi et al. [13] used the CFD method to analyze the thermal stratification and self-pressurization of the liquid phase in a cylindrical liquefied natural gas tank due to thermal boundary. Liu et al. [14] established a CFD model of cryogenic liquid oxygen tanks, studied the self-pressurization and thermal stratification in cryogenic tanks, and gave the temperature distribution and pressure change in cryogenic liquid oxygen tanks. To study the self-pressurization behavior in a liquid hydrogen tank under a normal gravity environment, Barsi [15] established a gas-liquid two-phase CFD model in the tank and gave the effect of filling rate on the self-pressurization behavior in a liquid hydrogen tank. Lopez et al. [16,17] established a CFD model to study the pressure change and pressure control in spherical liquid hydrogen tanks under different gravity environments. Using a two-dimensional geometric model with rotational symmetry and a 3.5 W/m^2 heat flow boundary, Kassemi et al. [18] simulated and examined the self-pressurization process in a zero-evaporation liquid hydrogen tank under normal gravity conditions. Perez-Raya et al. [19] utilized user-defined functions in the Ansys-Fluent program to accurately describe heat transport, mass transfer, and phase change processes at the gas-liquid interface. To investigate the impact of microgravity and heat leaks on the self-pressurization and thermal stratification in the liquid oxygen tank, Wang et al. [20] developed a 3D CFD model based on the VOF model. Bouhal et al. [21] simulated and predicted the heat and mass transfer process using the gas-liquid two-phase VOF model for the closed geometric model.

According to the above papers show, first of all, in most simulation processes, the gas phase is treated as an ideal gas. The liquid phase density has the following two treatment methods: the first is considered an incompressible fluid, and its density is a constant value under atmospheric pressure; the second is to use the Boussinesq approximation method, which allows it to change linearly with temperature within a certain range. The ideal gas assumption of gas phase and the fixed value treatment of liquid phase density are different from the actual situation. Secondly, the main factors affecting the pressurization rate of cryogenic propellant are as follows: wall heat flux, initial liquid filling rate, and gravity environment. Finally, due to the limited simulation and experimental research on the self-pressurization laws of cryogenic liquid krypton tanks, it is necessary to study the simulation analysis of the thermodynamic process in cryogenic liquid krypton tanks. Based on the study of the three mentioned issues, in this paper, the Redlich-Kwong-Soave real gas model and the Boussinesq approximation density model are chosen as the gas phase and liquid phase material models of cryogenic krypton. The cryogenic liquid krypton tank self-pressurization process was simulated using Ansys-Fluent. The distribution nephograms and correlation curves of the cryogenic liquid krypton tank's temperature, pressure, and gas volume fraction at various times were obtained and analyzed and summarized the laws of several condition parameters on the cryogenic liquid krypton tank about self-pressurization rate. The purpose is to provide data reference for the design and safety management of a cryogenic liquid krypton tank.

2. Model building

2.1. Physical model

In this paper, a disc-shaped tank model is used. The height of the column section is 50 mm. The semi-major axis is 300 mm and the semi-minor axis is 100 mm. Ten temperature monitoring points were set on the axial position of the cryogenic liquid krypton tank, as shown in Figure 1. To simplify the calculation model, the simulation analysis uses a two-dimensional rotational symmetry model.

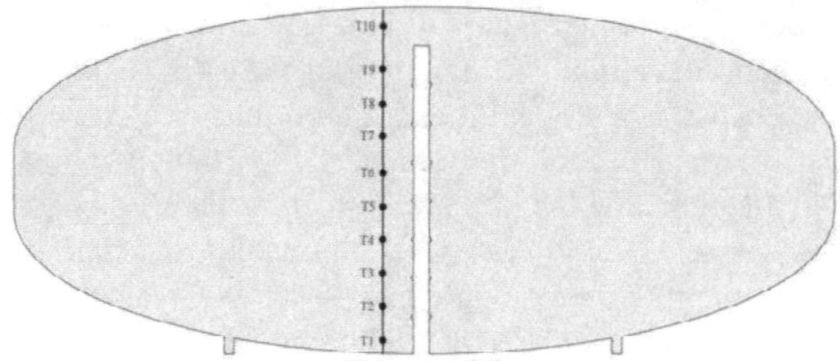

Figure 1 Two-dimensional cryogenic liquid krypton tank model

The material model for the krypton gas and liquid phases were built in the Fluent material library. The material-related property parameters are shown in Table 1, and the curves for the gas phase and liquid phase isobaric-specific heat capacity versus temperature are shown in Figure 2.

Table 1 Material parameters of krypton

Vapor			
Molar Mass /(kg/mol)	83.798	Thermal Conductivity	0.004
Critical pressure /MPa	5.525	Acentric factor	-0.000894
Critical temperature/K	209.48	Standard state entropy /J/(kg•K)	94535
Critical specific volume /(m^3/kg)	0.1101	Dynamic Viscosity /kg/(m•s)	1.04e-5
Cp/J/(kg•K)	Cp Satisfy the curve relation with T in Fig. 2		
Liquid			
Reference density (kg/m^3)	2413.4381	Thermal expansion coefficient /kg/(m^3•K)	0.008
Latent heat/(J/kg)	106910	Surface tension/(N/m)	0.015
Thermal Conductivity	0.11033	Dynamic Viscosity /kg/(m•s)	4.05e-4
Cp/J/(kg•K)	Cp Satisfy the curve relation with T in Fig. 2		

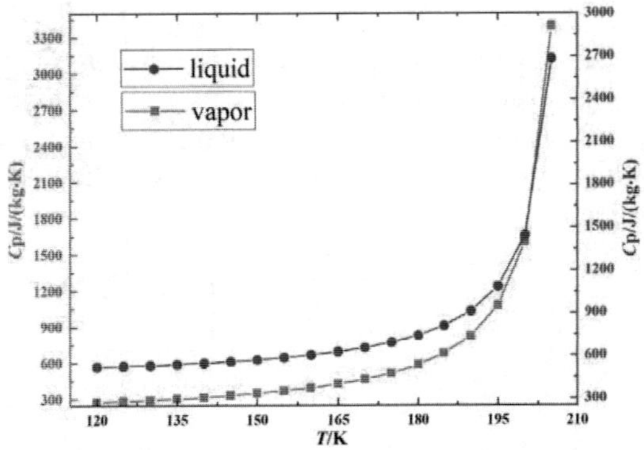

Figure 2 The curve of Krypton Cp with temperature

2.2. Numerical models

2.2.1. Governing equations

The fluid flow and heat transfer process in a cryogenic tank is usually described by continuity, momentum, and energy. In the following expression, the subscript "L" represents the liquid phase, the subscript "G" represents the gas phase, and the subscript "I" represents the gas-liquid interface:

$$\frac{\partial \rho}{\partial t} + \nabla(\rho \vec{v}) = S_C \tag{1}$$

$$\frac{\partial(\rho \vec{v})}{\partial t} + \nabla(\rho \vec{v} \vec{v}) = -\nabla p + \nabla[\mu_{\text{eff}}(\nabla \vec{v} + \nabla v^T)] + \rho \vec{g} + \vec{F}_{\text{vol}} \tag{2}$$

$$\frac{\partial(\rho E)}{\partial t} + \nabla[\vec{v}(\rho E + p)] = \nabla(k_{\text{eff}} \nabla T) + S_h \tag{3}$$

where V denotes velocity in m/s; ρ is fluid density; t is time; p is pressure; T is temperature; $\vec{g}$ is gravity, taken as 9.8 m/s^2; S_C denotes the mass source term in the continuity equation; μ_{eff} denotes the effective kinetic viscosity; k_{eff} denotes the effective heat transfer coefficient; S_h denotes the energy source term in the energy equation; α denotes the volume fraction term, where the variables in the cell grid are determined by the volume fraction: $\rho = \alpha_G \rho_G + \alpha_L \rho_L$, $\mu_{\text{eff}} = \alpha_G \mu_{\text{eff}_G} + \alpha_L \mu_{\text{eff}_L}$, $k_{\text{eff}} = \alpha_G k_{\text{eff}_G} + \alpha_L k_{\text{eff}_L}$.

The energy source term E is treated as a mass-averaging term:

$$E = \frac{\alpha_G \rho_G E_G + \alpha_L \rho_L E_L}{\alpha_G \rho_G + \alpha_L \rho_L} \tag{4}$$

Since the surface tension effect of cryogenic fluids in tanks under normal gravity and microgravity cannot be ignored, the surface tension at the interface layer (including the interface layer in the gas phase and the interface layer in the liquid phase) is often added to the volume force in the momentum equation through the Continuum Surface Force (CSF) model. Therefore, the volume force term can be expressed as:

$$\vec{F}_{\text{vol}} = \sigma_{LG} \frac{\alpha_G \rho_G \kappa_G \nabla \alpha_L + \alpha_L \rho_L \kappa_L \nabla \alpha_G}{\frac{1}{2}(\rho_G + \rho_L)} \tag{5}$$

where σ_{LG} represents the surface tension coefficient, κ_G and κ_L represent the curvature of the gas-liquid phase interface, respectively, which can be obtained from the gradient around the surface normal:

$$\kappa = \nabla \cdot \vec{n} \tag{6}$$

2.2.2. Gas-liquid phase change model

Due to the obvious interface between gas and liquid phases, there is no fusion and mixing phenomenon. Therefore, the VOF model for free surface and stratified flow is more applicable to the phase change process of propellant mass and heat transfer in the cryogenic tank. Based on the VOF model, the continuity equation, mass source term, and energy source term of gas-liquid two-phase can be expressed as:

$$\frac{1}{\rho_G}\left[\frac{\partial(\alpha_G\rho_G)}{\partial t}+\nabla\cdot(\alpha_G\rho_G\vec{v_G})\right]=S_{\alpha G} \tag{7}$$

$$\frac{1}{\rho_L}\left[\frac{\partial(\alpha_L\rho_L)}{\partial t}+\nabla\cdot(\alpha_L\rho_L\vec{v_L})\right]=S_{\alpha L} \tag{8}$$

$$S_C=\vec{\dot{m}}_I\cdot A_I=\begin{cases}S_{\alpha G},T>T_{sat}\\ S_{\alpha L},T<T_{sat}\end{cases} \tag{9}$$

$$S_h=S_C\cdot\gamma \tag{10}$$

where γ is the latent heat of vaporization value; $\vec{\dot{m}}_I$ is the evaporation rate at the gas-liquid phase interface; A_I is the area of the gas-liquid phase interface layer; T_{sat} is the saturation temperature value; $S_{\alpha G}$ and $S_{\alpha L}$ represents the mass source terms for evaporation and condensation:

$$S_{\alpha G}=\dot{m}_G=-\dot{m}_L=r\alpha_L\rho_L\frac{T-T_{sat}}{T_{sat}},T>T_{sat} \tag{11}$$

$$S_{\alpha L}=\dot{m}_L=-\dot{m}_G=r\alpha_G\rho_G\frac{T_{sat}-T}{T_{sat}},T<T_{sat} \tag{12}$$

where r in Equations (2-11) and (2-12) denotes the evaporation and condensation coefficients, the default value of 0.1 was taken for the experiments for faster verification of the conclusions. The energy source term, the mass source term of the gas and liquid phase, and the saturation temperature mentioned above are written by user-defined functions and explained in the simulation software to realize the gas-liquid phase change process.

2.3. Finite element model

In this paper, the Governing equations and Gas-liquid phase change equations are solved by Fluent, and the gas-liquid two-phase flow and heat transfer process in a cryogenic liquid krypton tank are simulated. The Redlich-Kwong-Soave real gas model is selected for the gas phase, and the Boussinesq approximation density model is selected for the liquid phase. According to NIST Refprop 9.1, the gas phase pressure theoretical value and simulation value of krypton are compared, as shown in Figure 3. ICEM generates the computational domain mesh. The column section adopts regular tetrahedral mesh, and the upper and lower ellipsoid head area mesh adopts 'C' type segmentation. The mesh refinement is performed on the near-wall surface, and 8 boundary layers are set. The boundary layer growth rate is 1.2, and the number of 2 D meshes is 5528. The finite element model is shown in Figure 4. The wall adopts a uniform heat flux boundary. The turbulence model selects the standard k-e model according to the initial Reynolds number. The solver uses PISO, the pressure and momentum relaxation factors are 0.2 and 0.5, and the energy relaxation factor is 0.8. The rest takes the default value, the step size is variable step size setting, and the calculation convergence is good.

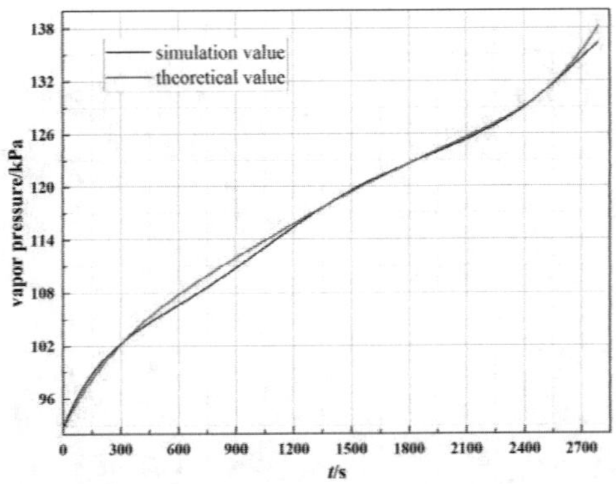

Figure 3 Comparison of the simulation value of the gas-liquid phase model with the theoretical value

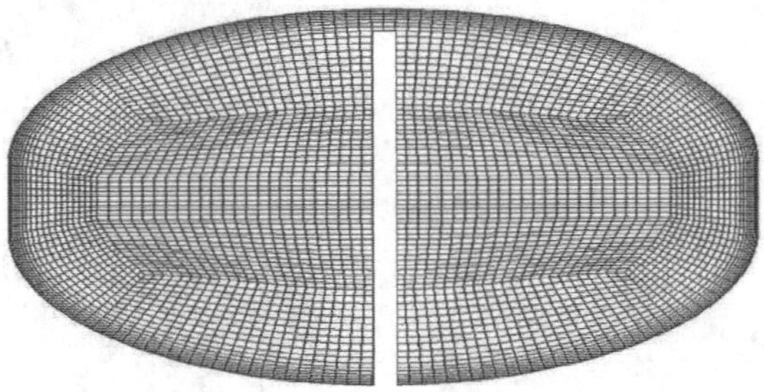

Figure 4 Finite element model of a cryogenic liquid krypton tank

3. Results and analysis

The fundamental reason for the formation of the self-pressurization phenomenon is the evaporation of the liquid phase and the thermal expansion of the gas phase caused by heat leakage from the wall. In addition, many factors affect the self-pressurization in the cryogenic liquid krypton tank, among which the liquid phase filling rate and gravity factors have a more prominent effect on the pressurization rate. Therefore, this paper focuses on the influence of wall heat leakage, liquid filling rate, and gravity environment on self-pressurization.

3.1. Effect of wall heat leakage on the self-pressurization rate

The wall heat leakage is the main factor of self-pressurization in cryogenic liquid krypton tanks. Based on the normal gravity environment and the initial liquid filling rate of 50 %, this paper compares the flow pattern and parameter changes in a cryogenic liquid krypton tank with a heat flux boundary of 5 w/m^2, 10 w/m^2, and 20 w/m^2. As shown in Figure 5, a1, b1 and c1 are the comparison of liquid surface positions at different times. When the initial filling rate is 50%, it takes the 1400 s, 2100 s, and 2800 s for the liquid surface to drop to the same position under the three heat flux boundaries, respectively. The greater the heat flux, the longer it takes to reach the same gas phase volume fraction. Mainly because the greater the heat flux, the gas phase thermal expansion rate increases, which in turn affects the rate of pressurization rate also becomes larger, the gas phase pressure becomes larger so that the saturation temperature of the phase change also becomes larger, according to the phase change formula of the evaporation process, it is obvious that the evaporation process is slower.

The magnitude and distribution of temperature and pressure as the liquid level drops to the same position at different heat flow boundaries can be observed in Figure 5 (a2), (b2), and (c2). It is clear that as the heat flow density increases, the temperature and pressure parameters corresponding to the same liquid level height are increasing. The gas phase pressures are 4223 Pa, 14240 Pa, and 34960 Pa, and the maximum gas phase temperatures are. 123.3 K, 127.1 K, 133.1 K.

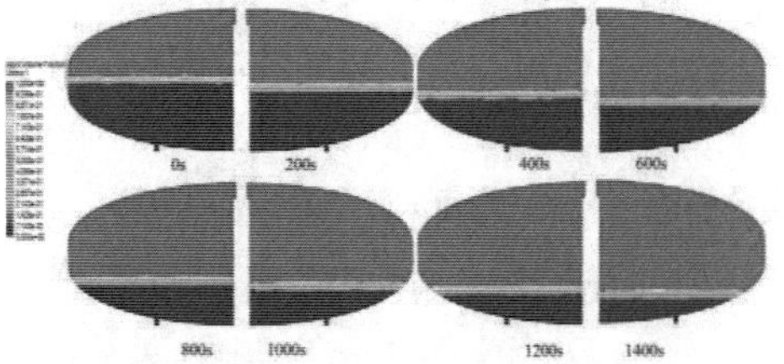

(a1) Liquid level change of 5 w/m^2

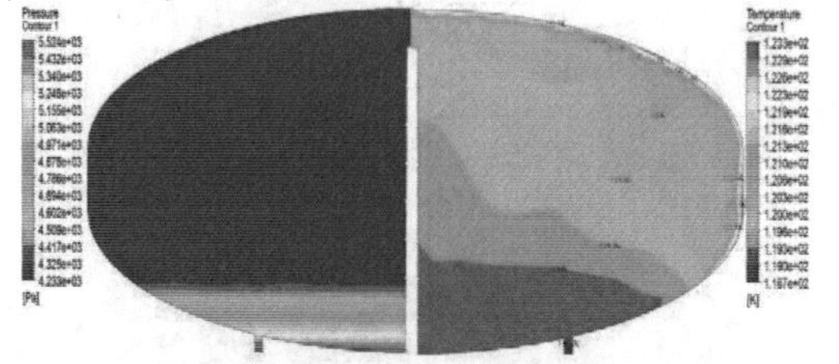

(a2) 1400s pressure and temperature contour

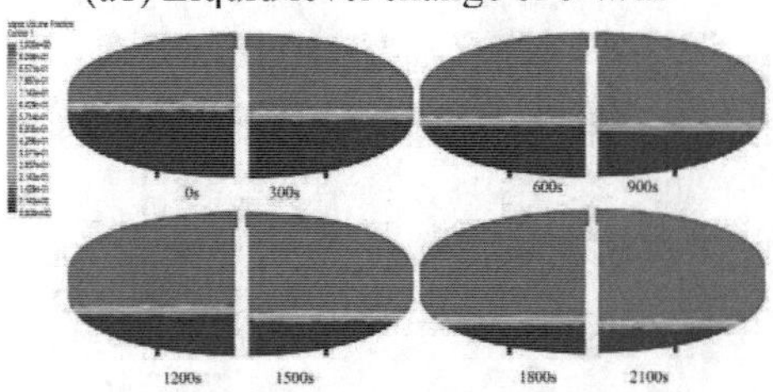

(b1) Liquid level change of 10 w/m^2

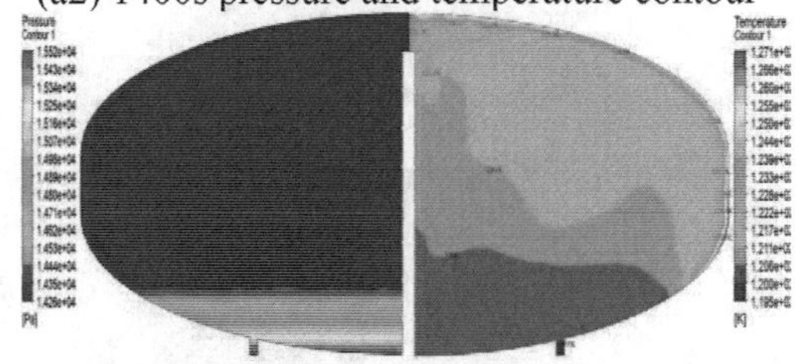

(b2) 2100 s pressure and temperature contour

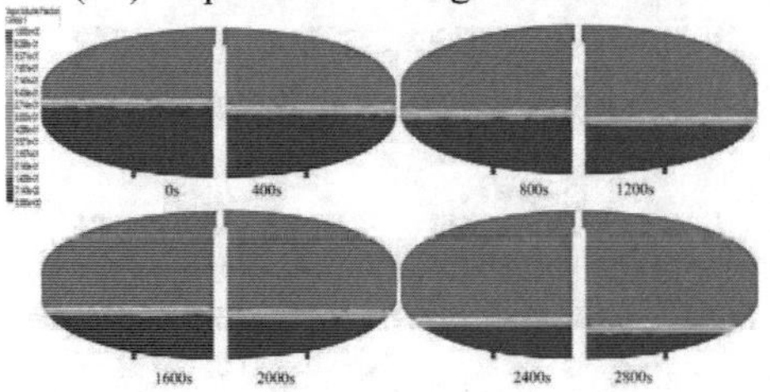

(c1) Liquid level change of 20 w/m^2

(c2) 2800 s pressure and temperature contour

Figure 5 Liquid level change and thermostatic cloud map at different heat flux densities

Secondly, the axial temperature change curves with time shown in Figure 6 (a), (b), and (c) were obtained by monitoring 10 temperature measurement points. The following conclusions were obtained by comparing the axial temperature curves under different heat flow densities: (1) When reaching the same liquid level height, the smaller the heat flow density, the smaller the temperature change in the overall process; and (2) The axial temperature curve corresponding to the same heat flux gradually shifts to the right with the increase of time, that is, the temperature in the tank continues to increase; and (3) From the analysis of each temperature curve alone, it is obvious that there is thermal stratification in the cryogenic liquid krypton tank during the change from point T1 to point T10.

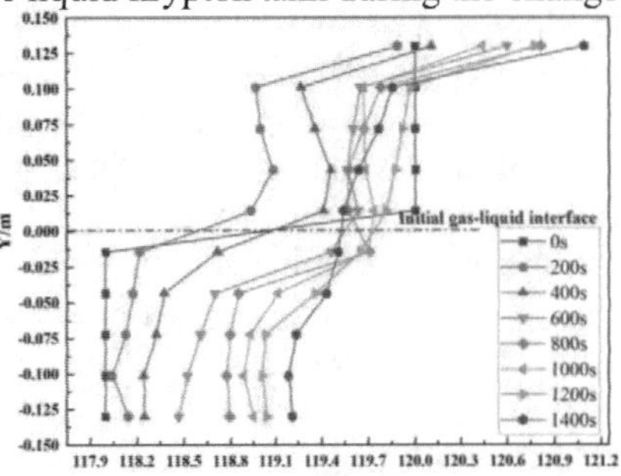

(a) 5 w/m^2 Curve of axial temperature changing with time

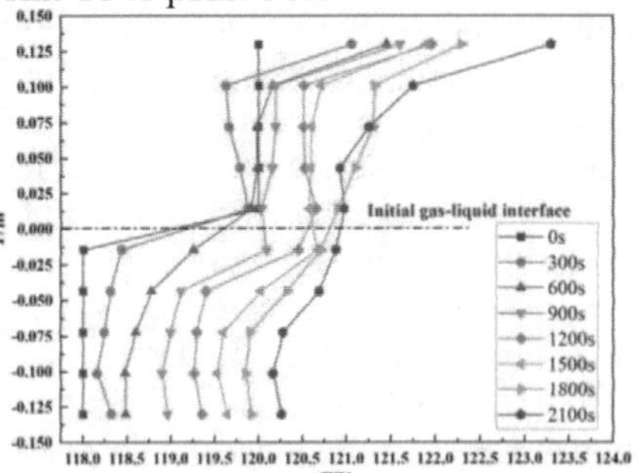

(b) 10 w/m^2 Curve of axial temperature changing with time

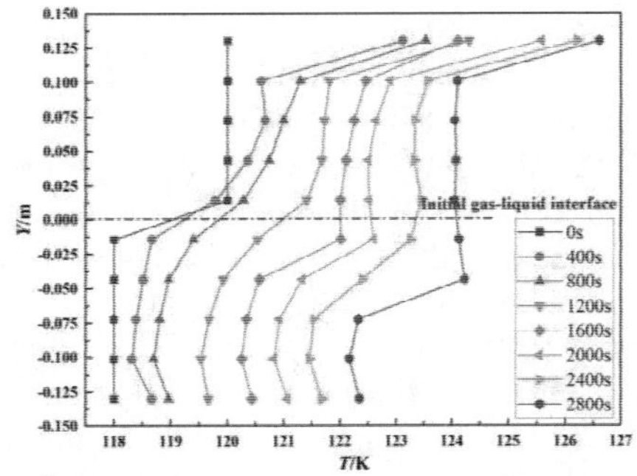

(c) 20 w/m^2 Curve of axial temperature changing with time

Figure 6 Axial temperature change curve over time at different heat flux densities

Both the parameter cloud and the axial temperature curve illustrate that the greater the heat flow density at the wall, the greater the range of parameter variation within the cryogenic liquid krypton tank. According to the convective heat transfer equation, the temperature inside the cryogenic liquid krypton tank varies more when the heat flow density rises and the heat transfer coefficient stays the same. When at the same liquid level, the volume fraction of the gas phase is equal and according to the ideal gas equation of state, and the change in gas phase pressure is also greater. To get a clearer view of the heat transfer process in the cryogenic liquid krypton tank due to the natural convection between the gas and liquid phases, a cloud diagram was drawn for the heat transfer process between the heat flow microclusters in the gas and liquid phases, as shown in Figure 7.

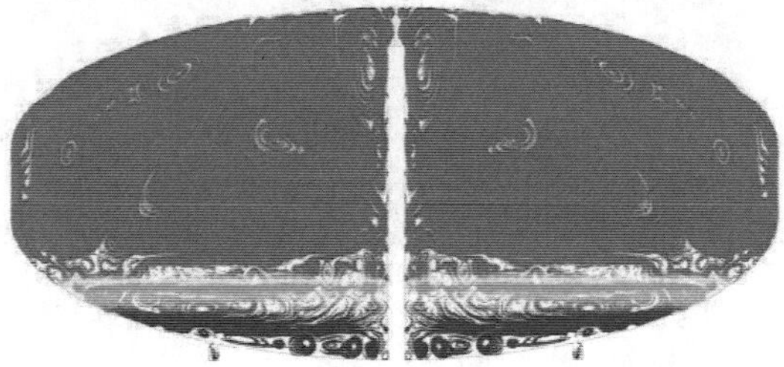

Figure 7 Heat transfer process at the interface layer of gas-liquid phase microclusters

Finally, the gas phase pressure variation curves at 5 w/m^2, 10 w/m^2, and 20 w/m^2 heat flow densities were compared, as shown in Figure 8, and the corresponding gas phase pressure variation values were 10 kPa, 14 kPa, and 20 kPa for a pressurization time of the 1200 s. With the increase of wall heat flux, the self-pressurization rate in the cryogenic liquid krypton tank increases gradually, and each corresponding self-pressurization curve shows a nonlinear growth trend.

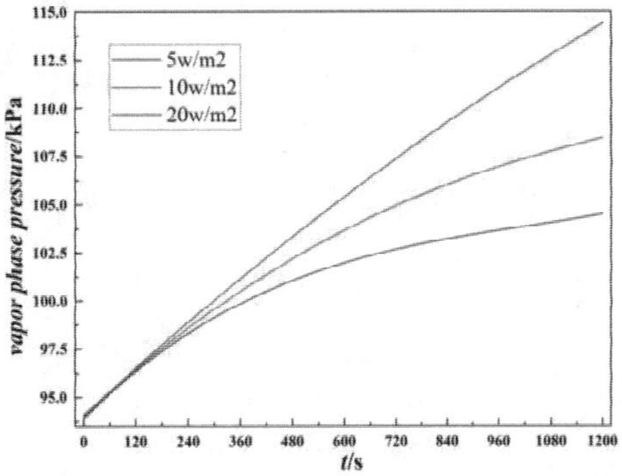

Figure 8 Comparison of self-pressurization curves with different heat flux densities

3.2. Effect of liquid phase filling rate on self-pressurization rate

The liquid phase filling rate determines the proportion of gas phase volume in the cryogenic tank, which will directly affect the self-pressurization rate of the cryogenic propellant. To verify this conclusion, the cryogenic krypton working fluid is used to simulate and analyze the three filling rate conditions of 37%, 50%, and 83% by using 5 w/m^2 uniform heat flow boundary under a normal gravity environment, and the curve results are shown in Figure 9. In the early and late stages of self-pressurization, the smaller the initial liquid filling rate, the greater the self-pressurization rate, which is consistent with the conclusions of most cryogenic propellant simulation studies. The initial liquid filling rate is inversely proportional to the self-pressurization rate for the entire liquid krypton evaporation process.

The main reasons for the above phenomenon are analyzed as follows: (1) As the initial liquid phase filling rate of cryogenic liquid krypton propellant increases, the greater the latent heat value of liquid krypton in the tank after the same uniform heat flow boundary continues to heat, the corresponding evaporation rate is significantly reduced; (2) In the short term, the change of liquid phase temperature is not obvious, so the contribution of liquid phase expansion caused by temperature rise to the pressurization of liquid krypton tank is very small, which is mainly caused by the thermal expansion of gas phase. But as the liquid filling rate increases, the corresponding gas phase percentage lowers, and the contact area between the gas phase and the wall surface reduces the gas phase's expansion rate declines.

However, the initial filling rate of the liquid phase cannot be increased constantly. Suppose the filling rate is increased to 100%. In that case, all the internal energy of the liquid phase volume expansion caused by heat leakage from the tank wall will be changed into mechanical energy to do work externally in the form of pressure acting directly on the tank wall. At that point, the safety of the cryogenic liquid krypton tank would be dramatically reduced.

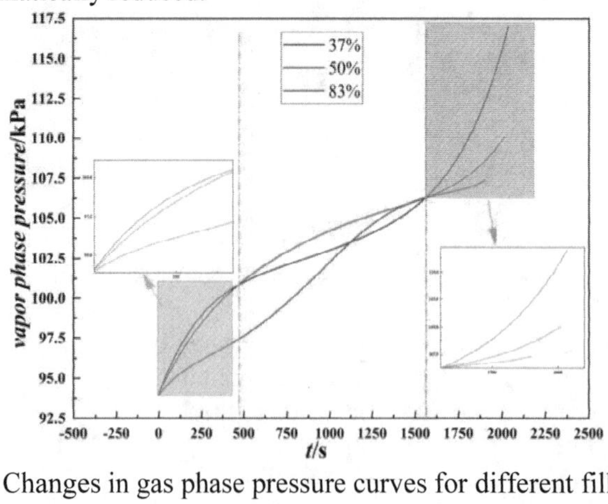

Figure 9 Changes in gas phase pressure curves for different filling rates

3.3. Effect of gravity conditions on the self-pressurization rate

The simulation was carried out for gravity environments of 1 g, 0.1 g, and 0.001 g using cryogenic krypton working fluid, respectively, with a liquid phase filling rate of 50% and an initial pressure of 94 kPa in the tank. The velocity clouds corresponding to the gas phase pressure in the tank reaching 115 kPa are shown in (a), (b), and (c) in Figure 10. The liquid-liquid phase tends to be diffusely distributed. When the pressure inside the cryogenic liquid krypton tank reaches 115 kPa from 94 kPa, the three gravity environments require 2336.3 s, 6231 s, and 16105 s, respectively, while the corresponding maximum velocities are 0.1092 m/s, 0.0389 m/s and 0.00583 m/s. Combined with the comparison of the gas phase pressure change curves for different gravities in Figure 11, it is clear that as the gravity decreases, the self-pressurization rate in the cryogenic liquid krypton tank decreases significantly.

The above conclusions are analysed from the following two aspects: (1) Cryogenic propellants' evaporation and condensation process is mainly carried out through the interfacial layer. Under normal gravity, the gas-liquid phase has a clear interface. The gas-phase heating by natural convection is relatively stable and efficient, so the gas-phase pressure curve shows a regular and stable upward trend. As the gravity decreases, the gas-liquid phase in the cryogenic liquid krypton tank gradually tends to disperse, and the gas-liquid phase interface is constantly reconfigured over time. At this time, the heat transfer efficiency is low and the heating is not uniform, so the phenomenon of slowing down the self-pressurization rate occurs. (2) With the reduction of gravity, the role of surface tension in the cryogenic propellants is gradually highlighted so that the inner wall of the cryogenic tank begins to be in a wetted state, that is, the phenomenon of liquid phase covering the gas phase. The flow rate within the gas phase can be seen from the reduction in gravity leads to a slower natural convective heat transfer process within the cryogenic liquid krypton tank, so there is a tendency for the liquid film on the wall to be locally overheated and the internal gas phase pressure change rate to decrease.

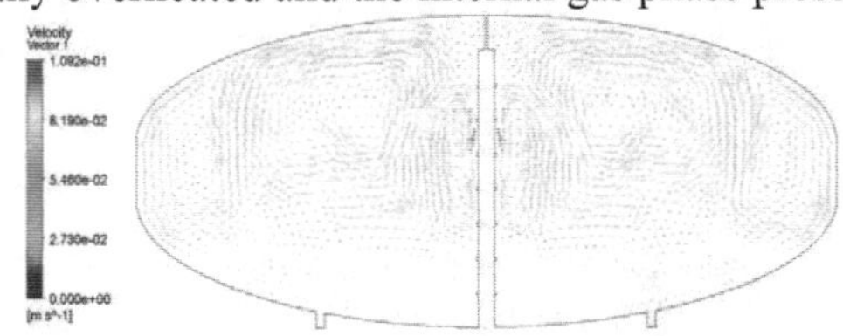

(a) 1g velocity contour

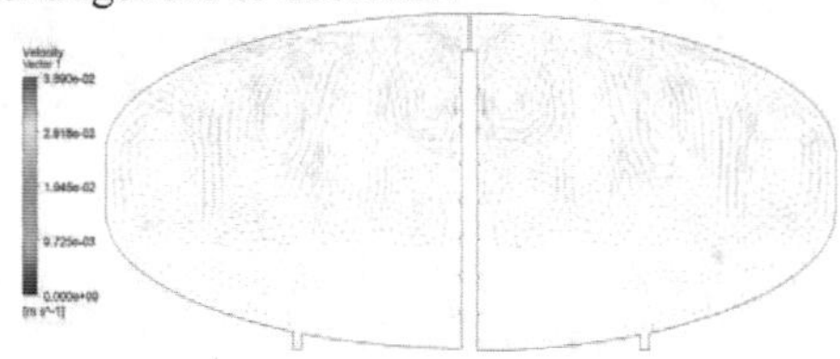

(b) 0.1g velocity contour

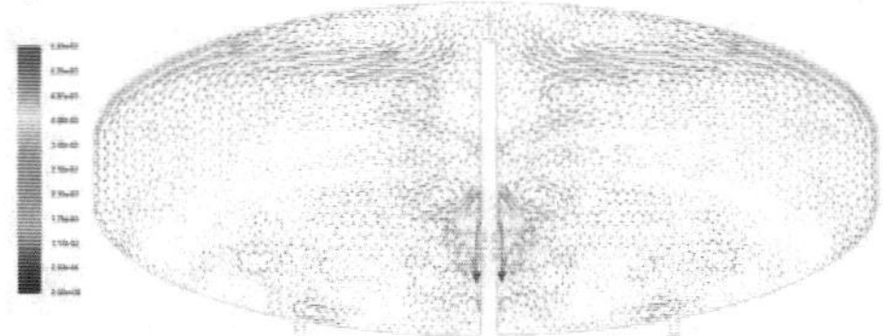

(c) 0.001g velocity contour

Figure 10 Comparison of flow velocities under different gravity

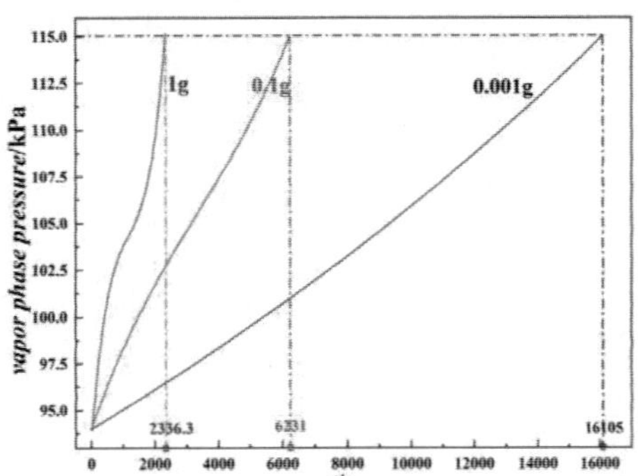

Figure 11 Changes in gas phase pressure curves under different gravity

4. Conclusion

In this paper, a two-dimensional axisymmetric tank model is used to simulate the self-pressurization phenomenon in a cryogenic liquid krypton tank using Fluent under different conditions of wall heat leakage, different filling rates, and different gravitational forces. By analyzing the evolution of the liquid level in the cryogenic liquid krypton tank with time, the distribution of pressure, temperature, and velocity vectors and the gas phase pressure variation curve, the influence of different influencing factors on the self-pressurization process was investigated. The main conclusions are as follows: maintaining 1 g and 50% liquid phase filling rate, the greater the wall heat flow density, the greater the time required when the same liquid level is reached, the greater the corresponding temperature and pressure parameters in the tank at the final moment, and the greater the corresponding gas phase pressurization rate. Maintaining 1 g and a heat flow density of 5 w/m^2, as the filling rate decreases, the volume fraction of the gas phase increases, the contact area with the wall increases, the gas phase is more fully heated and the expansion rate increases, which in turn leads to an increase in the gas phase pressurization rate at the same uniform heat flow boundary at the wall. By maintaining a 50% filling rate and heat flow density of 5 w/m^2, the natural convective heat transfer process in the tank is significantly weakened due to the reduced flow velocity of the gas phase as a result of the reduced gravity, resulting in a greater reduction of the inhomogeneous heat transfer and the gas phase pressurization rate.

The cryogenic liquid krypton tank will go through several stages in the application process, including ground parking, rising, and on-orbit operation. Different stages correspond to the different thermal and gravity environments, and the initial liquid filling rate is also an important factor in determining the pressure rise in the tank. Therefore, it is of great significance to study the variation of self-pressurization rate in cryogenic liquid krypton tanks with different wall heat flux, different liquid filling rates, and different gravity environments for the design of cryogenic liquid krypton tank and safety management in the life cycle.

Acknowledgments
This work was supported by the Key Laboratory Fund for Vacuum Technology and Physics (B202003).

References

[1] Choueiri E Y. A critical history of electric propulsion: The first 50 years (1906-1956) [J]. Journal of propulsion and power, 2004, 20(2): 193-203.
[2] Jacobson D, Manzella D. 50 kW class krypton Hall thruster performance[C]. 39th AIAA/ASME/SAE/ASEE Joint Propulsion Conference and Exhibit, 2003: 4550.
[3] Welle R. Xenon and krypton availability for electric propulsion - An updated assessment[C]. 29th Joint Propulsion Conference and Exhibit, 1993: 2401.
[4] Duchemin O, Valentian D, Cornu N. Cryostorage of propellants for electric propulsion[C]. 45th AIAA/ASME/SAE/ASEE joint propulsion conference & exhibit, 2009: 4912.
[5] Welle R P. Propellant storage considerations for electric propulsion[C]. 22nd International Electric Propulsion Conference, 1991: 91-107.
[6] Guo Z V, Ju Y L. Current status and problems of cryogenic liquid hydrogen storage[J]. Cryogenics and Superconductivity, 2019, 47: 21-29.
[7] Chen J D, Zhu J B, Zhao J P, et al. Analysis of key technologies for liquid krypton cryogenic propellant tanks for electric propulsion systems[J]. Vacuum and Cryogenics, 2022, 28: 1-11.
[8] Fu J, Sunden B, Chen X, et al. Influence of phase change on self-pressurization in cryogenic tanks under microgravity[J]. Applied Thermal Engineering, 2015, 87: 225-233.
[9] Fu J, Sunden B, Chen X. Analysis of self-pressurization phenomenon in a cryogenic fluid storage tank with VOF method[C]. ASME International Mechanical Engineering Congress and Exposition, 2013: V001T01A020.
[10] Sun D, Xu J, Chen Q. Modeling of the evaporation and condensation phase-change problems with FLUENT[J]. Numerical Heat Transfer, Part B: Fundamentals, 2014, 66(4): 326-342.
[11] Sun D-L, Xu J-L, Wang L. Development of a vapor-liquid phase change model for volume-of-fluid method in FLUENT[J]. International Communications in Heat and Mass Transfer, 2012, 39 (8): 1101-1106.
[12] Choi S W, Lee W I, Kim H S. Numerical analysis of convective flow and thermal stratification in a cryogenic storage tank [J]. Numerical Heat Transfer, Part A: Applications, 2017, 71 (4): 402-422.
[13] Ovidi F, Pagni E, Landucci G, et al. Numerical study of pressure build-up in vertical tanks for cryogenic flammables storage [J]. Applied Thermal Engineering, 2019, 161: 114079.
[14] Liu Z, Li Y, Jin Y. Pressurization performance and temperature stratification in cryogenic final stage propellant tank[J]. Applied Thermal Engineering, 2016, 106: 211-220.
[15] Barsi S, Kassemi M. Investigation of tank pressurization and pressure control-part II: numerical

modeling [J]. Journal of Thermal Science and Engineering Applications, 2013, 5 (4): 041006.
[16] Lopez A, Grayson G, Chandler F, et al. Cryogenic pressure control modeling for ellipsoidal space tanks[C]. 43rd AIAA/ASME/SAE/ASEE Joint Propulsion Conference & Exhibit, 2007: 5552.
[17] Lopez A, Grayson G, Chandler F, et al. Cryogenic pressure control modeling for ellipsoidal space tanks in reduced gravity[C]. 44th AIAA/ASME/SAE/ASEE Joint Propulsion Conference & Exhibit, 2008: 5104.
[18] Kassemi M, Kartuzova O, Hylton S. Validation of two-phase CFD models for propellant tank self-pressurization: crossing fluid types, scales, and gravity levels[J]. Cryogenics, 2018, 89: 1-15.
[19] Perez-Raya I, Kandlikar S G. Numerical modeling of interfacial heat and mass transport phenomena during a phase change using ANSYS-Fluent [J]. Numerical Heat Transfer, Part B: Fundamentals, 2016, 70 (4): 322-339.
[20] Wang B, Qin X, Jiang W, et al. Numerical simulation on interface evolution and pressurization behaviors in cryogenic propellant tank on orbit[J]. Microgravity Science and Technology, 2020, 32 (1): 59-68.
[21] Bouhal T, Agrouaz Y, Kousksou T, et al. Performance optimization of a two-phase closed thermosyphon through CFD numerical simulations [J]. Applied Thermal Engineering, 2018, 128: 551-563.

Numerical Calculation of Hydrodynamic Characteristics of Individuals in Coral Nurseries

Can Liu[1], Zhiyong Dong[1]*, Yun Pan[2]

[1.] College of Civil Engineering, Zhejiang University of Technology, Hangzhou 310023, China;

[2.] School of Naval Architecture and Maritime, Zhejiang Ocean University, Zhoushan 316022, China

* Correspondence: Zhiyong Dong, E-mail address: dongzy@zjut.edu.cn

Abstract: Coral nurseries play a critical role in solving the technical and ecological problems that hinder the large-scale and long-term culture of coral populations, which has become an effective tool for the coral culture and the restoration of coral reefs. We need to conduct a profound study of the hydrodynamic characteristics of individuals in coral nurseries to ensure that the coral nurseries achieve long-term, industrialized, and batch coral cultivation. In this paper, the calcified coral skeletons of *Acropora cervicornis* are selected, followed by refined statistical analysis of the characteristic dimensions of individual corals to calculate the average density, and fitted equations for the horizontal projected area and volume. In addition, we also analyze the force of the individual coral, which lays a foundation for studying the hydrodynamic characteristics of corals under wave action.

1. Introduction

Mass bleaching has resulted in significant losses of live coral. The Status of Coral Reefs of the World [1]: 2020, published by the Global Coral Reef Monitoring Network (GCRMN), stated that global bleaching events could become the norm in the coming decades under the rising global temperature. The worldwide coral reef will die at the end of the 21st century unless we can rapidly reduce carbon emissions. Therefore, accelerated ecological restoration in coral reefs has become one of the most urgent problems for marine environmental protection to solve in China. The coral gardening concept and underwater nurseries based on active coral reef restoration can help solve this problem. The path for ecological restoration of the "coral gardening" approach is shown in Figure.1 [2].

Coral nurseries are subjected to the combined action of both waves and currents in the ocean as marine structures. The hydrodynamics of oceans have long been known to play a critical role in shaping the growth, diversity, and distribution of corals, and understanding the forces of waves on individual corals from multiple perspectives could help in crafting more targeted and effective responses for reef restoration.

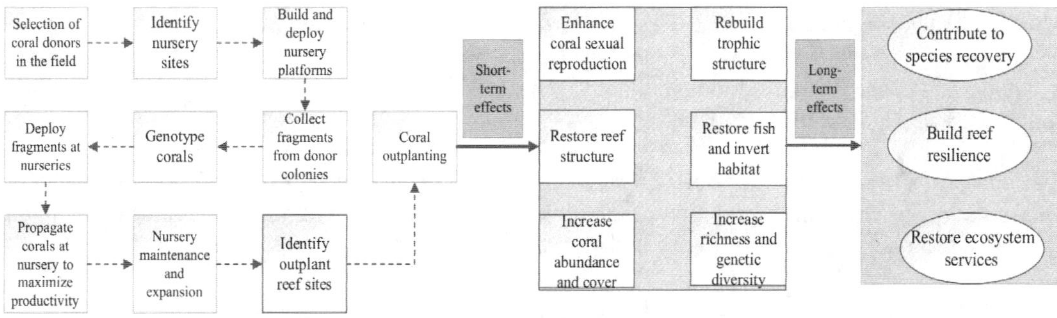

Figure 1. Implementation Path of Ecological Restoration of Gardening Coral Reefs

Robinson (1954) [4] first proposed that the energy dissipated by waves on shallow forereefs and reef crests facilitates the rapid exchange of nutrients and other metabolites necessary to support the robust growth of reef communities in otherwise nutrient-poor environments. This theory would stimulate modern research aimed at understanding how hydrodynamic processes in the benthic boundary layer control the uptake of dissolved nutrients by coral communities. Later, Darwin et al. (1962) [3] hypothesized that the surrounding seawaters provide a major source of nutrition to corals and that corals grow best at the edges of reefs. Lenihan et al. (2015) [5] suggested that hydrodynamic conditions can influence the distribution and abundance of aquatic species in many ways, including directly through disturbance and resource delivery and indirectly by altering environmental cues and species interactions. Hydrodynamic enhancement of coral growth via simultaneous direct and indirect effects can be important for the persistence of coral reefs as well as their recovery following disturbance.

In coral nurseries, the flow of fluid around the coral body causes some of the momenta of the fluid to be transferred to the interior of the coral body. The hydrodynamic forces (lift, drag, and inertia) generated by this momentum transfer create stresses on the coral body, which will cause the coral to break if the applied load exceeds the mechanical strength of the coral form [6]. The quantitative analysis of the relationship between wave forces and individual coral morphology can be targeted for improving coral body survival, decreasing shedding rates during coral cultivation and accelerating reef restoration.

2. Statistics of Characteristic Size of Individual Corals

With a faster growth rate and resistance to heat and high temperature, *Acropora cervicornis* has become the selection of preferred species for artificial transplantation recovery and restoration of coral reefs in China. Seven individual corals with natural breakage were selected, and the external dimensions were measured with a digital vernier caliper,as shown in Figure 2. We used a method based on Archimedes' Principle to measure the volume of each coral.

Figure 2. Seven Individual Corals with Natural Breakage

The morphology of a colony or branch was defined according to a shape index expressed as:

$$\lambda = \frac{h}{d} \tag{1}$$

where h is the height and d is the diameter. According to the shape index, the coral morphology can be divided into three different categories: 1. $\lambda > 1$, the shape index greater than 1, that is, taller than its width; 2. $\lambda < 1$, the shape index less than 1, that is, short and broad; 3. Multiple branches exist consistently in a longitudinal direction.

2.1 Calculation of projected area and volume of individual corals with $\lambda > 1$ and multiple branches.
For an individual coral, the calculation of the projected area (S) and volume (V) could be divided into two parts: the main trunk and its branches. The main formula can be reached as follows:

$$S=\sum_{i=1}^{n} s + s_i \tag{2}$$

$$V=\sum_{i=1}^{n} v + v_i \tag{3}$$

where s and s_i are the main trunk and its branches parts' projected area of the individual coral, respectively; v and v_i are the main trunk and its branches parts' volume of the individual coral, respectively, and the expressions are shown in Equations (4) to (7).

$$s = \begin{cases} a_2 \times l & a_2 / b_2 < 1.5 \\ (a_2 + b_2) \times l / 2 & a_2 / b_2 \geq 1.5 \end{cases} \tag{4}$$

$$s_i = l \times a_1 \tag{5}$$

$$v = \begin{cases} \pi \times (b_2/2)^2 \times l & b_1 / b_2 < 1.8 \\ \pi \times (b_1 + b_2/4)^2 \times l & b1 / b_2 \geq 1.8 \end{cases} \tag{6}$$

$$v_i = \pi \times (b_2 / 2)^2 \times l \tag{7}$$

where l is the length of the main trunk or branch; a_1, a_2, b_1, and b_2 are the upper and lower width and thickness of the main trunk or its branches, respectively.

2.2 Calculation of volume and projected area of individual corals for $\lambda < 1$.
When $\lambda < 1$, the calculation of the projected area (S) and volume (V) could be divided into two parts: the upper and lower parts. The main formula can be reached as follows:

$$S=\sum S_{up} + S_{down} \tag{8}$$

$$V=\sum V_{up} + V_{down} \tag{9}$$

where S_{up} and S_{down} are the upper and lower parts' projected area, $S_{up}=b_1 \times l$, and $S_{down}=b_2 \times l$; V_{up} and V_{down} are the upper and lower parts' volume, $V_{up}=\pi \times (a_1/2)^2 \times b_1$, and $V_{down}=\pi \times (b_2/2)^2 \times l$.

3. Linear Regression Analysis of the Projected Area and Volume

We can solve the density of coral skeletons by measuring the individual coral of the total mass and drainage volume. The projected area and volume equation of individual corals were fitted by linear regression analysis as a theoretical foundation for the force analysis of individual corals under wave action.

3.1 Calculation of density
The total mass of each individual coral was measured using an electronic balance with a 0.01-g accuracy. Based on the drainage volume of individual corals that have been measured, the density of each

individual coral can be calculated through the density formula of $\rho = m / v$, and the calculation results are shown in Figure 3. It can be seen that individual corals with different morphologies have very small fluctuations in density; therefore, the density can be taken as 2000 kg/m^3.

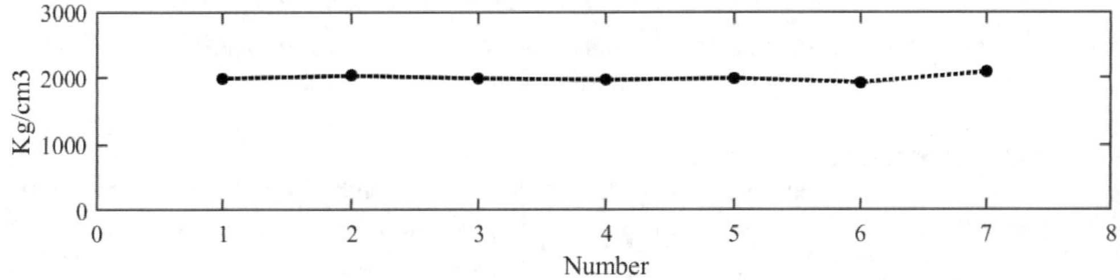

Figure 3. Density Variation of Individual Corals

3.2 Projection area formula of individual corals

The shape index of individual corals can be divided into three different categories and generalized as a cylinder. The projected area of individual corals in the vertical direction with the water flow is rectangular. We measured the projected area using the Image J software, which is in the public domain and does not require specific devices or software packages. The measurement results of the projected area are shown in Table 1.

Table 1. Projected Area of Individual Corals

Number	Measuring area (S')	Main trunk area (s)	Branch area (s_1)	Branch area (s_2)	Branch area (s_3)
1	67.54	41.37	8.81	7.63	6.75
2	97.4	72.29	18.78	11.42	0
3	60.7	44.69	14.85	0	0
4	29.83	25.07	4.27	0	0
5	21.04	16.19	2.75	0	0
6	45.18	39.02	4.98	0	0
7	73.91	55.47	9.13	5.03	7.19

The projected area equation of individual corals was fitted by linear regression analysis using the SPSS software, as shown in Equation (10).

$$S = 0.801s + 1.285s_1 + 0.909s_2 + 1.314s_3 + 5.45 \tag{10}$$

where s_1, s_2, and s_3 represent the projected area of branches arranged from large to small, respectively; s is the projected area of the main trunk. The calculation method is shown in Equations (4) to (5).

3.3 Volume formula of individual corals

The volumes of the main trunk and its branches were calculated using Equations (6), (7), and (9), as shown in Table 2.

Table 2. Volume of Each Part of the Coral Body

Numb er	Measured volume(V')	trunk volume(v)	Branch volume(v1)	Branch volume(v2)	Branch volume(v3)
1	102	82.46	4.95	11.81	8.59
2	102	53.83	46.28	8.82	0
3	87	74.78	24.42	0	0
4	45	46.79	2.89	0	0
5	22	21.98	1.36	0	0
6	68	59.33	21.34	0	0
7	110	49.36	34.61	5.52	9.94

The volume equation of individual corals was fitted by linear regression analysis using the SPSS software, as shown in Equation (11).

$$V = 0.717v + 1.084v_1 + 0.806v_2 + 2.664v_3 + 5.687 \tag{11}$$

where v_1, v_2, and v_3 represent the volumes of branches arranged from large to small, respectively; v is the volume of the main trunk. The calculation method is shown in Equations (6) to (7).

4. Force Analysis of Individual Corals

It will cause the coral to break or detach from the substrate when the hydrodynamic forces exceed the mechanical strength of individual corals. In order to simplify the force calculations, the forces of the individual coral under the wave action that are mainly maintained in balance by their gravity and physical strength can be generalized as shown in Figure 4. The morphology and forces of individual corals can be simplified in three ways [7]: a) Individual corals or very densely branched coral forms can be generalized to cylindrical shapes of comparable size. b) The skeletal structure and strength are uniform throughout the individual coral form, so the stress generated by water flow will be greatest at the base of the coral, where it attaches to the substrate. c) The position of the center of dynamic pressure and the center of mass for the coral coincide at the midpoint of the individual coral.

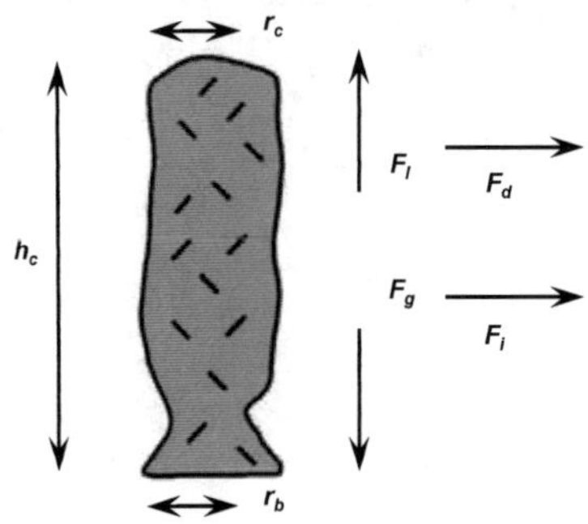

Figure 4. Diagram of Forces on a Coral Body. (F_g, F_l, F_d, and F_i are the coral's weight and the wave-induced lift, drag, and inertial forces, respectively.)

4.1 Force analysis of individual corals

As water flows over the coral, the flow is disturbed due to flow lines diverging, and the fluid temporarily accelerates over the coral. This acceleration causes a lifting force. The lift force (F_l) on the coral is defined as:

$$F_l = \frac{1}{2}\rho_f C_l \bar{u}^2 S\beta \tag{12}$$

where u is the horizontal velocity of the water particle near the bottom of the seabed, calculated on the basis of Stokes' second-order wave theory, and the lift coefficient is taken as C_l=0.15. β is a function of coral species, varying from 0–1. S is the projected area of individual corals calculated by Equation (10).

Moving fluid imposes the drag force on any structure projecting into and disrupting the flow. The drag force (F_d) on the coral was defined as:

$$F_d = \frac{1}{2}\rho_f C_d \bar{u}^2 S\beta \tag{13}$$

where C_d is the drag coefficient, based on the range of 0.7–1.0 determined for corals and rough cylinders. When the individual coral is generalized to a cylinder, C_d *is* approximated 1.

The inertial forces are present due to the acceleration and deceleration of the flow field around the coral canopy under wave action. The inertial force F_i is defined as:

$$F_i = \frac{51}{24}\rho_f Va \tag{14}$$

where a is the water particle acceleration around the coral; S is the volume of individual corals calculated by Equation (10).

Torsional stress is absent in the model of the individual coral and its branches, whose projected areas are balanced on the central vertical axis and thus ignored. The coral skeleton and the attached calcified matrix are very similar to concrete and limestone, which are superior in strength when subjected to compressive and shearing loads but with limited hydrodynamic forces. If levels of the hydrodynamic forces exceed the strength of the coral skeleton, breakage will occur.

4.2 Analysis of results

We numerically calculate the hydrodynamic force on the No. 1 individual coral under the action of the regular wave with a wave height of H=2.0 m, a wave period of T_{wave}=10 s, and a water depth of h=20 m. The calculation results are shown in Figure 5.

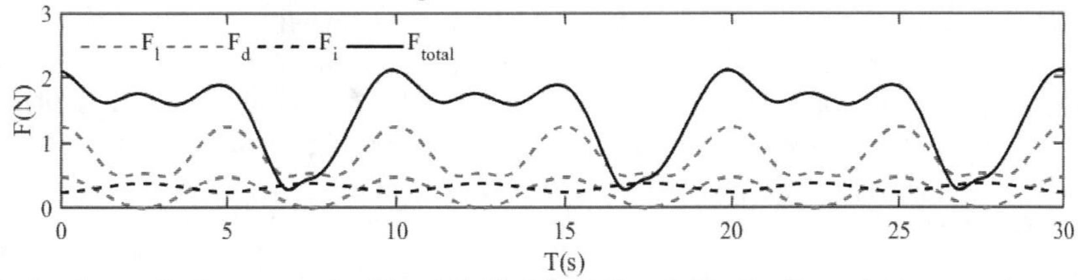

Figure 5. Hydrodynamic Force on the No. 1 Individual Coral (F_l, F_d, F_i, and F_{total} are the wave-induced lift, drag, inertial forces, and total force, respectively)

As can be seen from Figure 5, under the action of regular waves, various hydrodynamic forces on the individual coral show a periodic variation consistent with the wave period, and the drag force is more than twice the inertial force. When the water particle velocity is 0, the No. 1 individual coral is subjected to the smallest lift and drag force and the largest inertial force, and the maximum hydrodynamic force is about 2 N.

We also numerically calculate the hydrodynamic force on the No. 1–7 individual corals under the action of the regular wave with a wave height of H=2.0 m, a wave period of T_{wave}=10 s, and a water depth of h=20 m. The calculation results are shown in Figure 6.

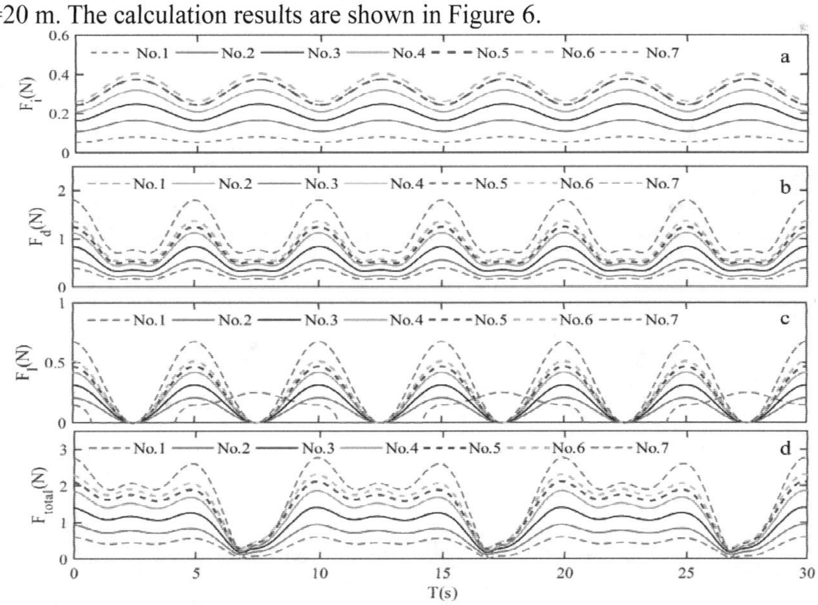

Figure 6 Hydrodynamic Forces on Different Individual Corals

F_l, F_d, F_i, and F_{total} are the wave-induced lift, drag, inertial forces, and total force, respectively, as shown in Figure 6 (a), (b), (c) and (d). The magnitude of the hydrodynamic force on the individual coral with different morphologies are proportional to their projected area. From Table 2, we can see that No. 5 coral body and No. 7 coral body have the same volume, but the projected area of coral body No. 7 is larger. Except for the lifting force in Figure 6(a), which is basically the same, the inertial force and drag force on coral body No. 7 are larger than those on coral body No. 5. Therefore, it can be seen from Figure 6 that the projected area has a more significant influence on the hydrodynamic force on the individual coral under the action of regular waves than the volume does.

5. Conclusion

In the past decade, coral nurseries have shown very high seedling survival and growth rates, which have developed into an ecological engineering tool for coral mariculture in modern marine ranching. Nonetheless, their wave resistance and mechanisms are not obvious, and the hydrodynamic effects are unclear. Therefore, we need to profoundly study the hydrodynamic characteristics of individuals in coral nurseries to ensure that coral nurseries can achieve long-term, industrialized, and batch coral cultivation. In this paper, the skeletal density of the calcified individual corals is calculated to be about 2000 kg/m^3. The equations for the projected area and volume of individual corals are fitted by linear regression analysis. And the projected area has a more significant influence on the forces of waves than the volume does. We numerically calculate the hydrodynamic force on the individual coral under the action of the regular wave, which lays a theoretical foundation for studying the hydrodynamic characteristics of individual corals.

References

[1] Global Coral Reef Monitoring Network, 2020. Status of Coral Reefs of the World: 2020[R/OL]. [2021-10-5].https://icriforum.org/events/gcrmn-2020-report-launch/

[2] Lirman D, Schopmeyer S. Ecological solutions to reef degradation: optimizing coral reef restoration in the Caribbean and Western Atlantic[J]. PeerJ, 2016, 4: e2597.

[3] Darwin C, Bonney T G, 1962. The Structure and Distribution of Coral Reefs[M]. California: University of California Press:60–71.

[4] Robinson M K, 1954. Bikini and nearby atolls, Marshall Islands; oceanography (physical): Adjustment of Bikini atoll to ocean waves[J]. Center for Integrated Data Analytics Wisconsin Science Center.260:275–291.

[5] Lenihan H S, Hench J L, Holbrook S J, *et al.*,2015. Hydrodynamics influence coral performance through simultaneous direct and indirect effects. Ecology,96:1540–1549.

[6] Madin J S, 2005. Mechanical limitations of reef corals during hydrodynamic disturbances[J]. Coral Reefs, 24(4): 630–635.

[7] Graus R R, Chamberlain J A, Boker A M,1977. Structural modification of corals in relation to waves and currents[J]. The American Association of Petroleum Geologists:135–153.

Experimental research on sound quality evaluation of aircraft cabin

Hongnian Li[1]

[1]Shanghai Aircraft Design and Research Institute, No. 5188 Jinke Road, Shanghai, China
lihongnian@comac.cc

Abstract: This paper proposes a construction method for the sound quality evaluation model of the aircraft cabin. By combining the subjective with the objective method, the relationship between the subjective results and the psychoacoustic parameters is established, and the sound quality evaluation model of the sound field in the passenger cabin is formed. The model takes the characteristic acoustic parameters that are more sensitive to the human ear, such as loudness and sharpness, as input, which can efficiently obtain the subjective result of sound quality and accurately reflect the subjective quality feeling of human ear on the current sound field.

1. Introduction

With the improving requirements for aircraft flight comfort, cabin comfort has become an important indicator for evaluating the performance of civil aircraft. As one of the main ways to improve comfort, cabin noise optimization has been paid more and more attention by aircraft manufacturers. In the past, improving cabin noise was mainly to reduce the A-weighted sound pressure level based on meeting the current noise standards and regulatory requirements. But passengers still feel that the noise is more intrusive, and it is increasingly expensive and difficult to rely on the noise reduction method to reduce the A-weighted sound pressure level alone. Reasonable evaluation of the quality of the sound field in the cabin enables manufacturers to improve the sound environment from the perspective of sound design and gradually becomes an effective means to guide noise optimization.

In recent years, the method of subjective sound quality evaluation, consisting of a graded scoring, semantic segmentation, and paired comparison method, has been used more frequently in sound field evaluation [1]. However, people's subjective evaluation will be affected by various factors such as race, subjective initiative, mental state, experience, etc., which makes it lack certain objectivity as a verification method. A new method for evaluating the sound quality of the sound field in the cabin of a passenger aircraft is proposed. By combining subjective with objective evaluation method, the relationship between the subjective results and the psychoacoustic parameters is established, and the evaluation model of the sound field in the passenger cabin is formed [2]. The evaluation model eliminates the influence of individual differences in personnel, has high evaluation efficiency, and can effectively evaluate and verify the noise optimization effect in the passenger cabin.

The construction of the sound field quality evaluation model in the passenger aircraft cabin is based on the objective psychoacoustic parameter test of the sound field sound quality in the cabin and the subjective quality evaluation test. The specific construction method is divided into the following six procedures [3]:

(1) We establish a sample set of acoustic signals in the cabin of passenger aircraft under multiple flight conditions, as follows:

Step 1: We arrange microphones at different positions in the cockpit or cabin of the passenger aircraft to collect acoustic signals at different positions during the complete flight phase of the passenger aircraft.

Step 2: We cut the collected acoustic signals according to different flight stages to construct a sample set of acoustic signals in the multi-flight condition cabin of the passenger aircraft. The duration of each sample after the acoustic signal of each measuring point is cut and controlled within 5-10 s. Sufficient subjective evaluation time must be given to the participants to ensure the accuracy of the evaluation.

(2) We take the acoustic signal sample set constructed in procedure (1) as input and use the cabin noise simulation system to accurately reproduce the cabin sound field inside the engineering sample section. Based on this environment, using the pairwise comparison method, the participants were organized to conduct a subjective evaluation of the sound field in the cabin of the passenger aircraft under multiple flight conditions, and the evaluation results were counted. If two or more sets of acoustic signal samples are cut out for each flight condition, the sample sets must be screened by the following methods to eliminate outliers:

Step 1: We disorder all the acoustic signal sample sets as the input of the noise simulation system, repeat the sound field in the engineering sample section in turn, and invite evaluators to use the graded scoring method to evaluate the sound quality subjectively.

Step 2: We calculate the variance of the subjective results of each acoustic signal sample set under the same working condition and remove a group of acoustic signal sample sets with large variance. If there are still two or more acoustic signal sample sets for each flight condition, we go to step 3. Otherwise, the screening ends.

Step 3: We eliminate a group of acoustic signal sample sets with a large degree of deviation of the subjective evaluation results under each flight condition, and the deviation degree index H is shown in the following formula:

$$H = \left(\bar{x}_i - \bar{x}\right)^2 \tag{1}$$

$$\bar{x} = \frac{\sum_{i=1}^{n} \bar{x}_i}{n} \tag{2}$$

In the formula, $\bar{x}_i$ represents the average value of evaluation results of the i-th sample set of a certain working condition; n represents the number of remaining sample sets of the working condition. If there are still 2 or more acoustic signal sample sets for each flight condition at this time, then arbitrarily select a set as the acoustic signal sample set for the formal test of the working condition. Otherwise, the screening ends.

(3) Based on the acoustic signal sample set screen in procedure (2), we extract objective psychoacoustic parameters: loudness, roughness, sharpness, semantic clarity, and jitter;

(4) If there are two types of objective psychoacoustic parameters, the objective psychoacoustic parameters are directly used as input items, and the subjective results obtained in the procedure (2) are used as output items to establish a multiple linear regression model; if the types of objective psychoacoustic parameters are more than two types, we will perform Pearson correlation on the subjective results and various psychoacoustic parameters, record each correlation coefficient r, and perform the procedure (5).

$$r = \frac{\sum_{i=1}^{n}\left(X_i - \bar{X}\right)\left(Y_i - \bar{Y}\right)}{\sqrt{\sum_{i=1}^{n}\left(X_i - \bar{X}\right)^2}\sqrt{\sum_{i=1}^{n}\left(Y_i - \bar{Y}\right)^2}} \tag{3}$$

In the formula, X_i represents the i_{th} value of a psychoacoustic parameter; represents the average value of the psychoacoustic parameter; represents the i_{th} result of subjective evaluation; represents the average value of subjective results; n represents the number of flight conditions. We eliminate a psychoacoustic parameter with the smallest absolute value of correlation coefficient r.

(5) Taking the remaining several objective psychoacoustic parameters as input items and the subjective evaluation results obtained in procedure (2) as output items, a multiple linear regression model is established, and the coefficient of determination is recorded.

(6) We repeat procedures (5)-(6) until the remaining two objective psychoacoustic parameters are completed, and the model corresponding to maximum value of the coefficient of determination is selected as the sound quality evaluation model.

2. Construction of sound quality evaluation model

This chapter will introduce the construction method of the sound quality evaluation model. The sound quality evaluation model is a mathematical relationship established by linear regression or curve fitting based on the results.

2.1. Sound quality evaluation test

2.1.1. Objective parameter test

The loudness and sharpness were selected as objective psychoacoustic parameters, and an artificial head was used to conduct binaural tests for 15 seconds each, as shown in Figure 1 [4]. The test results of loudness and sharpness are shown in Figure 2 and Figure 3.

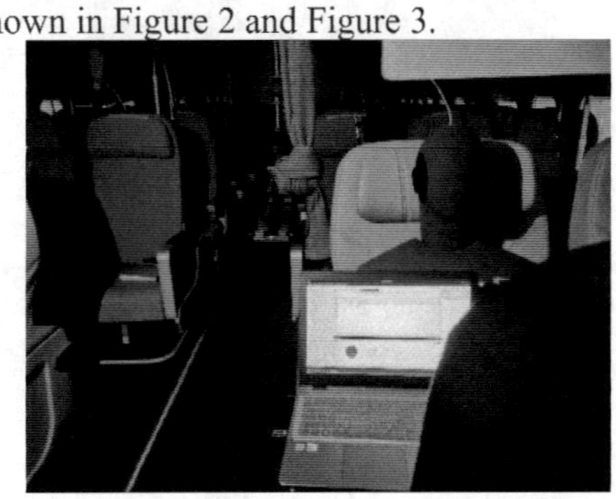

Figure 1. Objective parameter test of sound quality

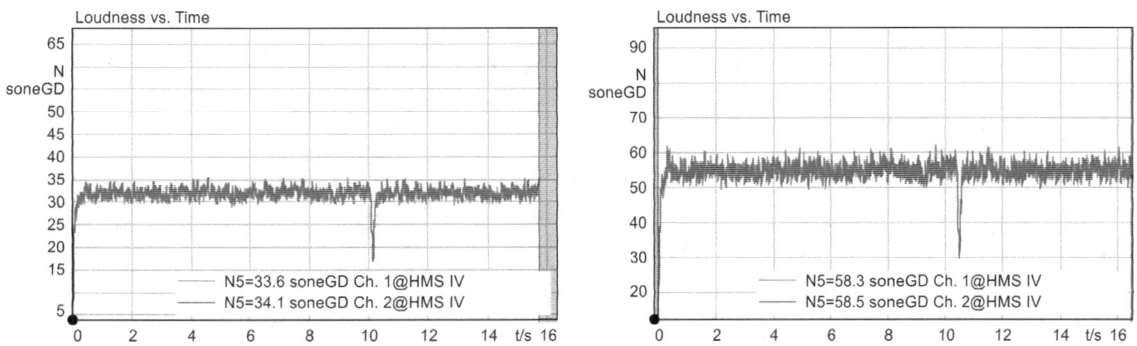

a) Ground taxiing b) Take off

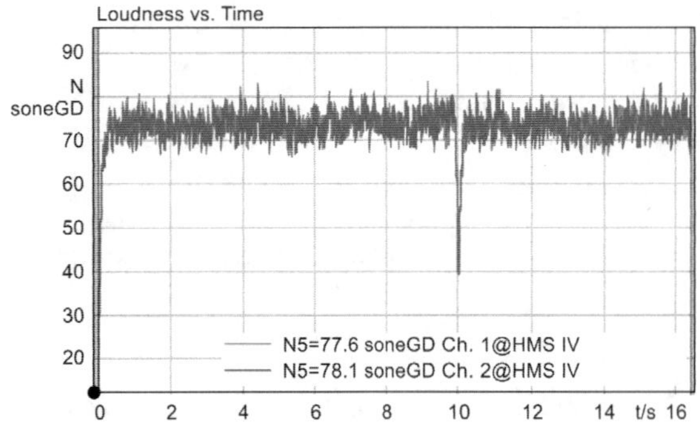

c) Cruise

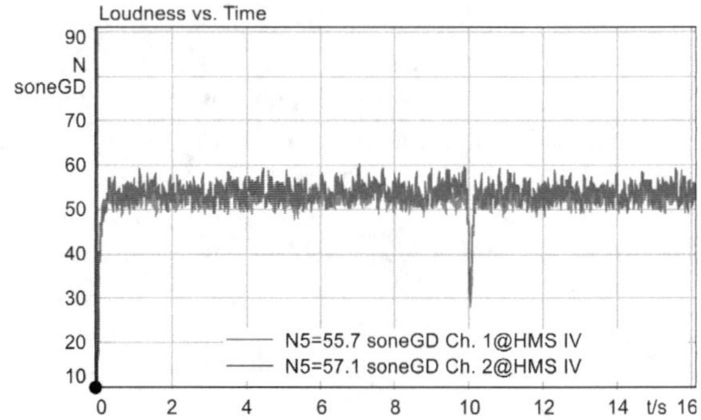

d) Acceleration

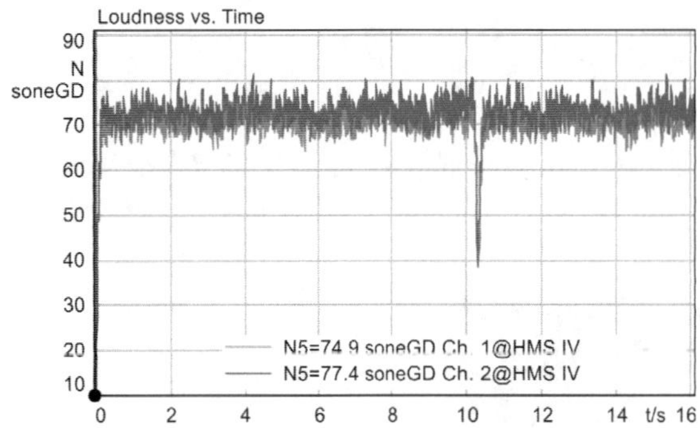

e) High-altitudeude cruise

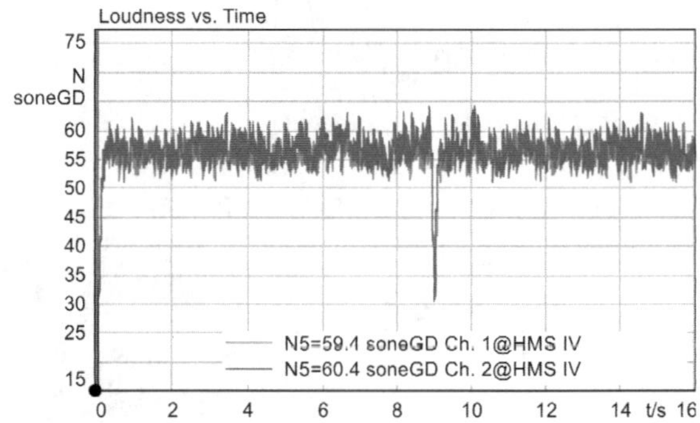

f) Landing

Figure 2. Business class loudness test

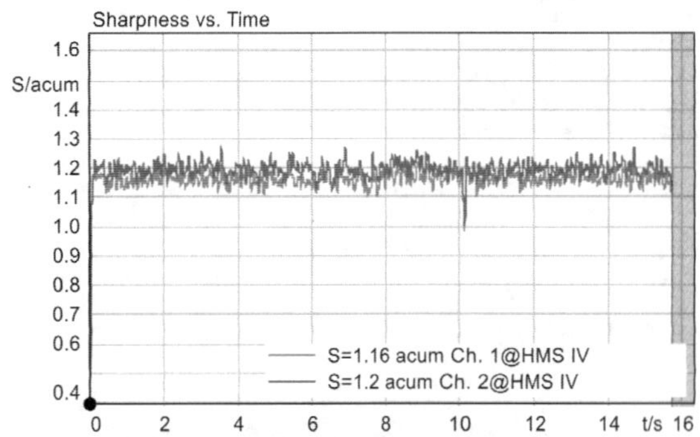

a) Ground taxiing

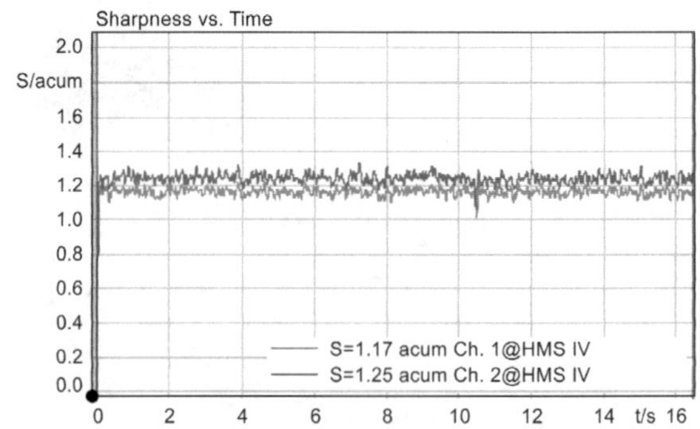

b) Take off

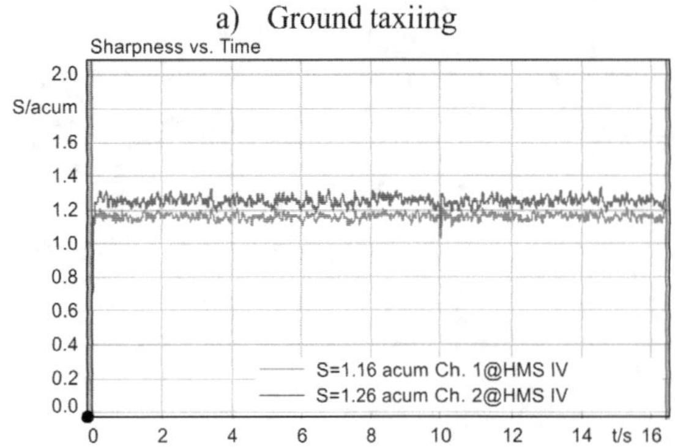

c) Cruise

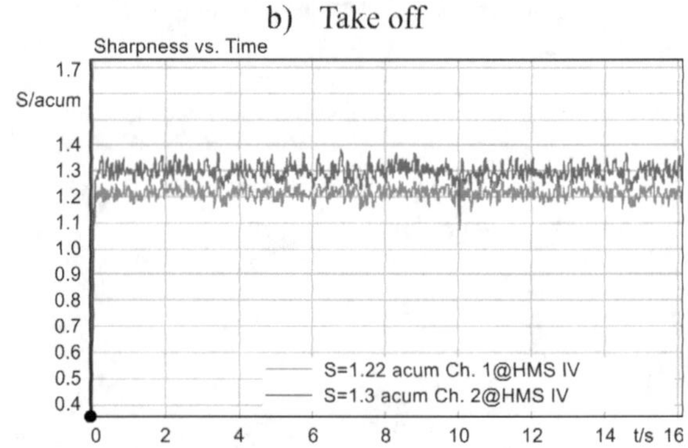

d) Acceleration

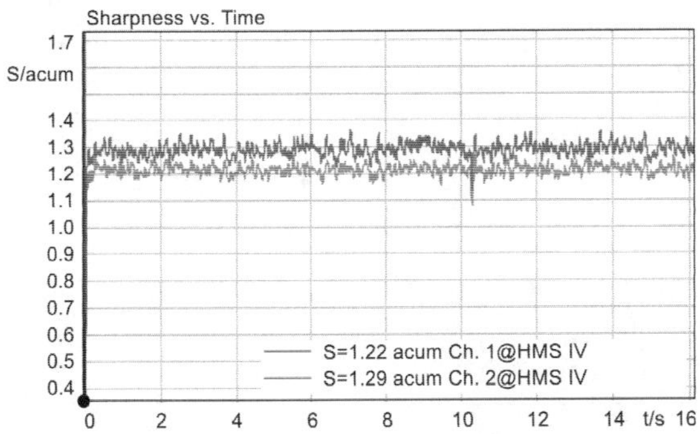

e) High-altitudeude cruise

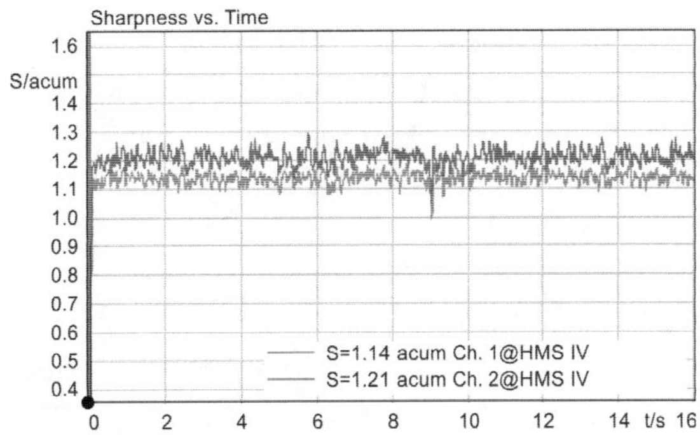

f) Landing

Figure 3. Business class sharpness test

The above test results were carried out synchronously and independently by the left ear and right ear of the artificial head, respectively, the average value of two ears was taken as the objective parameter test result of the corresponding sound field, as shown in Table 1.

Table 1. Business class psychoacoustics objective parameter table

Working conditions	Ground taxiing	Take off	Cruise	Acceleration	High altitude cruise	Landing
Loudness/sone	33.85	58.4	77.85	56.4	76.15	59.9
Sharpness/acum	1.18	1.21	1.21	1.26	1.255	1.175

2.1.2. Subjective evaluation test

The paired comparison method was adopted, with two acoustic samples in each group, and the acoustic samples of 6 working conditions were compared in pairs, forming a total of 15 groups of tests[5][6][7]. The evaluators compare the two samples in each group and score the evaluation, as shown in Figure 4.

Figure 4. Subjective evaluation test of sound quality

The evaluator gave a judgment after listening to a set of sound samples A and B: for example, if "A is better than B", A will be scored 2 points, and B will be scored 0 points. "B is better than A" is the opposite. "A and B are almost the same" means each scored 1 point, and finally, the scores of each sample are summarized, as shown in Table 2.

Table 2. Composition of Subjects

Category	Composition	Number of people	Percentage
Sex	Male	13	76.47%
	Female	4	23.53%
Profession	Engineer	14	82.35%

	Service workers	2	11.77%
	Construction worker	1	5.88%
	20~30	12	70.59%
Age	30~50	4	23.53%
	50~60	1	5.88%

According to the above test method, the subjective test was carried out. The scores of all questionnaires were counted, as shown in Table 3.

Table 3. Results under different flight conditions

Sort	Flight conditions	Score
1	Ground taxiing	166
2	Acceleration	111
3	Take off	91
4	Decline	80
5	Cruise	28
6	High altitude cruise	20

2.2. Sound quality evaluation model

The model is to establish the mathematical relationship between the subjective evaluation results of sound quality and objective parameters through linear regression or curve fitting [8][9]. There are two objective parameters and six conditions in this experiment. The following models can be constructed [10][11]:

$$M = a + b*L + c*S + d*L^2 + e*S^2 + f*L*S \qquad (4)$$

In the above formula, M is the subjective rating value, L is loudness value, S is sharpness value, and a, b... f are the coefficients, which can be obtained from Table 1 and Table 3, respectively:

$L = [33.85 \quad 58.40 \quad 77.85 \quad 56.40 \quad 76.15 \quad 59.90]^T$,

$S = [1.18 \quad 1.21 \quad 1.21 \quad 1.26 \quad 1.26 \quad 1.18]^T$,

$M = [166 \quad 111 \quad 91 \quad 80 \quad 28 \quad 20]^T$,

and substituting into the above formula, the coefficients can be obtained as:

$$[a \quad b \quad c \quad d \quad e \quad f]^T = \begin{bmatrix} -77623.22 \\ 103.85 \\ 122226.31 \\ 0.18 \\ -47363.39 \\ -104.87 \end{bmatrix} \qquad (5)$$

We substituting various parameters into Equation (4), the evaluation model can be obtained:

$$M = -77623.22 + 103.85*L + 122226.31*S + 0.18*L^2 - 47363.39*S^2 - 104.87*L*S \qquad (6)$$

3. Conclusion

A construction method is proposed for the sound evaluation model of the aircraft cabin. Based on this method, a sound evaluation model is developed with the results of the subjective test of the sound field in the aircraft cabin and the objective psychoacoustic parameter test results. The model takes the characteristic acoustic parameters that are more sensitive to the human ear, such as loudness and sharpness, as input, which can efficiently obtain the subjective result and accurately reflect the subjective feeling of the human ear on the current sound field. In the follow-up, the sound field test results of more aircraft models and various working conditions can be combined to improve the evaluation model of sound quality to make it more universal.

References

[1]Schiffbanker H, Brandl F K. Development and application of an evaluation technique to assess the subjective character of engine noise [C]. Traverse City: SAE Noise and Vibration Conference and Exhibition, 911081. [1] Vargas Jose V C, Bejan A 2001 Integrative thermodynamic optimization of the environmental control system of an aircraft *INT J HEAT MASS TRAN* 44 3907

[2] Sung-Jong Kim, Sang-Kwon Lee. Objective Evaluation for the Passenger Car During Acceleration Based on the Sound Metric and Artificial Neural Network [J]. Noise and Vibration Conference and Exhibition St. Charles, Illinois May 15-17, 2007.

[3]Bodden M. Instrumentation for sound quality evaluation [J]. Acta Acusticaunited with Acustica, 1997,83:775-783

[4] Lilis G N, Angelosante D, Giannakis G B. Sound field reproduction using the lasso. IEEE Transactions on Audio, Speech, and Language Processing, 2010, 18(8): 1902-1912.

[5] Gauthier P A, Berry A. Adaptive wave field synthesis with independent radiation mode control for active sound field reproduction: Theory. The Journal of the Acoustical Society of America, 2006, 119(5): 2721-2737.

[6] Naoe M, Kimura T, Yamakata Y, et al. Performance evaluation of 3D sound field reproduction system using a few loudspeakers and wave field synthesis: Proceedings of International Symposium on Universal Communication (ISUC), IEEE, 2008: 36-41.

[7] Liao X N, Cheer J, Elliott S, Zheng S F. Design of a loudspeaker array for personal audio in a car cabin. The Journal of the Audio Engineering Society, 2017,65(3):226-238.

[8]Tang RJ, Research on prediction and control technology of vehicle interior sound quality [D] University of Jilin, 2013

[9] Frank E C, Pickering D J, Raglin C. In-vehicle tire sound quality prediction from tire noise data [C]. SAE Noise and Vibration Conference and Exhibition, 2007-01-2253

[10]Lifang Yang, Fanyu Meng, Tao Wang. Sound Quality Evaluation Analysis on the Interior Noise of High-speed Train [J]. Advances in Applied Acoustics. 2013.

[11]Noumura K, Yoshida J. Perception modeling and quantifying sound quality in cabin. SAE, 2003.

Numerical investigation of submerged impinging jet based on different turbulent models

Jie Ge[a,b], Chuan Wang[a,c], Yulong Yao*[c], Xinxin Chen[c]

[a]College of Mechatronics Engineering, Hainan Vocational University of Science and Technology, Haikou, Hainan China 571126;

[b]SHIMGE Pump Co., Ltd., Taizhou, Zhejiang China 317525;

[c] College of Hydraulic Science and Engineering, Yangzhou University, Yangzhou, Jiangsu China 225009

* Corresponding author: yyl199876@126.com

ABSTRACT. In today's world, the loss of effective storage capacity caused by reservoir siltation has reached 50 billion cubic meters per year. As a high impact and simple water jet technology, submerged jet impact can have a powerful washout effect on submerged sediment. On this article, the submerged impinging jet is numerically computed with importing the Wray–Agarwal (W-A) turbulence models in FLUENT. Then, the streaming field analysis of the jet under different conditions is carried out and compared to the PIV test outcomes. To validate the effectiveness of the W-A turbulence models for predicting the streaming field of submerged impinging jet. The results show that, relative to other turbulence models, the calculated findings of the Wray–Agarwal turbulence model are in much closer accordance with the test data.

1. INTRODUCTION

Today's economy and water construction are developing rapidly, reservoir siltation problem is getting worse. The resulting loss of effective reservoir capacity worldwide is approximately 50 billion cubic meters per year [1-2]. Thus, the study of reservoir dredging is of major practical relevance. Jet can be divided into different types according to different characteristics (nozzle shape, flow pattern, impinging angle, surrounding boundary and jet motive force) [3]. This paper investigates the flow of a submerged impinging jet in a water tank, it can be considered as incompressible iso-density circular turbulent submerged impinging jet, referred to as submerged impinging jet.

In recent years, a lot of research has been done on the experimental device and flow mechanism of impinging jet [4-7]. Among them, the jet pump has been extensively studied as a typical jet device. In terms of flow mechanism [8-13], Cooper et al. [14] measured the mean velocity, pulsation velocity and turbulent shear stress at various radial locations in the vertical impinging jet flow field for different impinging heights and different Reynolds numbers (mean Reynolds number at the jet outlet) using a hot-wire anemometer (HWA). Hassan et al. [15] used PIV and polarographic analysis to study the vortical kinetic properties and the clipping stress at the wall of a vertical impinging jet, and a high correlation between the wall shear stress and large surface vortices was found. Naib et al. [16] adopted LDV to evaluate the submerged impact jet flow field at different impinging angles($45^{\circ} \leq \theta \leq 90^{\circ}$), and found that the velocity distribution was highly consistent with the theoretical analysis results.

In summary, the internal flow in impinging jets has been studied extensively by researchers domestically and internationally with different methods The LDV technology of single-point measurement method is basically used, which makes it difficult to gain a characteristic in the structure of large-scale vortices in a flow pattern. Hence, in the present paper, the submerged impinging jet is numerically computed by importing the W-A turbulence models into FLUENT, and the flow field analysis of the jet under various conditions is carried out. Finally, the calculated results are checked against the PIV test findings to verify the reliability of the W-A turbulence models for predicting the flow pattern of submerged impinging jets.

2. TEST RIG DESIGN AND EXPERIMENTAL SET-UP

In this paper, the PIV technology is mainly used to obtain the velocity field of the submerged impinging jet, and a pressure sensor is installed to survey the impinging pressure on the impinging wall. The relevant test set-up is depicted in Fig. 1 with the major geometrical measurements given in Table 1:

Table 1. Geometric parameters of the test set.

Test tank length	5000mm	Outer diameter of circular copper tube d	22mm
Test tank width	300mm	Impinging angle θ	0°~90°
Test tank height	400mm	Impinging height H	20~160mm
Inner diameter of circular copper tube D	20mm	Jet outlet velocity V_b	0.59/1.17/1.76m/s

The jet velocity is measured with a 2D PIV system. The tracer particles are insulating glass globules with a size of 10-15μm in diameter and a relative frequency of 1.05-1.15. The particle imagery is documented using a CCD video camera with a capacity of 1600×1200 megapixels. The pressure is obtained from the side of a pressure transducer mounted on the impinging plane, which measures from 0 to 30 mbar and is accurate to ±0.3%.

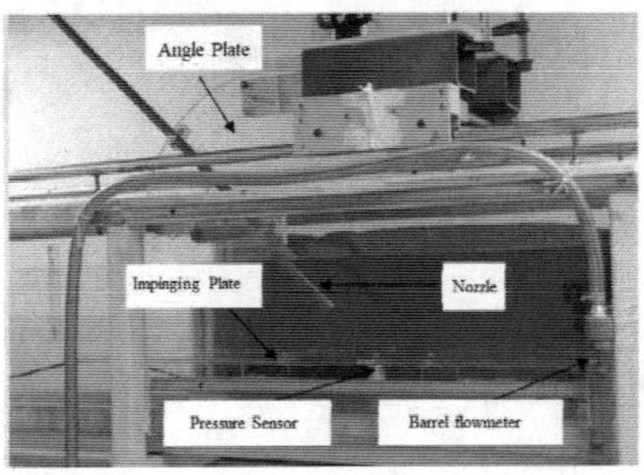

Figure 1. Experimental rig of submerged impinging jets.

3. W-A TURBULENCE MODEL AND NUMERICAL METHOD

3.1 W-A Turbulence Model

The Wray-Agarwal model is a one-formula turbulence account using k-ω closure model, which is a linear eddy-viscous model based on the Boussinesq assumption. In the W-A models, the ω formula was adopted to a counter-diffusion clause. And retains references for the k-ε model, which raises the prediction level of accuracy of balanced movements and the ability to explicate unbalanced flows.

Eddy viscosity:

$$\mu_{\tau}=f_{\mu}\rho R \tag{1}$$

$$R=\kappa / \omega \tag{2}$$

R transport equation:

$$\frac{\partial \rho R}{\partial t}+\frac{\partial \rho \mu_j R}{\partial x_j}=\frac{\partial}{\partial x_j}\left[(\partial_R \mu_\tau+\mu)\frac{\partial R}{\partial x_j}\right]+\rho C_1 RS+\rho f_1 C_{2k\omega}\frac{R}{S}\frac{\partial R}{\partial x_j}\frac{\partial S}{\partial x_j}-(1-f_1)\rho C_{2k\omega}\left(\frac{R\frac{\partial S}{\partial x_j}}{S}\right) \quad (3)$$

Where: f_μ represents damping function; f_1 represents switch function; ρ represents fluid density, kg/m^3; κ represents the turbulent pulsatile kinetic energy, J; ω represents the specific dissipation ratio; t represents the time, s; S represents the mean stress, 1/s.

3.2 Numerical Methods

The 3D unconfined inundated impinging jet model can be observed in Fig. 2(a). The jet is injected diagonally through a circular nozzle into a regular water tank. The bottom surface is the impinging plane, the uppermost is the surface of the free liquid, the exit boundary is on the left and right side, and the anterior and posterior sides are set as fixed wall borders. Internal diameter of spray pipe D=20mm, external diameter d=22mm. In this article all lengths are quantified in D. The quality of the mesh has a vital impact on the quantitative calculations. In this article, the ICEM meshing application is applied to the computational region for meshing hexahedra. The overall number of computational field meshes is 4.26 million, as depicted in Fig. 2(b).

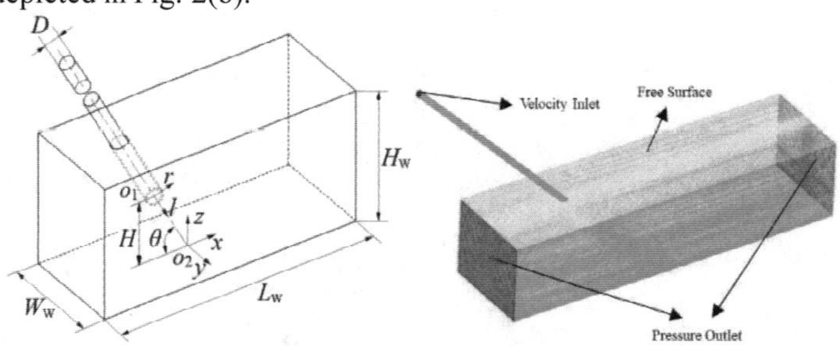

Figure 2. (a) 3-D submerged impinging jet. (b) Computational mesh of 3-D submerged impinging jet.

In boundary condition settings: the inlet adopts the velocity inlet, and the velocity is uniformly distributed, which is set to V_b = 1.76m/s (the corresponding jet nozzle outlet average Reynolds number Re = 35100); The outlet adopts the pressure outlet, and the pressure value is set to 0Pa; The free surface adopts the steel cover assumption and is set as a symmetrical surface, The wall is set as a fixed no-slip wall.

4. RESULTS AND DISCUSSION

Fig. 3(a) illustrates the comparison of the V/V_{max} curve obtained from the calculation with the experience equation $V/V_{max} = (1\text{-}2r/D)^{1/n}$ to the turbulent velocity distribution in a completely evolved circle tube. When n = 7, this equation is considered to correspond approximately to the adequately evolved circular tube turbulent velocity profile. As is evident from the graphs, the application of the various turbulence models resulted in a relatively good agreement with the empirical formulae. This represents a completely fully evolved round tube turbulence at the injection outlet. Of these, the findings obtained from the Wray–Agarwal turbulence models are nearest to the empirical equation distribution pattern.

Fig. 3(b) illustrates the profile of the jet's axial speed V/V_b along the axis for various turbulence patterns. It is observed that the axial speed stays essentially constant in the region of the free-jet (approx. $0\leq l/D \leq 3$) and reduces swiftly on impact zone (approx. $3\leq l/D \leq 4.24$). On the area of free-jet, the axial speeds computed with the Wray–Agarwal models are higher compared to the speeds counted by the alternative ones, while in the impact zone the velocities are about equal in magnitude. In the beginning of the free jet zone (approximately $0\leq l/D \leq 0.8$), the test results are lower than those calculated by the Wray–Agarwal turbulence models, but in the zone $1\leq l/D \leq 1.5$ the test data in agreement with the computed outcomes. A comprehensive comparison of the speed distributions in the free-zone and impact zones shows that the Wray–Agarwal turbulence models calculations agree well with the test outcomes than other turbulence models.

The direction of flow at either cross section (*oxz* section) in the jet can be expressed in terms to the flow angle of the submerged impacting jet, and its magnitude is defined for $\varphi=arctan(-V_z/V_x)$, as depicted in Fig. 4(a). The spread of flow angles on the shaft line is illustrated in Fig. 4(b). In the region of the free jet (approx. $0\leq l/D \leq 3$), the flow rate remains at around 45°. However, in the impacting zone (approx. $3\leq l/D \leq 4.24$) it quickly drops to zero. As a result, the calculated flow angles from the various patterns of turbulence agree well with the experimental outcomes, except for the near-wall region.

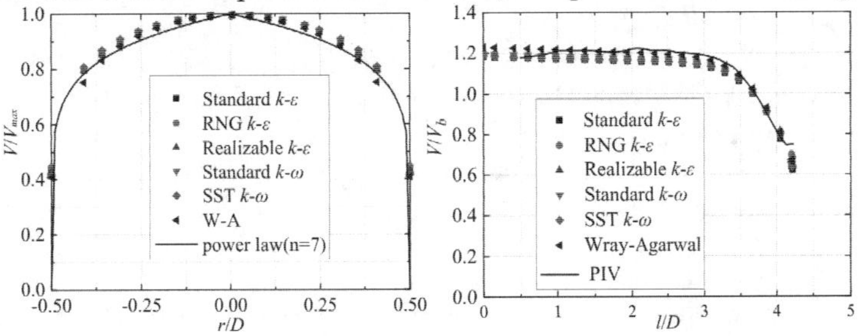

Figure 3. (a) Profiles of the standardized axial velocity V/V_{max} on the jet outlet. (b) standardized average axial speed V/V_b at the midline of the jet.

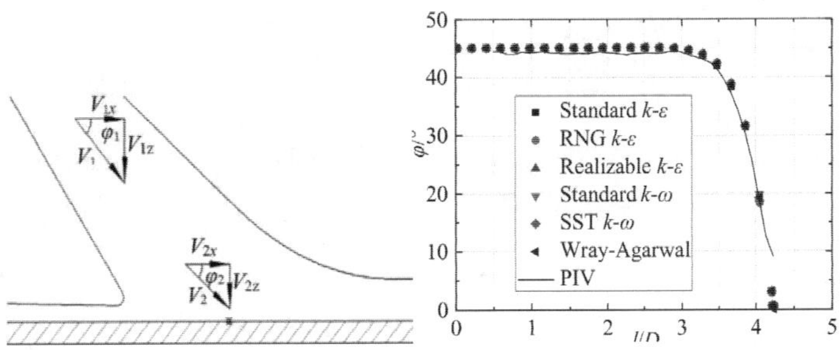

Figure 4. (a) Definition of angle of flow φ, (b) angle of flow φ at the midline of the jet.

5. CONCLUSION

This paper introduces the submerged impinging jet PIV test device and some basic theories of jet hydrodynamic numerical simulation. The focus is on the Wray–Agarwal turbulence models and the setting for free liquid levels and boundary conditions. In addition to this, this paper uses different turbulence models to numerically calculate the submerged impinging jet at a fixed impinging angle and impinging height, and compares the results with the PIV test results and finds that: the Wray–Agarwal turbulence models can better predict the flow field for submerged impact jets.

For the completely developed circular tube turbulent velocity distribution the experience equation $V/V_{max}=(1-2r/D)^{1/n}$, which is computed by the Wray–Agarwal turbulence models are the nearest to the empirical equation pattern of distribution. A comprehensive comparison of the speed distributions in the free-zone and impact-zone disclosures that the results of the Wray–Agarwal turbulence models are in much greater accordance with the test outcomes than those of other turbulence models.

REFERENCES

[1] Jian H. M., Ni J. C., Liu X. F., "Sediment Removal of Reservoir, "Yellow River Water Conservancy Press, Zhengzhou, (2004).

[2] Qin L., Yang X., Tian H. J., "Difficulties and Prospects of Mechanical Desilting in Reservoir Desilting," China International Dredging Technology Development Conference, 156-161 (2011).
[3] Dong Z. Y., "Jet Mechanics," Science Press, (2005).
[4] Chen X. X., Wang C., Shi W. D. and Zhang Y. C., "Numerical simulation of submerged impinging water jet at different impact angles," Journal of Drainage and Irrigation Machinery Engineering, 38(7), 658-662 (2020).
[5] Wang X., Wang X. K., Fan E. D., Xu S. R. and Xue Z. L., Optimization of structural parameters of capillary jet tee and experimental validation," Journal of Drainage and Irrigation Machinery Engineering, 38(9), 966-972 (2020).
[6] Zhang C. X., Wang X. K., Jin B. B., Yao J. C., Xue Z. L. and Wang X., "Variation of hydraulic performance of drip irrigation belts connected by jet tee," Journal of Drainage and Irrigation Machinery Engineering, 39(3), 300-305 (2021).
[7] Wang X. Y., Wu S. B., Wang Z. T., Wang J. F. and Wang D. B., "Instability of charged jet in liquid-liquid system," Journal of Drainage and Irrigation Machinery Engineering, 38(6), 596-601 (2020).
[8] Zhu R., Wang Y., Wang Q., Tang H. T. and Yu H. Q., "Effects of water baffle on self-priming performance of ejecting self-priming centrifugal pump," Journal of Drainage and Irrigation Machinery Engineering, 38(5), 469-475 (2020).
[9] He X. H., Yang H. and Yang S., "Multi-field coupling simulation of valveless piezoelectric micropump with wall attached jet," Journal of Drainage and Irrigation Machinery Engineering, 39(5), 439-444 (2021).
[10] Li M. Q., Chen Y, L., Hua Y. M. and Xu Y., "Experimental study on liquid jet pump based on cyclic symmetrical aeration," Journal of Drainage and Irrigation Machinery Engineering, 40(3), 223-229 (2022).
[11] Zhao K., Cheng H. Y., Long X. P., Ji B. and Lai X., "FTLE evaluation method for mixing capacity of jet pump," Journal of Drainage and Irrigation Machinery Engineering, 40(6), 556-562 (2022).
[12] Zou C. H., Li H., Harvey T., Zindine K. and Xiang Q. J., "Integrated method of optimizing jet pump anti-wear performance," Journal of Drainage and Irrigation Machinery Engineering, 38(2), 109-114 (2020).
[13] Wu W., Wang Y. and Han Y. W., "Effect of jet centrifugal pump with different throat lengths on performance of self-priming," Journal of Drainage and Irrigation Machinery Engineering, 36(12), 1258-1263 (2018).
[14] Cooper D., Jackson D. C., Launder B. E. and Liao G. X., "Impinging jet studies for turbulence model assessment-I. Flow-field experiments," International Journal of Heat & Mass Transfer, 36(05), 2675-2684 (1993).
[15] Hassan M. E., Assoum H. H., Sobolik V., Ve´tel J., Abed-Meraim K., Garon A. and Sakout A., "Experimental investigation of the W-All shear stress and the vortex dynamics in a circular impinging jet," Experiments in Fluids, 52(6), 1475-1489 (2012).
[16] Naib S. A. and Sanders J., "Oblique and Vertical Jet Dispersion in Channels," Journal of Hydraulic Engineering, 123, 456-462 (2014).

AUTHOR INDEX